PESTICIDE FACT HANDBOOK

PESTICIDE FACT HANDBOOK

U.S. Environmental Protection Agency

NOYES DATA CORPORATION
Park Ridge, New Jersey, U.S.A.

Copyright © 1988 by Noyes Data Corporation
Library of Congress Catalog Card Number 87-31528
ISBN: 0-8155-1145-0
Printed in the United States

Published in the United States of America by
Noyes Data Corporation
Mill Road, Park Ridge, New Jersey 07656

10 9 8 7 6 5 4 3 2 1

Library of Congress Cataloging-in-Publication Data

Pesticide fact handbook.

   Includes index.
   1. Pesticides--Handbooks, manuals, etc. I. United
States. Environmental Protection Agency.
SB951.P396 1988   632'.95'0212   87-31528
ISBN 0-8155-1145-0

Foreword

This book contains 130 Pesticide Fact Sheets issued by the U.S. Environmental Protection Agency and announced in the *Federal Register* through December, 1987. The Pesticide Fact Sheets include a description of the chemical use patterns and formulations, scientific findings, a summary of the Agency's regulatory position/rationale, and a summary of major data gaps. The Fact Sheets cover more than 550 trade-named pesticides.

The Fact Sheets are issued if one of the following regulatory actions occurs: (1) a Registration Standard has been issued, (2) a significantly different use pattern has been registered, (3) a new chemical is registered, or (4) a Special Review determination document has been issued.

Fact Sheets have been prepared for Registration Standards issued since June 1982 and for new chemicals and for chemicals with significantly changed use patterns registered since January 1984. Fact Sheets have also been issued for Special Review final determinations since June 1983.

Noyes has reproduced these fact sheets directly from copies of EPA original material and, because of their expected usefulness, bound them in a durable, hard cover edition at a fraction of their cost if purchased separately ($9.95 per fact sheet, or $1,293.50).

The table of contents is organized alphabetically and provides easy access to the information contained in the book. A Glossary and a Numerical List of Pesticide Fact Sheets, as well as Indexes of Common Names, Generic Names, and Trade Names, can be found at the end of the book.

> Advanced composition and production methods developed by Noyes Data Corporation are employed to bring this durably bound book to you in a minimum of time. Special techniques are used to close the gap between "manuscript" and "completed book." Due to this method of publication, certain portions of the book may be less legible than desired.

NOTICE

The information in this book was prepared by the U.S. Environmental Protection Agency. On this basis the Publisher assumes no responsibility nor liability for errors or any consequences arising from the use of the information contained herein. Mention of trade names or commercial products does not constitute endorsement or recommendation for use by the Agency or the Publisher.

Final determination of the suitability of any information or procedure for use contemplated by any user, and the manner of that use, is the sole responsibility of the user. The reader is warned that caution must always be exercised when dealing with pesticides and pesticide residues, and expert advice should be sought at all times.

Contents

Actellic 1
Aldicarb 6
Aldoxycarb 19
Aldrin 24
Aliette 30
Amitrole 34
Anilazine 43
Arosurf 48
Arsenal 52
Arsenic Acid 57
Arsenic Trioxide 64
Avermectin 67
Azinphos-Methyl 72

Bentazon and Sodium Bentazon 80
Brominated Salicylanilide 89
Bronopol 91
Butylate 95

Cadmium Pesticide Compounds 101
Calcium Arsenate 106
Captafol 109
Captan 120
Carbaryl 130
Carbofuran 140
Carbon Tetrachloride 149
Carbophenothion 152
Chlordane 157
Chlorimuron Ethyl 164

Contents

Chlorobenzilate 170
Chlorothalonil 174
Chlorpyrifos. 180
Chlorpyrifos-Methyl. 188
Clipper (Paclobutrazol). 192
Command Herbicide 197
Copper Sulfate 200
Cryolite. 207
Cyanazine 211
Cyhexatin 220
Cyromazine 226

Daminozide 232
Dantochlor. 236
DCNA. 239
Demeton 243
Diazinon 247
3,5-Dibromo. 252
Dicamba 254
Dichlobenil 260
2,4-Dichlorophenoxyacetic Acid 268
1,3-Dichloropropene 273
Dicofol 281
Diflubenzuron 286
Dinocap. 292
Dinoseb. 294
Diphenamid 298
Dipropetryn 307
Disulfoton 314
Diuron 324
Dodine 329

EPN 336
EPTC 346
Ethalfluralin. 351
Ethoprop. 354
Ethylenethiourea (ETU) 360

Fenaminosulf 364
Fenbutatin-Oxide 367
Fenoxycarb 373
Fensulfothion. 378
Fluchloralin 385
Fluometuron 392
Fluridone. 400
Fluvalinate. 405
Fonofos. 413
Formetanate Hydrochloride 417

Glycoserve 422

Heliothis NPV. 428
Heptachlor. 433
Hybrex 440

Imazaquin 447
Isazophos. 453
Isomate-M 458

Lactofen 462
Lead Arsenate. 469
Linalool. 473
Lindane. 475
Linuron. 484

Mancozeb 491
Methiocarb. 500
Methyl Bromide 506
Methyl Parathion. 513
Metolachlor 522
Metribuzin. 531
Metsulfuron Methyl. 541
Monocrotophos. 547

Nabam 555
Naled 563
Naptalam. 571
Nitrapyrin 578
Norflurazon 583

Oxamyl. 590

Paraquat 596
Parathion. 606
Pendimethalin. 614
Perfluidone 625
Phorate 632
Phosmet 637
Picloram 646
Potassium Bromide 651
Potassium Permanganate. 655
Prometryn 658
Pronamide 666
Propachlor 675
Propargite 683
Propham 693

x Contents

Simazine 703
Sodium and Calcium Hypochlorites 708
Sodium Arsenate 712
Sodium Arsenite 715
Sodium Omadine 719
Sodium Salt of Fomesafen 722
Sulfuryl Fluoride 731

Tebuthiuron 735
Terbufos 744
Terbutryn 750
Thiodicarb 759
Thiophanate Ethyl 769
Thiophanate-Methyl 773
Thiram 784
TPTH 789
Trimethacarb 798

Vitamin D_3 802

Wood Preservatives 806

GLOSSARY 811

NUMERICAL LIST OF PESTICIDE FACT SHEETS 812

COMMON NAME INDEX 816

GENERIC NAME INDEX 818

TRADE NAME INDEX 821

ACTELLIC

Date Issued: June 30, 1985
Fact Sheet Number: 59

1. **DESCRIPTION OF CHEMICAL**

    Generic Name: $\underline{O}$-[2-(diethylamino)-6-methyl-4-pyrimidinyl] $\underline{O},\underline{O}$-dimethyl phosphorothioate

    Common Name: Pirimiphos-methyl

    Trade Name: Actellic

    EPA Shaughnessy Code: 108102

    Chemical Abstracts Service (CAS) Number: 29232-93-7

    Year of Initial Registration: 1984

    Pesticide Type: Insecticide

    Chemical Family: Organophosphate

    U.S. and Foreign Producers: ICI Americas, Inc., Imperial Chemical Industries, PLC., United Kingdom

2. **USE PATTERNS AND FORMULATIONS**

    Application Sites: Stored grain products: corn, rice, wheat, and grain sorghum intended for export only

    Types of Formulations: Emulsifiable concentrate

    Types of Methods of Application: Sprays

    Application Rates: 0.006 to 0.015 lbs. a.i. per 1000 lbs. of grain

    Usual Carriers: petroleum solvents

3. SCIENCE FINDINGS

Summary Science Statement

Pirimiphos-methyl is an organophosphorothioate compound with moderate acute toxicity. This chemical has demonstrated adverse chronic effects. It is also a moderate toxicant to wildlife species, however the registered use precludes exposure to wildlife.

This chemical is toxic to fish, and other wildlife. The proposed use precludes any impact on endangered species.

In case of a significant chemical spill call (800) 426-9300 (CHEMTREC).

Chemical Characteristics

Physical State: Liquid

Color: Amber

Odor: Putrid- a typical organophosphorothioate odor. Odorless when pure

Molecular weight: 305 ($C_{11}H_{20}N_3O_3PS$)

Melting point: 15-18°C

Boiling point: Decomposes above 100°C

Vapor Pressure: 1.1×10^{-4} tor at 30°C

Flash Point: not reported

Solubility in various solvents: Solubility in water: 5 ppm at 30°C. Miscible in all proportions with methanol, ethanol, chloroform, acetone, benzene, toluene and xylene.

Toxicology Characteristics

Acute Oral: 2050 mg/kg, Toxicity Category III

Acute Dermal: 1505 mg/kg, Toxicity Category II

Primary Dermal Irritation: No irritation, Toxicity Category III

Primary Eye Irritation: Corneal opacity persisted to 14 days. Toxicity Category III

Skin Sensitization: Not a sensitizer

Acute Inhalation: Uncharacterized. The use pattern precludes inhalation exposure.

Neurotoxicity: Not an acute delayed neurotoxic agent at doses up to 10 mg/kg/day for 90 doses.

Oncogenicity: Not shown to be an oncogen in rat or mouse studies at dose levels up to 300 and 500 ppm (highest dose tested), respectively.

Teratogenicity: The Agency has determined that this chemical is not teratogenic at levels up to 16 mg/kg/day, however, an additional study in a second species (rats) is still required.

Reproduction-3 generation: Two studies adequately demonstrate that pirimiphos-methyl does not produce reproductive effects. No effects were demonstrated at dose levels up to 100 ppm.

Metabolism: The studies suggest that pirimiphos-methyl is rapidly excreted and no evidence of bioaccumulation was noted.

Mutagenicity: This chemical has been determined to be non-mutagenic in all three required studies.

Physiological and Biochemical Behavioral Characteristics

Mechanism of Pesticidal Action: An insecticide which is active by contact, ingestion, and vapor action and causes phosphorylation of the acetylcholinesterase enzyme of tissues, allowing accumulation of acetylchloline at cholinergic neuro-effector junctions (muscarinic effects), and at skeletal muscle myoneural junctions and autonomic ganglia. Poisoning also impairs the central nervous system function.

Symptoms of poisoning include: headache, dizziness, extreme weakness, ataxia, tiny pupils, twitching, tremor, nausea, slow heatbeat, pulmonary edema, and sweating. Continual absorbtion at intermediate dosages may cause influenza-like illness which includes symptoms like weakness, anorexia, and malaise.

Metabolism and Persistence in Plants and Animals:

The metabolism of pirimiphos-methyl in plants and animals is not, at this time, adequately understood in order to establish a tolerance for grain crops.

Environmental Characteristics

Uncharacterized. This use pattern preclude exposure to the environment.

Ecological Characteristics

Avian oral:
    Mallard duck--76.6 mg/kg

    Ring necked pheasant--17.7 mg/kg

Avian dietary:
    Mallard duck--633 ppm
    Bobwhite quail--207 ppm

Freshwater Fish:
    Coldwater fish (rainbow trout)--0.40 ppm
    Warmwater fish (bluegill sunfish)--2.9 ppm

Acute Freshwater Invertebrates:
    Daphnia--0.21 ppb

4. Tolerance Assessment

   No domestic tolerances exist for this chemical. Tolerances for small grains have been proposed to support domestic consumption of treated grains. There is an established tolerance of 5 ppm for imported kiwifruit.

5. SUMMARY OF MAJOR DATA GAPS

   ° A second mamalian species (rat) teratology study.

6. CONTACT PERSON AT EPA

   Jay S. Ellenberger
   Product Manager (12)
   Insecticide-Rodenticide Branch
   Registration Division (TS-767C)
   Office of Pesticide Programs
   Environmental Protection Agency
   401 M Street, S. W.
   Washington, D. C.   20460

   Office location and telephone number:
   Room 202, Crystal Mall #2
   1921 Jefferson Davis Highway
   Arlington, VA   22202
   (703) 557-2386

DISCLAIMER: The information presented in this Chemical Information Fact Sheet is for informational purposes only and may not be used to fulfill data requirements for pesticide registration and reregistration.

ALDICARB

Date Issued: March 30, 1984
Fact Sheet Number: 19

1. ## DESCRIPTION OF CHEMICAL

   Generic Name: 2-methyl-2-(methylthio) propionaldehyde
   0-(methylcarbamoyl) oxime

   Common Name: Aldicarb

   Trade Name: Temik®

   EPA Shaughnessy Code: 098301

   Chemical Abstracts Service (CAS) Number: 116-06-3

   Year of Initial Registration: 1970

   Pesticide Type: insecticide, acaricide and nematicide

   Chemical Family: carbamate

   U.S. and Foreign Producers: Union Carbide Corporation

2. ## USE PATTERNS AND FORMULATIONS

   ### Application Sites and Rates:

   ### Terrestrial Food Use

Sites	Rates (lbs. active ingredient)
Beans, dried	0.5 - 2.0
Cotton	0.3 - 6.0
Grapefruit	5.0 - 10.0
Lemons	5.0 - 10.0
Oranges	5.0 - 10.0
Peanuts	1.0 - 3.0
Pecans	0.7 - 10.0
Potatoes	1.0 - 3.0
Sorghum	0.5 - 1.0
Soybeans	0.7 - 3.0
Sugar beets	1.0 - 6.0
Sugar cane	2.0 - 3.0
Sweet potatoes	1.5 - 3.0

Terrestrial Non-Food Uses

Sites	Rates (lbs. active ingredient)
Birch	5.0 - 10.0
Dahlias	5.0 - 8.0
Holly	5.0 - 10.0
Lilies, bulbs	5.0 - 7.0
Roses	7.0 - 10.0

Commercial Greenhouse Uses

Sites	Rates (lbs. active ingredient)
Carnations	7.5 - 10.0
Chrysanthemum	7.5 - 10.0
Easter lilies	5.0 - 7.5
Gerbera	5.0 - 10.0
Orchids	7.5 - 10.0
Poinsettia	7.5 - 10.0
Roses	5.0 - 10.0
Snapdragons	5.0 - 10.0

Types of Formulations and Method of Application:

Aldicarb is a soil incorporated pesticide commercially formulated into a 15% granular formulation, two (2) 10% granular formulations and a 5% granular formulation in a mixture with the fungicides pentachloronitrobenzene and 5ethoxy-3-(trichloromethyl) 1,2,4-thiadiazole.

3. ### SCIENCE FINDINGS

Summary Science Statement:

Aldicarb has a high acute toxicity as an acetylcholinesterase inhibitor via the oral, inhalation and dermal routes of exposure. Neither aldicarb not its metabolites have been shown to be neurotoxic, oncogenic, teratogenic or mutagenic in studies that have been reviewed to date.

Aldicarb has been found to leach in fine to coarse textured soils, including those soils with a high organic matter content. The wide natural variability in soil types, weather patterns and aquifer characteristics make it impossible to specify a precise set of circumstances under which aldicarb will not reach ground water. In an effort to reduce further ground water contamination, stringent label/use restrictions have been established by the guidance document. An expedited "Special Review" of the aldicarb contamination problem is being initiated to determine whether aldicarb products can be labeled any practical way that would both permit the continued use of the chemical and preclude ground contamination of unacceptable amounts. Under this review we will also be determine what level will be acceptable in ground water. Aldicarb is highly toxic to wildlife organisms. From the available data, only limited exposure is expected to large animals, estuarine/ marine organisms and freshwater organisms as a result of the current label uses. Data suggest that application of this pesticide may result in some mortality to certain avian species. Additional data are being required to fully assess the impact on avian and small mammal populations.

<u>Chemical Characteristics</u>:
Technical aldicarb is a white crystalline solid with a melting point of 98-100° C (pure material). Under normal conditions, aldicarb is a heat-sensitive, inherently unstable chemical and

must be stabilized to obtain a practical shelf-life.

Toxicology Characteristics:

Aldicarb is a carbamate insecticide which causes cholinesterase inhibition (ChE) at very low exposure levels. It is highly toxic by the oral, dermal and inhalation routes of exposure (Toxicity Category I). The oral LD_{50} value for technical aldicarb is 0.9 mg/kg and 1.0 mg/kg for male and female rats, respectively.

The acute dermal LD_{50} for aldicarb in rats is 3.0 mg for males and 2.5 mg for females. High mortality was evident in rats, mice and guinea pigs (6.7 mg/M^3), a 15 minute exposure period was not lethal; however, 5 of the 6 test animals died during a 30-minute exposure. Exposure of rats for eight hours to air that had passed over technical aldicarb or granular aldicarb produced no mortality. Aldicarb applied to the eye of rabbits at 100 mg of dry powder caused ChE effects and lethality.

The toxicity data base for aldicarb is nearly complete; however, additional mutagenicity tests are being required by the aldicarb guidance document. The toxicity data base includes a 2-year rat feeding/oncogenicity study with a no-observed effect level (NOEL) of 0.3 mg/kg bw/day for effects other than cholinesterase inhibition and was negative for oncogenic effects at the level tested (0.3 mg/kg bw/day); a 2-year rat oncogenicity study which was negative for oncogenic effects at the levels tested (0.1 and 0.3 mg/kg bw/day); a 100-day dog feeding study and a 2-year dog feeding study with NOELs of 0.7 and 0.1 mg/kg bw/day, respectively, for effects other than cholinesterase inhibition (highest levels tested (HLT)); an 18-month mouse feeding/oncogenicity study with a NOEL of 0.7 mg/kg bw/day and was negative for oncogenic effects at the levels tested (0.1, 0.3 and 0.7 mg/kg bw/day); a 2-year mouse oncogenicity study which was negative for oncogenic effects at the levels tested (0.3 and 0.9 mg/kg bw/day); a 6-month rat feeding study using aldicarb sulfoxide with a NOEL of 0.125 mg/kg bw/day for cholinesterase inhibition; a 3-generation rat reproduction study with a 0.7 mg/kg bw/day NOEL; a rat teratology study which was negative for teratogenic effects at 1.0 mg/kg bw/day (HLT); a rabbit teratology study which was negative for teratogenic effects at 0.5 mg/kg bw/day (HLT); and a hen neurotoxicity study which was negative at up to 4.5 mg/kg bw/day.

Physiological and Biochemical Behavior Characteristics:
Aldicarb and its metabolites are absorbed by plants from the soil and translocated into the roots, stems, leaves and fruit. The available data indicate that the metabolism of aldicarb in plants and small animals are similar.

Aldicarb is metabolized rapidly by oxidation to the sulfoxide metabolite and followed by slower oxidation to the sulfone metabolite. Both the sulfoxide and sulfone are subsequently hydrolyzed and degraded further to yield less toxic entities. Sufficient information is not available to adequately define the metabolism of aldicarb in ruminant animals. However, at present the major residues of concern in plants and animals appear to be the parent compound aldicarb and its sulfoxide and sulfone metabolites.

Environmental Characteristics:

Sufficient data are not available to fully assess the environmental fate of aldicarb. Additional data are being required on soil metabolism, soil and aquatic dissipation, leaching and volatility. However, from the available data, aldicarb has been determined to be mobile in fine to coarse textured soils, even including those soils with high organic matter content and may reach ground water. Aldicarb is not expected to move horizontally from a bare, sloping field. Therefore, accumulation of aldicarb in aquatic nontarget organisms is expected to be minimal. This is further supported by an octanol/water partition coefficient of 5 and an ecological magnification value of 42.

Ecological Characteristics:

Aldicarb is highly toxic to mammals, birds, estuarine/marine organisms and freshwater organisms. LC_{50} values for the bluegill sunfish and rainbow trout have been reported as 50 ug/liter and 560 ug/liter, respectively. A LC_{50} of 410.7 ug/liter was reported for the Daphnia magna. Studies on the toxicity of aldicarb to the mallard duck and bobwhite quail indicate LD_{50} values of 1.0 and 2.0 mg/kg, respectively.

Limited exposure to mammals is expected from a dietary standpoint. However, data from field studies and the use history of aldicarb provide sufficient information to suggest that application of this pesticide may result in some mortality, if not local population reductions of certain avian species. Whether these effects are excessive, long lasting, or likely to diminish wildlife resources cannot be stated with any degree of certainty. Therefore, additional field studies are being required to further quantify the impact on avian and small mammal populations.

Aldicarb has also been found to pose a threat to the endangered species, Attwater's Greater Prairie Chicken, residing in or near aldicarb treated fields. Accordingly, all aldicarb products are required to bear labeling restrictions prohibiting the use of the product in the Texas counties of Aransas, Austin, Brazoria, Colorado, Galveston, Goliad, Harris, Refugio and Victoria if the Attwater's Greater Prairie Chicken is located in or immediately adjacent to the treatment area.

Tolerance Assessments:

The Agency is unable to complete a full tolerance assessment of aldicarb tolerances at this time because of the lack of 1) a large animal metabolism study which adequately identifies and quantifies residues in tissue; 2) analyses of treated cotton foliage and; 3) processing studies for instant coffee, potato granules and dried potatoes and soybean processing fractions. Additionally, there are some concerns over the appropriate Acceptable Daily Intake (ADI) for aldicarb.

In 1983, the Assistant Administrator for the EPA formed an Aldicarb Review Committee to evaluate the available toxicity data for aldicarb and determine the appropriate ADI. A formal report is not available at this time; however, one of the suggestions of the committee was the use of an "intermediate uncertainty factor of 32". The resulting ADI, if adopted, would be 0.0038 mg/kg/day. An independent evaluation of the toxicity data base by the World Health Organization (1982) and the Institute for Comparative and Environmental Toxicology (Cornell University) further supports the ADI used by the Office of Pesticide Programs (OPP). Although there has been and continues to be much discussion on the subject, the OPP considers it prudent, at this time, not to alter its established ADI of 0.003 mg/kg/day.

Based on the below listed tolerances for aldicarb and its sulfoxide and sulfone metabolites, the current theoretical maximum residue contribution (TMRC) is 0.1120 mg/day for a 1.5 kg diet which utilizes 62.14 percent of the ADI.

Tolerances for Residues

Raw Agricultural Commodities	Parts Per Million
Bananas	0.3
Beans (dry)	0.1
Beets, sugar	0.05
Beets, sugar, tops	1.0
Cattle, fat	0.01
Cattle, mbyp	0.01
Cattle, meat	0.01
Coffee beans	0.1
Cottonseed	0.1

Tolerances for Residues (cont.)

Raw Agricultural Commodities	Parts Per Million
Goats, fat	0.01
Goats, mbyp	0.01
Goats, meat	0.01
Grapefruits	0.3
Hogs, fat	0.01
Hogs, mbyp	0.01
Hogs, meat	0.01
Horses, fat	0.01
Horses, mbyp	0.01
Horses, meat	0.01
Lemons	0.3
Limes	0.3
Milk	0.002
Oranges	0.3
Peanuts	0.05
Peanut, hulls	0.5
Pecans	0.5
Potatoes	1.0
Sheep, fat	0.01
Sheep, mbyp	0.01
Sheep, meat	0.01
Sorghum, fodder	0.5
Sorghum, grain	0.2
Soybeans	0.02
Sugarcane	0.02
Sugarcane, fodder	0.1
Sugarcane, forage	0.1
Sweet potatoes	0.1

Problems Which are Known to Have Occurred With Use of Aldicarb:

In 1979, aldicarb residues were found in drinking water wells located near aldicarb treated potato fields in Suffolk County, Long Island, New York at levels <200 parts per billion (ppb). Aldicarb residues have since been found in drinking water wells at levels above 10 ppb in other states including Wisconsin, Florida, Maine, Virginia, Connecticut, Delaware, Maryland, New Jersey and Rhode Island. The Agency's Office of Drinking Water has established a Health Advisory Level (HAL) of 10 parts per billion (ppb) for residues of aldicarb in drinking water. However, since the HAL is derived from the ADI and the ADI may undergo changes in the future, the HAL may be revised. Such a change would likely be an increase in the HAL.

The Pesticide Incident Monitoring System (PIMS) reports on aldicarb from 1966 through 1982, contained 165 incidents associated with human injury, and 6 incidents each involving animals, environmental contamination and non-target plants and crops. Most of the human incidents alleged that aldicarb was the cause of the problem, but there was insufficient evidence to support such a conclusion. Those incidents involving confirmed aldicarb poisonings appeared to be the result of failure to use label recommended safety equipment while applying aldicarb. Other incidents resulted from accidental spillage, ingestion of aldicarb, or consumption of food commodities improperly treated with aldicarb.

4. SUMMARY OF REGULATORY POSITION AND RATIONALE:

In 1982, Union Carbide Corporation voluntarily classified all aldicarb products "Restricted Use" in an effort to minimize further ground water contamination. However, since aldicarb is highly toxic by the oral, dermal and inhalation r⁻ tes of exposure (Toxicity Category I), the Agency is requiring that all aldicarb products be classified "Restricted Use".

As the result of the aldicarb contamination of drinking water wells, Union Carbide Corporation excluded the use of aldicarb products in Suffolk County, Long Island, New York. The company also limited the use of aldicarb products on potatoes to once every two years and only after plant emergence in the state of Maine and Wisconsin and the counties of Hartford in Connecticut, Kent and New Castle in Delaware, Franklin and Hampshire in Massachusetts, Worchester in Maryland, Atlantic, Burlington, Cumberland, Monmouth and Salem in New Jersey, Newport and Washington in Rhode Island and Accomack and Northampton in Virginia. Aldicarb may be applied at planting at the 1 lb. active ingredient/acre rate for for aphid control in the state of Maine.

Based on concerns for ground water contamination, product labeling must be further revised to include the following statements:

> "This product is usually decomposed into harmless residues. However, a combination of sandy and acidic soil conditions, moderate to heavy irrigation and/or rainfall, use of 3 or more pounds active ingredient per acre, and soil temperature below 50°F at the time of application, tend to reduce degradation and promote movement of residues to ground water. If this describes your local use conditions and ground water in your area is used for drinking, do not use this product without first contacting (company name)."

> "Do not apply this product in Del Norte County, California."

"Do not apply more than 5 lbs. active ingredient per acre in the state of Florida."

"Application to citrus fruits in the state of Florida may be made between January 1 and April 30 only."

In the absence of adequate soil dissipation data and dermal exposure data, the Agency is imposing an interm 24-hour re-entery interval.

5. SUMMARY OF MAJOR DATA GAPS

The data requirements represent major data gaps for aldicarb. These data are required to be submitted to the Agency within 4 years from the data of the issuance of the registration document.

- Ruminant metabolism study
- Anaerobic soil metabolism study
- Aerobic and anaerobic aquatic metabolism studies
- Photodegradation in water study
- Soil (field) dissipation study
- Leaching and Adsorption/Desorption Studies
- Volatility study
- Field monitoring data

6. CONTACT PERSON AT EPA

Jay Ellenberger,

Product Manager (PM) 12,

Registration Division (TS-767C),

Office of Pesticide Programs,

Environmental Protection Agency,

401 M St., SW.,

Washington, D.C. 20460.

Office location and telephone number:

Rm. 202, CM #2,

Jefferson Davis Highway,

Arlington, VA 22202,

(703-557-2386).

DISCLAIMER: The information presented in this Chemical Information Fact Sheet is for informational purposes only and may not be used to fulfill data requirements for pesticide registration and reregistration.

ALDOXYCARB

Date Issued: January 16, 1987
Fact Sheet Number: 115

1. DESCRIPTION OF CHEMICAL

    Generic Name: 2-methyl-2-(methylsulfonyl)propanal-0-(methylamino carbonyl oximel)

    Common Name: Aldoxycarb

    Trade Name: Standak

    Other Names: Sulfocarb, Aldicarb sulfone

    EPA Shaughnessy Code: 110801

    Chemical Abstracts Service (CAS) Number: 1646-88-4

    Year of Initial Registration: 1986

    Pesticide Type: Insecticide/Nematicide

    Chemical Family: Carbamate

    U.S. Producer: Union Carbide Agricultural Products Co., Inc.
    No other producer at this time

2. USE PATTERNS AND FORMULATIONS

    Application Sites: Containerized honey locust trees (Commercial Use Only) to control honey locust gall midge

    Type of Formulations: Insecticide/fertilizer spike; registered to International Spike, Inc.

    Method of Application: Spike is inserted into soil in container

    Application Rate: One or more spikes of 1% active ingredient per container, depending on size of container or plant.

    Usual Carriers: Formulation is a pressed mixture of aldoxycarb and fertilizer chemicals.

3. SCIENCE FINDINGS

Summary Science Statement

Technical aldoxycarb has high mammalian acute toxicity. It has not been shown to cause oncogenic, mutagenic, teratogenic, delayed neurotoxin or reproduction effects. Aldoxycarb is a known degradate/metabolite of aldicarb produced by the oxidation of the thio-moiety.

Sufficient data are available to characterize aldoxycarb from an environmental fate and ecological standpoint. Aldoxycarb is extremely toxic to wildlife. Use precautions are being imposed to reduce potential hazards. Although aldoxycarb has the potential to contaminate groundwater under certain environmental conditions, the proposed containerized plant use will preclude any measurable contamination.

A tolerance assessment is not needed because the registered use pattern is for an ornamental plant. There are no data gaps.

Chemical Characteristics of Technical Aldoxycarb

Physical State: Crystalline powder at 20°C

Color: White

Odor: Slightly sulfurous

Melting Point: 140-142° C

Vapor Pressure: 9×10^{-5} mm Hg at 25° C

Density: 1.35 g/cm^3 at 20° C

ph: 3-6

Toxicology Characteristics of Technical Aldoxycarb

Acute oral: 21.4 mg/kg, Toxicity Category I

Acute dermal: 1000 mg/kg, Toxicity Category II

Primary Dermal Irritation: No irritation, Toxicity Category III

Primary Eye Irritation: No irritation, Toxicity Category IV

Acute Inhalation: 0.209 mg/l, Toxicity Category II.

Neurotoxicity: Not an acute delayed neurotoxic agent at doses up to 250 mg/kg (highest dose tested (HDT)).

Oncogenicity: Two studies, rat and mouse. Both are acceptable and are negative for oncogenic effects up to 9.6 mg/kg/day (HDT).

Teratogenicity: Two teratology studies, rat and rabbit have been evaluated to determine the teratogenic potential of aldoxycarb. Both studies were negative for teratogenic effects at levels up to 9.6 mg/kg (HDT).

Reproduction/3-generation: No effects on reproduction at levels up to 9.6 mg/kg (HDT).

Metabolism: The metabolism of aldoxycarb is adequately understood. It is metabolized by hydrolysis of the carbamate ester to form the oxime. Other reactions of the oxime occur.

Mutagenicity: Adequate studies are available to demonstrate aldoxycarb is not a mutagen.

Physiological and Biochemical Behavioral Characteristics

Mechanism of Pesticidal Action: A systemic insecticide/nematicide which causes reversible carbamylation of the acetocholinesterase enzyme of tissues, allowing accumulation of acetylcholine at cholinergic neuroeffector junctions (muscarinic effects) and at skeletal muscle myoneural junctions and in autonomic ganglia (nicotinic effects). The central nervous system is also impaired.

Symptoms of poisoning include: headache, dizziness, weakness, ataxia, pinpoint pupils, blurred or dark vision, muscle twitching, nausea, vomiting, diarrhea, convulsions and death. The onset of these symptoms is rapid and their severity depends on the dose. The immediate cause of death is usually respiratory failure.

Metabolism and Persistence in Plants and Animals:

Acceptable studies have been submitted which show aldoxycarb is metabolized in plants and animals by hydrolysis of the carbamate ester to form the oxime. Further reactions of the oxime yield aldoxycarb aldehyde, aldoxycarb alcohol, aldoxycarb acid and aldoxycarb nitrile. The oxime and alcohol metabolites are easily conjugatd to form water soluble glycosides, sulfates, and other compounds.

Environmental Characteristics

In soil and water, aldoxycarb is very stable under acidic conditions, stable at neutral conditions and very unstable to hydrolysis at alkaline conditions. It is rapidly hydrolyzed to sulfocarb oxime which in turn rapidly degrades to methane sultonic acid and 2-hydroxy isobutyraldehyde oxime. Aldoxycarb is rapidly degraded to a variety of materials under both aerobic and anerobic conditions. In certain soils, such as those with a sandy loam texture, it has a half-life of 2-4 weeks. Aldoxycarb is mobile in certain soil types and does have the potential to contaminate groundwater under certain situations. Soil types of high sand content and organic matter will promote leaching of parent and degradation products, which are of lower toxicity than that of the parent compound.

However, it is believed the containerized ornamental plant use will not result in groundwater contamination because of the fact that application is made to soils in containers.

Ecological Effects of Technical Aldoxycarb

Avian oral: Mallard duck - 33.5 mg/kg

Avian dietary: Waterfowl species (Mallard duck) - >10,000 ppm
Upland game species (Bobwhite quail) - 5,706 ppm

Freshwater fish: Coldwater species (rainbow trout) - 42.0 ppm
Warmwater species (bluegill sunfish) - 53.0 ppm

Acute Freshwater Invertebrates: Daphnia - 0.176 ppb

Precautionary language would be required for outdoor terrestrial use for hazards to wildlife.

4. **SUMMARY OF REGULATORY POSITION AND RATIONALE**

The Agency has determined to register aldoxycarb for containerized ornamental plants because, adequate studies are available to assess the toxicological and environmental characteristics of aldoxycarb and its potential effects to humans from this use. The Agency concludes from this studies that this use pattern will not pose any unreasonable adverse effects to humans or the environment. None of the criteria for unreasonable adverse effects listed in section 162.11(a) of Title 40 of the U.S. Code of Federal Regulations have been met or exceeded for this use.

5. **SUMMARY OF MAJOR DATA GAPS**

There are no data gaps.

6. CONTACT PERSON AT EPA

Jay S. Ellenberger
Product Manager (12)
Insecticide-Rodenticide Branch
Registration Division (TS-767C)
Office of Pesticide Programs,
Environmental Protection Agency,
401 M St., S.W.
Washington, D.C. 20460.

Office location and telephone number:
Rm. 202, Crystal Mall Bldg. 2
1921 Jefferson Davis Highway,
Arlington, VA 22202,
(703) 557-2386.

DISCLAIMER: The information presented in this Chemical Information Fact Sheet is for informational purposes only and may not be use to fulfill data requirements for pesticide registration and reregistration.

ALDRIN

Date Issued: December, 1986
Fact Sheet Number: 108

1. DESCRIPTION OF CHEMICAL

    Generic Name: 1,2,3,4,10,10-hexachloro-1,4,4a,5,8,8ahexahydro-exo-1,4-endo-5,8-dimethanonaphtalene.

    Common Name: Aldrin

    Trade and Other Names: Aldrine, HHDN, Aldrex®, Aldrex 30®, Aldrite®, Aldrosol®, Altox, Bangald®, Drinox®, Octalene®, Rasayaldrin®, Seedrin® Liquid, Entoma 15949 and Compound 118.

    EPA Shaughnessy Code: 045101

    Chemical Abstracts Service (CAS) Number: 309-00-2

    Year of Initial Registration: 1949

    Pesticide Type: Insecticide

    Chemical Family: chlorinated cyclodiene

    U.S. and Foreign Producers: Shell International Corp.

2. USE PATTERNS AND FORMULATIONS

    Application Sites: soil surrounding wooden structures for termite control

    Types of Formulations: 2 and 4 lb/gal emulsifiable concentrate

    Types and Methods of Application: trenching, rodding, subslab injection, low pressure spray

    Application Rates: 0.25 to 0.5% emulsion

3. SCIENCE FINDINGS

    Summary Science Statement

    Aldrin is a chlorinated cyclodiene with high acute toxicity.

The chemical has demonstrated adverse chronic effects in mice (causing liver tumors). Aldrin may pose a significant health risk of chronic liver effects to occupants of structures treated with aldrin. The Agency is continuing to evaluate the potential risk from the termiticide use of aldrin to determine whether further regulatory action may be warranted. Aldrin is extremely toxic to aquatic organisms and birds. Aldrin is persistent and bioaccumulates. Aldrin may have a potential for contaminating surface water; thus, a special study is required to delineate this potential. Applicator exposure studies are required to determine whether exposure to applicators may be posing health risks. Special subacute inhalatioon testing is required to evaluate the respiratory hazards to humans in structures treated with aldrin. Data available to the Agency show a pattern of misuse and misapplication of aldrin. The Agency is requiring restricted use classification of all end-use products containing aldrin. Application must be made either in the actual physical presence of the Certified Applicator, or if the Certified Applicator is not physically present at the site, each uncertified applicator must have completed a State approved training course and be registered in the State in which the uncertified applicator is working.

Chemical Characteristics of the Technical Material

Physical State: Crystalline solid
Color: Tan to dark brown
Odor: Mild chemical odor
Molecular weight and formula: 364.93 - $C_{12}H_8Cl_6$
Melting Point: 104 to 104.5 °C
Boiling Point: Decomposes at 1 atm.
Vapor Pressure: 6.6×10^{-6} mmHg at 25 °C
Solubility in various solvents: Very soluble in most organic solvents; practically insoluble in water
Stability: Stable with alkali and alkaline-oxidizing agents; not stable with concentrated mineral acids; acid catalysts, acid-oxidizing agents, phenols, active metals

Toxicology Characteristics

Acute Oral: Data gap
Acute Dermal: Data gap
Primary Dermal Irritation: Data gap
Primary Eye Irritation: Data gap
Skin Sensitization: not a sensitizer
Acute Inhalation: Data gap

Major routes of exposure: Inhalation exposure to occupants of treated structures; dermal and respiratory exposure to termiticide applicators.

Delayed neurotoxicity: does not cause delayed neurotoxic effects.

Oncogenicity: This chemical is classified as a Group B_2 (probable human oncogen). Rat oncogenicity study is a data gap. There are three long-term carcinogenesis bioassays of aldrin in mice which were independently conducted by investigators affiliated with the National Cancer Institute and the Food and Drug Administration. These studies were found to produce significant tumor responses in three different strains of mice (C_3H, CF_1, and $B6C3F_1$) in males and females with a dose-related increase in the proportion of tumors that were malignant.

Available data from seven existing carcinogenesis bioassays in rats are inadequate and inconclusive and a well-designed study in rats is needed to determine the carcinogenic potential of aldrin in this species.

Chronic Feeding: Based on a rat chronic feeding study, a Lowest Effect Level (LEL) of 0.025 mg/kg/day has been calculated.

Metabolism: In biological systems, aldrin is readily epoxidized to dieldrin.

Teratogenicity: Data gap

Reproduction: Data gap

Mutagenicity: Aldrin does not possess mutagenic activity in bacteria. Further testing is required to assess the mutagenic potential of aldrin in eukaryotes.

Physiological and Biochemical Characteristics

The precise mode of action in biological systems is not known. In humans, signs of acute intoxication are primarily related to the central nervous system (CNS), including hyperexcitability, convulsions, depression and death.

Environmental Characteristics

Available data are insufficient to fully assess the environmental fate of aldrin. Data gaps exist for all applicable studies. However, available supplementary data indicate general trends of aldrin behavior in the environment. Aldrin degrades readily to dieldrin, which is persistent in the environment. Reports on leaching and field studies suggest that aldrin/dieldrin would be unlikely to leach to underground aquifers. However, additional data are necessary to assess the potential for ground-water contamination as a result of the termiticide use of aldrin.

Ecological Characteristics

Avian oral toxicity: 6.59 mg/kg in bobwhite quail; 52 mg/kg in mallard ducks.

Avian dietary toxicity: 34 and 155 ppm in Japanese quail and mallard duck, respectively.

Freshwater fish acute toxicity: 5 and 53 ppb in largemouth bass and channel catfish, respectively (warmwater species), 2.6 and 8.2 ppb in rainbow trout and chinook salmon, respectively (coldwater species).

Freshwater invertebrate toxicity: 18 ppb in a species of seed shrimp; 32 ppb in a species of water flea.

4. Required Unique Labeling and Regulatory Position Summary

° EPA is currently evaluating the potential human health risks of 1) non-oncogenic chronic liver effects, and 2) oncogenic effects from exposure to aldrin. Following the completion of Registration Standards on alternative chlorinated cyclodiene termiticides (chlordane and heptachlor) EPA will determine whether the risks posed by the termiticide use of aldrin warrant further regulatory action.

° In order to meet the statutory standard for continued registration, retail sale and use of all end-use products containing Aldrin must be restricted to Certified Applicators or persons under their direct supervision. For purposes of Aldrin use, direct supervision by a Certified Applicator means 1) the actual physical presence of a Certified Applicator at the application site during application, or 2) if the Certified Applicator is not physically present at the site, each uncertified applicator must have completed a State approved training course and be registered in the State in which the uncertified applicator is working; the Certified Applicator must be available if and when needed.

° In order to meet the statutory standard for continued registration, Aldrin product labels must be revised to provide specific Aldrin disposal procedures, and to provide fish and wildlife toxicity warnings.

° The Agency is requiring a special monitoring study to evaluate whether and to what extent surface water contamination may be resulting from the use of Aldrin as a termiticide.

° Special product-specific subacute inhalation testing is required to evaluate the respiratory hazards to humans in structures treated with termiticide products containing Aldrin.

° The Agency is requiring the submission of applicator exposure data from dermal and respiratory routes of exposure.

° While data gaps are being filled, currently registered manufacturing use products and end use products containing Aldrin may be sold, distributed, formulated, and used, subject to the terms and conditions specified in the Registration Standard for Aldrin. Registrants must provide or agree to develop additional data in order to maintain existing registrations.

5. TOLERANCE REASSESSMENT

No tolerance reassessment for Aldrin is necessary since there are no food or feed uses.

6. SUMMARY OF MAJOR DATA GAPS

° Hydrolysis

° Photodegradation in Water

° Aerobic Soil Metabolism

° Anaerobic Soil Metabolism

° Leaching and Adsorption/Desorption

° Aerobic Aquatic Metabolism

° Soil Dissipation

° Reproductive Effects in Rats

° Rat Oncogenicity Study

° Mutagenicity Studies

° Teratology Studies

° Battery of Acute Toxicity Studies

° Special Surface Water Monitoring Studies

° Applicator Exposure Studies

° Special Guinea Pig Inhalation Study

o   All Product Chemistry Studies

7. CONTACT PERSON AT EPA

    George LaRocca
    Product Manager (15)
    Insecticide-Rodenticide Branch
    Registration Division (TS-767C)
    Office of Pesticide Programs
    Environmental Protection Agency
    401 M Street, S.W.
    Washington, D.C. 20460

    Office location and telephone number:

    Room 204, Crystal Mall #2
    1921 Jefferson Davis Highway
    Arlington, VA 22202

    (703) 557-2400

ALIETTE

Date Issued: June 30, 1983
Fact Sheet Number: 1

1. Description of the chemical

   * Generic name: Aluminum Tris (O-ethyl phosphonate)
   * Common name: Fosetyl-Al
   * Trade name: Aliette
   * EPA Shaughnessy code: 123301
   * Chemical Abstracts Service (CAS) number: 39148-24-8
   * Year of initial registration: 1983
   * Pesticide type: fungicide
   * Chemical family: Aluminum ester of alkyl phosphonates
   * U.S. and foreign producers: Rhone-Poulenc, Incorporated

2. Use patterns and formulations

   * Application sites: Greenhouses, field grown ornamentals, and pineapple seed pieces
   * Types of formulations: Wettable power
   * Types and Methods of applications: Dip and ground spray apparatus
   * Usual carriers: water

3. Science Findings

   Summary science statement: There are no extensive data gaps that exist for Fosetyl-Al. The toxicity from oral, dermal, and inhalation route of exposure is low. The chemical has been demonstrated to be a stong eye irritant. The chemical is considered to be an oncogen. Available data indicate that Fosetyl-Al has no teratogenic potential nor is it a mutagen.

   Fosetyl-Al is degraded rapidly in both moist and dry soils under aerobic conditions, with half-lives 1 to 1/2 half hours and 20 minutes respectively. The chemical is not expected to contaminate groundwater.

   Fosetyl-Al has a low toxicity for both fish and wildlife, and the chemical and its use patterns presents no problems to endangered species.

   Chemical characteristics:

   * Fosetyl-Al is a fine white, odorless powder. It is stable to heat, melts with decomposition at temperatures greater than 200° C. The chemical does not present any unusual handling hazards.

Toxicological characteristics:

- Fosetyl-Al is considered a strong eye irritant (Toxicity Category I).
- It demonstrates low toxicity from oral, dermal and inhalation routes of exposure (Toxicity Category IV, III, & III respectively).
- The chemical is not considered to be a skin irritant.
- Toxicology studies on Fosetyl-Al are as follows:

    - Oral LD_{50} in rats: 5.4 gm/kg body weight
    - Dermal LD_{50} in rats: >3 gm/kg
    - Inhalation LC_{50} in rats: >1.73 mg/l
    - Eye Irritation in rabbits: 2/6 animals revealed pannus of the cornea at 7 days and continued to show irritating effects at 21 days
    - Dermal Irritation in rabbits: not an irritant
    - Skin Sensitization in guinea pigs: Not a skin sensitizer
    - Teratology in rabbits: No observeable effect level (NOEL) > than 500 mg/kg/day
    - Teratology in rats: NOEL > than 1000 mg/kg/day
    - Multigeneration Reproduction study in rats: The reproduction NOEL is 6000 parts per million (PPM); the lowest effect level (LEL) is 12,000 ppm (crystalline deposits in urogenital system, lower overall weight gains in the F_{2b} generation, lower litter and mean pup weights in late lactation)
    - Chronic feeding/oncogenicity in rats: The chronic feeding/ oncogenicity NOEL is 8000 ppm; the LEL is 3000/4000 ppm (calculi and mineralization in 14/79 males in urogenital)
    - Chronic feeding in dogs: The chronic feeding NOEL is 10,000 ppm; the LEL is 20,000 ppm (presence of spermatocytic and or spermatidic giant cells within the semineferous tubules).
    - Oncogenicity in mice: No oncogenic effects were induced at any dose level under conditions of this test (highest dose tested 20,000/30,000 ppm).
    - Mutagenicity: Fosetyl-Al did not show any mutagenic activity in gene mutation, chromosomal aberration, or micronucleus tests.

Environmental Characteristics:

- The degradation of Fosetyl-Al in soil under aerobic conditions is quite rapid and is due to microbial action. Half-life is 1 to 1/2 hours in loamy sand, silt loam, and clay loam soils, and 20 minutes in sandy loam soil. The degradation proceeds through the hydrolysis of the ethyl ester bond resulting in the formation of phosphorous acid and ethanol. The ethanol is further degraded into carbon dioxide. The

phosphorous acid formed will form precipitates with calcium, aluminum, or iron in the soil or with aluminum from the Fosetyl-Al.
- Under field conditions, the chemical is not expected to leach.
- Bioaccumulation does not appear to be a factor.
- The use patterns are not expected to result in direct contamination hazards (via spray drift) to humans, livestock, or wildlife outside the application sites.

Ecological characteristics

- Avian oral LD_{50}: >8000 mg/kg (extremely low toxicity)
- Avian dietary LC_{50}: >20,000 ppm
- Fish LC_{50}: >150 mg/l (low toxicity)
- Aquatic invertebrate LC_{50}: 189 ppm
- The use patterns of the chemical do not present any problem to endangered species.

Tolerance assessment

- A tolerance of 0.1 ppm is required for pineapples and pineapple forage. Data developed with C^{14} labeled Fosetyl-Al indicate that residues of Fosetyl-Al or its metabolites could remain in harvested pineapples at or below the level of analytical method detection.

4. Summary of Regulatory Position and Rationale

- General Use
- No use, formulation, or geographical restrictions are required
- Because of oncogenicity potential of this compound a risk assessment was calculated for all uses involved. The risk calculations are based on overestimates of any actual risk exposure. The calculated risks, although low, are likely to be overestimates.
- Dietary risk for pineapples is $0.27-2.40 \times 10^{-8}$
- Oncogenic risk exposure ranged from $1.0-9.0 \times 10^{-9}$ to applicators to $0.7-6.0 \times 10^{-6}$ for pineapple planters wearing gloves.
- For the pineapple use, the following statement is required to reduce the exposure to seed piece planters: "Note to User: "Gloves impermeable to Fosetyl-Al must be worn during the handling and planting of pineapple crowns (seed pieces)."

5. Summary of Major Data Gaps:

- Exposure monitoring study for pineapple seed piece treaters.

6. Contact person at EPA

   Henry M. Jacoby
   EPA (TS-767-C)
   401 M St., S.W.
   Washington, D.C. 20460
   Phone (703) 557-1900

DISCLAIMER: The information presented in this Chemical Information Fact Sheet is for informational purposes only and may not be used to fulfill data requirements for pesticide registration and reregistration.

AMITROLE

Date Issued: May 14, 1984
Fact Sheet Number: 20

1. Description of chemical:

    Generic name: Amitrole

    Common name: Amitrole

    Trade names: Weedazole®, Amino Triazole Weed Killer®, Cytrol®, Amitrol T®, Domatol®, Vorox®, Amizole®, X-All®, Ustinex®, AT, ATA, Aminotriazole 90 and Chempar Amitrole

    EPA Shaughnessy code: 004401

    Chemical abstracts service (CAS) number: 61-82-5

    Year of initial registration: 1948

    Pesticide type: Herbicide

    U.S. and foreign producers: Not produced in U.S., major importers are Union Carbide, American Cyanamid and Aceto Chemical.

2. Use patterns and formulations:

    Application sites: Noncrop sites including rights-of-way, marshes, drainage ditches, ornamentals and around commercial, industrial, agricultural, domestic and recreational premises.

    Types of formulations: Technical (90%, 95%); wettable powder (15%, 25%); flowable concentrate (0.33 lb/gal, 0.44 lb/gal, 1%); soluble concentrate/liquid (0.3 lb/gal, 2 lb/gal); soluble concentrate/solid (50%, 90%) and pressurized liquid (0.36%, 1%)

    Types and methods of application: Applied as a spray for broadcast, spot or directed treatments using aerial or ground equipment.

    Application rates: 0.9 to 20 lbs a.i./A depending upon weed species and method of application

    Usual carriers: water

3. **Science findings:**

    Summary science statement:

    Extensive data gaps exist for Amitrole in product chemistry, toxicology, ecological effects and environmental fate. Amitrole has demonstrated oncogenic potential and is a candidate for Special Review. Because of this oncogenic risk, all use patterns and application techniques (except for homeowner uses) are classified as restricted.

    Chemical characteristics:

    Physical state: Crystalline powder

    Color: Transparent, colorless

    Odor: Odorless

    Melting point: 159° C

    Solubility: 28g/100g water, soluble in some polar solvents

    Stability: Stable in heat to 100° C. Amitrole sublimes under reduced pressure.

    pH: Aqueous solutions are neutral.

    Unusual handling characteristics: None

    Toxicology characteristics:

    Acute toxicology results:

    Acute oral LD_{50} (rat) > 4.08 gm/kg, Toxicity category III

    Acute dermal LD_{50} (rabbit): No mortalities reported, Toxicity category III

    Primary eye irritation (rabbit): Amitrole is slightly irritating, additional testing is required.

Chronic toxicology results:

>Feeding/Oncogenicity: Amitrole has an anti-thyroid effect in laboratory rats. Dogs fed Amitrole exhibited thyroid and pituitary changes.
>
>Reproduction: Amitrole does not cause reproductive effects.
>
>Teratology: Additional testing required.
>
>Mutagenicity: Amitrole does not cause mutagenic effects.
>
>Metabolism: Amitrole is rapidly eliminated from the body.

Major routes of exposure: Mixers, loaders and applicators would be expected to receive the most exposure via skin contact and inhalation.

Physiological and biochemical behavioral characteristics:

>Absorption and translocation: It is readily absorbed and rapidly translocated in the roots and leaves of higher plants.
>
>Mechanism of pesticidal action: Amitrole interferes with the metabolism of nucleic acid precursors, disrupts chloroplast development and regrowth from buds.

Environmental characteristics:

>Adsorption and leaching in basic soil types: Amitrole exhibits intermediate soil mobility.
>
>Microbial breakdown: Microbial metabolism is the expected major route of degradation.
>
>Resultant average persistance: Amitrole residues degrade with a half-life of <1 to 56 days in non-sterile aerobic soils. The soil dissipation rate is affected by moisture, temperature, cation exchange capacity and clay content, but is unaffected by soil pH. Amitrole is persistent in pond water and hydrosoil.

Ecological characteristics:

    Hazards to fish and wildlife:

        Avian dietary LC_{50}: Mallard duck > 5,000 ppm
                                     Ring-neck pheasant > 5,000 ppm

        Freshwater fish LC_{50}: Rainbow trout > 180 mg/l
                                     Bluegill sunfish > 180 mg/l

        Aquatic invertebrate LC_{50} > 10 ppm

Tolerance assessments: Temporary Maximum Residue Limits for Amitrole of 0.02 ppm have been established by FAO/WHO for those crops where residues are likely to occur. There are no established tolerances for Amitrole in the U.S., Canada, and Mexico. There are no food or feed uses in the U.S. and residues are not permitted on any food or water intended for irrigation, drinking, or other domestic purposes.

Problems known to have occurred from use: The Pesticide Incident Monitoring System (PIMS) listed eight incidents resulting from the use of Amitrole alone from 1972 to 1977. One incident involved illegal residue on apples and two others involved plant injury resulting from soil residues. The remaining five incidents involved pesticide applicators receiving medical attention after exposure. Symptoms included skin rash, vomiting, diarrhea and nosebleed. There were no reported fatalities. PIMS incidents are voluntarily reported, do not include detailed follow-ups and are not validated in any way.

4. <u>Summary of regulatory position and rationale:</u>

Use classification: Restricted (for all uses except for homeowner uses)

Use, formulation, geographical restrictions: Noncropland areas only

Unique label warning statements:

    Manufacturing-Use Products:

        Products intended for formulation into end-use products must bear the following statement:

            "For formulation only into end-use herbicide products intended for noncropland, outdoor use."

"The use of this product may be hazardous to your health. This product contains amitrole, which has been determined to cause cancer in laboratory animals. Products intended for formulation into restricted-use pesticides must require on their labeling that a respirator be worn during mixing and loading. Lightweight waterproof clothing (jumpsuit [or coverall], boots [or shoes], gloves, and a wide-brimmed plastic hardhat) must be worn when mixing and loading all products and when applying all products to control dense, tall vegetation. Workers applying this product in all other situations must wear lightweight waterproof gloves and boots (or shoes). Products intended for formulation into general-use pesticides must require on their labeling that waterproof gloves be worn while handling the product."

All products must bear the following statements:

"Each formulator is responsible for obtaining EPA registrations for its formulated product(s)."

"Do not discharge into lakes, streams, ponds, or public waters unless in accordance with NPDS permit. For guidance, contact your Regional Office of the EPA."

End-Use Products:

All restricted-use products must bear the following statements:

"Restricted Use Pesticide"

"For retail sale to and application only by certified applicators or personnel under their direct supervision."

"The use of this product may be hazardous to your health. This product contains amitrole, which has been determined to cause cancer in laboratory animals. Wear a respirator during mixing and loading of all products. Wear lightweight waterproof clothing (jumpsuit [or coverall], boots [or shoes], gloves, and a wide-brimmed plastic hardhat) when applying all products to control dense, tall vegetation. Workers applying this product in all other situations must wear lightweight waterproof gloves and boots (or shoes)."

All homeowner products must bear the following statements:

> "The use of this product may be hazardous to your health. This product contains amitrole, which has been determined to cause cancer in laboratory animals. Wear waterproof gloves when using this product."

All products intended for nonaquatic uses must bear the following statement on the label:

> "Do not apply directly to water or wetlands. Do not contaminate water by cleaning of equipment or disposal of wastes."

All products intended for aquatic uses must bear the following statement on the label:

> "Consult your state Fish and Game Agency before applying this product to public waters. Permits may be required before treating such waters."

All products must bear the following statements, regardless of classification:

> "Do not allow spray or spray drift to contaminate edible crops or water intended for irrigation, drinking or other domestic purposes."

> "Do not allow livestock to graze or feed in treated noncrop areas."

Summary of preliminary risk/benefit review:

Risks:

Amitrole is not used on food crops and there is no dietary exposure to amitrole. Dermal exposure is the major source of exposure, with inhalation furnishing only a minor contribution to the total body burden. Human exposure, in some circumstances, occurs at doses which resulted in antithyroid effects in laboratory animals.

Conservatively assuming 100% dermal penetration, the oncogenic risk associated with some use patterns and application techniques is high. Lightweight waterproof clothing and a respirator are expected to reduce exposure and risk for all uses except the power wagon application.

Benefits:

The largest use site by production volume, the highway rights-of-way site was selected for this limited analysis. Amitrole is not produced in the United States, with under 800 thousand pounds being imported by Union Carbide, American Cyanamid and Aceto Chemical. Amitrole, in combination with other chemicals, offers low cost, broad spectrum control of both newly emerged or established broadleaf weeds as well as seasonal control by residual chemicals with which it is mixed. Alternatives include contact herbicides and mechanical cutting.

5. Summary of major data gaps:

Generic data requirements:

Product chemistry: data due 6 months after receipt of Standard

Statement of composition
Discussion of formation of unintentional ingredients
Preliminary analysis
Density, bulk density, or specific gravity
Solubility
Vapor pressure
Dissociation constant
Octanol/Water partition coefficient
Submittal of samples

Toxicology:

Acute testing: data due 6 months after receipt of Standard

Primary eye irritation
Primary skin irritation
Dermal sensitization

Subchronic testing: data due 24 months after receipt of Standard

90-day dermal
90-day inhalation

Chronic testing: data due 24 months after receipt of Standard

Teratogenicity - 2 species

Special testing: data due 6 months after receipt of Standard

Dermal absorption study

Wildlife and aquatic organisms: data due 24 months after receipt of Standard

  Avian oral LD_{50}
  Freshwater fish LC_{50}
  Acute LC_{50} freshwater invertebrates
  Acute LC_{50} estuarine and marine organisms

Environmental fate:

  Data due 6 months after receipt of Standard:

    Hydrolysis studies
    Photodegradation studies in water
    Photodegradation studies on soil
    Leaching and adsorption/desorption
    Special exposure study - Protective clothing effectiveness

  Data due 24 months after receipt of Standard:

    Aerobic soil metabolism study
    Anaerobic aquatic metabolism study
    Aerobic aquatic metabolism study
    Soil dissipation study - field
    Aquatic (sediment) dissipation study - field
    Forestry dissipation study - field
    Soil, long-term dissipation study (field) - reserved, depending upon results of field dissipation study
    Accumulation studies - irrigated crops

Product specific data requirements for manufacturing-use products containing Amitrole:

  Product chemistry: data due 6 months after receipt of Standard

    Statement of composition
    Discussion of formation of unintentional ingredients
    Preliminary analysis
    Certification of limits
    Analytical methods for enforcement of limits
    Density, bulk density, or specific gravity
    pH
    Oxidizing or reducing action
    Flammability
    Explodability
    Storage stability

Toxicology:

   Acute testing: data due 6 months after receipt of Standard

   Primary eye irritation - rabbit
   Primary dermal irritation
   Dermal sensitization

6. Contact person at EPA:

   Robert J. Taylor
   Product Manager (25), TS-767C
   Environmental Protection Agency
   401 M Street, S.W.
   Washington, D.C. 20460
   (703) 557-1800

DISCLAIMER: The information presented in this Chemical Information Fact Sheet is for informational purposes only and may not be used to fulfill data requirments for pesticide registration and reregistration.

ANILAZINE

Date Issued: December 16, 1983
Fact Sheet Number: 12

1. Description of Chemical:
   Generic name: 2,4-dichloro-6-(O-chloroanilino)-s-triazine
   Common name: Anilazine
   Trade names: Dyrene, Direz, Kemate, Triasyn, B-622, Ent-26,058, HCl-008684, 4,6-dichloro-N-(2-chlorophenyl)-1,3,5-triazin-2-amine.
   EPA Shaughnessy Code: 080811
   Chemical Abstracts Service (CAS) number: 101-05-3
   Year of intitial registration: 1957
   Pesticide type: fungicide
   Chemical family: triazine
   U.S. and foreign producers: Mobay Chemical Corporation

2. Use Patterns and Formulations:
   Application sites: turf, tobacco, ornamentals, various fruits and vegetables
   Types of formulations: dust, wettable powder, granular
   Types and methods of application: foliar application, additive to vinyl plastics
   Application rates: 1.0 to 3.0 lb active ingredient per acre
   Usual carriers: water

3. Science Findings:

   Summary science statement:
   Extensive data gaps exist for Anilazine. No human toxicological hazards of concern have been identified in studies reviewed by the Agency for the standard. The Agency has no information that indicates continued use will result in any unreasonable adverse effects to man or his environment during the time required to develop the data.

   Chemical characteristics:
   Anilazine is a white to tan crystalline solid. It is stable in neutral to slightly acidic media and subject to hydrolysis. It has a melting point of 159-160 C. The chemical does not present any unusual handling hazards.

Toxicological characteristics:
  Anilazine is considered a skin sensitizer. It demonstrates low toxicity from oral routes of exposure. The chemical has been found to be highly toxic to fish and aquatic invertebrates.
  Acute toxicology results:
    Oral LD_{50} in rats: 2.71 g/kg body weight (28-33 days) (Tox Category III)
    Oral LD_{50} in rabbits: 460 mg/kg (Tox Category II)
    Oral LD_{50} in dogs: MLD > 7.1 g/kg (Tox Category IV)
    Oral LD_{50} in monkeys: MLD > 3.2 g/kg (Tox Category III)
    Dermal LD_{50} in rabbits: > 9.4 g/kg (intact skin) (Tox Category III) > 2.5 g/kg (abraded skin)
    Eye irritation in rabbits: Corneal opacity in 3/6 animals, iris irritation, redness, chemosis which persisted through day 21 (Tox Category I)

  Chronic toxicology results:
    Rat and mouse oncogenicity studies were negative at dose levels tested (500 and 1000 ppm). Clinical toxic signs were noted at both doses in second year.

  Major routes of exposure:
  dermal

Environmental characteristics:

  Microbial breakdown:

    Anilazine is degraded rapidly in both moist and dry soils under aerobic conditions, with half-lives of 0.5 and 2.5 days, respectively.

  Adsorption and leaching in basic soil types:

    Anilazine is classified to be of intermediate mobility in a sandy loam and of low mobility in agricultural sand, sandy clay loam, silt loam and silty clay soils based on soil TLC.

Ecological characteristics:

  Hazards to fish and wildlife:

    Avian dietary LC_{50}: Anilazine did not cause 50% mortality when birds were exposed to a diet containing 5000 ppm for > 10 to <100 days.
    Avian oral LD_{50}: > 2,000 mg/kg
    Fish LC_{50}: 0.14 to 0.326 ppm (highly toxic)
    Aquatic invertebrate LC_{50}: 0.270 ppm (highly toxic)

Tolerance assessment:

   Due to the absence of pertinent data, the Agency is unable to complete its reassessment of anilazine tolerances.

   List of present tolerances:

Crop	ppm
Blackberries	10.0
Blueberries	10.0
Celery	10.0
Cranberries	10.0
Cucumbers, including pickles	10.0
Dewberries	10.0
Garlic	1.0
Onions, dry bulb	1.0
Onions, green	10.0
Potatoes	1.0

Problems known to have occurred from use:

   The Pesticide Incident Monitoring System (PIMS) includes a report of dermatitis and delayed dermal hypersensitivity in several laborers hand harvesting anilazine treated strawberries and tomatoes in Tennessee. With treatment, recovery required at least one week, but symptoms recurred with increased severity upon re-exposure.

4. Summary of Regulatory Position and Rationale:

   Use Classification: General

   Unique label warning statements:

   All product labeling is required to bear the statement "Protective clothing should be worn during periods of exposure, such as, during application or when contacting treated foliage."

   Manufacturing-use labels must contain the statements "This pesticide is toxic to fish and aquatic invertebrates. Do not discharge into lakes, streams, ponds or public water unless in accordance with an NPDES permit. For guidance, contact your Regional Office of EPA."

   All end-use labels, except those for use on cranberries, must contain the statements "This pesticide is toxic to fish and aquatic invertebrates. Do not apply directly to water or wetlands. Drift and runoff from treated areas may be hazardous to aquatic organisms in neighboring areas. Do not contaminate water by cleaning of equipment or disposal of wastes."

All end-use labels for use on cranberries must contain the statements "This pesticide is toxic to fish and aquatic invertebrates. Movement from treated areas may be hazardous to aquatic organisms in neighboring areas. Do not contaminate water by cleaning of equipment or disposal of wastes."

All end-use labels must contain the following statements:
"Do not use on seed crops intended to be used for feed or forage. Do not graze treated areas."
"Do not reenter treated fields within 24 hours following application of this product."

5. <u>Summary of major data gaps</u>:

    Product Chemistry (Due August 1984)
        Product identity
        Analysis and certification of product ingredients.
        Physical and chemical characteristics

    Residue Chemistry (Due June 1986)
        Nature of residue and analytical method for plants and animal residues
        Storage stability data
        Crop field trials on all crops except cucumbers, summer squash, dewberries, loganberries, and raspberries
        Processed food/feed studies on potatoes and tomatoes

    Environmental Fate (Due December 1987)
        Hydrolysis
        Photodegradation
        Metabolism studies
        Mobility studies
        Soil and aquatic dissipation studies
        Accumulation studies

    Toxicology (Due December 1987)
        Inhalation LC_{50} - rat
        21-day dermal
        90-day dermal (vinyl additive use only)
        90-day inhalation - rat
        Chronic toxicity - rodent and non-rodent
        Teratogenicity - 2 species
        Reproduction - 2 generation
        Mutagenicity testing
        General metabolism

    Reentry Protection (Due December 1987)
        Foliar dissipation
        Dermal exposure

Wildlife and Aquatic Organisms (Due December 1987)
  Avian oral LD_{50}
  Avian dietary LC_{50} - upland game bird and waterfowl
  Acute LC_{50} - freshwater invertebrates
  Fish early life stage and aquatic invertebrate life-cycle - invertebrate and fish

6. <u>Contact person at EPA:</u>
Henry M. Jacoby
EPA (TS-767C)
401 M. St., S.W.
Washington, D.C. 20460
Phone (703) 557-1900

DISCLAIMER: The information presented in this Chemical Information Fact Sheet is for informational purposes only and may not be used to fulfill data requirements for pesticide registration and reregistration.

AROSURF

Date Issued: February 15, 1984
Fact Sheet Number: 17

1. Description of Chemical

    Generic name: Poly(oxy-1,2-ethanediyl), alpha-isooctadecyl-
          omega-hydroxy
    Common name: none
    Trade name: Arosurf® MSF (Monomolecular Surface Film)
    EPA Shaughnessy code: 124601
    Chemical abstracts service (CAS) number: 52292-17-8
    Year of initial registration: 1984
    Pesticide type: Insecticide (mosquito larvacide/pupacide)
    Chemical family: ethoxylated fatty alcohol (a nonionic
          surfacant
    U.S. producer: Sherex Chemical Company, Inc.

2. Use Patterns and Formulations

    Application sites: Semi-permanent or permanent fresh
       (including potable and irrigation sources) or salt
       water habitats. Semi-permanent or permanent polluted
       water habitats.
    Types of formulations: liquid
    Types/methods of application: ground and air spray
    Application rates: 3-5 gal/acre (i.e., 0.2-0.5 gal.
       a.i./acre
    Usual carriers: water

3. Science Findings

    Summary science findings:

    Based on submitted toxicity and fish and wildlife data,
    Arosurf® demonstrates low toxicity to humans, fish
    and wildlife. The product is conditionally registered
    pending the submission of additional fish and wildlife
    studies identified below.

    Chemical characteristics:

    Arosurf® MSF is a clear, light amber liquid at 77°F
    with a melting point of 19-27°F and a flash point of
    395°F. It is capable of reducing the surface tension
    of clean water to 28.2 dyns/cm at 77°F. As a surfactant,
    it forms a film over the surface of the water to which

it is applied. Its mode of action is considered to be physical rather than chemical in that it reduces the water surface tension resulting in suffocation of the larvae and pupae.

Toxicology Characteristics:

Arosurf® MSF demonstrates low toxicity (Tox Category III) from the following submitted studies:

- Two oral LD_{50} rat studies; 20 gm/kg and 5.0 gm/kg
- Eye irritation, rabbit; not an eye irritant
- Primary eye irritation, rabbit; not an eye irritant
- Acute LC_{50} inhalation, rat; >29 mg/L
- Salmonella microsomal Ames mutagenicity assay; negative
- Primary skin irritation, guinea pig; mild dermal irritant
- Primary skin irritation, rabbit; severe dermal irritant with reversibility of effect
- Primary skin irritation, human patch test; not a skin irritant
- 2-year dog study with a NOEL of 50,000 ppm (1250 mg/kg)
- Two 2-year dog studies with NOEL of 20,000 ppm (1,000 mg/kg)
- 10-month monkey study with a NOEL of 1,000 mg/day

Physiological and Biochemical Behavior Characteristics:

Foliar absorption: N/A
Translocation : N/A
Mechanism of pesticidal action: suffocation (physical)
Metabolism and persistence in plants and animals: N/A

Environmental Characteristics:

Adsorption and leaching in basic soil types: N/A
Microbial breakdown: data show that the chemical is degraded by unaclimated, mixed cultures from natural sources by the shake culture method.
Loss from photodecomposition and/or volatization: Arosurf® MSF is stable to hydrolysis and aqueous photolysis in the absence of photosensitizers.
Bioaccumulation: Similar alcohol ethoxylates degrade under field conditions with the major route of degradation being hydrolysis at the ether linkage and subsequent oxidation of the alkyl chain to lower molecular weight polyethylene glycol-like materials and ultimately CO_2 and H_2O.
Resultant average persistance: 2 to 10 days at recommended dosage rates.

Ecological Characteristics:

Hazards to fish and wildlife: the following data have been submitted:

- Avian acute oral LD_{50} (mallard duck); >2,000 mg/kg
- Avian dietary LC_{50} (bobwhite quail); >5,000 mg/kg
- Fish acute LC_{50} (bluegill); 290 mg/L: and rainbow trout; 98 mg/L
- Aquatic invertebrate LC_{50} (Daphnia); 1.9 mg/L

Efficacy review results:

Efficacy data not required.

Tolerance Assessments:

An exemption from the requirement of a tolerance for the pesticide in or on fish, shellfish, irrigated crops, meat, milk, poultry and eggs, when used as a mosquito control agent in aquatic areas has been established under 40 CFR 180.1078.

Studies on closely related fatty acid and fatty alcohol ethoxylates having various degrees of ethoxylation supported the clearance of the chemical as an inert ingredient and were also used in support of the exemption as an active ingredient.

4. **Summary of Regulatory Position and Rationale**

Use classification: General
Use, formulation, manufacturing process or geographical restrictions : None
Unique warning statements required on labeling: None
Summary of risk/benefit review: None conducted

It has been determined that (1) the use of Arosurf® MSF has met the public interest finding as required for the conditional registration of a new chemical under section 3(c)(7)(C) of FIFRA, as amended; and (2) that the use of the pesticide during the period in which the requested data is being generated will not cause any unreasonable adverse effect on the environment.

5. **Summary of Major Data Gaps**

- One avian dietary study on mallard duck (submitted and currently under review)
- One fish early lifestage study

- One aquatic invertebrate (Mysid shrimp) lifecyle study

These studies are to be submitted no later than February 1, 1987.

6. Contact person at EPA

William H. Miller
Product Manager (16)
Insecticide-Rodenticide Branch
Registration Division (TS-767)
Environmental Protection Agency
Washington, DC 20460

Tel. No. (703) 557-2600

DISCLAIMER: The information presented in this Chemical Information Fact Sheet is for informational purposes only and may not be used to fulfill data requirements for pesticide registration and reregistration.

ARSENAL

Date Issued: September 5, 1985
Fact Sheet Number: 63

1. Description of Chemical

    Code Names: Cl 243,997 and AC 243,997

    EPA Shaughnessy Code: 128821

    Chemical Abstracts Service (CAS) Number: 81334-34-1

    Year of Initial Registration: 1984

    Pesticide Type: Herbicide

    Chemical Family: Imidazolinone

    U.S. and Foreign Producers: American Cyanamid Company

2. Use Patterns and Formulations

    Application sites: Noncropland areas such as railroad, utility and pipeline rights-of-way, utility plant sites, petroleum tank farms, pumping installations, fence rows, storage areas, nonirrigation ditchbanks and other similar areas.

    Types of formulations: Aqueous solution of isopropylamine salt

    Types and methods of application: End-use product may be applied either pre-emergence or postemergence to the weeds. In most situations, the preferred method is postemergence application. Product is mixed with water and applied as a spray.

    Application rates: 0.5 to 1.5 lbs. acid equivalent per acre depending on weed species and degree of infestation.

    Usual carriers: Water. Drift control agent and foam reducing agent may be added if needed.

3. Science Findings

Summary science statement: Results of acute toxicity studies indicate toxicity category III. Chronic studies present no evidence of unacceptable health hazards resulting from proposed use. Ecological effects data indicate that the technical acid is practically nontoxic to avian and aquatic organisms. Additional fish and wildlife studies are required, on the salt formulation. Several data gaps exist in environmental fate.

Chemical characteristics:

Physical state:	Powder
Color:	White to off-white
Odor:	Slight acetic acid odor
Melting point:	169-173°C
Bulk density:	13.1 lb/cubic foot
Solubility:	1.0 to 1.5% in water at 25°C
Dissociation constant:	pk_1=1.9, pk_2=3.6

Octanol/Water partition coefficient: 1.3

pH:	3-3.5 as 1% solution in water at 25°C
Stability:	Stable at 25°C for at least 18 months

Toxicology characteristics:

Acute toxicology results:

Acute oral toxicity (rat): greater than 5,000 mg/kg
Toxicity category III

Acute dermal toxicity (rabbit): greater than 2,000 mg/kg
Toxicity category III

Primary eye irritation (rabbit): Toxicity category III

Primary skin irritation (rabbit): Toxicity category IV

Acute inhalation toxicity (rat): greater than 5.1 mg/l (nominal)
greater than 1.3 mg/l (gravimetric)
Toxicity category III

Dermal sensitization (guinea pig): Technical material is not a skin sensitizer.

Chronic toxicology results:

21-day dermal (rabbit): Systemic NOEL is 400 mg/kg/day (HDT). The NOEL for skin is 400 mg/kg/day (HDT).

Teratology (albino rat): NOEL for teratogenicity and fetotoxicity is 1,000 mg/kg/day. The NOEL for maternal toxicity is 300 mg/kg/day. The LEL is 1,000 mg/kg/day with salivation occurring in 6 of 22 females.

Teratology (albino rabbit): Test material was not teratogenic or fetotoxic at dosages up to 400 mg/kg/day (HDT). The maternal toxic NOEL is 400 mg/kg/day.

Mutagenicity: Technical material was not mutagenic in the Ames assay.

Metabolism (rat): The half-life of the technical was less than one day. There was no significant radiolabeled compound detected in tissue residues.

Major routes of exposure: Mixers, loaders and applicators would be expected to receive the most exposure via skin contact and inhalation.

Physiological and biochemical behavioral characteristics:

Foliar absorption: The product is absorbed by roots and foliage.

Translocation: The product translocates readily following absorption and is distributed between roots and foliage.

Mechanism of pesticidal action: The herbicide prevents the production of the amino acids; valine, leucine and isoleucine. Once the levels of these amino acids decrease, protein synthesis slows down and growth stops. The growing points of target plants die first. Mature, green tissue is not as rapidly affected.

Environmental characteristics:

Adsorption and leaching in basic soil types: The technical material has a moderate leaching potential. The adsorption coefficient (K) ranges from 1.7 in a clay loam soil (4.6% organic matter) to 4.9 on a silt loam soil (4.0% organic matter). The adsorbed material will desorb from soil.

Loss from photodecomposition: Technical material will photodegrade in aqueous solution with a half-life of 2.5 to 5.3 days (12 hours of sunlight/day).

Ecological characteristics:

Fish acute toxicity (Rainbow trout): greater than 100 mg/l

Fish acute toxicity (Bluegill sunfish): greater than 100 mg/l

Fish acute toxicity (Channel catfish): greater than 100 mg/l

Aquatic invertebrate toxicity (Daphnia magna): greater than 100 mg/l

Avian acute oral toxicity (Bobwhite quail): greater than 2,150 mg/kg

Avian acute oral toxicity (Mallard ducks): greater than 2,150 mg/kg

Avian dietary toxicity (Bobwhite quail): greater than 5,000 ppm

Avian dietary toxicity (Mallard ducklings): greater than 5,000 ppm

4. Summary of Regulatory Position and Rationale:

Use, formulation, manufacturing process or geographical restrictions:

Do not use on food or feed crops. Do not apply to ditches used to transport irrigation water. Do not apply where runoff water may flow onto agricultural land as injury to crops may result. Keep from contact with fertilizers, insecticides, fungicides and seeds. Do not apply or drain or flush equipment on or near desirable trees or other plants, or on areas where their roots may extend, or in locations where the chemical may be washed or moved into contact with their roots. Do not use on lawns, walks, driveways, tennis courts, or similar areas. Prevent drift of spray to desirable plants. Do not use in California.

Unique label warning statements:

PHYSICAL AND CHEMICAL HAZARDS

Spray solutions of Arsenal should be mixed, stored and applied only in stainless steel, fiberglass, plastic and plastic-lined containers. Do not mix, store or apply Arsenal or spray solutions of Arsenal in unlined steel (except stainless steel) containers or spray tanks.

5. Summary of Major Data Gaps:

   Ecological Effects:

   The following studies are to be done with the end-use product:

   Avian single-dose oral LD_{50} (1 test)
   Avian dietary LC_{50} (2 tests - waterfowl and upland game bird)
   Acute toxicity test for freshwater fish (2 tests - warm water and cold water species)
   Acute toxicity test for freshwater aquatic invertebrates (1 test)

   A waiver of above studies will be considered if data are supplied that fulfill: one avian dietary study, one freshwater fish study, an aquatic invertebrate study and demonstrate equivalent toxicity levels to the technical material.

   Environmental Fate:

   Nonguideline study - Application of ^{14}C Arsenal to weeds
   Fish accumulation study
   Field dissipation study (if field crop uses are proposed in the future)

6. Contact Person at EPA:

   Robert J. Taylor
   Product Manager (25), TS-767C
   Environmental Protection Agency
   401 M Street, S.W.
   Washington, D.C. 20460
   Phone: (703) 557-1800

DISCLAIMER: The information presented in this Pesticide Fact Sheet is for informational purposes only and may not be used to fulfill data requirements for pesticide registration and reregistration.

ARSENIC ACID

Date Issued: September, 1986
Fact Sheet Number: 91

1. DESCRIPTION OF CHEMICAL

    Chemical Names: Arsenic acid, Orthoarsenic acid
    Common Name: Arsenic acid
    Trade Names: Desiccant L-10®, Hi Yield® H-10, Poly Brand Desiccant, Hi Yield® Synergized H-10®
    EPA Shaughnessy Code: 006801
    Chemical Abstracts Service (CAS) Number: 7778-39-4
    Pesticide Types: Desiccant, Wood preservative
    Chemical Family: Inorganic arsenicals
    U. S. and Foreign Producers: Pennwalt Corporation
    Voluntary Purchasing Corporation

2. USE PATTERNS AND FORMULATIONS

Application Sites: Terrestrial crop use on machine and stripper harvested cotton as a desiccant; Non-food use on seed crop okra (Arizona only) as a desiccant.

Types of formulations: 75% soluble concentrate formulation intermediate, also used as end use product.

Types and Methods of Application: Foliar spray (single application at least 4 and 10 days prior to harvest of cotton and okra, respectively.)

Application Rates: Cotton — 2.94 to 4.42 pounds active per acre

Okra — 4.42 pounds active per acre

(Note: 1 quart of 75% liquid = about 2.94 pounds active ingredient)

Usual Carriers: Water

3. SCIENCE FINDINGS

Science summary: Arsenic acid is a form of inorganic arsenic. Such compounds are acutely toxic to humans by ingestion. Inorganic arsenical compounds have been classified as Class A oncogens, demonstrating positive oncogenic effects based on sufficient human epidemiological evidence. The weight of evidence indicates that inorganical arsenical compounds are also mutagens. Although there is teratogenic and fetotoxic potential based on intravenous and intraperitoneal routes of exposure, there is insufficient evidence by the oral route to confirm arsenic acid's teratogenic or fetotoxic effects. Neurotoxic effects have been demonstrated after acute, subchronic and chronic exposures. The metabolism of arsenical compounds in humans is well-documented, but animal studies are not adequate to determine no observed effect levels (NOELs) and acceptable daily intakes (ADIs).

The environmental fate of arsenic acid is not well documented. Studies to demonstrate its fate must take into account the fact that inorganic arsenicals are natural constituents of the soil, and that the forms of inorganic arsenic may change depending on environmental conditions. Based on very limited data, arsenic acid is not predicted to leach significantly. Although elevated levels of arsenic have been found in groundwater in Texas, the source of the arsenic cannot be determined; non-pesticide sources may have been the cause.

Arsenic acid is moderately toxic to birds, slightly toxic to fish and moderately toxic to aquatic invertebrate species.

Chemical Characteristics:

Physical state - Aqueous solution
Oxidation state - pentavalent (As^{+5})
Color - Pale yellow to pale green
Odor - None
Boiling point - Not available
Specific gravity - 1.884 at 20°C
Solubility - Readily soluble in water, forming various As salts
Stability - Most stable under conditions favoring oxidation and at high pH; under reduction conditions or low pH, pentavalent form may convert to trivalent arsenic

Unusual handling characteristics: Reacts with fabric, galvanized metals, black iron and certain other metals resulting in deterioration, corrosion, or liberation of toxic gases (e.g., hydrogen, arsine).

Forms of inorganic arsenic referenced in the Registration Standard

Arsenic acid - H_3AsO_4, containing arsenic in a +5 oxidation form
Sodium arsenate (Na_3AsO_4) - the sodium salt of arsenic acid, also +5
Arsenic trioxide (As_2O_3) - an oxide of arsenic, containing arsenic in a +3 oxidation form
Sodium arsenite ($NaAsO_2$) - a sodium compound related to arsenic trioxide, also +3

Arsenic Acid 59

Toxicological Characteristics:

Acute toxicity. Although arsenic is known to be highly toxic by ingestion, few animal studies are available on the active ingredient, or on the formulated products of arsenic acid. Moreover, the toxicity of arsenic compounds may vary widely depending on the type of formulation and the form of inorganic arsenic in the product.

- Oral (rat) - 40-100 mg/kg. Rats are not a good test species, however, since, alone among animal test species, they retain arsenic in their bodies without significant excretion. Humans are known to be more sensitive to acute arsenic effects than rats.

- Dermal - Undetermined

- Inhalation - Undetermined

- Eye and Skin Irritation - Undetermined

- Dermal Sensitization - Undetermined

Chronic toxicity.

Oncogenicity: Arsenic compounds, including arsenic acid, have been classified as Class A oncogens. Epidemiological studies on workers in copper smelting and pesticide manufacturing, and on populations exposed to excess levels of arsenic in well water in Taiwan are the basis for this classification. Inhalation exposure leads to lung cancers, and ingestion exposure has shown a correlation with development of skin cancers.

The lifetime inhalation oncogenic risks to workers from the cotton use have been estimated at negligible for applicators, and 10^{-4} to 10^{-5} for mixer/loaders.

Dermal and oral oncogenic risks have not been calculated because the risk models are still undergoing Agency review. Completion of this review is expected in late 1986 or early 1987.

Mutagenicity: The sodium salt of arsenic acid (sodium arsenate) and the sodium salt of arsenous acid (a related form of arsenic) have been found to be mutagenic, that is, to interact with DNA to cause heritable effects. Numerous assays have been conducted on cells *in vitro*. Other observed effects include interference with DNA repair mechanisms, direct toxicity to mammalian gonads, and positive effects in microbial systems. Other evidence suggests that similar effects may occur *in vivo*. Sodium arsenite is a more potent mutagen than sodium arsenate.

Teratogenicity/fetotoxicity: Sodium arsenate has been shown to produce teratogenic or fetotoxic effects in hamsters (15-25 mg/kg intravenously); mice (40-45 mg/kg intraperitoneally); and rats (20-50 mg/kg interperitoneally). Similar results have been obtained with sodium arsenite at lower dosages. These results have not been demonstrated using an oral route of exposure, or have been found only at dosages that also cause significant maternal mortality. Because the effects have been shown only using routes of exposure that are not likely to occur with pesticide use, and because the studies were not adequate to establish no-observed effect levels (NOELs) the Agency will require an oral teratogenicity study in two species other than the rat.

Reproductive effects: No data that meet Agency standards are available. A reproduction study on a species other than the rat will be required.

Neurotoxicity: Subchronic and chronic exposure to arsenic compounds causes peripheral and central nervous system neuropathy, the effects of which vary from slight to severe depending on the level and duration of exposure.

Other subchronic and chronic effects: Inorganic arsenic compounds have been observed to cause cardiovascular, skin, blood, and liver and kidney effects in humans. The same effects have been observed in experimental animals. The NOEL for blood effects in dogs is 50 ppm (1.25 mg/kg). The NOEL for liver effects in rats of arsenites is 62.5 ppm and of arsenates is 125 ppm.

Metabolism: The metabolism of inorganic arsenic compounds in animals is well known. The pentavalent form, such as arsenic acid, is metabolized by reduction into the trivalent form, followed by transformation into organic forms which are excreted within several days via the urine. All mammals exhibit this metabolism except rats, which retain arsenic in their bodies for up to 90 days.

Physiological and Behavioral Characteristics: Mechanism of Pesticide Action -- Protein denaturation and enzyme inactivation, resulting in desiccation of plant foliage and stems

Environmental Characteristics: Few data are available on the environmental fate of arsenic acid. Arsenic is a naturally occurring compound that is ubiquitous and exists in different forms (species) depending on environmental conditions. Arsenic acid rapidly dissolves in water. The arsenic moiety of the residue cannot be distinguished from natural arsenic in the soil. Special environmental fate data are required to be submitted. Studies on environmental fate of arsenic acid must be designed to differentiate between natural and pesticide sources of arsenic.

Groundwater concerns: Arsenic has been detected in groundwater underlying areas of arsenic acid use. However, the source of this contamination cannot be determined. Limited information currently available suggests that arsenic acid will not leach significantly. Additional data are required to further evaluate leaching potential.

Arsenic Acid   61

Ecological Characteristics:

Avian acute toxicity:	No data available
Avian dietary toxicity:	Mallard duck - 1606 ppm Bobwhite quail - 168 ppm
Freshwater fish toxicity:	Bluegill sunfish - 66.8 ppm Rainbow trout - 53.1 ppm
Aquatic invertebrates:	Daphnia magna - 6.5 ppm

Based on limited data, the Agency characterizes arsenic acid as moderately toxic to birds by ingestion in the diet. Arsenic acid is slightly toxic to fish and moderately toxic to aquatic invertebrates.

Endangered Species: Cotton desiccant use may pose a potential hazard to the Attwater's Greater Prairie Chicken in three Texas counties (Victoria, Refugio, and Fort Bend). The Agency has referred arsenic acid to the Office of Endangered Species, U.S. Department of Interior, for review as part of a group of cotton pesticides, and will require labeling to protect this endangered species.

Tolerance Assessment

A tolerance of 4.0 ppm in cottonseed oil has been established for arsenic acid, expressed as arsenic trioxide (As_2O_3) (40 CFR 180.180). The use on okra seed crop is a non-food use for which no tolerance is required.

Because the metabolism and chronic effects of inorganic arsenic in humans are well-known, the Agency is not requiring the submission of chronic feeding studies on arsenic acid per se. For regulatory purposes, and until teratogenicity and reproduction studies are submitted, the Agency has calculated a provisional acceptable daily intake (PADI) based upon studies using sodium arsenate.

Based upon a dog study having a NOEL of 50 ppm (1.25 mg/kg of actual arsenic) and a safety factor of 100, the PADI is 0.0165 mg/kg/day of As_2O_3, and the maximum permissible intake for a 60 kg person is 0.99 mg/day.

Available residue data indicate that the maximum residue that will theoretically occur in cottonseed from use of arsenic acid is 0.009 mg/day. The maximum residue therefore uses 0.009/0.99 of the maximum permissible intake, or 0.9%.

Reported Pesticide Incidents

In the period from 1966-1981, 8 incidents were reported to the Agency concerning arsenic acid related to its cotton use. Among these, one involved one human fatality and two persons hospitalized, three involved cattle, and four involved crop damage from spray drift of arsenic acid from nearby areas.

4. SUMMARY OF REGULATORY POSITION AND RATIONALE

--Arsenic acid is currently ungoing Agency Special Review, based upon its oncogenic and teratogenic effects. Products will remain registered until the conclusion of this review. New uses, however, will not be accepted.

--Arsenic acid products will be restricted to use by certified applicators, because of its acute toxicity and oncogenicity.

--Use restrictions based upon groundwater concerns are not warranted at the present time.

--Reentry intervals are not required because use as a desiccant does not lead to significant exposure to field workers.

--Protective clothing is specified for mixer/loaders and applicators, because of the acidic properties of arsenic acid, and its potential oncogenic risks to mixer/loaders and applicators.

--Endangered species labeling statements are required because of potential hazard to the Attwater's Greater Prairie Chicken in Texas. An avian residue monitoring study is required to determine actual levels of arsenic acid in avian feed items.

--Tolerances will be reassessed based upon residue and metabolism studies to be submitted. A rotational crop restriction or tolerances may be needed if followup crops take up arsenic acid residues from the soil.

5. SUMMARY OF MAJOR DATA GAPS

Product chemistry data - arsenic acid	Feb. 1987
Residue chemistry data - arsenic acid --Plant and animal metabolism --Analytical methodology for residues --Magnitude of residues in cotton	Feb. 1988
Environmental fate studies --Metabolism in soil --Leaching --Laboratory volatility --Soil dissipation --Rotational crop studies	 Dec. 1988 Aug. 1987 Aug. 1987 Dec. 1988 Dec. 1989
Toxicology studies --Acute toxicity --Teratology (rabbit and mouse or hamster) --Reproduction (rodent other than rat) --Dermal penetration --Glove permeability	 Feb. 1987 Dec. 1987 Dec. 1989 Aug. 1987 Feb. 1987

Ecological effects studies
--Avian acute toxicity May 1987
--Residue monitoring study on avian food items Dec. 1987
--Aquatic invertebrate early life stage Dec. 1987

6. CONTACT PERSON AT EPA

Richard F. Mountfort
U.S. Environmental Protection Agency
TS-767C
401 M St., SW
Washington, D.C. 20460
703-557-1830

DISCLAIMER: The information presented in this Pesticide Fact Sheet is for informational purposes only and may not be used to fulfill data requirements for pesticide registration and reregistration.

ARSENIC TRIOXIDE

Date Issued: December 1986
Fact Sheet Number: 110

1. DESCRIPTION OF CHEMICAL

    Common Name: Arsenic Trioxide

    Chemical Name: Arsenious oxide - As_2O_3

    Trade Names: Refined arsenic trioxide is known as white arsenic

    EPA Shaughnessy Code: 007001

    Chemical Abstracts Service (CAS) Number: 1327-53-3

    Year of Initial Registration:

    Pesticide Type: Antifoulant agent, Herbicide, Insecticide and Rodenticide.

    Chemical Family: Inorganic Arsenicals

    U.S. and Foreign Producers: Grant Laboratories

2. USE PATTERNS AND FORMULATIONS

    Approximately 85% of the pesticidal use of arsenic trioxide is as a liquid rodenticide bait to control rats and mice. The remaining 15% is used to kill moles and pocket gophers, as an insecticide bait to control ants and as an antifoulant agent in boat paints. Arsenic trioxide is currently registered as a noncrop herbicide; however, there is no known usage at this time.

    ° Methods of Application: Rodenticide baits are applied where rats, mice, moles and gophers are seen. As an insecticide it is used in bait stations and as an antifoulant it is applied by painting.

    ° Application Rates: Rodenticide- These baits contain 1.14% arsenic.
    Antifoulant- 2.4% arsenic/gallon batch

    Types of Formulations: Pellets, wettable powders, pastes, ready to use solutions, impregnated material, granular and as a formulation intermediate.

3. SCIENCE FINDINGS

- **Chemical Characteristics**

    Arsenic trioxide is a form of inorganic arsenic. It normally exists either as transparent crystals or an amorphous white powder with no discernible odor. Arsenic trioxide contains 76% arsenic and is slightly soluble in water and other solvents which do not promote chemical transformation. However, the compound dissolves in acidic or alkaline aqueous media to yield either a free acid or salt which are soluble in a number of solvents. As_2O_3 sublimes at 193°C, the density is 3.865 and the molecular weight is 197.84. The technical chemical contains between 90% and 99.5% and the formulations contain from 0.25% to 25.0% arsenic trioxide. Arsenic trioxide is produced as a by-product of copper smelting operations. It is the base compound from which all other arsenicals are produced.

- **Toxicological Characteristics**

    Inorganic arsenical compounds have been classified as Class A oncogens, demonstrating positive oncogenic effects based on sufficient human epidemiological evidence.

    Inorganic arsenicals have been assayed for mutagenic activity in a variety of test systems ranging from bacterial cells to peripheral lymphocytes from humans exposed to arsenic. The weight of evidence indicates that inorganic arsenical compounds are mutagenic.

    Evidence exists indicating that there is teratogenic and fetotoxic potential based on intravenous and intraperitoneal routes of exposure; however, evidence by the oral route is insufficient to confirm arsenic trioxide's teratogenic and fetotoxic effects.

    Inorganic arsenicals are known to be acutely toxic. The symptoms which follow oral exposure include severe gastrointestinal damage resulting in vomiting and diarrhea, and general vascular collapse leading to shock, coma and death. Muscular cramps, facial edema, and cardiovascular reactions are also known to occur following oral exposure to arsenic.

- **Environmental Characteristics:** The environmental fate of arsenic trioxide is not well documented. Studies to demonstrate its fate must take into account the fact that inorganic arsenicals are natural constituents of the soil, and that forms of inorganic arsenic may change depending on environmental conditions.

- Ecological Characteristics: Arsenic trioxide is moderately toxic to birds, slightly toxic to fish and moderately toxic to aquatic invertebrate species.

- Metabolism: The metabolism of inorganic arsenic compounds in animals is well known. The pentavalent form is metabolized by reduction into the trivalent form, followed by transformation into organic forms which are excreted within several days via the urine. All animals exhibit this metabolism except rats, which retain arsenic in their bodies for up to 90 days.

- Reported Pesticide Incidents: The Agency's Pesticide Incident Monitoring System (PIMS) contains many recorded incidents of accidental poisonings from the use of arsenic trioxide baits. Between 1966 and 1979, 72 incidents were reported; ten of these incidents resulted in child fatalities.

4. SUMMARY OF REGULATORY POSITION AND RATIONALE

The Agency is proposing to cancel all existing nonwood registrations of arsenic trioxide. Based upon the risk of acute toxicity poisonings and the other toxicological characteristics described above, the Agency has determined that in light of the limited benefits for nonwood uses of arsenic trioxide the risks of continued use outweigh the benefits.

- Benefits Analysis: Nationwide, user costs would be expected to increase by approximately five thousand dollars if arsenic trioxide were cancelled for moles and pocket gophers, resulting in no measurable impact. For all other uses negligible impact is expected.

5. CONTACT PERSON

Douglas McKinney
Special Review Branch, Registration Division
Office of Pesticide Programs (TS-767C)
401 M Street, S.W.
Washington, D.C. 20460
(703) 557-5488

DISCLAIMER: The information presented in this Pesticide Fact Sheet is for informational purposes only and may not be used to fulfill data requirements for pesticide registration or reregistration.

AVERMECTIN

Date Issued: April 18, 1986
Fact Sheet Number: 89

1. DESCRIPTION OF CHEMICAL

   Generic Name: Avermectin B_1 [A mixture of avermectins containing $\geq$ 80% avermectin B_1a (5-O-dimethyl-avermectin A_1a) and $\leq$ 20% avermectin B_1b (5-O-demethyl-25-de(1-methylpropyl)-25-(1-methylethyl)avermectin A_1a)]

   Common Name: None assigned.

   Trade Name: Affirm™

   EPA Shaughnessy Code: 0122804

   Chemical Abstracts Service (CAS) Numbers: 65195-55-3 and 65195-56-4

   Year of Initial Registration: 1986

   Pesticide Type: Insecticide

   Chemical Family: Avermectins (macrocylic lactones isolated from soil organism Streptomyces avermitilis).

   U.S. Producers: Merck Sharp & Dohme Research Laboratories

2. USE PATTERNS AND FORMULATIONS

   Application Sites: Turf, lawns and other non-crop wide outdoor areas.

   Type of Formulations: 0.011% insecticide bait.

   Method of Application: Bait broadcast (ground or air application) and individual mound to mound treatment.

   Application Rates: 50 mg active ingredient (A.I.) per acre (1 pound of product per acre).

   Usual Carriers: Pregelled defatted corn grit carrier.

   Limitations: Do not use in pastures, rangeland, or croplands.

3. SCIENCE FINDINGS

Summary Science Statement:

Technical avermectin exhibits high mammalian acute toxicity. It is not considered to be mutagenic nor teratogenic and does not sensitize skin. It is not readily absorbed by mammals and the majority of the residue is excreted in the feces. The results of the acute toxicity on the bait formulation indicates that it is of low toxicity. Chronic feeding and oncogenicity studies and a three-generation reproduction study are currently in progress and will be required in support of food/feed crop uses.

Sufficient data are available to characterize avermectin from an environmental fate and ecological stand point. Avermectin is extremely toxic to fish and aquatic invertebrates and highly toxic to birds. However, because of its low use rate and rapid rate of photolysis no adverse acute or chronic effects to aquatic, estuarine or endangered species are expected. The degradation products are less toxic than the parent and tend to become less toxic as they continue to degrade. Avermectin undergoes rapid photolysis, is readily degraded by soil microorganisms and, due to its binding properties and low water solubility, is expected to exhibit little or no potential for leaching.

A tolerance assessment is not needed because the registered use pattern is for non-crop/non-food use. There are no data gaps.

A. Chemical Characteristics:

Physical State: Crystalline powder.

Color: Yellowish-white.

Odor: Odorless.

Melting Point: 155 - 157°C.

Vapor Pressure: Being tested, expected to be extremely low.

Density: 1.16 $\pm$ 0.05 at 21°C.

Solubility: Insoluble in water ($\leq$ 5 ug/ml), readily soluble in organic solvents.

pH: NA. The avermectin molecule has neither acidic nor basic functional groups.

Octanol/Water Partition Coefficient: 9.9×10^3

B. Toxicological Characteristics:

   Technical Avermectin

   Acute Oral: 1.52 mg/kg. Toxicity Category I.

   Acute Dermal: LD_{50} >380 mg/kg. Toxicity Category II.

   21-day Dermal: Noel is 125 mg/kg/day

   Dermal Sensitization: Negative for skin sensitization.

   Acute Inhalation: 1.62 mg/l. Toxicity Category II.

   Teratogenicity: Three teratology studies (rat, rabbit, and mouse) have been evaluated to determine the teratogenic potential of avermectin. Teratogenic effects were negative for the rat up to 1.6 mg/kg/day and for the rabbit up to 2.0 mg/kg/day. Avermectin was positive in the mouse at 0.4 mg/kg/day and the NOEL for these effects was 0.2 mg/kg/day. However, the margin of safety between the NOEL and exposure to applicators and persons upon re-entry is estimated to be greater than 10,000.

   Mutagenicity: Adequate studies are available to demonstrate that avermectin is not a mutagen. Avermectin was not mutagenic in the Ames assay and in vivo bone marrow cytogenetics. In rat hepatocytes, Avermectin caused an induction of single strand DNA breaks in vitro. No effect was observed when this same assay was carried out in hepatoocytes from rats dosed in vivo at the LD_{50} (10.6 mg/kg). In the mammalian cell mutagenic assay, Avermectin was not mutagenic for V-79 cells.

   Metabolism (rats): The metabolic T 1/2 in rats is 1.2 days. Avermectin does not bioaccumulate in rats.

   Affirm Fire Ant Formulation

   Oral LD_{50} in rats: LD_{50} > 5.0 gm/kg. Toxicity Category III.

   Dermal LD_{50} in rats: LD_{50} > 2.0 gm/kg. Toxicity Category III.

   Acute inhalation LC_{50}: Not required due to large particle size and low vapor pressure of technical.

Primary eye irritation: Toxicity Category III.

Primary skin irritation: No irritation. Toxicity Category III.

C. Physiological and Biological Characteristics:

Foliar absorption: not absorbed.

Translocation: not translocated.

Mechanism of Pesticide Action: Avermectin is γ-aminobutyric acid (GABA) agonist, and thus acts in arthropods by inhibiting nervous signal transmission at the neuromuscular juncture causing paralysis. No effect on any cholinergic nervous systems have been demonstrated.

D. Environmental Characteristics:

Avermectin is not expected to hydrolyze in the environment. It photodegrades rapidly in water and soil with half-lives less than 1 day. Soil metabolism studies conducted in darkness indicate degradation does occur with a half-life of 2 weeks to 2 months under aerobic conditions. Anaerobic degradation is slower. It is not expected to accumulate in fish. Avermectin's solubility in water is determined to be 7.8 ppb. The field dissipation study indicates that avermectin, when applied in the bait formulation directly to the soil, dissipates with a half life of about a week but may persist longer if the bait is shaded. Avermectin and its degradates do not leach into the soil. There are no concerns at this time in regard to ground water contamination. Due to low application rates, it is unlikely that there would be any significant exposure to humans and nontarget organisms.

E. Ecological Characteristics:

Avian Oral: Bobwhite quail - LD_{50} > 2000 mg/kg.

Bobwhite quail - LC_{50} = 3102 ppm.

Avian Dietary: Mallard duck - LC_{50} = 383 ppm.

Freshwater Fish: Bluegill - LC_{50} = 9.6 ppb.

Rainbow trout - LC_{50} = 3.2 ppb.

Acute Freshwater Invertebrate: Daphnia - LC_{50} = 0.22 ppb.

Acute Estuarine Invertebrate: Shrimp, mysid - LC_{50} = 0.2 ppb.

Estuarine Fish: Fathead minnow - LC_{50} = 15 ppb.

Oyster Embryo Larvae: LC_{50} = 430 ppb.

4. SUMMARY OF REGULATORY POSITION AND RATIONALE

The Agency has determined that it should allow the registration of avermectin for non-crop, wide area general outdoor use for control of the imported fire ant. Adequate studies are available to assess the acute toxicological effects of avermectin to humans. Since the proposed use is a non-food/food use a tolerance assessment is not necessary. No reentry interval is necessary. Available data are sufficient to characterize the environmental fate of avermectin. Although technical avermectin is highly toxic to fish, birds, and invertebrates, the 50 mg/acre rate plus rapid degradation should minimize this potential hazard. The effectiveness of the bait formulation has been extensively tested by USDA and also by the registrant under an EPA-approved experimental use permit from 1982-1985. None of the criteria for unreasonable adverse effects listed in section 162.11(a) of Title 40 of the U.S. Code of Federal Regulations have been met or exceeded for this use.

5. SUMMARY OF MAJOR DATA GAPS

There are no data gaps for the present use pattern.

6. CONTACT PERSON AT EPA

George T. LaRocca
Product Manager 15
Insecticide-Rodenticide Branch
Registration Division (TS-767C)
401 M Street SW.
Washington, DC 20460
(703) 557-2400

DISCLAIMER: The information presented in this Chemical Information Fact Sheet is for informational purposes only and may not be used to fulfill data requirements for pesticide registration and reregistration.

AZINPHOS-METHYL

Date Issued: September 30, 1986
Fact Sheet Number: 100

1. Description of Chemical

-Generic Name: O,O-dimethyl-s-[(4-oxo-1,2,3-benzotriazin-3(4H)-yl)methyl] phosphorodithioate
-Common Name: Azinphos-methyl
-Trade Names: metiltriazotion, carfene, cotion-methyl, gusathion, gusation-M, guthion, Bay 9027, Bay 17147, R-1582
-EPA Shaughnessy Number: 058001
-Chemical Abstracts Service (CAS) Number: 86-50-0
-Year of Initial Registration: 1956
-Pesticide Type: Insecticide, Acaricide, Molluscicide
-Chemical Family: Organophosphate
-U.S. and Foreign Producers: Bayer AG (Federal Republic of Germany), Makhteshim Chemical Co.(Israel), and Mobay Chemical Corporation (U.S.) and Aceto Chemical Company.

2. Use Patterns and Formulations

-Application sites: fruit and field crops, vegetables, tobacco, ornamentals and forest trees
-Types and methods of application: foliar (by ground or aerial equipment), ULV foliar application, soil application, transplant water application
-Application rates: 0.047 to 10.35 lb a.i./A
-Types of formulation: technical, formulation intermediate, dust, granular, wettable powder, emulsifiable concentrate, flowable concentrate, soluble concentrate
-Usual carriers: Confidential Business Information
-All liquid formulations with concentrations of 13.5% or greater are currently classified as restricted use chemicals

3. Science Findings

Chemical Characterisitics:

Technical azinphos-methyl is a yellow-brown solid with a melting point at around 63°C, that decomposes at elevated temperatures and is soluble in most organic solvents. The empirical formula is $C_{10}H_{12}N_3O_3PS_2$ and the molecular weight is 317.1.

Toxicology Characterisitics

- Acute oral: 4.4 mg/kg (rat), Toxicity Category I, highly toxic
- Acute dermal: 200 mg/kg (rat), Toxicity Category I, highly toxic
- Primary Eye Irritation: Data gap.
- Acute Inhalation: Data gap.
- Primary Skin Irritation: Data gap.
- Skin Sensitization: Data gap.
- Major Routes of Exposure: Handling of the concentrated pesticide and airblast application produce the largest exposure per pound of azinphos-methyl handled. Human exposure to azinphos-methyl from handling, application, and re-entry

operations would be minimized by the use of respirators and
protective clothing.

-Oncogenicity: A mouse oncogenicity study showed that no
increased incidence of tumors could be attributed to exposure
to azinphos-methyl. A rat oncogenicity study, that was
found to be uacceptable upon re-evaluation, showed that male
rats developed tumors of the pancreatic islets and of the
follicular cells of the thyroid. Because of the wide range of
spontaneous incidence of these tumors in Osborne-Mendel rats,
the Agency concludes that there is no clear link between the
development of tumors to the administration of azinphos-methyl,
the Agency is requiring that a rat oncogenicity study be
conducted and submitted.
-Metabolism: Data gap.
-Teratology: Data gap.
-Reproduction: Data gap.
-Mutagenicity: Data gap.

Physiological and Biochemical Characteristics

-Mechanism of Pesticidal Action: Cholinesterase inhibition
by all routes of exposure.
-Metabolism and persistence in plants and animals:
Radiolabeled azinphos-methyl residues are absorbed through
the roots and translocated to the shoot of hydroponically
treated plants. Available data indicate that azinphos-methyl
is relatively persistent on leaf surfaces and is gradually
degraded to polar metabolites. Although the submitted
studies are not adequate to assess the nature of azinphos-methyl
in animals, metabolites of azinphos-methyl were identified
in chicken excreta and rat urine.

Environmental Characteristics:

-Available data indicate that azinphos-methyl exhibits both
low soil mobility and low leaching potential in a variety of soil
types. Leaching studies indicate a low potential for leaching
of azinphos-methyl residues to groundwater. In sandy soils
azinphos-methyl dissipates in the field fairly rapidly and is
metabolized by microorganisms.

The half-life of azinphos-methyl in a non-sterile soil is
21 days under aerobic conditions and 68 days under anaerobic
conditions. It is more stable in sterile conditions with a
half-life of 355 days. Under aerobic conditions the following
azinphosmethyl metabolites were detected: oxygen analog
residues, mercaptomethyl benzazimide, benzazimide, hydroxymethyl
benzazimide, and bis-methyl benzazimide sulfide. Benzazimide
and hydroxymethyl benzazimide, were identified as the major
degradation products resulting from hydrolysis. Available
photodegradation data are inadequate to assess the
photodecomposition of azinphos-methyl.

-Re-entry: Treated areas may not be re-entered for at least
24 hours unless protective clothing is worn.

Ecological Characteristics

-Acute avian oral toxicity: Supplemental data indicate an LD_{50}
of 136 mg/kg body weight for mallard ducks, 60-120 mg/kg for
bobwhite quail, 74.9 mg/kg for pheasant, and 84.2 mg/kg
for chukar. (Moderately toxic for all species tested).
-Avian dietary toxicity: LC_{50} = 488 ppm for bobwhite quail
(highly toxic), LC_{50} = 639 ppm for Japanese quail (moderately toxic
and LC_{50} = 1940 ppm for mallard duck and 1821 ppm for ring-necked
pheasant (slightly toxic).
-Freshwater fish acute toxicity (LC_{50}): Available data
place azinphos-methyl in the range from very highly toxic to
moderately toxic [LD_{50} values of 0.36-4,270 ug/l, depending
on species tested with most values in the very highly
toxic range (less than 100 ug/l)].
-Freshwater aquatic invertebrate toxicity: 0.13-56 ug/l, depending
on the species tested (very highly toxic).

Tolerance Assessment

-Tolerances have been established for residues of azinphos-methyl
in a variety of raw agricultural commodities in meat, fat and
meat byproducts (refer to 40 CFR 180.154 and 40CFR 180.154a. for
listing of tolerances), and in processed food (21 CFR 193.150) and
and feed (21 CFR 561.180).
-The Agency is unable to complete a full tolerance assessment for
the established tolerances due to residue chemistry and
toxicology data gaps.
-No Mexican tolerances have been established for azinphos-methyl.

Commodity	U.S.	Canadian	Codex
Alfalfa	2.0	0.1	2.0
Alfalfa,hay	5.0	-	-
Almonds	0.3	-	0.2
Almonds,hulls	10.3	-	10.0
Apples	2.0	2.0	1.0
Apricots	2.0	2.0	2.0
Artichokes	2.0	0.5	0.5
Barley,grain	0.2	0.1	0.2
Barley,straw	2.0	0.1	-
Beans(dry)	0.3	2.0	0.5
Beans(snap)	2.0	2.0	0.5
Birdsfoot Trefoil	2.0	0.1	-
BirdsfootTrefoil,hay	5.0	0.1	-
Blackberries	2.0	2.0	1.0
Blueberries	5.0	2.0	1.0
Boysenberries	2.0	2.0	1.0
Broccoli	2.0	1.0	1.0
Brussels Sprouts	2.0	1.0	1.0
Cabbage	2.0	1.0	0.5

Commodity	U.S.	Canadian	Codex
Cattle,fat	0.1	-	-
Cattle,meat by product	0.1	-	-
Cattle,meat	0.1	-	-
Cauliflower	2.0	0.5	0.5
Celery	2.0	0.5	2.0
Cherries	2.0	1.0	1.0
Citrus fruits	2.0	2.0	2.0
Clover	2.0	0.1	-
Clover,hay	5.0	-	-
Cottonseed	0.5	-	0.2
Crabapples	2.0	-	1.0
Cranberries	2.0	-	-
Cucumbers	2.0	0.5	0.5
Eggplant	0.3	-	0.5
Filberts	0.3	-	-
Goats,fat	0.1	-	-
Goats,meat by product	0.1	-	-
Goats,meat	0.1	-	-
Gooseberries	5.0	-	-
Grapes	5.0	5.0	4.0
Grass,pasture	2.0	0.1	-
Grass,pasture,hay	5.0	-	-
Horses,fat	0.1	-	-
Horses,meat by products	0.1	-	-
Horses,meat	0.1	-	-
Kiwi fruit	10.0	0.4	4.0
Loganberries	2.0	2.0	1.0
Melons	2.0	-	2.0
Nectrines	2.0	-	1.0
Nut,Pistachio	0.3	-	-
Oats,grain	0.2	0.1	0.2
Oats,straw	2.0	-	-
Onions(green)	2.0	1.0	0.5
Parsley(leaves)	5.0	-	0.5
Parsley(roots)	2.0	-	0.5
Peaches	2.0	2.0	4.0
Pears	2.0	2.0	1.0
Peas,black-eyed	0.3	-	0.5
Pecans	0.3	-	-
Peppers	0.3	0.2	0.5
Plums	2.0	1.0	1.0
Potatoes	0.3	0.1	0.2
Quinces	2.0	2.0	1.0
Raspberries	2.0	2.0	1.0
Rye,grain	0.2	0.1	0.2
Rye,straw	2.0	0.1	-
Sheep,fat	0.1	-	-
Sheep,meat by product	0.1	-	-
Sheep,meat	0.1	-	-
Soybeans	0.2	0.2	-

Commodity	U.S.	Canadian	Codex
Spinach	2.0	2.0	0.5
Strawberries	2.0	1.0	1.0
Sugarcane	0.3	-	-
Tomatoes	2.0	1.0	0.5
Walnuts	0.3	0.1	-
Wheat, grain	0.2	0.1	0.2
Wheat, straw	2.0	0.1	-
Milk	0.04	-	-
Soybean oil	1.0	-	-
Dried citrus	5.0	-	-
Sugarcane bagasse	1.5	-	-

The data for azinphos-methyl residues in or on potatoes, parsley, pistachios, sugarcane bagasse and sugarcane are adequate to support the respective established tolerances.

Data are not adequate to support the established tolerance for residues in or on onions, celery, spinach, broccoli, Brussels sprouts, cabbage, cauliflower, beans, soybeans, peppers, tomatoes, cucumbers, melons, citrus fruit, apples, apricots, cherries, nectarines, peaches, plums, blackberries, blueberries, cranberries, grapes, strawberries, almonds, almond hulls, filberts, pecans, walnuts, wheat grain, wheat straw, pasture grass, alfalfa, clover, artichokes, cottonseed, kiwi fruit, eggplant, crabapples, pears, quices, boysenberries, loganberries, raspberries, oat grain, rye grain, barley grain, oat straw, rye straw, barley straw, birdsfoot trefoil, soybean oil, and dried citrus pulp.

A Provisional Limiting Dose (PLD) based on a 2 year dog feeding study has been established using a uncertainty factor of 100. The Maximum Permissible Intake (MPI) is 0.075 mg/kg for a 60 kg. person. The Theoretical Maximum Residue Contribution (TMRC) to the human diet from the existing tolerances is 0.6678 mg for a 1.5 kg diet which is 899% of the MPI. (This value is greatly inflated when compared to actual dietary residues expected in practice, and is due solely to the additional 10-fold uncertainty factor applied for the reduced data base and two-fold increase in sensitivity of the dog compared to that of the rat). When the required chronic feeding studies are submitted, the PLD and TMRC will be re-evaluated.

-Reported Pesticide Incidents: The Pesticide Incident Monitoring System (PIMS) reported on azinphos-methyl from 1966 through 1979. The 1981 report is the latest one available, which cites 71 incidents associated with human injury. There was one fatality which was a confirmed suicide. The PIMS report cited one bird kill and 29 incidents involving fish and other aquatic organisms. For the period 1978-1985, seven additional fish kills, citing azinphos-methyl, alone, were obtained at the Agency's Office of Water.

4. Summary of Regulatory Position

   A. The Agency is requiring extensive field monitoring data to better define the extent of exposure and hazard to wildlife.

   B. No new tolerances or new food uses will be considered until the Agency has received data sufficient to assess existing tolerances for azinphos-methyl.

   C. The Agency is concerned about the potential for human poisonings (cholinesterase inhibition) from the use of azinphos-methyl. The Agency will continue to restrict all liquid formulations of azinphos-methyl with greater than 13.5% a.i. In addition, registrants must either classify all products for restricted use (due to acute toxicity) or submit dermal and inhalation data to support the appropriate toxicity category.

   D. Revised Protective clothing statements are required to be included on the labels of azinphos-methyl end-use products.

   E. The Agency has concluded that data are not adequate to determine the oncogenic potential of azinphos-methyl and is requiring numerous toxicology studies mentioned below.

   F. The Agency is requiring that endangered species labeling be added to labels for certain azinphos-methyl uses.

   G. A 24 hour re-entry interval, previously established under 40 CFR 170.3 (b)(2) will remain in effect.

   H. The Agency is imposing an interim rotational crop restriction of 6 months for root crops and 30 days for all other registered crops until the required crop rotation data are submitted.

5. Summary of Major Data Gaps

-Additional crop residue studies on various commodities and plant and animal metabolism studies are required to support existing tolerances.
-The following studies are required to assess the toxicological characteristics of technical azinphos-methyl:
Acute inhalation, Delayed Neurotoxicity, Oncogenicity/ Chronic testing in rat, Teratology, Reproduction, Mutagenicity, and General Metabolism.
-The following data are required to fully characterize azinphos-methyl's environmental fate: Re-entry data (foliar dissipation, soil dissipation, dermal exposure, inhalation exposure

Special Testing-Glove Permeability, Photodegradation, Anaerobic Aquatic*, Forestry Dissipation*, Rotational Crops (Confined), Irrigated Crops Accumulated Study*, Accumulation in Non-target Organisms*
-Additional data are required to assess the impact on wildlife from the use of azinphos-methyl: acute avian oral toxicity, wild mammal toxicity, avian reproduction, field testing for mammals and birds, acute toxicity to freshwater invertebrates, acute toxicty to estuarine and marine organisms, fish early life stage, simulated or actual field testing for aquatic organisms and honey bee toxicity.
-Product chemistry and acute toxicity data are required.

*Data not needed if label clarifications are made.

Contact person at EPA:

Lawrence J. Schnaubelt
Product Manager (12)
Insecticide-Rodenticide Branch
Registration Division (TS-767C)
Office of Pesticide Programs
Environmental Protection Agency
401 M. Street, S.W.
Washington,D.C. 20460

Office location and telephone number:
Room 202, Crystal Mall Building #2
1921 Jefferson Davis Highway
Arlington,Va. 22202
703-557-2386

Disclaimer: The information presented in this pesticide Fact Sheet is for informational purposes only and may not be used to fufill data requirements for pesticide registration and reregistration.

BENTAZON AND SODIUM BENTAZON

Date Issued: September 30, 1985
Fact Sheet Number: 64

1. Description of chemical

Common Names: Bentazon

   Sodium bentazon (= sodium salt of bentazon)

Chemical Name: 3-(1-methylethyl)-1H-2,1,3-benzothiadiazin-4(3H)-one 2,2-dioxide

   3-(1-methylethyl)-1H-2,1,3-benzothiadiazin-4(3H)-one 2,2-dioxide, sodium salt of

Trade names for Federal Section 3 Registered Products:

   Manufacturing-use product: "Bentazon Manufacturer's Concentrate" (EPA Registration No. 7969-42)

   End-use product (single active ingredient): "Basagran" Postemergence Herbicide" (EPA Registration No. 7969-45)

   End-use product (combined with atrazine): "Laddock" Postemergence Herbicide" (EPA Registration No. 7969-54)

NOTE: All three of the above products have sodium bentazon as the active ingredient.

EPA Chemical Code: Bentazon: 275200

   Sodium bentazon: 103901

Chemical Abstracts Service (CAS) Registry Number:

   Bentazon: 50723-80-3

Year of Initial Registration: 1975

Pesticide Type: Heterocyclic nitrogen herbicide

Producer Marketing in U.S.: BASF Wyandotte Corporation

2. Use patterns and formulations

   Application sites:

   soybeans, rice, corn, sorghum, peanuts, beans (dry or succulent), peas (dry or succulent), established peppermint and spearmint, established ornamental turf.

The only additional site under Section 24(c) (special local need) registration is Bohemian chili peppers.

Percent of Particular Crops Treated with the Pesticide as of 1984:

soybeans	20%
rice	10%
dry beans and peas	5%
peanuts	2%
corn	<1%

Percent of Pesticide Applied to Particular Crops as of 1984:

soybeans	96%
corn	2%
rice	1%
dry beans and peas	<1%
peanuts	<1%

Types and Methods of Application:

Sodium bentazon is applied, by ground or air, as a broadcast foliar spray after the crop and weeds have emerged from the soil. It is used to control selected broadleaf weeds and sedges only.

Application Rates:

Rates, from 0.75 lb active ingredient (a.i.)/acre to 2 lb a.i./acre, vary by crop, geographic region, target species, and site conditions.

Types of Formulations:

The manufacturing-use product is a 46% sodium bentazon liquid. Basagran™ Postemergence Herbicide is a "soluble liquid concentrate" containing 42% sodium bentazon. Laddock™ Postemergence Herbicide is a "flowable liquid concentrate" containing 18.52% sodium bentazon and 16.96% atrazine.

Usual Carriers:

The usual carrier is water. An oil concentrate is used under certain conditions.

3. Science Findings

Chemical Characteristics:

Technical grade bentazon (isolated before formation of sodium bentazon) is an odorless, non-volatile solid with a melting point of 137-139°. It's solubility (g/100 g solvent, 20°C) is 0.05 in water and 150.7 in acetone. Sodium bentazon is considerably more soluble in water than bentazon, with a solubility of 230 g/100 g in water.

Mammalian Toxicology Characteristics:

Acute toxicity:

Sodium bentazon is in Toxicity Category III (defined in 40 CFR 162.10), based upon acute oral toxicity in the rat and acute dermal testing in the rat. An acute inhalation study is supplementary (valid but does not meet EPA guideline requirements) and will need to be repeated.

Subchronic toxicity:

Subchronic data are supplementary (90-day rat feeding study and 13-week dog study; bentazon) or invalid (21-day dermal study; sodium bentazon), and will require replacement. Compound-related effects were seen at 300 and 3000 ppm test levels in the dog study (numerous effects at 3000 ppm; prostatitis at 300 ppm was basis of lowest-effect-level of 300 ppm and no-effect level of 100 ppm). A one year dog study is required for hazard assessment.

Chronic Toxicity:

All chronic toxicity data for bentazon are supplementary or invalid, and will require replacement. 24-month rat and 18-month mouse studies have been invalidated and a "for cause" laboratory audit requested, due to substantial deficiencies. In a supplementary mouse oncogenicity study, no specific clinical or pathological symptoms could be associated with bentazon exposure.

A rat 3-generation reproduction study found no compound-related effects for bentazon up to a dietary level of 180 ppm. However, without effects at the highest dose, the dose selection is considered inadequate.

The available rat and rabbit teratogenicity studies are inadequate, but do not suggest that bentazon is a potent teratogen or fetotoxic agent. Terata were observed in one rat study at a dose of 200 mg/kg/day, but the utility and validity of these data are in question. Additional teratology studies in the rat and rabbit are required.

A variety of mutagenicity studies (analytical grade bentazon; or sodium salt) have been reviewed, but none are adequate for regulatory purposes. This is also the case for metabolism studies with labeled bentazon.

Physiological and Biochemical Behavioral Characteristics:

Mechanism of pesticidal action:

Foliar application of sodium bentazon results in photosynthesis inhibition in susceptible species. Visible injury to the treated leaf surface usually occurs in 4-8 hours, followed by plant death.

Foliar absorption and translocation:

Bentazon applied post-emergence to young plants may be absorbed and translocated from the site of application. The degree of translocation depends on the plant species. Whether translocated or not, bentazon is rapidly metabolized, conjugate and incorporated into natural plant components.

Metabolism in plants and animals:

The metabolism of bentazon in tolerant plants is partially understood. The 6-hydroxy and the 8-hydroxy metabolites are included with bentazon in the tolerance on crops. Further work is necessary to characterize other metabolites.

The metabolism in animals is not understood as the acid hydrolysis procedure was not run on any of the methanol extracts of eggs and poultry tissues. However in animals there is no hydroxylation of bentazon as in plants. Based on available information, the residues in animals consist of the metabolite AIBA and bentazon, and tolerances for residues in animal products should be expressed in terms of these combined residues.

Environmental Characteristics

Decomposition:

Based on available, validated data, bentazon appears to be stable to hydrolysis, but photodegrades in water with a half-life of <24 hours. It also photodegrades on soil. Under aerobic conditions in lab and field, bentazon degrades with a half-life of <1 month in soil.

Bioaccumulation:

Bentazon residues accumulate in the tail meat and viscera of crayfish with bioconcentration factors of <10X.

Surface and ground water contamination concerns:

Bentazon is very mobile in soil but the relatively rapid degradation is expected to prevent groundwater contamination. An aged leaching study is needed to determine the potential for metabolites to contaminate groundwater. Bentazon does have the potential to contaminate surface water because of 1) its mobility in runoff water, for all crops and 2) its rice use pattern that involves either direct application to water or application to fields prior to flooding.

Ecological Effects

Hazards to fish and wildlife:

Technical bentazon is considered slightly toxic to birds based on subacute dietary testing. Formulated bentazon is considered slightly toxic to birds based on acute oral testing with a 50% a.i. wettable powder. Avian reproduction testing did not show effects up to the highest dietary level tested, but the studies were found to lack vital information and do not presently meet EPA guidelines.

Technical bentazon is characterized by EPA as practically nontoxic to both coldwater and warmwater fish, and slightly toxic to aquatic invertebrates, based on review of acute testing. Formulated bentazon is considered practically nontoxic to coldwater and warmwater fish based on acute testing with a 48% a.i. liquid product.

Applications of bentazon on registered use sites are considered unlikely to result in acute hazard to most nontarget organisms because of its generally low toxicity, based on available data, and low application rates. However, a final risk assessment is deferred due to lack of critical environmental chemistry data and certain ecological effects data.

Potential problems related to endangered species:

Biological opinions (covering all registered pesticides) have been received from the U.S. Fish and Wildlife Service Office of Endangered Species (OES) for three of the crops for which sodium bentazon is registered: corn, sorghum, and soybeans To avoid jeopardy to the Valley Elderberry Longhorn Beetle (Desmocerus californic dimorphus), OES indicates that herbicides should be prohibited from designated areas in California in order to protect the host plant, elderberry (Sambucus spp.). Labeling developed by EPA to implement this prohibition is described below.

The above opinions also stated that "...to avoid jeopardy to Solano grass...the use of any herbicides toxic to graminoides should be prohibited within..." certain defined geographic areas of California, because of concerns with spray drift and runoff from agricultural areas. Solano grass (Tuctoria (= Orcuttia) mucronata) is an endangered plant species found in a vernal lakebed in Solano County, California. Since no grass species are claimed on existing labeling to be controlled with sodium bentazon and the herbicide is used on various grass crops, Ecological Effects Branch (with informal consultation with OES) did not consider there to be a threat to Solano grass from the registered use of this chemical. Subsequently, information has been located indicating that sodium bentazon may affect certain germinating grass species with direct exposure at full dosage rates, but that it does not affect grasses after germination. Solano grass germinates in March and April. The earliest planting date for corn, sorghum, or soybeans is for corn, which can be planted as early as April 15. Since sodium bentazon is applied postemergence to the weeds, there may be little opportunity for an application that could affect Solano grass. Further consultation with OES will be initiated. Label restrictions, as above, and/or plant protection data requirements (under 40 CFR § 158.150) may be imposed.

An oyster study and further environmental chemistry data are required, in part, to evaluate whether there is any hazard to endangered mussel species from sodium bentazon use.

The Agency is not aware of any other data which would suggest that the risk criteria of § 162.11 have been met or exceeded for the uses of sodium bentazon at the present time.

TOLERANCES

In the United States, tolerances are currently established in 40 CFR §180.355 for combined residues of bentazon (3-isopropyl-1H-2,1,3-benzothiadiazin-4(3H)-one-2,2-dioxide) and its 6- and 8-hydroxy metabolites in or on agricultural commodities as follows:

Commodity	parts per million
Beans (except soybeans), dried	0.05
Beans (exc. soybeans), dried, vine hays	3
Beans (exc. soybeans), forage	3
Beans, lima (succulent)	0.05
Beans, succulent	0.05
Bohemian chili peppers	0.5*
Corn, fodder	3
Corn, forage	3
Corn, grain	0.05
Corn, fresh (incl. sweet K+CWHR)	0.05
Mint	1
Peanuts	0.05
Peanuts, hay	3
Peanuts, hulls	0.3

Commodity	parts per million
Peanuts, forage	3
Peas (dried)	0.05
Peas (dried), vine hays	3
Peas, forage	3
Peas, succulent	0.5
Rice	0.05
Rice, straw	3
Sorghum, fodder	0.05
Sorghum, forage	0.20
Sorghum, grain	0.05
Soybeans	0.05
Soybeans, forage	3
Soybeans, hay	0.3

and

b) combined residues of bentazon (3-isopropyl-1H-2,1,3-benzothiadiazin-4(3H)-one-2,2-dioxide) and its metabolite 2-amino-N-isopropyl benzamide in raw agricultural commodities as follows:

Commodity	parts per million
Cattle, fat	0.05
Cattle, mbyp	0.05
Cattle, meat	0.05
Eggs	0.05
Goats, fat	0.05
Goats, mbyp	0.05
Goats, meat	0.05
Hogs, fat	0.05
Hogs, mbyp	0.05
Hogs, meat	0.05
Milk	0.02
Poultry, fat	0.05
Poultry, mbyp	0.05
Poultry, meat	0.05
Sheep, fat	0.05
Sheep, mbyp	0.05
Sheep, meat	0.05

*The misprinted tolerance of 0.5 ppm for Bohemian chili peppers will be corrected to read 0.05 ppm.

Tolerance reassessment cannot be conducted because of toxicology and residue chemistry data gaps.

International Tolerances:

There are 0.1 ppm Canadian tolerances for bentazon on soybeans, beans, peas, corn, rice, and peanuts. Presently, there are no Mexican or Codex Alimentarius tolerances for bentazon.

4. Summary of Regulatory Position and Rationale

The available data do not indicate that any of the risk criteria listed in § 162.11 (a) of Title 40 of the U.S. Code of Federal Regulations have been met or exceeded for the uses of sodium bentazon at the present time. However, substantial data gaps exist (see below). Under FIFRA §3(c)(2)B, the registrant must provide or agree to develop this data to maintain the existing MP registration or to permit new registrations of substantially similar sodium bentazon MPs.

The Agency will complete its hazard evaluation, or determine what further data are necessary to do so, upon review of the data being required under this Registration Standard. The Agency will determine at that time if such data will affect the registrations of bentazon. If such review determines that criteria for determinations of unreasonable adverse effects are met or exceeded (as specified under §162.11), a rebuttable presumption shall arise that a a notice of intent to cancel registration(s) pursuant to FIFRA §6(b)(1) (or a notice of intent to hold a hearing to determine whether the registration(s) should be cancelled) may be issued. If, at any time, review of the data indicates that an imminent hazard (as defined by FIFRA § 2(1)) is posed by continued bentazon use, immediate suspension procedures may be initiated as per FIFRA §6(c)(1).

No new uses of sodium bentazon will be permitted until the data base is adequate to complete a hazard assessment. As per Conditional Registration Interim Final Regulations (FR Vol. 44, No. 93, May 11, 1979), no new uses or new products may be registered without "data sufficient to allow the Agency to determine that approval of the application would not cause a significant increase in the risk of unreasonable adverse effects on the environment". For bentazon, the entire subchronic and chronic toxicology data base (required to evaluate hazards to humans/domestic animals from existing uses) is invalid or otherwise inadequate, and thus totally insufficient to evaluate any new uses. The Agency is unable to complete a tolerance reassessment of bentazon because of these gaps, as well as residue chemistry data gaps.

The Agency is unable to fully assess potential human exposure, potential for groundwater contamination, or complete an ecological effects hazard assessment of existing sodium bentazon uses because of exposure assessment/environmental chemistry data gaps. There are also certain product chemistry, wildlife/aquatic organism, and nontarget insect data gaps that prevent full assessment of existing use.

Clarification or verification of all test materials used in studies submitted by the registrant is required. The Agency reserves the right to impose additional testing of either bentazon or sodium bentazon following review of this information and/or review of new studies submitted to fulfill data gaps identified in this Standard.

All manufacturing-use and end-use products containing sodium bentazon must bear appropriate labeling as specified in 40 CFR §162.10. The following statements are also required. All labeling changes must appear on all products released for shipment by September, 1986. All labeling changes must appear on all product in channels of trade by September, 1987.

Manufacturing-Use Products

"Do not discharge effluent containing this product into lakes, streams, ponds, estuaries, oceans, or public waters unless this product is specifically identified and addressed in a National Pollutant Discharge Elimination System (NPDES) permit. Do not discharge effluent containing this product to sewer systems without previously notifying the sewage treatment plant authority. For guidance, contact your State Water Board or Regional Office of the U.S. Environmental Protection Agency."

When citing the bentazon equivalent, the chemical name for bentazon should be written as "3-(1-methylethyl)-1H-2,1,3-benzothiadiazin-4(3H)-one 2,2 dioxide".

End-Use Products

"Do not graze treated corn fields for at least 12 days after the last sodium bentazon treatment."

"Do not graze treated peanut fields for at least 50 days after the last sodium bentazon treatment."

"Do not rotate crops used for food or feed, which are not registered for use with sodium bentazon, on areas previously treated with this chemical."

"Do not use sodium bentazon on rice fields in which the commercial cultivation of catfish or crayfish is practiced."
"Do not use water containing bentazon or sodium bentazon residues from rice cultivation to irrigate crops used for food or feed unless sodium bentazon is registered for use on these crops."

For all uses except rice: "Do not apply directly to water or wetlands. Do not contaminate water by cleaning of equipment or disposal of wastes."

For rice use: "Do not contaminate water by cleaning of equipment or disposal of wastes."

For corn, soybean, and sorghum uses:
"Notice: It is a violation of federal laws to use any pesticide in a manner that results in the death of an endangered species or adverse modification of their habitat.

"The use of this product may pose a hazard to certain federally designated endangered species known to occur in specific areas within the CALIFORNIA counties of Merced, Sacramento, and Solano. Before using this product in these counties you must obtain the EPA Endangered Species Bulletin specific for these areas. The bulletin (EPA/ES-85-6) is available from either your County Agricultural Extension Agent, the Endangered Species Specialist in your State Wildlife Agency Headquarters, or the Regional Office of the U.S. Fish and Wildlife Service (Portland, Oregon). THIS BULLETIN MUST BE REVIEWED PRIOR TO PESTICIDE USE. THE USE OF THIS PRODUCT IS PROHIBITED IN THESE COUNTIES UNLESS SPECIFIED OTHERWISE IN THE BULLETIN."

When citing the bentazon equivalent, the chemical name for bentazon should be written as "3-(1-methylethyl)-1H-2,1,3-benzothiadiazin-4(3H)-one 2,2 dioxide".

Summary of Major Data Gaps:

There are numerous generic data gaps in the areas of product chemistry, residue chemistry, environmental fate, and ecological effects.

Generic toxicology data gaps and time frames for submittal are as follows:

Acute inhalation toxicity—rat	9 months
90-day feeding—rodent	15 months
—non-rodent (dog)	18 months
21-day dermal—rabbit	12 months
Chronic toxicity—rodent	50 months
—non-rodent (dog)	50 months
Oncogenicity—rat (preferred)	50 months
—mouse (preferred)	50 months
Teratogenicity—rat	15 months
—rabbit	15 months
Reproduction—rat (2-generation)	39 months
Mutagenicity—gene mutation (Ames Test)	9 months
—structural chromosomal aberration	12 months
Mutagenicity—other genotoxic effects	12 months
General metabolism (using bentazon and sodium bentazon)	24 months
Dermal Penetration	12 months

Product-specific data on the manufacturing-use product include product chemistry data and two acute toxicology studies.

BROMINATED SALICYLANILIDE

Date Issued: December 1985
Fact Sheet Number: 133

1. Description of Chemical

   Generic Name: 3,4',5 Tribromosalicylanilide
   Common Name: Tribromsalan, TBS
   Trade Name: Temasept IV
   EPA Shaughnessy Code No: 077404
   Chemical Abstracts Service No: 87-10-5
   Generic Name: 3,5 Dibromosalicylanilide
   Common Name: N/A
   Trade Name: N/A
   EPA Shaughnessy Code No: 077405
   Chemical Abstracts Service No: 2577-72-2
   Generic Name: 4',5 Dibromosalicylanilide
   Common Name: N/A
   Trade Name: N/A
   EPA Shaughnessy Code No: 077402
   Chemical Abstracts Service No: 87-12-7
   Year of Initial Registration: July 6, 1964
   Pesticide Type: Antimicrobial, Preservative
   Chemical Family: Brominated Salicylanilides
   U.S. Producers: Hexcel Corporation, Sherwin Williams Chemicals

2. Use Patterns and Formulations

   Application Sites: Hard Surfaces, Laundry, Textiles, Manufactured Products
   Types of Formulations: Solid, Solutions, Sprays
   Application Rates: N/A
   Usual Carriers: Soap

3. Science Findings

   Chemical Characteristics:

   Physical State: Powder
   Color: White
   Odor: None
   Melting Point: 227-228°
   Density: No Data Available
   Solubility: Soluble (0.5-4.0%) in methanol, ethanol, isopropanol, ethyl ether and benzene; sparingly soluble (0.01-0.5%) in chloroform and cyclohexane; insoluble in water and petroleum ether.

Toxicity Characteristics:

There are no available data to assess the potential toxicity of the brominated salicylanilides.

Physiological/Biochemical Behavioral Characteristics: N/A

Environmental Characteristics: Indoor uses only; limited exposure is possible based on the registered uses of these products as disinfectants, laundry additives, textile preservatives and manufactured products.

Ecological Characteristics: Indoor uses only; limited exposure is possible based on the registered uses of these products as disinfectants, laundry additives, textile preservatives and manufactured products.

Efficacy Data: Product specific microbiological efficacy data will be required for all health related uses.

4. Summary of Regulatory Position and Rationale

No data are available to establish whether or not the risk criteria listed in Section 162.11(a) of Title 40 of the U.S. Code of Federal Regulations have been met or exceeded.

Unique Warning Statements: None

Tolerance Assessments: N/A The registered uses of the brominated salicylanilides do not include direct application to a food or feed crop.

Problems which are known to have occurred with use of the chemical (e.g., PIMS): None.

5. Summary of Major Data Gaps:

Toxicology, environmental fate, fish and wildlife and product chemistry data are required to assess the hazard associated with the registered uses of these chemicals.

6. Contact person at EPA:

Jeff Kempter
Product Manager (32)
Disinfectants Branch
Registration Division (TS-767C)
Telephone: (703) 557-3964

DISCLAIMER: The information presented in this Chemical Information Fact Sheet is for informational purposes only and may not be used to fulfill data requirements for pesticide registration and reregistration.

BRONOPOL

Date Issued: August 2, 1984
Fact Sheet Number: 32

1. Description of Chemical

    Generic Name: 2-Bromo-2-nitropropane-1,3-diol
    Common Name: Bronopol - Boots
    Trade Name: Myacide B10, Myacide AS
    EPA Shaughnessy Code: 216400
    Chemical Abstracts Service (CAS) number: 52-51-7
    Year of initial registration: August 2, 1984 (EPA Reg. No. 33753-1; MUP)
    Pesticide Type: Bactericide
    Chemical Family: Alcohol
    Producers: U.S. _____ none
    Foreign ___ The Boots Co., PLC

2. Use Patterns and Formulations

    Application Sites: pesticide manufactering facilities.
    Types of Formulations: crystalline powder
    Types and Methods of application: the only registration is for manufacturing use only. N/A
    Application rates: N/A
    Usual carriers: water

3. Science Findings

    Chemical Characteristics:

    Physical state: white crystalline powder
    Melting point: approx. 130°C
    Bulk density: 1.10 mg/ml
    Solubility:
    in ethanol   50%
    in water (25°C)   25%
    in ethylene glycol   61%
    in chloroform   0.9%

Toxicity Characteristics:

Study	Effects	Category
Eye Irritation	5% solution strongly irritating	I
Skin Irritation	a severe irritant	I
Dermal LD_{50}	not specified	I
Oral LD_{50} (mg/kg):		II
orally administered	male 307 & female 342 (rat)	
	male 374 & female 327 (mouse)	
intraperitoneal	male 22.0 & female 30.2 (rat)	
	male 22.8 & female 34.7 (mouse)	
Inhalation LC_{50}	LC_{50} > 5 mg/L	III
Dermal Sensitization	not a skin sensitizer	
90 day Oral (rat)	no observable effect level: 20 mg/kg/day	
	lowest effect level: 80 mg/kg/day	
Teratology	no observable effect level: 3.3 mg/kg/day	
(rabbit)	lowest effect level: 10 mg/kg/day	

Physiological and Biochemical Behavioral Characteristics:

Foliar absorption: N/A
Translocation: N/A
Mechanism of pesticidal action: N/A
Metabolism and persistence in plants and animals: N/A

Environmental Characteristics:

Adsorption and leaching basic soil types: N/A
Microbial breakdown: N/A
Hydrolysis: bronopol appears to hydrolyze slowly at acidic or neutral pH conditions. Bronopol decomposes in aqueous solution on exposure to light. Increases in temperature increases decomposition. Label required to disallow any pesticide waste discharge into aquatic environments.
Bioaccumulation: N/A

Ecological Characteristics:

Mallard duck Acute Oral LD_{50} = 510 mg/kg
Fresh Water Invertebrate (<u>Daphnia</u>) 48 hr EC_{50} = 1.4 mg/L

Efficacy Data: N/A

Tolerance assessments: N/A, no crop applications.

Problems which are known to have occurred with use of the chemical (e.g., PIMS): none

4. Summary of Regulatory Position and Rationale:

Use Classification: General Use
Use, formulation, manufacturing process or geographical restrictions: do not discharge into aquatic environments unless in accordance with and NPDES permit.

Unique warning statements required of labels: none
Summary of risk/benefit review: none, risk assessment was not conducted.

5. **Summary of major data gaps**

    The following studies are to submitted by December 13, 1984:

    a. Avian Acute LC_{50}
    b. Fish Acute LC_{50} (Rainbow trout)

6. **Contact person at EPA:**   John H. Lee,
    Product Manager 31
    Disinfectants Branch
    Registration Division (TS-767C)
    Tel. (703) - 557-3675

DISCLAIMER: The information presented in this Chemical Information Fact Sheet is for informational purposes only and may not be used to fulfil data requirements for pesticide registration and reregistration.

BRONOPOL - BOOTS™

For Use in Formulating Indoor Products*

*Including Disinfectants, Sanitizers, and Preservatives.

Active Ingredient:

2-Bromo-2-nitropropane-1,3-diol   99%

Inert Ingredients: Total   1%
                                       100%

Net Contents:

Keep Out of Reach of Children

DANGER

Statement of Practical Treatment

In case of contact, immediately flush eyes or skin with plenty of water for at least 15 minutes. For eyes, call a physician. If swallowed, drink milk, egg whites, gelatin solution or if these are not available, drink large quantities of water. Call a physician.

Note to Physician

Probable mucosal damage may contraindicate the use of gastric lavage. Measures against circulatory shock, respiratory depression and convulsion may be needed.

See side panel for additional precautionary statements

MADE IN ENGLAND BY
THE BOOTS CO., PLC
NOTTINGHAM
N12 3AA

EPA Registration No. 33753-1
EPA Establishment No. 33753-EN-01

DIRECTIONS FOR USE

It is a violation of Federal Law to use this product in a manner inconsistent with its labelling.

For use by manufacturers in formulating disinfectants, sanitizers, and preservatives.

Formulators using this product are responsible for obtaining EPA registration for their formulated product and for providing appropriate data.

STORAGE AND DISPOSAL

-- Do not contaminate water, food, or feed by storage or disposal. Keep away from heat.
-- Pesticide wastes are acutely hazardous. Improper disposal of excess pesticide, spray mixture, or rinsate is a violation of Federal Law. If these wastes cannot be disposed of by use according to label instructions, contact your State Pesticide or Environmental Control Agency, or the Hazardous Waste representative at the nearest EPA Regional Office for guidance.
-- Completely empty liner by shaking and tapping sides and bottom to loosen clinging particles. Empty residue into application equipment. Triple rinse (or equivalent). Then offer drum for recycling or reconditioning, puncture. Dispose of drum and liner in a sanitary landfill, or by incineration, if allowed by State and local authorities. If burned, stay out of smoke.

Chemical and Physical Properties

Molecular Weight   200
Solubility
  in Water (25°C)   25% w/v
  in Ethanol   50% w/v
  in Ethylene glycol   61% w/v
  in Chloroform   0.9% w/v

Bulk Density   1.1 mg/ml
Melting Point   approx 130°C
Physical State   white, or almost
                         white crystals
                         or crystalline
                         powder.

Bronopol-Boots is compatible with non-toxic or cationic additives and with anionic additives which are neutral or acidic.

PRECAUTIONARY STATEMENTS

Hazards to Humans and Domestic Animals

DANGER: Corrosive. Causes eye and skin damage. Do not get in eyes or skin or on clothing. May be fatal if swallowed. Avoid breathing dust. Wear goggles or face shield and rubber gloves when handling.

ENVIRONMENTAL HAZARDS

Do not discharge into lakes, streams, ponds or public water unless in accordance with an NPDES permit. For guidance contact your Regional Office of the EPA.

BUTYLATE

Date Issued: March 22, 1984
Fact Sheet Number: 7

1. <u>Description of chemical</u>:

    Generic Name: Butylate
    Common name: butylate
    Trade name: Sutan
    EPA Shaughnessy Code: 041405
    Chemical Abstracts Service (CAS) Number: 2008-41-5
    Year of Initial Registration: 1967
    Pesticide Type: Herbicide
    Chemical family: Thiocarbamate
    U.S. and Foreign Producers: Stauffer Chemical Company, PPG Industries, Inc.

2. <u>Use patterns and formulations</u>:

    Application sites: Sweet corn, field corn, and popcorn
    Types of formulations: Granulars, emulsifiable concentrates, and encapsulated
    Types and Methods of Application: Soil incorporation, generally with discs or hooded powerdriven tillers, often in combination with atrazine and/or cyanazine herbicide. Center pivot irrigation systems can be used in some areas.
    Application Rates: 3.4- 6.7 lbs ai/A
    Usual carriers: Emulsifiable liquid formulations are diluted in water.

3. <u>Science Findings</u>:

    Summary science statement:

    Butylate appears to pose few, if any, acute toxicological hazards to humans or non-target wildlife. The only major concern is the lack of inhalation toxicity data. Such data has been requested in the standard.

Chemical characteristics:

Physical state:	Liquid
Color:	Yellow to amber
Odor:	Amine
Boiling point:	71° C at 10 mm Hg
Melting point:	Not applicable
Flash point:	(TOC) 110°C

Unusual handling characteristics: None. Non-corrosive, stable at normal ambient temps.

Toxicological characteristics:

Acute Effects:

Acute Oral LD_{50} - Low - (Tox Category III) (3.0 g/kg)
Acute Dermal LD_{50} - Low - (Tox Category III) (>2 g/kg)
Dermal Irritation - Not an irritant
Acute Inhalation Toxicity - data gap
Primary Eye Irritation - Caused permanent damage in 1/6 of unwashed eyes (Tox Category II)

Chronic Effects:

Oncogenicity - No dose-related effects at levels up to 320 mg/kg/day (HDT) (Highest Dose Tested) in a 24-month study.
Teratology - No effects at up to 24 mg/kg (HDT).
Reproductive Effects - No effects at up to 24 mg/kg (HDT).
Mutagenicity - data gap
Feeding Studies - 13-week study with dogs: No effects on behavior or bodyweight. No neurological opthalmological, hematological, nor blood chemistry effects. No effect on brain AChE, nor on gross organ appearance or weight. 56-week study with rats: No major effects at 10 and 30 mg/kg. HDT 180 mg/kg produced liver pericholagitis, uterine and testicular changes with focal hemorrhage. Blood clotting parameters were affected at lowest dose (10 mg/kg/day).

Major Routes of Exposure: Dermal, inhalation

Physiological and Biochemical Behavioral Characteristics:

   Foliar absorption: Absorbed by leaves, but not normally applied to the foliage.
   Translocation: Butylate is rapidly absorbed by the roots of the corn plant and translocated throughout the whole corn plant.
   Mechanism of pesticidal action: Unknown. Inhibits growth in the meristemic region of the leaves of grassy weeds.
   Metabolism and persistence in plants and animals: Metabolized rapidly to CO_2, diisobutylamine, fatty acids, conjugates of amines and fatty acids, and certain natural plant constituents. Disappears from the stems and leaves of corn plants 7 to 14 days after treatment.

Environmental Characteristics:

   Adsorption and leaching in basic soil types: In sandy dry soils, butylate leached about one-third the distance that 20 cm (8 inches) of water moved. Leaching decreased as clay and organic matter increased. In heavy clay soils, butylate leached slightly downward 2.5 to 7.6 cm (1 to 3 inches) with 20 cm of water.

   Microbial breakdown: Microbial breakdown plays an important role in the disappearance of butylate from soils.

   Loss from photodecomposition and/or volatilization: Butylate is lost by vaporization when applied to the surface of wet soils without incorporation. Very little loss occurs after application to dry soil surfaces.

   Bioaccumulation: Butylate has moderate potential for bioaccumulation in fish. After 28 days of exposure, bluegill sunfish had bioaccumulation ratios of 33X ambient in edible tissues, and 119-174X ambient in non-edible tissues.

Resultant average persistance: The half-life of butylate under crop growing conditions was 1.5 to 3 weeks in several soils. In a loam soil at 21 to 27° C (70 to 80° F) the half-life was 3 weeks.

Half-life in Water: Data not yet available.

Ecological characteristics:

Hazards to Birds: Minimal, owing to low toxicity and low exposure rates
Hazards to Aquatic Invertebrates: Minimal, owing to moderate toxicity and low exposure rates
Hazards to Fish: Not fully assessed yet. Butylate is at least moderately toxic to fish, but requested data might show a greater toxicity. If earlier studies prove accurate, and butylate is only moderately toxic to fish, the hazards to fish from the registered use patterns are low.
Potential Problems with Endangered Species: None anticipated.

Tolerance Reassessment:

List of crops and tolerances: Corn grain (including popcorn), fresh corn (including sweet corn – kernels plus cob with husk removed), and corn forage and fodder (including sweet corn, field corn, and popcorn) at 0.1 ppm.
List of food contact uses: All corn products listed above.
Results of tolerance assessment: Assuming 100% of all corn products to be treated with butylate, the dietary burden amounts to no more than .032% of the ADI. Reassessment has been conducted. No tolerance changes are needed at this time.
Problems known to have occured from use: None

4. **Summary of Regulatory Position and Rationale:**

Use Classification: General use classification.
Use, Formulation or Geographic Restrictions: Uses are limited to application to corn fields. No other restrictions.

Unique Label warning statements:

Manufacturing Use Products:

"Do not discharge into lakes, streams, ponds, or public waters unless in accordance with a NPDES permit. For guidance contact your regional office of the EPA."

End-Use Products:

"Harmful if swallowed. Avoid contact with skin, eyes and clothing. Aoid breathing spray mist. Wear goggles, rubber gloves and protective clothing. Wash skin with soap and water immediately after contact. Flush eyes with water.

"Do not apply directly to water or wetlands. Do not contaminate water by cleaning of equipment or disposal of wastes. Cover or incorporate spills."

Data gaps exist (see below), but the Agency will not cancel or withhold registration solely because of data gaps. The available toxicity and environmental fate data indicate that butylate use is unlikely to cause severe hazards to humans or wildlife, and although this analysis of hazard cannot be considered complete until the data gaps are filled, there is sufficient justification for continuing the registration of butylate products.

5. Summary of major data gaps

An inhalation LC_{50} for rats is needed, and has been required. Other data gaps exist, but none is considered major, or as important as the lack of an inhalation LC_{50}. The inhalation LC_{50} is to be supplied by April 1985. The mutagenicity data and other data are to be supplied by October 1987.

6. <u>Contact person at EPA</u>: Robert Taylor, TS-767-C, 401 M Street SW, Washington DC 20460
(703) 557-1800

DISCLAIMER: The information presented in this Chemical Information Fact Sheet is for informational purposes only and may not be used to fulfill data requirements for pesticide registration and reregistration.

CADMIUM PESTICIDE COMPOUNDS

Date Issued: September 1986
Fact Sheet Number: 103

1. Description of chemicals

    Chemical names: cadmium carbonate
    cadmium chloride
    cadmium sebecate
    cadmium succinate
    anilinocadmium dilactate

    Common names: same as above

    Trade names: none

    EPA Shaughnessy codes (respectively): 012901, 012902, 012903, 012904, 064601

    Chemical abstracts service (CAS) numbers (respectively): 134A, 135, 136A, 136B, 051D

    Years of Initial Registration: 1949--1952

    Pesticide type: fungicides

    Chemical family: cadmium salts

2. Use patterns and formulations

    Application sites: golf course and home lawn turf

    Types of formulations: wettable powders, dusts, granulars

    Types and methods of application: ground application by hand held sprayers and boom sprayers

3. **Science Findings**

   Physical and Chemical Characteristics of Cadmium:

   Physical state: solid
   Boiling point: 765° C
   Melting point: 321° C

   Human Toxicology Characteristics:

   Acute toxicity: moderate to moderately high (Toxicity Categories III and II); specific values are unavailable for each compound since there are no technical registrations and there are data gaps on formulated products

   acute effects to kidneys are formation of fatty bodies in the kidneys and degeneration of renal tubules

   Chronic toxicity:

   Oncogenic as demonstrated in laboratory animal and human epidemiological studies:

   rat chronic inhalation study--LOEL of 12.5 ug Cd chloride/m^3 (lowest dose tested) for lung tumors

   rat chronic injection study--3.6% Cd chloride (lowest concentration tested) caused testicular and pancreatic islet tumors

   epidemiological studies of factory workers chronically exposed to cadmium oxide and dust have shown statistically significant increases in the incidences of lung tumors

   Kidney effects of proteinuria, glucosuria, excretion of amino acids and decreased renal function:

   rat drinking water study (24 wks)--NOEL of 10 mg/L (lowest dose tested) for proteinuria

   epidemiological study of factory workers exposed to cadmium oxide and dust (50 yrs)-- LOEL of 2 ug/m^3 for renal tubular proteinuria

   Mutagenic effects from 36 studies on various cadmium compounds are equivocal; depending on protocol and end point examined, results vary

Teratogenic, fetotoxic and reproduction effects have been shown in laboratory animal studies however, the data do not support that cadmium would produce these types of effects in humans

4. Summary of regulatory position and rationale

The Agency initiated a Special Review of cadmium pesticide compounds in October 1977 based on data indicating that they may pose risks of oncogenicity, mutagenicity, teratogenicity and fetotoxicity from their application to golf courses and home lawns to control turf diseases.

During this time the Agency conducted further evaluations of toxicology studies of various cadmium compounds. Additionally, the Agency assessed new applicator exposure data and cadmium fungicide benefits information.

The Agency concluded that the toxicology data from laboratory animal and human epidemiological studies demonstrate a correlation between inhalation exposure to cadmium and an increased incidence of lung tumors. The Agency has classified cadmium as a "B1" or "probable" human carcinogen. Based on the derived potency value and estimates of applicator exposure, the estimates of oncogenic risk to applicators are 10^{-4} to 10^{-6} from applications to golf courses and 10^{-8} from applications to home lawns.

In regard to mutagenicity, the Agency reassessed the studies for which the Special Review was initiated and assessed additional studies that have become available since the initiation. From the results of the many studies, the Agency now concludes that there are conflicting results which cannot be readily resolved due to the many protocols and end points that have been used. Therefore, the Agency believes that the data no longer support the mutagenicity risk criteria for Special Review.

Since initiation of the Special Review, the Agency also reassessed cadmium's potential as a human teratogen and fetotoxin. Although data from some studies link cadmium with teratogenic and fetotoxic effects in laboratory animals, the Agency now concludes that the composit of data from studies do not support the risk criteria that cadmium is teratogenic and fetotoxic in humans.

Additional laboratory animal and human epidemiological studies that the Agency has reviewed since 1977 demonstrate that cadmium causes acute and chronic effects to the kidneys, including fatty body formation, renal tubular degeneration, proteinuria, glucosuria and amino acid excretion. In comparison of the lowest effect level for kidney effects with the estimated applicator exposure the Agency concludes that there are risks of applicators developing kidney effects from the use of cadmium fungicides on golf courses and home lawns. The risk to golf course applicators is much higher than to home lawn applicators. Therefore, the Agency has added this hazard with oncogenicity for the risk criteria for this Special Review.

The Agency examined the benefits from the uses of cadmium fungicides on golf courses and home lawns and the availability of alternate fungicides and their associated hazards. The Agency concludes that the benefits are low. Use of cadmium fungicides is very low in annual volume (30,000 lbs) and in percentage of golf course acreage treated (2 %) as compared to the eleven alternatives. Use on home lawns is negligible. Some of the alternatives are as effective and some are more expensive. Total substitution for cadmium fungicides could annually cost the golf courses nationwide as much as $240,000 or $500 each. Some of the alternatives are associated with chronic hazards while others are not.

The Agency received seven comments from the cadmium registrants during the comment period. These were arguments against the risk criteria for the initiated Special Review and additional benefits information. All comments and the benefits information have been considered and are addressed in the Technical Support Document.

The Agency has carefully evaluated all the information and also considered possible measures to reduce exposure and the risks of oncogenicity and kidney effects. The weight of the evidence leads the Agency to conlude that the risks to applicators outweigh the minor benefits and therefore cancellation is the prudent regulatory action to propose.

This proposed regulatory action along with a request for comments from the public, USDA and the FIFRA Science Advisory Panel will be published in the Federal Register in October 1986. The comment period will be 60 days. After that time the Agency will evaluate any received comments in consideration of their impact on the risk and benefit assessments and the proposed regulatory decision. The Agency will then complete the Special Review by publishing in the Federal Register a final decision.

5. Contact person at EPA: Valerie Meredith Bael
   EPA
   Office of Pesticide Programs
   Registration Division (TS-767C)
   401 M Street, S.W.
   Washington, D.C. 20460

DISCLAIMER: The information in this Chemical Information Sheet is for informational purposes only and may not be used to fulfill data requirements for pesticide registration or reregistration.

CALCIUM ARSENATE

Date Issued: December 1986
Fact Sheet Number: 111

1. DESCRIPTION OF CHEMICAL

    Common Name: Calcium Arsenate

    Chemical Name: Calcium Orthoarsenate - $Ca_3(AsO_4)_2$

    Trade Name: Turf-Cal

    EPA Shaughnessy Code: 013501

    Chemical Abstracts Service (CAS) Number: 7778-44-1

    Year of Initial Registration:

    Pesticide Type: Herbicide

    Chemical Family: Inorganic Arsenicals

    U.S. and Foreign Producers: Security Chemical Co.

2. USE PATTERNS AND FORMULATIONS

    Calcium arsenate is currently used as a herbicide on 18% of U.S. golf courses for controlling Poa annua, crabgrass, and other annual grasses. It has also been used as a crop herbicide and an insecticide; however, these uses are considered to be inactive.

    ° Methods of Application: The flowable (liquid suspension) formulation is generally applied by ground boom. Products which are currently suspended include application by hand-held sprayer or broadcast spreader.

    ° Application Rates: Turf- 4.5 lb ai/A

    Types of Formulations: The remaining active registered product is a flowable (liquid suspension). Products which are currently suspended include granular, wettable powder, and wettable powder/dust formulations.

3. SCIENCE FINDINGS

- **Chemical Characteristics**

    Calcium arsenate is a pentavalent form of inorganic arsenic. It normally exists as a colorless amorphic powder with no discernible odor. Calcium arsenate contains 38% arsenic and is slightly soluble in water and soluble in dilute acids. The melting point of calcium arsenate is 1455°C , the density is 3.62 and the molecular weight is 398.08.

- **Toxicological Characteristics**

    Inorganic arsenical compounds have been classified as Class A oncogens, demonstrating positive oncogenic effects based on sufficient human epidemiological evidence.

    Inorganic arsenicals have been assayed for mutagenic activity in a variety of test systems ranging from bacterial cells to peripheral lymphocytes from humans exposed to arsenic. The weight of evidence indicates that inorganic arsenical compounds are mutagenic.

    Evidence exists indicating that there is teratogenic and fetotoxic potential based on intravenous and intraperitoneal routes of exposure; however, evidence by the oral route is insufficient to confirm calcium arsenate's teratogenic and fetotoxic effects.

    Inorganic arsenicals are known to be acutely toxic. The symptoms which follow oral exposure include severe gastrointestinal damage resulting in vomiting and diarrhea, and general vascular collapse leading to shock, coma and death. Muscular cramps, facial edema, and cardiovascular reactions are also known to occur following oral exposure to arsenic.

- **Environmental Characteristics:** The environmental fate of calcium arsenate is not well documented. Studies to demonstrate its fate must take into account the fact that inorganic arsenicals are natural constituents of the soil, and that forms of inorganic arsenic may change depending on environmental conditions. Based on very limited data calcium arsenate is not predicted to leach significantly.

- **Ecological Characteristics:** Calcium arsenate is moderately toxic to birds, slightly toxic to fish and moderately toxic to aquatic invertebrate species.

- **Metabolism:** The metabolism of inorganic arsenic compounds in animals is well known. The pentavalent form, such as calcium arsenate, is metabolized by reduction into the trivalent form, followed by transformation into organic forms which are excreted within several days via the urine. All animals exhibit this metabolism except rats, which retain arsenic in their bodies for up to 90 days.

4. SUMMARY OF REGULATORY POSITION AND RATIONALE

The Agency is proposing to cancel all existing registrations of calcium arsenate, with the exception of the flowable formulation for use on turf. Measures to mitigate the inhalation risks including dust masks, respirators, which would be expected to reduce inhalation exposure by 80 and 90 percent, respectively, and restricting the use to certified applicators were considered by the Agency during the Special Review. The Agency has determined that these protective measures would not reduce risks to an acceptable level in light of the limited benefits. The Agency has further determined that the toxicological risks from all non-wood uses of calcium arsenate, except the aforementioned use on turf, outweigh the limited benefits. The flowable formulation for use on turf is being deferred pending further evaluation by EPA's Risk Assesment Forum of the carcinogenic potency of inorganic arsenic from dermal and dietary exposure.

- **Benefits Analysis:** No economic impact is expected as a result of cancellation of these uses. Viable alternatives are available.

5. CONTACT PERSON

Douglas McKinney
Special Review Branch, Registration Division
Office of Pesticide Programs (TS-767C)
401 M Street, S.W.
Washington, D.C. 20460
(703) 557-5488

DISCLAIMER: The information presented in this Pesticide Fact Sheet is for informational purposes only and may not be used to fulfill data requirements for pesticide registration or reregistration.

CAPTAFOL

Date Issued: October 1, 1984
Fact Sheet Number: 35

1. Description of Chemical

    Generic Name: cis-N-[(1,1,2,2-tetrachloroethyl)
    thio]-4-cyclohexene-1,2-dicarboximide

    Common Name: Captafol

    Trade Names: Difolatan, Folcid, Haipen, Merpafol, Ortho 5865, Sanspor and Sulfenimide

    EPA Shaughnessy Code: 081701
    Chemical Abstract Service (CAS) Number: 2425-06-1
    Year of Initial Registration: May 21, 1962
    Pesticide Type: Fungicide
    Chemical Family: Phthalimide
    U.S. Producer: Chevron Chemical Company

2. Use Patterns and Formulations

    Application Sites: Apples (Midwestern and Eastern states only), apricots, blueberries, cherries (unspecified-use in California and Oregon only), cherries (sour), cranberries, grapefruit, lemons, limes (Florida only), macadamia nuts (Hawaii), nectarines, oranges (Florida only), peaches, pineapples, plums, prunes, tangerines, corn (sweet, Florida only), cucumbers, melons, onions, potatoes, taro (wetland) (Hawaii only), tomatoes (mechanically harvested only), watermelons, peanuts, seed treatment uses [corn (field), cotton, peanuts, rice and sorghum (grain and forage)] and wood.

    Types of Formulations: Dusts, emulsifiable concentrates, flowable suspensions, wettable powders, and water dispersible granules.

    Types and Methods of Application: Dusting, spraying, misting, and dipping under pressure for wood treatment.

    Application Rates: See use patterns in USDA Compilations of Registered Uses of Fungicides and Nematicides, Part I, Pages C-09-95.01 through C-09-95.11.

    Usual Carriers: Clay, talc, silica, water.

3. **Scientific Findings**

   Chemical Characteristics:

Physical State:	Crystalline solid
Color:	White

   Odor: Slight characteristic pungent odor
   Melting Point: Range 156-161°C (162°C, pure compound)
   Stability: Stable under ordinary environmental conditions. Decomposes slowly at melting point. Very slight hydrolytic decomposition at acid or neutral pH at ambient temperature. Strong sodium and potassium hydroxide solutions may cause spontaneous ignition of concentrate forms. In neutral or weakly basic solution, captafol slowly decomposes at a rate depending upon pH and temperature, being very slow below 7.0 and rapid above 9.0.

Toxicity Characteristics:

Acute oral LD_{50} - Rat - males   6780 mg/kg, Category III
                             females   6330 mg/kg, Category III

Acute inhalation LC_{50} - Data gap

Dermal Irritation - Rabbit - moderate dermal irritation at 72 hours, Category III. Severe dermal sensitization.

Eye Irritation - Rabbit - corneal opacity, iris and conjunctive irritation present through day 21, Category I.

Teratology - Rat - no teratogenic effects seen. Highest dose tested: 100 mg/kg/day.
    Rabbit - recently submitted tests currently under review.
    Hamster - no effect level (NOEL) for fetotoxicity and possible teratogenic effects = 300 mg/kg. New study must be submitted by the registrant.

Oncogenicity - Mouse - Oncogenic lesions were observed at 1000 and 3000 parts per million (ppm) but not at 300 ppm. Information concerning the control groups is being requested from the registrant in order for a final risk assessment to be performed.
    - Rat - Oncogenic lesions were seen at all dose levels. Additional histopathology examinations of liver and mammary tissues at low (56 ppm) and mid (241 ppm) dose levels have been requested from the registrant. This information is necessary for a complete risk assessment to be performed.

Physiological and Biochemical Behavioral Characteristics:

Foliar Absorption: Captafol and/or its metabolites and degradrates is (are) absorbed by roots and shoots of plants.

Translocation: Captafol and/or its metabolites and degradrates is (are) translocated in plant tissue. Captafol and/or its metabolites and degradates is (are) absorbed and translocated within plants as a result of seed treatment, soil treatment and foliar application.

Mechanism of Pesticidal Action: Unknown

Metabolism and Persistence in Plants and Animals: The metabolism and accumulation of captafol in plants are not understood. The metabolism of captafol is understood for ruminants but not for avian species.

The following metabolites have been identified in animal tissues:

delta^4-tetrahydrophthalimide

phthalimide

4,5-epoxyhexahydrophthalimide

4,5-dihydroxyhexahydrophthalimide

3-hydroxy-delta^4-tetrahydrophthalimide

5-hydroxy-delta^3-tetrahydrophthalimide

delta^4-tetrahydrophthalamic acid

delta^4-tetrahydrophthalic acid

The above metabolites, with the exception of delta^4-tetrahydrophthalic acid, were found in milk. No parent captafol per se was detected in ruminant tissues or in milk.

Environmental Characteristics:

Absorption and Leaching in Basic Soil Types:
Captafol per se does not leach from basic soils. The leachability and persistence of its metabolites and degradates are not understood.

Microbial Breakdown: Inadequate data. Limited data indicate that captafol per se has a half-life of <3, 5, and 8 days in nonsterile organic, sandy and clay loam soils, respectively. The soil degradates and metabolites have not been identified.

Loss from Photodecomposition and/or Volatilization: Inadequate data.

Bioaccumulation: No data.

Resultant Average Persistence: Inadequate data on metabolites and degradates.

Ecological Characteristics:

Hazards to fish and wildlife

Rainbow trout: 96-hr LC_{50} = 0.027 - 0.190 ppm
Bluegill sunfish: 96-hr LC_{50} = 0.045 - 0.230 ppm
Characterized as "very highly toxic" to both cold water and warm water fish. Inadequate data for fish accumulation.

Avian Toxicity: LD_{50} = >2510 ppm
LC_{50} = >5620 ppm

Daphnia magna: 96-hr LC_{50} = 3.34 ppm

Moderately to very highly toxic to freshwater invertebrates.

Avian Reproduction: Strong potential for reproductive effect in birds (based on supplementary information).

Potential problems related to endangered species:

Additional data on exposure persistence, avian reproduction studies, and accumulation are required to complete the endangered species assessment for captafol.

Efficacy Review Results:

Captafol product registrations with EPA have been supported with efficacy data. Tolerant strains of target plant pathogens have not been reported.

Tolerance Assessments:

1. List of Crops and Tolerances:

The following table lists the present status for tolerances in parts per million (ppm) for residues of captafol:

	Parts Per Million in Captafol Residues			
Raw Agricultural Commodity	U.S.	Canada	Mexico	Codex
Apples	0.25	0.1(N)	—	5.0
Apricots	30.0	0.5	—	15.0
Blueberries	35.0	—	—	—
Cherries, sour	50.0	10.0	—	10.0
Cherries, sweet	2.0	2.0	—	2.0
Citrus fruit	0.5	—	—	—
Corn, Fresh (sweet K+CWHR)	0.1 (N)	—	—	—
Cranberries	8.0	—	—	8.0
Cucumbers	2.0	2.0	—	2.0
Macadamia nuts	0.1(N)	—	—	0.1 (N)
Melons	5.0	2.0	—	2.0
Nectarines	2.0	—	—	—
Peanuts, hulls	2.0	—	—	—*
Peanuts, meats (hulls removed)	0.05	—	—	0.05
Onions	0.1(N)	0.1 (N)	—	0.5
Peaches	30.0	15.0	—	15.0
Pineapples	0.1(N)	—	—	10.0
Plums (fresh prunes)	2.0	0.2	—	10.0
Potatoes	0.5	0.1(N)	—	0.5
Taro (corms)	0.02	—	—	—
Tomatoes	15.0	5.0	—	5.0
Carrots				0.5

*Codex MRL for whole peanuts is 0.5 ppm.

2. Seed Applications:

No tolerances have been established for captafol in or on any crop for which captafol is registered solely for seed treatment or plant propagule uses, because heretofore seed treatment and plant propagule uses have been considered to be nonfood uses. These crops include: corn (field), cotton, rice, sorghum (grain and forage), strawberries (propagating bed use), and sugarcane (seed piece use).

Results of Tolerance Assessment:

Sufficient data are available to determine that the currently established tolerances for residue of captafol in or on the following commodities are adequate: apricots, blueberries, cranberries, macadamia nuts, nectarines, peaches, peanut hulls, and taro corms (40 CFR 180.267). However, additional data concerning residues of metabolite THPI must be submitted for all commodities since the residues of concern in or on plants have recently been modified to include THPI. In addition to THPI data, other residue data are required to support the tolerance for captafol in or on the following commodities: apples, citrus fruits, corn (sweet), cucumbers, melons, onions, peanuts, pineapples, plums (prunes), potatoes, and tomatoes. THPI data and tolerance changes must be submitted for sweet and sour cherries. Residue data and tolerance proposals must be submitted for the following commodities: crops receiving only seed treatments (cotton, field corn, rice and sorghum), pears (state label use only), strawberries [propagating bed use: Section 24(c)], sugarcane (seed piece use), and taro foliage. Tolerances for the following commodities are pending: carrots, coffee beans, pecans, soybeans, and soybean forage and hay. Plant metabolism of captafol is not adequately understood, therefore tolerance for residues other than captafol and THPI may need to be sought if those residues are deemed to be of concern. A poultry metabolism study will be needed. Also, feeding studies are required for ruminants and poultry. Additional storage stability data are necessary to validate the residue data. No crop group tolerance may be established based on the available data.

The Theoretical Maximum Residue Contribution (TMRC) is 1.4579 mg/day based on a 1.5-kg diet. The change in the residue definition (by including the metabolite THPI), the requested tolerance proposals and the pending tolerances noted above will affect a change in the TMRC level.

The maximum permissible intake (MPI) for a 60 kg person is 1.68 mg/day based on an acceptable daily intake (ADI) of 0.028 mg/kg. The present TMRC represents 86.8% of the MPI. The inclusion of the major metabolite (THPI) in the tolerance expression may result in an increase in the TMRC and a greater percentage of the MPI utilized.

The ADI for captafol is based on a chronic toxicity study in rats. A NOEL was established at 56 ppm for non-oncogenic effects. In that study the next highest dose level caused cholangiectasis in liver, increase in hyperplasia of tubule epithelium, megalocytic cells and transitional cell hyperplasia in kidney, increased erosion/ulceration, hyperkertosis/acanthosis, ground substance in glandular mucosa, and dilated pits in stomach. The reported dosage level was corrected from 75 ppm to 56 ppm based on the instability of captafol in the diet. In the rat, 56 ppm is approximately equivalent to 2.8 mg/kg. Using a safety factor of 100 and a NOEL of 2.8 mg/kg/day, the ADI would be 0.028 mg/kg for humans. The only registrant of technical captafol, Chevron Chemical Company, has been requested to complete the histopathology examination of the low and mid-dose liver and mammary gland tissues; therefore, the NOEL for cholangiectasis in the liver may change. The NOEL of 56 ppm is for systemic effects excluding the issue of oncogenicity for which a risk assessment will be performed.

Captafol 115

Problems that are Known to Have Occurred with Use of the Chemical

1. Captafol is a skin sensitizer. Incidents of farmworkers being disabled from its effect have been reported. Restricting the use of captafol to mechanically harvested tomatoes, and labeling requiring use of gloves and protective clothing by citrus harvesters have reduced this problem.

4. Summary of Regulatory Positions and Rationale

Based on the oncogenic and fish toxicity potential risks of hazard, captafol is being placed under Special Review by the Environmental Protection Agency. The presumption of risk of hazards outweighing the derived benefits must be rebutted by the manufacturers of pesticide products that have captafol as an active ingredient. Based on the oncogenic potential and the fish toxicity potential, captafol is classified as a "Restricted Use" pesticide. All captafol products must be relabeled as "Restricted Use" pesticides.

The following labeling restrictions will be required for continued registration of captafol products:

All technical grade, manufacturing-use (MUP's) and end-use products (EUP's) must bear appropriate labeling as specified below and in 40 CFR 162.10.

1. Precautionary statements to be used on labeling of all end-use captafol products:

    a. All end-use product labels must reflect the following statements: RESTRICTED USE PESTICIDE. For retail sale to, and use only by Certified Applicators or persons under the direct supervision of a Certified Applicator, and only for those uses covered by the Certified Applicator's certification.
    "This product is classified as a "Restricted Use Pesticide" because it contains captafol which has been determined to cause tumors in laboratory animals."

    Exposure to captafol during mixing, loading, and application may be hazardous to your health.
    This product contains captafol which has been determined to cause tumors in laboratory animals.

    b. All end-use products must have the labeling precaution under "Direction for Use": Wear impervious gloves and full body clothing during handling and application".

c. All end-use products intended for crop use, except seed, seed piece, and plant propagule treatments must bear the following use restrictions:

   i. Do not rotate treated crop with crops other than those with registered captafol uses.

   ii. Do not allow persons to enter treated areas within 24 hours following application unless full body clothing is worn. Conspicuously post reentry information at sites of application.

   iii. Do not use captafol-treated rice seed in fields subsequently to be used for agricultural crops other than those with registered captafol uses

   iiii. Water from cranberry bogs, wetland taro fields (foliarly treated with captafol) and rice fields planted with captafol-treated rice seed must not be used for irrigation of crops other than those with registered captafol uses.

2. For products that bear label claims for use in greenhouses:

   a. Only the applicator is permitted in the greenhouse during application of captafol to soil. Open vents to greenhouse during application and for at least 1 hour after application.

   b. Workers planting in captafol treated soil in greenhouses must wear impervious gloves and full body clothing.

3. All products, manufacturing-use and end-use, must bear the labeling precautionary statements.

   DANGER

   Causes irreversible eye damage. Harmful if swallowed or inhaled. May cause allergic skin reactions. Do not get in eyes. Wear goggles or face shield when handling. Avoid contact with skin and clothing. Remove and separately launder clothing before reuse. <u>This product contains captafol which has been determined to cause tumors in laboratory animals</u>

In addition, the following specific environmental hazard labeling requirements apply to either technical, manufacturing-use or end-use product as indicated.

a. All technical and MUP's must bear the following precautionary statements:

> This pesticide is extremely toxic to fish. Do not discharge into lakes, streams, ponds, or public waters unless in accordance with an NPDES permit. For guidance contact your Regional Office of the Environmental Protection Agency. Do not discharge effluent containing captafol into sewage systems without obtaining permission from the sewage treatment authority.

b. All EUP's that allow for foliar applications except foliar application to cranberries, must bear the hazard precaution:

> This pesticide is extremely toxic to fish. Do not apply directly to water. Drift and runoff from treated areas may be hazardous to aquatic organisms in neighboring areas. Do not contaminate water by cleaning of equipment or disposal of wastes.

c. All EUP's that allow foliar application to cranberries must bear the following environmental hazard precautions:

> This pesticide is extremely toxic to fish. Fish may be killed at recommended application rates. Drift and runoff from treated areas may be hazardous to aquatic organisms in neighboring areas. Do not contaminate water by cleaning of equipment or disposal of wastes.

d. All EUP's that allow seed treatment must bear the following environmental hazard precaution:

> This pesticide is extremely toxic to fish. Do not contaminate water by cleaning of equipment or diposal of wastes.

4. Both MUP's and EUP's may need special labeling to protect endangered species. This will be determined after required environmental chemistry data are reviewed, and in conjunction with EPA's evaluation of the potential risk of hazards to endangered species.

118   Pesticide Fact Handbook

Based on historical use experience (human health effects reports) and the benefits from the pesticide uses, the Agency will allow the registration of captafol products to continue for existing use-patterns until the risk of hazards is better defined by the data submitted in response to the Captafol Registration Standard and the completion of a Special Review by the Agency; and provided that the required labeling under the Standard is submitted within the time specified.

5. <u>Summary of Major Data Gaps</u>

Product Chemistry Data (6)[a]
    Description of beginning materials and manufacturing process
    Discussion of Formation of Impurities
    Preliminary Analysis of Product Samples
    Density, Bulk Density, or Specific Gravity
    Vapor Pressure
    Dissociation Constant
    Octanol/Water Partition Coefficient
    pH
    Solubility

Toxicology
    Inhalation LC_{50} - Rat (6)
    90-day feeding - Rodent, Non-rodent (18)
    90-day Inhalation - Rat (18)
    Chronic Toxicity - 2 species: Rodent and Non-rodent (36)
    Oncogenicity study - 2 species: Rat and Mouse preferred (6)
    Teratogenicity 2 species - needs hamsters study using two test groups, one dosed on day 7, and the other dose on day 8 of gestation (12)
    Gene Mutation (12)
    Dermal Absorption (6)

Residue Chemistry (24)
    Analytical Method
    Plant Metabolism
    Animal Metabolism (poultry)
    Storage Stability
    Magnitude of Residues on all food-use crops

Environmental Fate
    Soil Metabolism - Aerobic soil (24)
                       - Anaerobic aquatic (24)
                       - Aerobic aquatic (24)
    Degradation - Hydrolysis (6)
                - Photodegradation in water, soil and air (6)
    Dissipation - Soil (24)
              - Aquatic (sediment) (24)
    Reentry (24)

[a]/ Data to be submitted in number of months after receipt of the Guidance for the Reregistration of Pesticide Products Containing Captafol, EPA Case Number 116.

Mobility Studies (6)[a]
   Leaching and Absorption/Desorption
Accumulation Studies (24)
   Rotational Crops
Wildlife and Aquatic Organisms
   Acute LC_{50} Estuarine and
     Marine Organisms (6)

6. Contact Person

    Eugene M. Wilson

    EPA
    Office of Pesticide Programs
    Registration Division (TS-767C)
    Crystal Mall #2
    1921 Jefferson Davis Hwy.
    Arlington, VA
    Telephone (703) 557-1900

DISCLAIMER: The information in this Chemical Information Fact Sheet is for informational purposes only and may not be used to fulfill data requirements for pesticide registration or reregistration.

a. Data to be submitted in number of months after receipt of the Guidance for the Reregistration of Pesticide Products Containing Captafol, EPA Case Number 116.

CAPTAN

Date Issued: March 6, 1986
Fact Sheet Number: 75

1. Description of Chemical

    Generic Name: N-trichloromethylthio-4-cyclohexene-1, 2-dicarboximide

    Common Name: Captan

    Trade Names: Merpan, Orthocide, SR-406, and Vancide 89

    EPA Shaughnessy Code: 081301
    Chemical Abstract Service (CAS) Number: 133-06-2
    Year of Initial Registration: 1951
    Pesticide Type: Fungicide
    U.S. Producer: Chevron Chemical Company
    Stauffer Chemical Company
    Makteshim Beer Sheva Chemical Works, Ltd.
    Calhio Chemicals, Inc.

2. Use Patterns and Formulations

    Application Sites: Captan (N-trichloromethylthio-4-cyclohexene-1,2-dicarboximide) is a fungicide federally registered for use on almonds, apples, apricots, asparagus, avocados, beans, beets, blackberries, blueberries, broccoli, Brussels sprouts, cabbage, cantaloupes, carrots, cauliflower, celery, cherries, corn (sweet), cotton, cranberries, cucumbers, dewberries, eggplants, grapefruits, grapes, honeydew melons, kale, lemons, lettuce, limes, mangoes, mustard, nectarines, onions, oranges, peaches, pears, peas, peppers, pineapples, plums, potatoes, pumpkins, quinces, raspberries, rhubarb, rutabagas, soybeans, spinach, squach, strawberries, tangelos, tangerines, taro, tomatoes, turnips, and watermelons. The following crops may be seed-treated: Alfalfa, asparagus, barley, beans, beans (lima), beets (table), bluegrass, broccoli, Brussels sprouts, cabbage, cantaloupes, carrots, cauliflower, clover, collards, conifers (Douglas fir, Red pine, Scotch pine, Norway spruce), corn (field and sweet), cotton, cowpeas, crucifers, cucumbers, eggplants, flax, forage grasses, kale, lentils, lespedeza, millet, milo, muskmelons, mustard, oats, okra, onions, peanuts, peas, peppers, pineapples, pumpkins, radishes, rape, rice, rutabagas, rye, safflower, sesame, small-seeded legumes, sorghum, soybeans, spinach, squash, sugar beets, sunflowers, Swiss chard, tomatoes, trefoil, turnips, watermelons, and wheat. Ornamental crop uses include: foliar applications to azaleas, begonias (tuberous), camellias, carnations, chrysanthemums, dichondra, grasses (ornamental in nonpastered areas only), grasses (lawn seedbeds), hollyhocks, lilacs, snapdragons, spireas, roses, and stocks; for soil in plant beds and on green house benches used for the culture of flowers, roses shrubs and trees; and as a dip application to gladiolus corms, to begonia tubers, and to azalea cuttings. Captan may be applied to packing boxes for use in storage and shipping of fruits and vegetables and to soil used in greenhouses for culture of ornamentals and vegetables.

Household Uses Include: application to fruit, vegetable and ornamental gardens, house plants and lawns, in paints (oil based), on surfaces (awnings, blankets, boats, closets, clothing, draperies, floors, leather goods, luggage, mattresses, rugs, shoes, storage rooms, upholstery, walls, workshops, and on other articles. Industrial uses include incorporation into lacquers, paints (oil based), paper, paste (wallpaper flour), plasticizers, polyethylene, rubber stabilizer, textiles, vinyl, and vinyl resins.

Types of Formulations: Dusts, wettable powders, aqueous suspensions, and granules.

Types and Methods of Application: Dusting, spraying, misting, dipping, mixing, and low pressure bomb aerosols.

Application Rates: See use patterns in USDA Compilations of Registered Uses of Fungicides and Nematicides, Part I, Pages C-10-00-01 to C-10-00.21.

Usual Carriers: Clay, talc, silica, and water.

3. Scientific Findings

Chemical Characteristics:

Physical State:	Pure is white crystals, technical is white to buff colored amorphous powder.
Color:	White to buff
Odor:	Pure is odorless, the technical is pungent.
Melting Point:	158-164 °C
Vapor Pressure:	Less than 10^{-6} mm Hg at 25 °C.
Solubility:	Practically insoluble in water, soluble in acetone, ethanol, kerosene, xylene, chloroform, and benzene.
Stability:	Regarded as stable. Decomposes slowly at the melting point. In solution captan decomposes rapidly depending on the pH and temperature, being slower at pH 4 and rapid at pH above 10.

Toxicity Characteristics:

Acute Oral LD_{50} - Rat - 9 gm/kg, Category IV

Acute Inhalation LD_{50} - Rat - males   5.8 mg/L, Category III
                                  - females > 8.9 mg/L, Category III

Eye Irritation - Rabbit - corneal opacity, iris and conjunctive irritation present through day 21, Category I.

Dermal Sensitization - Moderate sensitizer.

Subchronic Rodent - Filled by the 2 year chronic feeding study in rats.

Teratogenicity - Rabbit - not teratogenic at 6, 12, 25, or 60 mg/kg/day. Maternal toxicity observed as weight loss at high dosage.

- Hamster - dosed at 50, 200, and 400 mg/kg/day. Severe maternal weight loss at 400 mg/kg/day. Incidence of skeletal abnormalities; i. e., fused ribs, was increased at high dosage. These lesions were considered to be within normal background incidence.

Reproduction and Fertility Effects (feeding)

a. Three Generation Reproduction Study - Rats: Rats were fed 25, 100, 250, and 500 mg/kg/day for three generations. Body weight reductions occurred at 100, 250, and 500 mg/kg/day and a reduction of food consumption occurred at 100, 250, and 500 mg/kg - in F_1 males and F_2 females. Pup litter weights were decreased in all dosage groups.

b. One Generation Reproduction Study - Rats: Rats were fed a diet of 6, 12.5, and 25 mg/kg/day. No treatment related effects due to captan were seen. (The NOEL for these studies (a and b) when combined, is 12.5 mg/kg/day.)

Mutagenicity

i. Gene Mutation
   o In vivo somatic mutation assay with mice - no mutations were observed.
   o Captan was mutagenic to various strains of S. typhimurium in saline. Decreased mutagenicity was observed when captan was incubated with blood or urine. Captan was not mutagenic in the host mediated assay in mice or rats with S. typhimurium hisG46 or TA1950. Negative findings were also obtained in vitro with blood or urine of captan-treated mice or rats.

   o Captan was mutagenic for point mutations in E. coli and

S. typhimurium. However, it did not produce heritable chromosome aberrations in vivo (this was the conclusion of a working group chaired by W. M. Generoso.)

o Captan was not mutagenic for the dominant lethal test using C3H male and SLR-ICR female mice. No significant increases in chromosomal aberrations were observed in human fibroblasts in vitro or in bone marrow cells of Wistar rats in vivo.

ii. Chromosomal Aberrations

o Chinese hamster V79 cells were treated with captan technical at concentrations up to 6.0×10^{-5} M. Chromosomal aberrations were observed at 4.5×10^{-5}. Increased frequency of sister chromatid exchanges were observed at 1.5×10^{-5} M and above.

iii. Unscheduled DNA Synthesis (UDS)

o WI-38 cells were incubated with captan technical and tritiated thymidine for 3 hr without or 1 hr with S-9 activation. The DNA was extracted and the incorporated labeled thymidine counted. Captan did not induce UDS in WI-38 cells.

Chronic Toxicity (feeding)

a. Rat

In a rat feeding study at 0, 25, 100 and 250 mg/Kg/day treatment-related neoplasms (renal tubular adenomas and carcinomas) in males were observed at 100 and 250 mg/Kg/day. The LEL was 100 mg/kg/day based on hepatocellular hypertrophy, increased kidney weight (male and female) and decreased body weight (male and female). The NOEL was 25 mg/Kg/day. This study partially satisfies the chronic testing requirement for registration.

Oncogenicity (feeding)

a. Rat

See a above, chronic toxicity in the rat.

b. Mouse (high dose study)

In a CD-1 mouse study technical captan at dietary concentrations of 6,000, 10,000, and 16,000 ppm induced both benign and malignant duodenal tumors in both males and females.

c. Mouse (low dose study)

In a second CD-1 mouse study there was an increased incidence of focal hyperplasia, adenoma/polyp(s) and primary carcinomas in the gastrointestinal track of both male and female mice at the highest level tested (6,000 ppm) and a possible increase at the lower dosage levels.

Physiological and Biochemical Behavioral Characteristics:

Foliar Absorption: Captan and/or its metabolites and degradates is (are) absorbed by roots and shoot of plants.

Translocation: Captan and/or its metabolites and degradates is (are) translocated in plant tissue. Captan and/or its metabolites and degradates is (are) absorbed and translocated within plants as a result of seed treatment, soil treatment and foliar application.

Mechanism of Pesticidal Action: Unknown

Metabolism and Persistence in Plants and Animals: The metabolism and accumulation of captan in plants are not understood. The metabolism of captan is understood for ruminants but not for avian species.

Environmental Characteristics: Data gap

Ecological Characteristics:

Hazards to Fish and Wildlife

Bluegill sunfish   96 hr LC_{50} = 0.047 - 0.111 ppm
Rainbow trout   96 hr LC_{50} = 0.066 - 0.080 ppm
Characterized as "very highly toxic" to both cold water and warm water fish.

Daphnia magna: 48 hr LC_{50} = 7.06 - 9.96 ppm
Moderately toxic to aquatic invertebrates.

Avian Toxicity: LC_{50} = quail > 2400 ppm

Avian Reproduction: Available data indicate that captan does not impair avian reproduction.

Potential problems related to endangered species. The Agency has made a preliminary finding that the use of captan as a fungicide in rice, cranberries, and citrus may affect the status of endangered birds, fish, and insect species. An endangered bat may be affected by the use of captan on taro in Hawaii. The Agency will seek the opinion of the U.S. Fish and Wildlife Service in these matters, to better determine what, if any, actions are necessary to protect these species.

Efficacy Review Results:

Captan product registrations with EPA have been supported with efficacy data. Tolerant strains of target plant pathogens have not been reported.

Tolerance Assessments:

1. List of Crops and Tolerances:

The following table lists the present status for tolerances in parts per million (ppm) for residues of captan:

Raw Agricultural Commodity	Part Per Million in Captan Residues			
	U.S.	Canada	Mexico	Codex
ALMOND, HULLS	100.0 I*	-	-	-
ALMONDS	2.0 I	-	-	-
APPLES	25.0**	5.0	25.0	25.0
APRICOTS	50.0	5.0	-	20.0
AVOCADOS	25.0	-	-	-
BEANS, DRY	25.0 I	-	25.0	-
BEANS, SUCCULENT	25.0 I	-	25.0	10.0
BEETS, GREENS	100.0	-	-	-
BEETS, ROOTS	2.0	-	-	-
BLACKBERRIES	25.0	-	-	-
BLUEBERRIES (HUCKLEBERRIES)	25.0	5.0	-	20.0
BROCCOLI	2.0	-	2.0	-
BRUSSELS SPROUTS	2.0	-	-	-
CABBAGE	2.0	-	2.0	-
CANTALOUPS	25.0	-	25.0	-
CARROTS	2.0	-	2.0	-
CATTLE, FAT	0.05	-	-	-
CATTLE, MBYP	0.05	-	-	-
CATTLE, MEAT	0.05	-	-	-
CAULIFLOWER	2.0	-	2.0	-
CELERY	50.0	-	50.0	-
CHERRIES	100.0	5.0	-	50.0
COLLARDS	2.0	-	-	-
CORN, SWEET (K+CWHR)	2.0	-	2.0	-
COTTON, SEED	2.0	-	2.0	-
CRABAPPLES	25.0	5.0	25.0	25.0
CRABERRIES	25.0	5.0	-	10.0
CUCUMBERS	25.0	-	25.0	10.0
DEWBERRIES	25.0	-	-	-
EGGPLANT	25.0	-	-	-
GARLIC	25.0	-	25.0	-
GRAPEFRUIT	25.0 I	-	-	15.0
GRAPES	50.0	5.0	-	-

*I Interim tolerance pending evaluation (under Special Review) of transfer of captan residues to meat, milk, and eggs from feeding the raw agricultural commodity or their byproducts.
** Established tolerance under regulation, Section 180.103, 40 CFR.

Continued

Raw Agricultural Commodity	Part Per Million in Captan Residues			
	U.S.	Canada	Mexico	Codex
HOGS, FAT	0.05	-	-	-
HOGS, MBYP	0.05	-	-	-
HOGS, MEAT	0.05	-	-	-
HONEYDEW MELONS	25.0	-	25.0	-
KALE	2.0	-	-	-
LEEKS	50.0	-	-	-
LEMONS	25.0 I	-	-	15.0
LETTUCE	100.0	-	100.0	10.0
LIMES	25.0 I	-	-	15.0
MANGOES	50.0	-	50.0	-
MUSKMELONS	25.0	-	25.0	-
MUSTARD, GREENS	2.0	-	-	-
NECTARINES	50.0	-	-	-
ONIONS, DRY BULB	25.0	-	25.0	-
ONIONS, GREEN	50.0	-	50.0	-
ORANGES	25.0 I	-	-	15.0
PEACHES	50.0	5.0	40.0	15.0
PEARS	25.0	5.0	25.0	25.0
PEAS, DRY	2.0	-	2.0	-
PEAS, SUCCULENT	2.0	-	2.0	-
PEPPERS	25.0	-	-	10.0
PIMENTOS	25.0	-	-	10.0
PINEAPPLES	25.0 I	-	25.0	-
PLUMS (FRESH PRUNES)	100	5.0	-	10.0
POTATOES	25.0 I	-	2.0	15.0
PUMPKINS	25.0	-	-	-
QUINCES	25.0	-	-	-
RASPBERRIES	25.0	5.0	-	10.0
RHUBARB	25.0	-	-	15.0
RUTABAGAS, ROOTS	2.0	-	-	-
SHALLOTS	50.0	-	-	-
SOYBEANS, DRY	2.0	-	2.0	-
SOYBEANS, SUCCULENT	2.0	-	2.0	-
SPINACH	100.0	-	100.0	20.0
SQUASH, SUMMER	25.0	-	-	-
SQUASH, WINTER	25.0	-	-	-
STRAWBERRIES	25.0	5.0	25.0	20.0
TANGERINES	25.0 I	-	-	15.0
TARO (CORN)	0.25	-	-	-
TOMATOES	25.0	5.0	-	15.0
TURNIPS, GREENS	2.0	-	-	-
TURNIPS, ROOTS	2.0	-	-	-
WATERMELONS	25.0	-	25.0	-

A feed additive regulation (§561.65, Title 21, Code of Federal Regulations; Parts 500 to 599) permits residues of captan at 100.0 ppm remaining on corn seed from its intended use as a seed protectant after detreatment. Detreated corn seed can be used only as a feed for cattle and hogs up to 14 days prior to slaughter.

A food additive regulation (§193.40, Title 21, Code of Federal Regulations Parts 170 to 199) permits 50.0 ppm residues of captan in or on washed raisins when present as a result of fungicidal treatment by preharvest application to grapes and postharvest application during the drying process.

No tolerances have been established for captan residues in or on any crop commodity for which captan is registered solely for seed or plant propagule application, because heretofore seed or plant propagule applications were considered as nonfood uses. Seed use sites are listed under Use Patterns and Formulations.

Data Gaps in Residue Chemistry

o Available plant metabolism data are not completely adequate for identifying the metabolites that may result from the maximum uses and necessary to support the established tolerances.

o Available animal metabolism data are not adequate to support the tolerances in meat; and to establish tolerances in milk, and poultry and eggs.

o For enforcement purposes, FDA's Pesticide Analytical Manual, Method I, Vol. II, Pesticide Regulation Section 180.103 is acceptable for plant commodities. No validated method is available for enforcement of tolerances for residues of captan in animal commodities.

o Inadequate data are available on the storage stability of residues of captan in animal commodities or in or on plant commodities.

o The following uses need tolerances to allow continued registrations:

   California, Special Local Needs registration CA780027 - Use of captan as a seed treatment or as a root dip in the culture of asparagus.

   Washington, Special Local Needs registration WA800035 - Use of captan as a seed treatment or as a soil treatment in the culture of kohlrabi.

   Use-patterns for treatment of soil and greenhouse benches in which vegetables (without tolerances) are grown.

o The data are insufficient to assess the established tolerance for residues in or on detreated seed corn because no data were submitted depicting residues resulting from detreated seed that originally had been treated at the maximum allowable rate. A mechanism must be implemented to prevent the feeding of detreated seed corn which contains residues of pesticides in addition to those of Captan.

o Heretofore, seed treatments and plant propagule treatments have been considered nonfood uses. Available plant metabolism data indicate that residues of captan may be taken up into mature plants from treated seed. Therefore, seed treatments are uses for which residue data and requests for EPA Pesticide Petitions for proposed tolerances must be submitted.

o Processing studies are required for the following commodities: potatoes, beans, soybeans, tomatoes, oranges, plums, sweet corn, and cottonseed.

o Captan may be used as a component of paper and paper board that may come in contact with aqueous and fatty foods [21 CFR 176.170(c)]. Residue data to support this regulation are required to support the EPA registered use-pattern. Alternatively, label amendments are required to restrict the use of captan-treated packing boxes for fruits and vegetables having tolerances for residues of captan.

o The theoretical maximum residue contribution (TMRC) from established tolerances is 12 mg/day based a 1.5 kg diet. The changes in the residue definition, the requested tolerance proposals, and the pending tolerances noted above will all affect a change in the TMRC level. The data requirements to support established tolerances as listed in 40 CFR 180.103 are identified.

o The Provisional Maximum Permissible Intake (PMPI) for a 60 kg person is 0.75 mg/day based on a Provisional Acceptable Daily Intake (PADI) of 0.0125 mg/kg. The present TMRC represents 1600 percent of the PMPI. The inclusion of the major metabolite (THPI) in the tolerance expression may result in an increase in the TMRC and a greater percentage of the PMPI utilized.

o The PADI for captan is based on a reproductive toxicity study in rats. A no-observed-effect level (NOEL) was established at 12.5 mg/kg/day for decreased pup weights. A safety factor of 1000 is used to derive the PADI because there was only chronic data on one species. The PADI will be changed to an ADI when chronic data on a second species (nonrodent) are submitted and found adequate. The data from the most sensitive species and a safety factor of 100 will be used. The NOEL of 12.5 mg/kg/day was based on reproductive toxicity excluding the issue of oncogenicity for which a risk assessment has been made.

4. Summary of Regulatory Positions and Rationale

The Agency has concluded that studies conducted with mice and rats have shown statistically significant increases in incidences of certain tumors. Use of captan results in dietary and environmental exposure that may pose unreasonable risks to human health unless certain steps are taken. Accordingly, the Agency proposed in the Federal Register of June 21, 1985 (50 FR 25884) to cancel or deny federal registrations of products containing captan for use on food crops with the proviso that in the final decision EPA would continue any use on food where data submitted demonstrate that captan residues on food are sufficiently lower than EPA's estimates or that alternative application methods will sufficiently reduce dietary exposure to captan. EPA also proposed in its preliminary determination that protective clothing and/or equipment be worn or used for specific non-food agricultural and non-agricultural uses of captan and that revised labeling be required on products intended for non-food uses. Extensive dietary data are due from registrants in May, 1987. A decision document, Position Document Number 4, is expected to be issued in August, 1988.

Required labeling reflects use restrictions needed to reduce human exposure to captan. The following areas of labeling will be required within 90 days from the receipt of the Standard:

1. Ingredient Statements
2. Precautionary Statements
3. Environmental Hazards Statements
4. Use Precaution Statements

Summary of Risk/Benefit Review

An EPA document entitled "Intent to Cancel Registration of Pesticide Products Containing Captan; Availability of Position Document 2/3 (50 FR 25884-25899, June 21, 1985) discusses the risks and benefits of captan.

5. Summary of Major Data Gaps

    Product Chemistry
    Toxicology
        Acute Testing
        Subchronic Testing
        Chronic Testing
        Special Testing
    Environmental Fate
        Photodegradation
        Metabolism Studies - Laboratory
        Mobility Studies
        Dissipation Studies Field
        Accumulation Studies
        Subdivision K, Reentry Studies
    Wildlife and Aquatic Organisms
        Aquatic Organism Testing
        Nontarget Insect Testing - Aquatic Insects
    Residue Chemistry

6. Contact person at EPA

    Eugene M. Wilson
    Assistant Product Manager (21)
    Fungicide/Herbicide Branch
    Registration Division (TS-767C)
    Office of Pesticide Programs
    Washington, D. C. 20460

CARBARYL

Date Issued: March 30, 1984 - Revised September 5, 1985
Fact Sheet Number: 21

1. DESCRIPTION OF CHEMICAL

    Generic Name: 1-Napthyl N-methylcarbamate

    Common Name: Carbaryl

    Trade Name: Sevin

    EPA Shaughnessy Code: 056801

    Chemical Abstracts Service (CAS) Number: 63-25-2

    Year of Initial Registration: 1958

    Pesticide Type: Insecticide

    Chemical Family: Carbamates

    U.S. and Foreign Producers: Union Carbide
    Makteshim Chemical Works, Inc.

2. USE PATTERNS AND FORMULATIONS

    Application Sites: Citrus, pome, stone and berry fruits, forage, field and vegetable crops, nuts, lawns, forests, ornamental plants, rangeland, shade trees, poultry and pets, indoor use

    Types of Formulations: Baits, dusts, granules, wettable powders, flowables and aqueous dispersions

    Types of Methods of Application: Ground and aerial

    Application Rates: Range from 0.53 lbs. a.i./A to 6.4 lbs. a.i./A

    Usual Carriers: Synthetic clays, talc, various solvents

3. SCIENCE FINDINGS

    Summary Science Statement

    Carbaryl has moderate to low mammalian toxicity. It is not considered to be an oncogen. It is a weak mutagen. Available data indicates that carbaryl has only low teratogenic potential. Long term dietary studies in rats and dogs and a short term study in humans (highest dose only) demonstrate an apparent effect on renal function.

No reentry interval is necessary for carbaryl. The Agency is requesting data to determine if carbaryl will contaminate groundwater. Data are insufficient to assess the environmental fate of carbaryl.

Carbaryl is extremely toxic to aquatic invertebrates and certain estuarine organisms. It is extremely toxic to honey bees. It is moderately toxic to both warmwater and coldwater fishes and has only low toxicity to birds.

A full tolerance reassessment cannot be completed. A one year dog feeding study is required as well as residue data on numerous processed commodities.

Chemical Characteristics

Physical State: Crystalline solid

Color: White

Odor: Essentially odorless

Melting point: 142° C

Vapor Pressure: < 0.005 mm Hg at 26°C

Flash Point: 380°F

Toxicology Characteristics

Acute Oral LD_{50}: 255 mg/kg, Toxicity Category II

Acute Dermal LD_{50}: > 2 g/kg, Toxicity Category III

Primary Dermal Irritation: No irritation, Toxicity Category IV

Primary Eye Irritation: Conjunctival irritation at 24 hours. Cleared at 48 hours. Toxicity Category III

Acute Inhalation LC_{50}: Data gap

Oncogenicity: Ten studies. Each study classified as "supplemental." Collectively these studies provide sufficient evidence that carbaryl is not oncogenic in experimental animals. Eighteen month mouse study was negative at 400 ppm. A 2 year rat feeding and oncogenicity study was negative at 200 ppm.

Teratogenicity: Twenty-four studies have been evaluated to determine the teratogenic potential of carbaryl. In evaluating these studies some were found to be flawed. Other studies demonstrated no teratogenicity or maternal toxicity. There are studies which demonstrate teratogenic effects although the doses also caused maternal toxicity. Two studies produced teratogenic effects in the beagle dog. These two studies are the primary reason carbaryl was made a candidate for RPAR in 1976.

The Agency has concluded (45 FR 81869) that carbaryl does not constitute "a potential human teratogen or reproductive hazard under proper environmental usage". The Agency has determined that the dog is a poor model to use for teratogenicity testing. The Agency has determined that a label precaution stating not to use carbaryl on pregnant dogs is warranted

There have been proposals that there are differences in the metabolism of carbaryl between the dog and man. These differences, however, have never been demonstrated. Therefore, a metabolism study in the beagle dog versus the rat is being required. This metabolism study should allow us to determine if there are meaningful differences between the dog and other mammalian species.

Reproduction: A rat three generation study was negative at 200 mg/kg.

Mutagenicity: Carbaryl is characterized as a weak mutagen. The Agency has determined that carbaryl does not pose a mutagenic risk. No additional data are being requested.

1-year Dog Feeding Study:
2-year Rat Feeding Study: Demonstrated an apparent effect on renal function. A kidney effect was also noted in a short-term human study. A one year dog feeding study using carbaryl is being requested in order to determine the effects of carbaryl on kidney dysfunction. The results of these may necessitate a re-evaluation of the ADI for carbaryl.

<u>Physiological and Biochemical Behavioral Characteristics</u>

Mechanism of Pesticidal Action: A contact insecticide which causes reversible carbamylation of the acetylcholinesterase enzyme of tissues, allowing accumulation of acetylchloline at cholinergic neuroeffector junctions (muscarinic effects), and at skeletal muscle myoneural junctions and autonomic ganglia. Poisoning also inpairs the central nervous system function.

Metabolism and Persistence in Plants and Animals: Carbaryl is rapidly excreted in animals, mainly in the urine. Residues in animals are carbaryl, 1-naphthol and hydroxycarbaryl. The hydroxy metabolites are found mainly as glucuronide and sulfate conjugates. Carbaryl is slowly taken up into plants, after which it is metabolized. The disappearance of carbaryl residue from plant surfaces is attributed to mechanical attribution, volatization and uptake into plant. Photochemical degradation does not appear to be a factor. 1-Naphthol is the major metabolite.

Environmental Characteristics

Available data are insufficient to fully assess the environmental fate of carbaryl.

Adsorption and Leaching in Basic Soil Types: The Agency is requesting data to determine if carbaryl will contaminate groundwater.

Microbial Breakdown: Carbaryl is degraded by fungi. The soil fungi attack carbaryl by hydroxylation of the side chain and ring structure.

Loss from Photodecomposition: Data gaps. Data are required

Bioaccumulation: Preliminary data indicate that there may be a potential for carbaryl and its residue(s) to accumulate in catfish, crayfish, snail, duckweed and algae. Additional data are requested.

Resultant Average Persistance: Carbaryl is metabolized by pure and mixed cultures of bacteria, fungi, and to some exby other soil and water organisms. The half-life appears to range from 7 to 28 days in aerobic and anaerobic soils, respectively.

Ecological Characteristics

Avian oral LD_{50} -
Mallard duck: > 2179 mg/kg
Ring necked pheasant: > 2000 mg/kg

Avian dietary LC_{50} -
Mallard duck: > 5000 ppm

Ring-necked pheasant: > 5000 ppm
Bobwhite quail: > 5000 ppm

Freshwater Fish LC_{50} -
Coldwater fish: rainbow trout - 1.95 ppm
Warmwater fish: bluegill sunfish - 6.76 ppm

Acute LC_{50} Freshwater Invertebrates -
Daphnia pulex - 6.4 ppb

Acute LC_{50} Estuarine and Marine Organisms: Data gap. Data being requested.

Freshwater Fish Early Life-Stage -
Fathead minnow - Maximum Acceptable Theoretical Concentration (MATC) - >0.21<0.68 ppb

No precautionary language is required for birds or fish. However, carbaryl is highly toxic to aquatic invertebrates. There is insufficient information to characterize the chronic toxicity of carbaryl to aquatic invertebrates.

Tolerance Assessments

The Agency is unable to complete a full tolerance reassessment because of certain residue chemistry and toxicology data gaps, namely a one year dog feeding study and the need for residue data on various processed food commodities.

Tolerances:

Commodity	Parts Per Million
Alfalfa	100
Alfalfa	100
Almonds	1
Almonds, hulls	40
Apples	10
Apricots	10
Asparagus	10
Bananas	10
Barley, grain	0
Barley, green fodder	100
Barley, straw	100
Beans	10
Beans, forage	100
Beans, hay	100
Beets, garden (roots)	5
Beets, garden (tops)	12

Tolerances:

Commodity	Parts Per Million
Birdsfoot trefoil, forage	100
Birdsfoot trefoil, hay	100
Blackberries	12
Blueberries	10
Boysenberries	12
Broccoli	10
Brussels sprouts	10
Cabbage	10
Carrots	10
Cauliflower	10
Celery	10
Cherries	10
Chestnuts	1
Chinese cabbage	10
Citrus fruits	10
Clover	100
Clover, hay	100
Collards	12
Corn, fresh (including sweet) Kernel(K) + Corn with husk removed(CWHR)	5
Corn, fodder	100
Corn, forage	100
Cotton, forage	100
Cottonseed	
Cowpeas	5
Cowpeas, forage	100
Cowpeas, hay	100
Cranberries	10
Cucumbers	10
Dandelions	12
Dewberries	12
Eggplants	10
Endive (escarole)	10
Filberts (hazelnuts)	1
Flax, seed	5
Flax, straw	100
Grapes	10
Grass	100
Grass, hay	100
Horseradish	5
Kale	12
Kohlrabi	10
Lentils	10
Lettuce	10
Loganberries	12

Tolerances:

Commodity	Parts Per Million
Maple sap	0.5
Melons	10
Millet, proso, grain	3
Millet, proso, straw	100
Mustard greens	12
Nectarines	10
Oats, fodder, green	100
Oats, grain	0
Oats, straw	100
Okra	10
Olives	10
Oysters	0.25
Parsley	12
Parsnips	5
Peaches	10
Peanuts	5
Peanuts, hay	100
Pears	10
Peas (with pods)	10
Peavines	100
Pecans	1
Peppers	10
Pistachio nuts	1
Plums (fresh prunes)	10
Poultry, fat	5
Poultry, meat	5
Potatoes	0.2(N)
Prickly pear cactus, fruit	12.0
Prickly pear cactus, pads	12.0
Pumpkins	10
Radishes	5
Raspberries	12
Rice	5
Rice, straw	100
Rutabagas	5
Rye, fodder, green	100
Rye, grain	0
Rye, straw	100
Salsify (roots)	5
Salsify (tops)	10
Sorghum, forage	100
Sorghum, grain	10
Soybeans	5
Soybeans, forage	100
Soybeans, hay	100
Spinach	12
Squash, summer	10
Squash, winter	10

Tolerances:

Commodity	Parts Per Million
Strawberries	10
Sugar beets, tops	100
Sunflower seeds	1
Sweet potatoes	0.2
Swiss chard	12
Tomatoes	10
Turnips, roots	5
Turnips, tops	12
Walnuts	1
Wheat, fodder, green	100
Wheat (grain)	3
Wheat, straw	100
Cattle, fat	0.1
Cattle, kidney	1
Cattle, liver	1
Cattle, meat	0.1
Cattle (mbyp)	0.1
Goats, fat	0.1
Goats, kidney	1
Goats, liver	1
Goats, meat	0.1
Goats (mbyp)	0.1
Horses, fat	0.1
Horses, kidney	1
Horses, liver	1
Horses, meat	0.1
Horses (mbyp)	0.1
Sheep, fat	0.1
Sheep, kidney	1
Sheep, liver	1
Sheep, meat	0.1
Sheep (mbyp)	0.1
Swine, fat	0.1
Swine, kidney	1
Swine, liver	1
Swine, meat	0.1

Based on established tolerances the theoretical maximum residue contribution (TMRC) for carbaryl residues in the human diet is calculated to be 5.48 mg/day. The acceptable daily intake (ADI) of carbaryl is 0.1 mg/kg/day. The maximum permissible intake (MPI) is 6 mg/day. To provide for conformity between U.S. tolerances for carbaryl and tolerances established by the Codex Alimentarius, Canada and Mexico, the expression of the U.S. tolerances for carbaryl will be changed to omit reference to 1-naphthol.

A one year dog feeding study is being requested in order to determine the effects of carbaryl on kidney dysfunction. The results of these data may require that the ADI for carbaryl be recalculated.

U.S. tolerances for most raw agricultural commodities are supported by current residue chemistry data. In some cases, however, more data are required.

4. SUMMARY OF REGULATORY POSITION AND RATIONALE

The Agency has determined that it should continue to allow the registration of carbaryl. Adequate studies are available to assess the acute toxicological effects of carbaryl to humans. None of the criteria listed in section 162.11(a) of Title 40 of the U.S. Code of Federal Regulations have been met or exceeded. However, because of gaps in the data base a full risk assessment of carbaryl cannot be completed.

A full tolerance reassessment cannot be completed because of certain residue chemistry and toxicology data gaps, namely a one year dog feeding study and the need for residue data on various processed commodities.

No federal or state reentry intervals have been established for carbaryl or will be established.

Available data are insufficient to fully assess the environmental fate of carbaryl. The Agency is requesting data to determine if carbaryl will contaminate groundwater.

5. SUMMARY OF MAJOR DATA GAPS

Residue data on various processed commodities
One year dog feeding study
Hydrolysis study
Photodegradation studies
Soil metabolism studies
Mobility studies
Dissipation studies
Accumulation studies
Metabolism study in dog versus rat

6. CONTACT PERSON AT EPA

Jay S. Ellenberger
Product Manager (12)
Insecticide-Rodenticide Branch Registration Division (TS-767C)
Office of Pesticide Programs
Environmental Protection Agency 401 M Street, S. W.
Washington, D. C.   20460

Office location and telephone number:
Room 202, Crystal Mall #2
1921 Jefferson Davis Highway
Arlington, VA   22202
(703) 557-2386

DISCLAIMER: The information presented in this Chemical Information Fact Sheet is for informational purposes only and may not be used to fulfill data requirements for pesticide registration and reregistration.

CARBOFURAN

Date Issued: June 25, 1984
Fact Sheet Number: 24

1. Description of Chemical

   Generic name: 2,3-dihydro-2,2-dimethyl-7-benzofuranyl methylcarbamate
   Common name: carbofuran
   Trade Names and code numbers: Furadan, Curaterr, Yaltox, Bay 78537, D 1221 ENT 27164, FMC 10242, and NIA 10242
   EPA Shaughnessy code: 090601
   Chemical Abstract Service (CAS) number: 1563-66-2
   Year of initial registration: 1969
   Pesticide type: insecticide, nematicide
   Chemical family: carbamate
   U. S. producer: FMC Corporation

2. Use Patterns and formulations

   Application sites: Fruit and field crops, vegetables, tobacco, ornamentals and forest trees

   Types and methods of applications: aerial and ground application as a granular or spray

   Application rates: 0.5 to 10 lbs a.i./acre

   Usual carriers: (Confidential Business Information)

3. Science Findings:

   Summary:

   Chemical Characteristics:

   Technical carbofuran is a white crystalline solid with a melting point of 153 - 154°C and a vapor pressure of 2×10^{-5} mm Hg at 33°C. The empirical formula is $C_{12}H_{15}NO_3$ and the molecular weight is 221.3. Solubility in water is 700 ppm. Other solubilities include 30% in N-methyl-2-pyrrolidone, 25% in dimethyl sulfoxide, 15% in acetone, 14% in acetonitrile, 12% in methylene chloride, 9% in cyclohexanone, and 4% in benzene.

Toxicological characteristics:

Current available acute toxicological studies on carbofuran show the following:

- Acute oral toxicity: rat, LD_{50} 3.8-34.5 mg/kg; mouse, LD_{50} 14.4 mg/kg cat, 2.5 - 3.5 mg/kg; dog, LD_{50} 15-18.9 mg/kg (Tox Category I)

- Acute dermal toxicity: rabbit, LD_{50} in isopropanol <46.4; LD_{50} in water >10,250; 75WP formulation LD_{50} in water, 3400 (Toxicity Category I)

- Acute inhalation: (dust) rat. 1hr. LD_{50} 0.80 - 0.108 mg/L; rat 4 hr. LD_{50} 0.075-0.108 mg/L; rat. 1hr. LD_{50} >0.026 mg/L; rat, LC_{50} 4 hr. 0.017 - 0.047 mg/L (Toxicity Category I)

- Primary eye irritation and primary dermal irritation not done since this is a category I material precluding condition of meaningful studies. (Assumed to be Toxicity Category I)

Major routes of exposure: Application by ground and aerial spray equipment increases the potential for exposure of humans, livestock, and wildlife due to spray drift. Human exposure to carbofuran from handling, application, and reentry operations would be minimized by the use of approved respirators and other protective clothing.

Physiological and Biochemical characteristics

Mechanism of pesticidal action: Cholinesterase inhibition following contact with treated surfaces or soil and/or ingestion of treated plant tissue.

Metabolism in plants and animals is similar, carbofuran® is systemic in plants. Like other carbamates, metabolized rapidly in animals into less toxic and finally non-toxic metabolites.

Environmental Characteristics

Carbofuran degrades fairly slowly in non-sterile, neutral or acid aerobic soils with half-lives ranging from 1-8 weeks. It is more stable in sterile soil and unstable under alkaline conditions. Under anaerobic conditions carbofuran is more stable and may take twice as long to degrade. The major degradates of concern are the 3-hydroxy carbamate and 7-phenol products resulting from hydrolysis. The metabolites of carbofuran are less toxic than the parent compound.

Carbofuran is mobile in soil, particularly sandy soil with high percolation rate. It has been found in shallow aquifers under or near treated fields in three states at levels up to 50ppb.

Ecological Characteristics

Acute avian oral toxicity; LD_{50} 90-5 mg/kg
Avian subacute dietary toxicity; LD_{50} 16-1104 ppm
Freshwater fish acute toxicity; LC_{50} 94-2859 ppb
Acute toxicity for freshwater invertebrates; 9.8-38.6 ppb

Based on studies available to assess hazards to wildlife and aquatic organisms carbofuran is characterized as very highly toxic to cold water and warm water fish, highly toxic to freshwater invertebrates and very highly toxic to birds. Label precautions reduce the hazard to wildlife and aquatic organisms.

Monitoring studies are being required to allow a better assessment of the actual effects of carbofuran use on non-target organisms under field conditions. Primary emphasis will be on effects of use of carbofuran on bird populations.

Efficacy Review Results: none conducted.

Tolerance Assessment:

The previously established tolerances for residues of carbofuran and its cholinesterase - inhibiting metabolites are published in 40 CFR 180.254, 21 CFR 193.43, and 21 CFR 561.67. A summary of these tolerances follows:

U. S. Tolerances

Crop	Maximum Residue Limits (PPM)	
	Carbamate	Total
alfalfa forage	5.0	10.0
alfalfa hay	20.0	40.0
bananas	-	0.1
barley, grain	0.1	0.2
barley, straw	1.0	5.0
cattle, (fat, meat, and meat byproducts)	0.02	0.05
coffee beans	-	0.1
corn, field and popcorn-grain	0.1	0.2
corn, fresh-kernels and cob (husk removed)	0.2	1.0
corn fodder and forage, field, pop-, and sweet	5.0	25.0
cottonseed	0.2	1.0
cranberries	0.3	0.5

U. S. Tolerances (continued)	Maximum Residue Limits (PPM)	
Crop	Carbamate	Total
cucumbers	0.2	0.4
goats, (fat, meat, and meat byproducts)	0.02	0.05
grapes	0.2	0.4
grapes, dried pomace	1.5	2.0
hogs (fat, meat, and meat byproducts)	0.02	0.05
horses, (fat, meat, and meat byproducts)	0.02	0.05
melons	0.2	0.4
milk	0.02	0.1
oats, grain	0.1	0.2
oats, straw	1.0	5.0
peanuts	1.5	4.0
peanut hulls	8.0	10.0
peanut soapstock	3.0	24.0
peppers	0.2	1.0
potatoes	1.0	2.0
pumpkins	0.6	0.8
raisins	1.0	2.0
raisin waste	3.0	6.0
rice, grain	-	0.2
rice, straw	0.2	1.0
sheep, (fat, meat, and meat byproducts)	0.02	0.05
sorghum, (grain)	-	0.1
sorghum, (grain), forage & fodder	0.5	3.0
soybeans	0.2	1.0
soybean, forage and hay	20.0	35.0
soybean, soapstock	1.0	6.0
squash	0.6	0.8
strawberries	0.2	0.5
sugar beets	-	0.1
sugar beet tops	1.0	2.0
sugarcane	-	0.1
sunflower	0.4	0.8
sunflower seed hulls and meal	0.5	1.0
sweet potato	0.6*	2.0*
tobacco	N/A	N/A
wheat, grain	0.1	0.2
wheat, straw	1.0	5.0

*Proposed
N/A not applicable - nonfood use

The data for carbofuran residues in or on the following agricultural commodities are adequate to fill the residue data requirements: sugar beet tops, soybean, soybean soapstock, soybean forage and hay, cucumbers, melons, pumpkins, squash, cranberries, raisins, strawberries, barley grain, oat grain, wheat grain, barley straw, oat straw, wheat straw, cottonseed, peanuts, the fat, meat and meat-by-products' of cattle, goats, hogs, horses, and sheep, milk, poultry, and eggs. However, although additional data are not required to to support the tolerances for milk, peanuts, peanut hulls, and fatty acids of peanut soapstock, the tolerances for these agricultural commodities must be changed as follows:

Commodity	Present Tolerance		Change Tolerance To	
	Carbamate	Total	Carbamate	Total
milk	0.02 ppm	0.1 ppm	0.05 ppm	0.2 ppm
peanuts	1.5 ppm	5.0 ppm	0.2 ppm	0.6 ppm
peanut hulls	8.0 ppm	10.0 ppm	0.3 ppm	1.5 ppm
peanut soapstone	3.0 ppm	24.0 ppm	1.0 ppm	6.0 ppm

Additional residue data are required for the following commodities (and their processed products, if applicable): potato, sugar beets, peppers, grapes (including dried pomace and raisin waste), corn (grain, forage and fodder), rice (grain and straw), sorghum (grain, forage and fodder), alfalfa forage and hay, bananas, coffee, sugarcane and tobacco. No crop groupings can be established at this time because of residue chemistry data gaps. Compatability between Codex MRLs and U.S. tolerances will be assessed when data have been submitted and evaluated.

The previously established Acceptable Daily Intake (ADI) for carbofuran is 0.005 mg/kg/day and the Theoretical maximum Residue Concentration (TMRC), based on the established tolerances for residues of carbofuran as cited under 40 CFR 180.254, is 0.3415 mg/day for a 1.5 kg food diet for a 60 KG person. The TMRC is 113.84% of the ADI. Actual concentrations of carbofuran are likely to be substantially lower than the ADI since the 113% calculation assumes 100% of all crops on the label were treated.

4. Summary of Regulatory Position and Rationale

Use classification: Restricted
(This and other label revisions to appear on all products released for shipment after September 1, 1985 and on all products in channels of trade after September 1, 1986)

Use restrictions: Do not use on Long Island, New York

Unique warning statement required on labels: Labels for all manufacturing use products (MPs) containing carbofuran must bear statements reflecting the hazards to man and the environment [40 CFR 162.10]. Carbofuran is in Toxicity Category I on the basis of acute toxic effects.

° Based on the data reviewed by the Agency, the environmental hazards statement below is required to appear on all MPs containing Carbofuran.

"This pesticide is toxic to fish, birds and other wildlife. Do not discharge into lakes, streams, ponds, estuaries, oceans or public water unless this product is specifically identified and addressed in an NPDES permit. Do not discharge effluent containing this product to sewer systems without previously notifying the sewage treatment plant authority. For guidance, contact your State Water Board or Regional Office of the Environmental Protection Agency.

Labels for all end-use products (EPs) containing carbofuran must bear a statement reflecting the hazard to man and the environment [40 CFR 162.10]

"Restricted Use Pesticides"

"For retail sale to and use only by certified applicators or persons under their direct supervision and only for those uses covered by the certified applicator's certification."

The "Directions for Use" section of the label must include the following statements:

Georgaphical Use Restrictions:

"Do not use this product on Long Island, N. Y."

Rotational Crops:

"Do not plant any crop other than those with registered carbofuran uses in carbofuran-treated soil sooner than 18 months after last application"

Reentry:

"If prolonged intimate contact with corn and/or sorghum foliage will result, do not re-enter treated fields within 14 days of application without wearing proper protective clothing. For all other situations, do not re-enter fields less than 24 hours following application."

º Based on data reviewed by the Agency, the following human and environmental hazards statements are required to appear on the EP products.

The following precautions must be included in the "Hazards to humans and Domestic Animals" section of the product label:

"Poisonous if swallowed. May be fatal or harmful as a result of skin or eye contact or by breathing dust. Causes cholinesterase inhibition. Warning symptoms of poisoning include weakness, headache, sweating, nausea, vomiting, diarrhea, tightness in chest, blurred vision, pinpoint eye pupils, abnormal flow of saliva, abdominal cramps, and unconsciousness. Atropine sulfate is antidotal."

"In case of skin contact, wash skin immediately with soap and water. Remove contaminated clothing and wash before reuse. In case of swallowing, drink 1 or 2 glasses of water and induce vomiting by touching back of throat with finger. Do not induce vomiting or give anything by mouth to an unconscious person."

"Wear long-sleeved clothing and protective gloves when handling. Wash hands and face before eating or smoking. Bathe at the end of the work day. Change clothing daily and wash before reuse.

The following statements must appear under the heading "Environmental Hazards."

i. "Carbofuran is known to leach through soil, and has been found in groundwater as a result of agricultural use. Users are advised not to apply in areas where soils are permeable, i. e., well drained, and which overlie shallow aquifers, particularly those currently being used for drinking water. Consult with the pesticide state lead agency for information regarding soil permeability and aquifer locations

ii. Granular products except for use on rice: "This product is toxic to fish, birds and other wildlife. Birds feeding on treated areas may be killed. Cover or incorporate granules in spill areas. Runoff from treated areas may be hazardous to fish in neighboring areas. Do not apply directly to water or wetlands. Do not contaminate wells, wetlands or any body of water by cleaning of equipment or disposal of waste."

iii. Granular products used on rice:
Same as for the granular products except delete "Do not apply directly to water" and substitute "Fish may be killed at recommend rates."

iv. Non-granular products:
"This product is toxic to fish, birds and other wildlife. Birds feeding on treated areas may be killed. Drift and runoff from treated areas may be hazardous to fish in neighboring areas. Do not apply directly to water. This pesticide is highly toxic to bees exposed to direct treatment or residues on crops. Do not apply this product or allow it to drift to blooming crops or weeds if bees are visiting the treatment area."

v. Non-granular products used on potatoes and/or alfalfa:
"For waterfowl protection, do not apply immediately before or during irrigation, or on fields in proximity of waterfowl nesting areas, or on fields where waterfowl are known to repeatedly feed."

5. Summary of Major Data Gaps

Product Chemistry: Data on product formation of ingredients, preliminary analysis and certification of limits are the major product chemistry gaps.

Residue Chemistry: Additional data are required to support the tolerances for a variety of crops. Residue chemistry data requirements are outlined in detail in the tolerance assessment in Section 3 of this Fact Sheet.

Wildlife and aquatic organisms: The primary concern is monitoring bird populations following commercial application to determine extent of adverse effects, particularly on songbirds and raptors.

Environmental Fate: In addition to regular environmental chemistry requirements, monitoring studies are needed to find the extent and level of groundwater contamination and the conditions under which groundwater contamination occurs.

All data must be submitted no later than June 1987. Interim reports on groundwater and avian monitoring must be submitted at the end of each season.

6. Contact person at EPA

>   Jay Ellenberger
>   Product Manager (12)
>   Insecticide-Rodenticide Branch
>   Registration Division (TS-767)
>   Environmental Protection Agency
>   Washington, DC 20460
>   Telephone No.: (703) 557-3286

DISCLAIMER: The information presented in this Chemical Information Fact Sheet is for informational purposes only and may not be used to fulfill data requirements for pesticide registration and reregistration.

CARBON TETRACHLORIDE

Date Issued: September, 1986
Fact Sheet Number: 102

1. Description of chemical

    Generic name: Carbon Tetrachloride

    Common name: same

    Trade name: Benzinoform, Carbona, CAS 56-235, Dowfume 75, ENT 4705, Flukoide, Halon 104

    EPA Shaughenssy Code (CAS) number: 016501

    Chemical Abstracts Service (CAS) number: 56-23-5

    Year of Initial Registration: 1956

    Pesticide Type: Fumigant

    Chemical family: Chlorinated hydrocarbon

2. Use patterns and formulations

    Application sites: Harvested grains throughout storage transfer, milling, distribution, and processing phases, fumigation of museum specimens.

    Type of formulations: Gas

    Types and methods of application: Fumigation

3. Science Findings

   Physical and Chemical Characteristics-

   Physical state: Liquid

   Color: Colorless

   Odor: Strong ether-like odor

   Boiling Point: 76.75°C

   Freezing Point: -23°C

   Vapor Density: 5.32 (air=1)

   Melting Point: -23°C (-9°F)

4. Regulatory Position

   ° Position Document 1 issued October 15, 1980.

   ° OPP issued a Data-Call-In Notice in March 1984 requiring submission of reproduction (1 species), teratogenicity (2 species), residue chemistry data, updated Confidential Statements of Formula and product chemistry data.

   ° On April 23, 1986 a notice announcing the availability of the draft PD 2/3/4 was published in the Federal Register.

   ° The Agency is issuing a Notice of Intent to Cancel all pesticide products containing carbon tetrachloride based on oncogenic risk and other effects in laboratory animals. This notice cancels all carbon tetrachloride registrations already suspended for non-compliance with the above mentioned Data-Call-In Notice. The use on encased museum specimens, which was not subject to the March 1984 Data Call-In, will be allowed to continue because the current label instructions are sufficient to reduce applicator exposure so that the benefits outweighed the risks.

5. Concerns

   ° Significant data are available which suggest that carbon tetrachloride poses significant health and environmental hazards.

   ° Has been shown to pose acute and subacute poisoning risks.

- Has been shown to cause oncogenic effects in laboratory animals.

- May contribute to the breakdown of the atmosphere's ozone layer.

6. <u>Contact person at EPA</u>: Douglas G. McKinney
　　　　　　　　　　　　　　　　EPA
　　　　　　　　　　　　　　　　Office of Pesticide Programs
　　　　　　　　　　　　　　　　401 M Street, S.W.
　　　　　　　　　　　　　　　　Washington, D.C. 20460

DISCLAIMER: The information in this Chemical Information Sheet is for informational purposes only and may not be used to fulfill data requirements for pesticide registration or reregistration.

CARBOPHENOTHION

Date Issued: June 30, 1984
Fact Sheet Number: 25

1. DESCRIPTION OF CHEMICAL

    Chemical Name: S-[[(p-chlorophenyl)thio]methyl] O,O-diethyl phosphorodithioate

    Common Name: Carbophenothion

    Trade Name: Trithion®

    EPA Shaughnessy code: 058102

    Chemical Abstracts Service (CAS) Number: 786-19-6

    Pesticide Type: Insecticide and Acaricide

    Chemical family: Organophosphate

    U.S. and foreign producers: Stauffer Chemical Co.

2. USE PATTERNS AND FORMULATIONS

    ° Registered for use on a wide variety of vegetable, fruit, nut, forage, ornamental, and forestry sites.
    ° Majority of pesticide use is on citrus.
    ° Commercially available as dust, granular, pelleted, wettable powder, and emulsifiable concentrate formulations.
    ° Applied as foliar applications using either ground or aerial equipment. Dormant and delayed dormant applications are made to some fruit and nut trees. There are also limited uses as a seed treatment, dip, and soil insecticide.
    ° See also EPA Index Entry for carbophenothion.

3. SCIENCE FINDINGS

    Carbophenothion has data gaps in areas of toxicology, environmental fate, and ecological effects. A summary of the science findings based on the available data is provided below:

    Chemical Characteristics

    Physical State: Liquid
    Odor: Mild mercaptan
    Color: Yellow-brown
    Empirical formula: $C_{11}H_{16}ClO_2PS_3$
    Molecular weight: 342.9

Carbophenothion 153

Vapor pressure 0.008 u at 25°C
solubility in water: 0.34 ppm at 20°C
specific gravity: 1.274 at 20°C
pH: 2.43
boiling point: 82°C at 0.01 mmHg
miscibility: miscible with most organic solvents such as petroleum
ether, benzene, toluene, xylene, ethers, alcohol and ketones.

TOXICOLOGY

Acute Oral Toxicity: 0.02 ml/kg in male rats

Reproduction: A rat three year generation study had a NOEL of 10 ppm

Acute Delayed Neurotoxicity: not neurotoxic at 330 mg/kg

2-Year Dog Feeding Study: NOEL of 5 ppm

Adequate studies are unavailable to assess the acute toxicological effects of carbophenothion. Preliminary data indicate that carbophenothion is in Toxicity Category I on the basis of acute oral effects. Carbophenthion is a cholinesterase inhibitor. It is not adequately tested for acute toxicology, chronic toxicity, oncogenicity or teratology.

ECOLOGICAL EFFECTS

Freshwater Fish Acute Toxicity
Coldwater fish: rainbow trout - 56 ppb
Warmwater fish: bluegill sunfish - 13 ppb

Avian Acute Oral Toxicity
Bobwhite quail - 320 mg/kg

Acute Toxicity to Freshwater Invertebrates
adult Palaemonetes - 1.2 ppb

Acute Toxicity to Marine and Estuarine Organisms
pink shrimp - 0.47 ppb
sheepshead minnow - 17 ppb

Chronic Toxicity for Marine and Estuarine Organisms
grass shrimp life cycle study - Maximum Acceptable Theoretical
Concentration (MATC) - >0.22<0.36 ppb
sheepshead minnow embryo/juvenile study - MATC >1.3<2.8 ppb

Carbophenothion is characterized as very highly toxic to freshwater and marine/estuarine organisms and highly toxic to upland gamebirds. See discussion under Section 4 [Summary of Regulatory Position and Rationale].

ENVIRONMENTAL CHEMISTRY

Available data are insufficient to assess the environmental fate of carbophenothion or to assess the potential exposure of humans and non-target organisms to carbophenothion. Preliminary data indicate that carbophenothion is relatively immobile in sandy loam soils. However, The Agency cannot more completely assess the potential for carbophenothion to contaminate groundwater until data are submitted. Preliminary data indicate that there may be a potential for carbophenothion to accumulate in spot and juvenile sheepshead minnows.

TOLERANCE REASSESSMENT

Tolerances for combined residues of the insecticide carbophenothion (S-[p-chlorophenylthio)methyl] O,O-diethyl phosphorodithioate) and its cholinesterase-inhibiting metabolites in or on raw agricultural commodities are established as follows:

10 parts per million in or on almond hulls.
5 parts per million in or on alfalfa (fresh), alfalfa (hay), bean straw, clover, (fresh), clover (hay), corn forage, sorghum forage, sugarbeets (roots), sugarbeets (tops).
4 parts per million in or on blueberries.
2 parts per million in or on grapefruit, lemons, limes, oranges, sorghum grain, tangerines.
0.8 part per million in or on apples; apricots; beans, snap (succulent form); beans, lima (succulent form); beets, garden (roots); beets, garden (tops); cantaloups; cherries; crabapples; cucumbers; eggplants; figs; grapes; nectarines; olives; onions (dry bulb); onions (green); peaches; pears; peas (succulent form); peppers; pimentos; plums (fresh prunes); quinces; soybeans (succulent form); spinach; strawberries; summer squash; tomatoes; watermelons.
0.2 part per million in or on corn (kernels plus cob with husks removed), undelinted cottonseed.
0.1 part per million in the fat of meat of cattle, goats, hogs and sheep.
0.1 part per million (negligible residue) in or on beans (dry) pecans, and walnuts.
Zero in milk.

The tolerances are published in 40 CFR 180.156. Tolerances for numerous raw agricultural commodities as well as processed products are not supported by available data.

No new crop groupings can be established at this time because of extensive residue chemistry data gaps. Compatibility between Codex MRL's and U.S. tolerances will be assessed when data gaps specified in Table A have been submitted and evaluated.

The Acceptable Daily Intake (ADI) for carbophenothion is 0.0125 mg/kg/day. This is based on an acceptable dog chronic feeding study with a No Observable Effect Level (NOEL) of 5.0 ppm and a safety factor of 10.

The Theoretical Maximum Residue Contribution (TMRC), based on relevant food factors and the tolerances cited in 21 CFR 193.50 and 40 CFR 180.156, is 0.5806 mg/day assuming a 1.5 kg diet. Accordingly, the percentage of the ADI used up is 77.42%.

4. SUMMARY OF REGULATORY POSITION AND RATIONALE

The Agency has identified concerns over the potential adverse effects of carbophenotion to aquatic and terrestial species. Based on acceptable aquatic acute toxicity studies, it is calculated that the expected concentration of carbophenothion following direct application to a 6-inch layer of water exceed 1/2 the acute toxicity level in aquatic species. Based on a scientifically sound subacute dietary study, it is calculated that the expected residues in avian foodstuffs following a single application of carbophenothion at a rate of 1 pound a.i. per acre exceed 1/5 the subacute dietary toxicity in avian species. In addition, although there is insufficient information on the granular formulations, the Agency expects that granular applications of carbophenothion would have an adverse impact on birds.

A total risk assessment cannot be made until gaps in the data base for terrestrial species and environmental fate are filled.

The Agency is unable to complete a full tolerance reassessment of carbophenothion because of extensive residue chemistry and toxicology data gaps. Future requests for tolerances will not be automatically rejected, but will be considered on a case-by-case basis.

California has established reentry intervals for carbophenothion of 14 days for citrus, peaches, nectarines, and grapes; and 2 days for all the other crops. A federal reentry interval of 2 days for carbophenothion has been established for all crops under 40 CFR 170. The Agency is now requiring 2 days for all crop uses of carbophenothion on an interim basis, and is requesting data for establishing permanent reentry interval(s). The Agency is also requiring an interim 24 hour reentry interval for the domestic outdoor usage on home lawns and ornamentals and requesting data to enable the Agency to make a risk assessment.

The Agency has determined that all products warrant restricted-use classification based on acute dermal toxicity. Registrants have the option of placing the restricted-use classification on the labeling, or submitting acute toxicity data to the Agency.

5. SUMMARY OF MAJOR DATA GAPS

158.130 Environmental fate

161-1 - Hydrolysis
161-2 - Photodegradation In Water
161-3 - Photodegradation On Soil

161-4 - Photodegradation In Air
162-1 - Aerobic Soil Metabolism Study
162-2 - Anaerobic Soil Metabolism Study
162-3 - Anaerobic Aquatic Metabolism Study
163-1 - Leaching and Adsorption/Desorption Mobility Studies
163-2 - Volatility (Lab) Mobility Studies
163-3 - Volatility (Field) Mobility Studies
164-1 - Soil Dissipation Studies
164-3 - Forestry Dissipation Studies
164-5 - Soil, Long-Term Dissipation Studies
165-1 - Rotational Crops Accumulation Studies (confined)
165-2 - Rotational Crops Accumulation Studies (Field)
165-4 - In Fish Accumulation Studies
165-4 - In Aquatic Non-Target Organisms Accumulation Studies

154.140 Reentry Protection

158.135 Toxicology

82-1 - 90-Day Subchronic Feeding - Rodent
83-1 - Chronic Toxicity - Rodent (rat)
83-2 - Oncogenicity - rat and mouse
83-3 - Teratogenicity - 2 species
84-2 - Gene Mutation
83-2 - Chromosomal Aberration
83-2 - Other Mechanisms of Mutagenicity

158.145 Wildlife and Aquatic Organisms

71-1 - Avian Acute Oral Toxicity
71-2 - Avian Subacute Dietary Toxicity
72-2 - Acute Toxicity To Freshwater Invertebrates

158.125 Residue Chemistry

see under Tolerance Reassessment

6. Contact person at EPA

William E. Miller
Product Manager (16)
Insecticide-Rodenticide Branch
Registration Division (TS-767C)
(703) 557-2600

DISCLAIMER: The information presented in this Chemical Information Fact Sheet is for informational purposes only and may not be used to fulfill data requirements for pesticide registration and reregistration.

CHLORDANE

Date Issued: December, 1986
Fact Sheet Number: 109

1. DESCRIPTION OF CHEMICAL

    Generic Name: 1,2,4,5,6,7,8,8-octachloro-2,3,3a,4,7,7a-
    (Chemical)   hexahydro-4,7-methanoindene

    Common Name: Chlordane

    Trade and   1,2,4,5,6,7,8,8-octachloro-3a,4,7,7a-tetrahydro-
    Other   4,7-methanoindan; Velsicol 1068; Velsicol 168;
    Names   M-410; Belt; Chlor-Kil; Chlortox; Corodane; Gold
                  Crest C-100; Kilex; Gold Crest C-50; Kilex;
                  Kypchlor; Niran; Octachlor; Synchlor; Termi-Ded;
                  Topiclor 20; Chlordan; Prentox; and Penticklor

    EPA Shaughnessy Code: 058201

    Chemical Abstracts Service (CAS) Number: 57-47-9

    Year of Initial Registration: 1948

    Pesticide Type: Insecticide

    Chemical Family: Chlorinated cyclodiene

    U.S. and Foreign Producers: Velsicol Chemical Corporation

2. USE PATTERNS AND FORMULATIONS

    Application Sites: subsurface soil treatment for termite
                       control; underground cables for termite
                       control; above ground structural
                       application for control of termites and
                       other wood-destroying insects

    Types of Formulations: emulsifiable concentrates; granular;
                           soluble concentrates

Types and Methods of Application: trenching, rodding, subslab injection, low pressure spray for subsurface termite control; brush, spray, or dip for applying to structural wood

Application Rates: 0.5 to 2.0% emulsion for termite control; 3.0 to 4.25% solution for above ground structural wood treatment

3. SCIENCE FINDINGS

Summary Science Statement

Chlordane is a chlorinated cyclodiene with moderate acute toxicity. The chemical has demonstrated adverse chronic effects in mice (causing liver tumors). Chlordane may pose a significant health risk of chronic liver effects to occupants of structures treated with chlordane for termite control. This risk may be determined to be of regulatory concern, pending further evaluation. Chlordane is highly toxic to aquatic organisms and birds. Chlordane is persistent and bioaccumulates. Chlordane may have a potential for contaminating surface water; thus, a special study is required to delineate this potential. Applicator exposure studies are required to determine whether exposure to applicators may be posing health risks. Special product-specific subacute inhalation testing is required to evaluate the short-term respiratory hazards to humans in structures treated with chlordane. An inhalation study of one-year duration using rats is required to assess potential hazards to humans in treated residences from this route of exposure. The Agency has been apprised of reported cases of optic neuritis associated with termiticide treatment of homes. To determine whether this is a significant health effect, the registrant must have eye tissue from the latest two-year rat oncogenicity study analyzed by neuro-pathologists specializing in optic tissue pathology. Data available to the Agency show an occurrence of misuse and misapplication of chlordane. The Agency is requiring restricted use classification of all end-use products containing chlordane. Application must be made either in the actual physical presence of a Certified Applicator, or if the Certified Applicator is not physically present at the site, each uncertified applicator must have completed a State approved training course in termiticide application meeting minimal EPA training requirements and be registered in the State in which the uncertified applicator is working.

Chemical Characteristics of the Technical Material

Physical State: Crystalline solid
Color: White
Odor: Chlorine odor
Molecular weight and formula: 409.8 - $C_{10}H_6Cl_8$
Melting Point: 95 to 96°C
Boiling point: 118°C at 0.66 mmHg (technical)
Density: 1.59 - 1.63 at 25°C
Vapor Pressure: 0.00001 mmHg at 25°C (technical)
Solubility in various solvents: Miscible with aliphatic and and aromatic hydrocarbon solvents, including deodorized kerosene; insoluble in water
Stability: Loses its chlorine in presence of alkaline reagents and should not be formulated with any solvent, carrier, diluent or emulsifier which has an alkaline reaction (technical)

Toxicology Characteristics

Acute Oral: Data gap

Acute Dermal: Data gap

Primary Dermal Irritation: Data gap

Primary Eye Irritation: Data gap (except for a 72% technical formulation)

Skin Sensitization: Not a sensitizer.

Acute Inhalation: Data gap

Subchronic Inhalation (2-week duration) using rats or guinea pigs: Data gap

Subchronic Inhalation (1-year duration) using rats: Data gap

Major routes of exposure: Inhalation exposure to occupants of treated structures; dermal and respiratory exposure to termiticide applicators.

Delayed neurotoxicity: does not cause delayed neurotoxic

effects.

Oncogenicity: This chemical is classified as a Group B_2 oncogen (probable human oncogen).

There are three long-term carcinogenesis bioassays of chlordane in mice which were independently conducted by investigators affiliated with the National Cancer Institute, the International Research and Development

Corporation, and the Research Institute for Animal Science in Biochemistry and Toxicology, Japan. Reported in these studies were significant tumor responses in three different strains of mice (IRC, CF_1, and $B6C3F_1$) in males and females with a dose-related increase in the proportion of tumors that were malignant. In Fischer 344 rats, significant tumor responses were reported in a study conducted by the Research Institute for Animal Science in Biochemistry and Toxicology.

Chronic Feeding: Based on a rat chronic feeding study with chlordane, a Lowest Effect Level (LEL) of 0.05 mg/kg/day for liver effects has been calculated.

Metabolism: Chlordane's major metabolite is oxychlordane. Oxychlordane has been found to be a major fat tissue residue in rats. Human fat samples frequently contain trans-nonachlor, a contaminant found in technical chlordane, as a major residue.

Teratogenicity: Data gap

Reproduction: Data gap

Mutagenicity: Data gap. Further testing is required in all three categories (gene mutation, structural chromosome aberrations and other genotoxic effects.

Physiological and Biochemical Characteristics

The precise mode of action in biological systems is not known. In humans, signs of acute intoxication are primarily related to the central nervous system (CNS), including hyperexcitabilty, convulsions, depression and death.

Environmental Characteristics

Available data are insufficient to fully assess the environmental fate of chlordane. Data gaps exist for all applicable studies. However, available supplementary data indicate general trends of chlordane behavior in the environment. Chlordane is persistant and bioaccumulates. Chlordane is not expected to leach, since it is insoluble in water and should adsorb to the soil surface; thus it should not reach underground aquifers. However, additional data are necessary to fully assess the potential for ground-water contamination as a result of the termiticide use of chlordane.

Ecological Characteristics

Avian acute toxicity: LD_{50} of 83.0 mg/kg in bobwhite quail

Avian dietary toxicity: 858 ppm in mallard duck; 331 ppm in
(8 day)   bobwhite quail; and 430 ppm in pheasant

Freshwater fish acute toxicity: 57 to 74.8 ug/L for bluegill;
(96 Hr. LC_{50})   42 to 90 ug/L for rainbow trout

Freshwater invertebrate toxicity: 15 to 590 ug/L for <u>Pteronarcys</u>
(48 hr. and 96 hr. EC_{50})   and <u>Daphnia</u>, respectively.

4. Required Unique Labeling and Regulatory Position Summary

○   EPA is currently evaluating the potential human health risks of 1) non-oncogenic chronic liver effects, and 2) oncogenic effects to determine whether additional regulatory action on chlordane may be warranted.

○   In order to meet the statutory standard for continued registration, retail sale and use of all end-use products containing chlordane must be restricted to Certified Applicators or persons under their direct supervision. For purposes of chlordane use, direct supervision by a Certitied Applicator means 1) the actual physical presence of a Certified Applicator at the application site during application, or 2) if the Certified Applicator is not physically present at the site, each uncertified applicator must have completed a State approved training course in termiticide application meeting minimal EPA training requirements and be registered in the State in which the uncertified applicator is working; the Certified Applicator must be available if and when needed.

° In order to meet the statutory standard for continued registration, chlordane product labels must be revised to provide specific chlordane disposal procedures, and to provide fish and wildlife toxicity warnings.

° The Agency is requiring a special monitoring study to evaluate whether and to what extent surface water contamination may be resulting from the use of chlordane as a termiticide.

° Special product-specific subacute inhalation testing is required to evaluate the respiratory hazards to humans in structures treated with termiticide products containing chlordane.

° Evaluation of eye tissue from the latest two-year rat oncogenicity study is required to determine whether chlordane's termiticide use may be causing optic neuritis in humans.

° The Agency is requiring the submission of applicator exposure data from dermal and respiratory routes of exposure.

° While data gaps are being filled, currently registerd manufacturing use products and end use products containg chlordane may be sold, distributed, formulated, and used, subject to the terms and conditions specified in the Registration Standard for chlordane, and any additional regulatory action taken by the Agency. Registrants must provide or agree to develop additional data in order to maintain existing registrations.

5. TOLERANCE REASSESSMENT

No tolerance reassessment for chlordane is necessary, since there are no food or feed uses. The Agency is proceeding to revoke all tolerances and replace them with action levels. The final rule is scheduled for publication in the Federal Register in early 1987.

6. SUMMARY OF MAJOR DATA GAPS

° Hydrolysis

° Photodegradation in Water

° Aerobic Soil Metabolism

° Anaerobic Soil Metabolism

- Leaching and Adsorption/Desorption
- Aerobic Aquatic Metabolism
- Soil Dissipation
- Chronic Toxicity Studies- Rodents and Non-rodents
- Teratogenicity
- Mutagenicity Studies
- Acute Toxicity Studies
- Optic Tissue Pathology
- Special Surface Water Monitoring Studies
- Applicator Exposure Studies
- Indoor Air Exposure Studies
- Special Product-Specific Subchronic Inhalation Study (two-week duration using guinea pigs or rats)
- Subchronic Inhalation Study (One-year duration using rats)
- All Product Chemistry Studies

7. <u>CONTACT PERSON AT EPA</u>

George LaRocca
Product Manager (15)
Insecticide-Rodenticide Branch
Registration Division (TS-767C)
Office of Pesticide Programs
Environmental Protection Agency
401 M Street, S. W.
Washington, D. C.   20460

Office location and telephone number:
Room 204, Crystal Mall #2
1921 Jefferson Davis Highway
Arlington, VA   22202
(703) 557-2386

DISCLAIMER: The information presented in this Chemical Information Fact Sheet is for informational purposes only and may not be used to fulfill data requirements for pesticide registration and reregistration.

CHLORIMURON ETHYL

Date Issued: April 4, 1986
Fact Sheet Number: 82

1. Description of Chemical

    Generic Name: Ethyl 2-[[[[(4-chloro-6-methoxyprimidin-2-yl)amino]carbonyl]amino]sulfonyl]benzoate

    Common Name: Chlorimuron ethyl; DPX-F6025
    Trade Name: DuPont Classic Herbicide
    EPA Shaughnessy Code: 128901
    Chemical Abstracts Service (CAS) Number: 90982-32-4
    Year of Initial Registration: 1986
    Pesticide Type: Herbicide
    Chemical Family: Sulfenylurea
    U.S. and Foreign Producers: E.I. du Pont de Nemours & Company

2. Use Patterns and Formulations

    Application sites: Soybeans
    Types and methods of application: Postemergence foliar by ground equipment.
    Application rates: 1/2 to 3/4 ounces active ingredient per acre (oz ai/A)
    Types of formulation: 25% dispersible granule
    Usual carrier: water

3. Science Findings

    Summary Science Statement:

    All data required for registration of this chemical is acceptable to the Agency.

    Chlorimuron ethyl has low acute toxicity (Category III) for acute dermal and primary eye irritation and Category IV for all other forms of acute toxicity. It was not oncogenic to mice or rats, not teratogenic to rabbits, and not mutagenic. The pesticide is foliarly absorbed and translocated within the plant. It works by inhibition of cell division in shoots and roots. The major degradation pathway is hydrolysis.

The pesticide will leach in some soils and has the potential to contaminate ground water at very low concentrations. Chlorimuron ethyl is practically nontoxic to birds and slightly toxic to fish and invertebrates. The nature of residues in plants is adequately understood and adequate methodology is available for enforcement of a tolerance of 0.05 part per million (ppm) on soybeans.

Chemical Characteristics:

Physical state: Solid
Color: Off-white to pale yellow
Odor: None
Melting point: 181 °C
Density: 1.51 gram/cubic centimeter (g/cc)
Solubility in various organic solvents at 25 °C

	g/100 ml
Acetone	7.05
Acetonitrile	3.10
Benzene	0.815
Ethyl acetate	2.36
Ethyl alcohol	0.392
n-hexane	0.006
Methyl alcohol	0.740
Methylene chloride	15.3
Xylenes	0.283

Solubility in water at controlled pH:

pH	Solubility milligram/liter (mg/l)
1.3	1.5
1.9	1.5
2.5	1.5
4.2	4.1
5.0	9.0
5.8	99
6.5	450
7.0	1200

Handling characteristics: Store product in original container only.

Toxicology Characteristics:

Acute Toxicity:

Acute Oral Toxicity (Rat): Greater than (>) 5000 milligrams/kilogram (mg/kg)
Toxicity Category IV[1]/

Acute Dermal Toxicity (Rabbits): > 2000 mg/kg
   Toxicity Category III[1]/

Primary Eye Irritation (Rabbit): Draize score = 1
   Toxicity Category III[1]/

Primary Dermal Irritation (Rabbit:) Primary irritation
   score from .13 to .63
   Toxicity Category IV[1]/

Primary Skin Irritation (Guinea Pig): Not irritating
   Toxicity Category IV[1]/

Dermal Sensitization (Guinea Pig): Not a sensitizer
   Toxicity Category IV[1]/

[1]/ Labeling statements required for: Toxicity Category
   III for acute dermal - "Harmful if absorbed
   through skin. Avoid contact with skin, eyes,
   or clothing. Wash thoroughly with soap and
   water after handling."
   Toxicity Category III for primary eye irritation -
   "Causes (moderate) eye injury (irritation). Avoid
   contact with eyes or clothing. Wash thoroughly
   with soap and water after handling."
   Toxicity Category IV - no precautionary statements
   required.

Major Routes of Exposure:

The major routes of exposure are via dermal and eye contact.

Chronic Toxicity

90-Day Feeding Study (Mouse) resulted in a no-observable-effect level (NOEL) of 18.75 mg/kg/day.

90-Day Feeding Study (Dog) resulted in a NOEL of 2.5 mg/kg/day.

One-Year Feeding Study (Dog) resulted in a NOEL of 6.25 mg/kg/day

18-Month Chronic Feeding/Oncogenicity Study (Mouse) resulted in a NOEL of 3.75 mg/kg/day and no oncogenic effects at 187.5 mg/kg/day for the highest dose tested (HDT).

Chronic Feeding/Oncogenic Study (Rat) resulted in a NOEL of 12.5 mg/kg/day and no oncogenic effects at 125 mg/kg/day (HDT).

2-Generation Reproduction Study (Rat) resulted in a maternal toxicity NOEL of 12.5 mg/kg/day and a fetotoxic NOEL of 1.25 mg/kg/day.

Teratology Study (Rats) resulted in a teratogenic NOEL of 150 mg/kg/day, a maternal toxicity NOEL of 30 mg/kg/day and a fetotoxic NOEL of 30 mg/kg/day.

Teratology Study (Rabbits) resulted in a maternal toxicity of 60 mg/kg/day and a fetotoxic NOEL of 15 mg/kg/day and no teratogenic effects at 300 mg/kg/day (HDT).

Mutagenicity (in vivo bone marrow test) - not mutagenic.

Mutagenicity (Ames test) - not mutagenic.

Mutagenicity (Unscheduled DNA Synthesis Activity) - not mutagenic.

Physiological or Biochemical Behavioral Characteristics

Foliar absorption: Rapid.

Translocation: Systemic after absorption through either the foliage or the roots.

Mechanism of pesticidal action: Inhibits plant cell division of rapidly growing tips of roots and shoots by inhibition of amino acid synthesis.

Metabolism in plants: Tolerant species metabolize the compound to nonherbicidal metabolites.

Persistance in plants: Does not persist in plants.

Environmental Characteristics:

Adsorption and leaching in basic soil types. Chlorimuron ethyl was very weakly absorbed on the two sandy loam soils and only weakly absorbed on the two silt loam soils. Absorbed radioactivity was readily desorbed from the sandy loam soils, but was more tightly retained on the silt loams. DPX-F6025 had low mobility on Keyport silt loam, intermediate mobility on Flanagan silt loam and Cecil sandy loam and high mobility on Woodstown sandy loam.

Microbial breakdown: Initial deactivation of the molecule is through hydrolysis followed by complete metabolism to low molecular weight compounds through normal soil microbial degradation.

Loss from photodecomposition and/or volatilization: Photodegradation of chlorimuron ethyl is not a major degradation pathway, but did proceed at twice the rate in exposed samples compared to the nonexposed samples.

Bioaccumulation: The octanol/water partition coefficient (K_{ow}) of 1.3 and available information show the hydrolysis products have a lower K_{ow}. Since the correlation between octanol/water partitioning and fish accumulation is only accurate within a factor of 100, chlorimuron ethyl and its degradation products have potential to accumulate in fish to levels 130 times higher than levels in water.

Resultant average persistance: Chlorimuron ethyl has a half-life of 7.5 weeks in soil.

Environmental fate and surface and ground water contamination concerns: Chlorimuron methyl has the potential to leach and contaminate ground water at very low concentrations.

Exposure of humans and nontarget organisms to pesticide or degradates. Human risk from exposure is minimal because of low acute toxicity (Category III and IV). Nontarget organism risk from exposure is minimal because maximum expected residues on soil and water do not approach the toxicity values for the organisms tested.

Ecological Characteristics:

Avian Acute Toxicity (Mallard duck):	> 2510 mg/kg
Avian Dietary Toxicity (Bobwhite quail):	> 5620 ppm
Avian Dietary Toxicity (Mallard duck):	> 5620 ppm
96-Hour Fish Toxicity (Rainbow trout):	> 12 mg/l
96-Hour Fish Toxicity (Bluegill sunfish):	> 10 mg/l
48-Hour Invertebrate Toxicity (_Daphnia magna_):	> 10 mg/l

Chlorimuron ethyl is practically nontoxic to birds on both an acute and dietary basis. It is slightly toxic to fish and invertebrates. It is not expected to adversely affect endangered/threatened species because of low toxicity and low application rate.

Tolerance Assessment:

The nature of the residue in plants is adequately understood. The residue of concern is chlorimuron ethyl. Adequate methodology (high pressure liquid chromatography

[HPLC] using a photoconductivity detector) is available for enforcement.

Tolerances are established for residues of the herbicide chlorimuron ethyl (ethyl 2-[[[[(4-chloro-6-methoxypyrimidin-2-yl)amino]carbonyl]amino]sulfonyl] benzoate in or on soybeans at 0.05 ppm.

The acceptable daily intake (ADI) based on the 1-year dog feeding study (NOEL of 6.25 mg/kg/day) and using a hundredfold safety factor is calculated to be .0625 mg/kg/day. The maximum permissible intake (MPI) for a 60 kg human is calculated to be 3.70 mg/kg/day. The theoretical maximum residue contribution (TMRC) for this tolerance for a 1.5 kg diet is calculated to be .000690 mg/day/1.5 kg. The current action will use .0184 percent of the ADI. There are no other published tolerances for this chemical.

4. Summary of Regulatory Position and Rationale

   Position: The Agency has registered chlorimuron ethyl for postemergence use on soybeans.

   Rationale: All data required for registration have been submitted. All submitted data fulfill guideline requirements and are acceptable to support registration.

5. Summary of Major Data Gaps

   None

6. Contact Person at EPA

   Robert J. Taylor
   Environmental Protection Agency
   Office of Pesticide Programs
   Registration Division (TS-767C)
   401 M Street, SW.
   Washington, DC 20460
   Phone: (703) 557-1800

Disclaimer: The information presented in this Pesticide Fact Sheet is for informational purposes only, and may not be used to fulfill data requirements for pesticide registration and reregistration.

CHLOROBENZILATE

Date Issued: December 30, 1983
Fact Sheet Number: 15

1. Description of the chemical:

    - Generic name: ethyl 4,4'-dichlorobenzilate
    - Common name: Chlorobenzilate
    - Trade name: Acaraben
    - EPA Shaughnessy Code: 028801
    - Chemical abstracts services number: 510-15-6
    - Year of initial registration: 1953
    - Pesticide type: Miticide
    - Chemical family: Organochlorine compound
    - U.S. and Foreign producers: Ciba-Geigy, Nippon Kayaku, Japan
      and Makhteshim Beer-Sheva, Israel

2. Use patterns and formulations:

    - Application sites: Citrus
    - Types of formulations: Four pound per gallon emulsifiable concentrate
    - Types and Methods of Application: Aerial and ground foliar sprays restricted to citrus use only in the states of Arizona, California, Florida and Texas for control of mites
    - Application rates: 0.75 lb/ acre
    - Usual carriers: water

3. Science findings:

    Summary of Science Statement:

    - Chlorobenzilate is classified RESTRICTED.

    - Chlorobenzilate met criteria for unreasonable adverse effects due to the oncogenicity "trigger".

    - Chlorobenzilate is suspected to be contaminated by DDT and/or its analogs at or near the limit of detection (by Thin Layer Chromatography). The Agency is requesting registrants to do further analyses with more sensitive analytical techniques to search for these impurities.

    Chemical charcteristics:

    - Chlorobenzilate, an organochlorine compound, is a brownish viscous liquid with boiling and melting points of 141-142 °C and 35-37 °C, respectively. The chemical is virtually insoluble in water and stable at room temperature.

Toxicological characteristics:

- Overall toxicity category III
- Acute Oral LD_{50} - 960 to 1220 mg/kg - Category III
- Acute Dermal LD_{50} - greater than 10,200 mg/kg - Category IV
- Acute Inhalation LC_{50} - Data Gap
- Primary Eye Irritation- Rabbit- Data Gap
- Primary Dermal Irritation - Data Gap
- Dermal Sensitization- Data Gap
- The major routes of exposure are believed to be dermal followed by inhalation.
-Chronic toxicity results:
  The chemical is an oncogen.
  Data gaps include, teratology and mutagenicity testing.
  Using the protective clothing and the use restriction limited to citrus as required in the Rebuttable Presumption Against Registration, the lifetime oncogenicity risk would be between 0.5 and 7.0×10^{-6} from dietary exposure and between 65 and 190×10^{-6} from applicator exposure. A risk-benefits analysis of this use pattern indicates the benefits derived from this use outweighes the risks involved.

Physiological and Biochemical Behavioral Characteristics:

-Foliar absorption remains a data gap.
-Metabolism and persistence in plants and animals are data gsps.
-Pesticidal action: nerve poison.

Ecological characteristics:

- No data available. Most studies reserved pending review of basic product chemistry, environmental fate, toxicology and residue chemistry data. Potential problems related to endangered species are unknown.

Environmental characteristics:

- An environmental assessment can not be made at this time because there is a lack of the data needed to make this assessment.

Tolerance Assessment:

- Because all uses of chlorobenzilate are cancelled except citrus, a recalculation of the Theoretical Maximum Residue Concentration (TMRC), Acceptable Daily Intake (ADI), and Maximum Permissible Intake (MPI) was undertaken. These values are 0.2859 mg/day/1.5 kg (3.81% of the ADI), 0.125 mg/kg/day and 7.50 mg/day/60 kg, respectively. These values are considered provisional at this time. Another reassessment and recalculation of the ADI/MPI will be made when the toxicology data gaps are filled.

- A food additive tolerance(s) is required for citrus oil.

- Established tolerances other than citrus and the associated meat, fat and meat byproducts will be revoked in 1984 because all other uses were cancelled by the RPAR action.

4. Summary of Regulatory Position and Rationale:

    - An interim 24-hour reentry interval on citrus crops has been established until the Agency receives reentry data.

    - The Agency will make a determination as to the continued registrability of this chemical when the data requested in the Registration Standard Guidance Document is submitted and reviewed.

    - The Agency determined in the Special Review process that chlorobenzilate end-use products be classified as "Restricted Use" to reduce exposure to loaders, mixers and applicators. The chemical will continue to be classified for "Restricted Use" until the Agency receives data to reevaluate its position.

    - The Agency has concluded, via the Special Review process, that by limiting the use of chlorobenzilate to citrus, classifying chlorobenzilate products for restricted use, and upgrading the protective clothing requirements, the exposure level and risk would be lowered to acceptable levels. The benefits are determined to exceed the risks and the chemical is allowed continued registration, provided the following data are submitted and they do not suggest new unacceptable toxicological and/or environmental properties of chlorobenzilate.

5. Summary of Data Gaps:

    The following data gaps were required to be submitted by January of 1987.

- Mutagenicity testing
- Rat inhalation LC_{50} study
- Primary eye irritation study (rabbit)
- Primary dermal irritation study
- Dissipation studies
- Mobility studies
- Hydrolysis studies
- Aerobic soil metabolism study
- photodegradation studies
- Fish accumulation studies
- Avian oral LD_{50} testing
- Avian dietary LC_{50}
- Avian reproduction studies
- Freshwater fish LC_{50}
- Acute LC_{50} freshwater invertebrates study
- Acute LC_{50} estuarine and marine organisms study
- Fish early life stage and aquatic invertebrate life-cycle study
- Fish life-cycle study

These data gaps are required to be submitted by April of 1984

- Identity of Ingredients
- Statement of Composition
- Discussion of Formation of Ingredients
- Preliminary Analysis
- Certification of Limits
- Analytical Methods for Enforcement of Limits
- Physical and Chemical Characteristics (except melting point & pH)

Teratology study- 2 species by October of 1984
Citrus Fractionation Study- Feb. of 1984 (promised April 1, 1984)
Feeding Citrus By-Products to Cattle- Feb. of 1984 (promised April 1, 1984)
Aerial Applicators Study- Feb. of 1984 (received March, 1984)
Citrus Picker Exposure Study- Feb. of 1984 (received March, 1984)

6. Contact person at EPA

    Jay S. Ellenberger
    Product Manager (12)
    Insecticide-Rodenticide Branch
    Registration Division (TS-767C)
    Office of Pesticide Programs
    Environmental Protection Agency
    401 M Street, S.W.
    Washington, D.C. 20460

CHLOROTHALONIL

Date Issued: September 30, 1986
Fact Sheet Number: 36

1. Description of the chemical:

    - Generic name: Tetrachloroisophthalonitrile
    - Common name: chlorothalonil
    - Trade name: Bravo®, Nopcocide® N-96, Daconil®, Chlorothalonil Technical Termil, Groutcide®, Mold-Ex®, Brovomil®
    - EPA Shaughnessy Code: 081901
    - Chemical Abstracts Service (CAS) Registry number: 1897-45-6
    - Year of initial registration: 1966
    - Pesticide Type: Fungicide
    - Chemical family: Chlorinated isophthalic acid derivative
    - U.S. and foreign producers: SDS Biotech Corporation and Griffin Chemical Company

2. Use patterns and formulations:

    - Application sites: chlorothalonil is a chlorinated isophthalic acid derivative registered for use as a fungicide to control a wide variety of fungal diseases on both crop and noncrop sites. Chlorothalonil is registered for use on numerous crop sites such as fruits, vegetables, and peanuts. In non-crop applications, chlorothalonil is used on ornamental sites, on turf, and as a paint and grout additive.
    - Types of formulations: chlorothalonil is available in wettable powder, granular flowable, pelleted/tableted, liquid suspension, and soluble concentrate formulations.
    - Types and methods of applications: chlorothalomil is applied as follows: broadcast, band, and soil surface using ground or aerial equipment.
    - Application rates: 0.75 lbs. a.i./A to 3.0 lbs. a.i./A on crop sites; and 1.125 lbs. a.i./A to 7.5 lbs. a.i./A on non-crop sites.
    - Usual carrier: Water.

3. Science Findings:

    a. Summary of science findings:

    - Chlorothalonil breaks down in aerobic soil with a half-life of 1 to 2 months with the formation of a major degradate, 4-hydroxy-2,5,6-trichloroisophthalonitrile, and several lesser degradates. This breakdown appears to be primarily the result of microbial degradation, since chlorothalonil is relatively stable to hydrolysis and photolysis. Chlorothalonil is immobile in most soil types, except sand in which it is moderately mobile. The 4-hydroxy degradate is moderately mobile in most soil types.

- Chlorothalonil is not toxic to birds. However, the major degradate is moderately toxic to birds. Additional studies are required for chlorothalonil and the 4-hydroxy degradate in order to determine whether residues of these compounds will produce an effect on the reproduction of birds.
- Chlorothalonil and its 4-hydroxy degradate are highly toxic to fish, aquatic invertebrates and marine/estuarine organisms. Low amounts of residues will affect the reproduction of these organisms. An additional field monitoring study is required that would record the effects of residues occurring in water resulting from registered uses of chlorothalonil.
- The metabolism of chlorothalonil in plants and animals is not adequately elucidated. However the major metabolite is the 4-hydroxy-2,5,6-trichloroisophthalonitrile. At low concentrations, chlorothalonil is expected to be excreted in the feces and urine within 24 hours. The amount of residues expected to be in or on raw agricultural commodities from use of chlorothalonil will be reassessed when the data on plant and animal metabolism are submitted along with the additional residue data.
- The impact of the manufacturing impurities of hexachlorobenzene and pentachlorobenzonitrile in chlorothalonil cannot be made until plant and animal residue data is submitted for these impurities.
- Chlorothalonil is a severe eye irritant. It is moderately toxic via inhalation and demonstrates low toxicity via dermal and oral exposures. Chlorothalonil is not considered to be mutagenic, teratogenic, or cause reproductive effects. Studies demonstrated that the chemical could have oncogenic potential, but additional data are needed to confirm this.
The major metabolite is not considered to be an oncogen or mutagen.

b. Chemical characteristics:

- Technical chlorothalonil is an odorless or slightly pungent, white, crystalline solid. It is stable towards moisture under conventional conditions and decomposes at 250-251° Celsius. The chemical does not exhibit any unusual handling hazards. Technical chlorothalonil contains low amounts of hexachlorobenzene (HCB) and pentachlorobenzonitrile (PCBN) as manufacturing impurities.

c. Toxicological characteristics:

- Technical chlorothalonil is a severe eye irritant, moderately toxic via inhalation, and of low toxicity via oral and dermal exposures. The toxicity categories assigned to chlorothalonil are I for eye irritation, II for inhalation, III for oral and dermal toxicities, and IV for dermal irritation.
- Toxicology studies on chlorothalonil are as follows:

- - Oral LD_{50} in rats: > 28.2 mg/kg body weight,

- - Dermal LD_{50} in rats: > 10 gm/kg body weight,

- - Skin irritation in rabbits: not an irritant

- - Eye irritation in rabbits: severe irritant

- - In a two-year rat chronic feeding study, the no-observed-effect-level was 60 ppm (3 mg/kg body weight). However, it was not adequate to address the evaluate oncogenic potential of chlorothalonil.

- - In a two-year feeding study in dogs, the no-observed-effect-level was 60 ppm.

- - In a 24-month mouse oncogenicity study, chlorothalonil produced tumors in renal cortical tubules and squamous tissue of the stomach, however there was no dose related response. The study demonstrates that chlorothalonil may have oncogenic potential. An additional two-year feeding study in rats is considered a data gap.

- - In a three-generation rat reproduction study, the no-observed-effect-level for chlorothalonil is < 0.15% of the diet based on reduced growth and renal and gastric effects. No teratogenic effects were observed at any dose level tested under the conditions of the study. Reproductive effects were observed at 0.15% of the diet.

- - In a rabbit teratology study and rat teratology study utilizing chlorothalonil and a rat and a rabbit teratology study utilizing the 4-hydroxy metabolite, no significant teratogenic effects were observed at any dose levels tested under the conditions of the studies.

- - Chlorothalonil did not demonstrate any mutagenic effects in a variety of mutagenicity tests.

- - The 4-hydroxy metabolite is not oncogenic in rats at doses of 20.0 mg/kg body weight/day [Highest Dose Tested (HDT)] or less and in mice at doses of 215 mg/kg body weight/day (HDT) or less under the conditions of the study.

- - The 4-hydroxy metabolite did not demonstrate any mutagenic affects in a variety of mutagenicity studies.

d. Physiological and Biochemical Behavioral Characteristics:

- Translocation: chlorothalonil is translocated in plants.
- Mechanism of pesticidal action: The mode of action of chlorothalonil involves its reaction with thiol groups of the pathogenic fungi's enzyme systems.
- Metabolism and persistence in plants and animals: The metabolism of chlorothalonil is not adequately delineated in both plants and animals. However, the major metabolite is the 4-hydroxy metabolite.

e. Environmental characteristics:

- The available environmental fate data are insufficient to fully assess the potential for exposure of humans and non-target organisms to chlorothalonil. When additional studies are submitted, a complete environmental exposure assesment can be made. However, chlorothalonil is moderately mobile in sand and the 4-hydroxy degradate is moderately mobile in most soils.

f. Ecological characteristics:

- Avian oral ID_{50}: chlorothalonil (>10,000 mg/kg) uses are not expected to affect avian wildlife. 4-hydroxy degradate (2000 mg/kg), is moderately toxic to birds. Reproduction studies using both mallard ducks and bobwhite quail for each, chlorothalonil and the 4-hydroxy degradate, are required.
- LC_{50} to fish: chlorothalonil (47-84 ppb) is highly toxic to fish. The 4-hydroxy degradate (16-45 ppm) is slightly toxic to fish.
- Fish reproduction: Chlorothalonil may affect fish populations at low levels (3 - 6.5 ppb).
- LC_{50} aquatic invertebrates: Chlorothalonil is highly toxic to aquatic invertebrates and may affect their reproduction at low concentrations (>79 ppb).
- An aquatic field monitoring study is required that reflects residue concentrations likely to occur under actual use conditions.

g. Tolerance assesments:

- A reassessment of the tolerances cannot be completed at this time. The residue data on the amounts of impurities in or on the commodities, the rat oncogenicity/chronic feeding study, and metabolism data, need to be submitted and evaluated before any assessment of the present tolerances can be made.

4. Summary of regulatory position and rationale:

- No use, formulation, or geographical restrictions are required.
- Chlorothalonil will be provisionally regulated as an oncogen. This position is based on a study in which chlorothalonil has exhibited renal adenomas and carcinomas in male mice and somewhat flawed NCI studies that indicate an oncogenic effect in the rat, but not in mice. This position will be reconsidered after the review of the ongoing rat chronic feeding/oncogenicity study in 1985. As a result of this evaluation, chlorothalonil may be referred to a Special Review.
- Skin sensitization study is a data gap. However, information received through the Pesticide Incident Monitoring System indicates that skin sensitization is a potential problem.
- Unique labeling statements:
    Under "Hazards to Humans and Domestic Animals", the words "Skin Sensitizer" must be added.
    Under "First Aid", the following statement must be added: "Note to Physician: Persons having an allergic reaction respond to treatment with antihistamines or steroid creams and/or systemic steroids."
    Under "Environmental Hazards", the following must be used: "This product is toxic to fish, aquatic invertebrates, and marine/estuarine organisms."

Under "Directions for Use", the following statements must be added:
"Note to User: This product may produce temporary allergic side effects characterized by redness of the eyes; mild bronchial irritation and redness or rash on exposed skin areas. Person having allergic reaction should contact a physician."
"Note to User: Wear long sleeve shirt, long pants, and gloves when mixing, loading and applying this product."
"Note to User: Do not enter treated areas to perform hand labor within 24 hours of application unless protective clothing is worn."
"Note to User: Do not rotate to crops other than those listed on the label within one year from last application. Leafy vegetables may be rotated after one year from last application."

5. Summary of major data gaps:

   a. The following toxicology data are required:

   - An oncogenic study in rats (currently underway) is required to be submitted within one year.
   - A skin sensitization study is required to be submitted within one year.
   - A 21-day subchronic dermal study is required to be submitted within one year.

   b. The following environmental fate data are required:

   - Hydrolysis test, (2)
   - Photodegradation test in water, (2)
   - Photodegradation test on soil, (2)
   - Photodegradation test in air, (2)
   - Metabolism test in anaerobic soil, (1)
   - Metabolism test in aerobic soil, (1)
   - Leaching study (2)
   - Mobility (volatility) test in the laboratory, (2)
   - Mobility (volatility) test in the field, (2)
   - Dissipation study in soil (1)

   (1) Long-term studies must be submitted within 15 months after receipt of the guidance package. (2) Short-term studies must be submitted within six months after receipt of the guidance package.

   c. The following ecological effects data are required:

   - A field study is needed to determine the effects of residues on fish and aquatic invertebrate occurring as a result of the terrestial use of chlorothalonil.
   - Mallard duck and bobwhite quail reproduction study with chlorothalonil and with it 4-hydroxy degradate are needed.

   These long-term studies must be submitted within 15 months after receipt of the guidance package.

d. The following residue chemistry data are required:

- Data related to the manufacturing process, formation of impurities, preliminary analysis of the technical product, certification of ingredient limits, analytical methodology, and physical/chemical characteristics. (2)
- Field residue data for several raw agricultural commodities.
- Studies pertaining to the metabolism of chlorothalonil residues in plants and animals. (1)
- Residues studies regarding the presence of hexachlorobenzane and pentachlorobenzonitrile in plants and in animals. (1)

(1) Long-term studies must be submitted within 15 months after receipt of the guidance package. (2) Short-term studies must be submitted within six months after receipt of the guidance package.

e. Contact Person at EPA:

> Henry M. Jacoby
> Product Manager (21)
> Environmental Protection Agency (TS-767C)
> 401 M Street, S. W.
> Washington, D. C. 20460
> (703) 557-1800

DISCLAIMER: The information presented in this Chemical Information Fact Sheet is for informational purposes only, and may not be used to fulfill data requirements for pesticide registration and reregistration.

CHLORPYRIFOS

Date Issued: September 30, 1984
Fact Sheet Number: 37

1. DESCRIPTION OF CHEMICAL

    Generic Name: 0,0-diethyl 0-(3,5,6-trichloro-2-pyridyl) phosphorothioate

    Common Name: Chlorpyrifos

    Trade Name: Dursban for household products, Lorsban for agricultural products

    EPA Shaughnessy Code: 059101

    Chemical Abstracts Service (CAS) Number: 2921-88-2

    Year of Initial Registration: 1965

    Pesticide Type: Insecticide

    Chemical Family: Organophosphate

    U.S. and Foreign Producers: Dow Chemical U.S.A.
    Makhteshim-Beer Shiva
    All India Medical Corp
    Planters Products, Inc.

2. USE PATTERNS AND FORMULATIONS

    Application Sites: Grain crops, nut crops, bananas, cole crops, citrus, pome and strawberry fruits, forage, field and vegetable crops, lawns and ornamental plants, poultry, beef cattle, sheep and dogs, livestock premise treatment, domestic dwellings, terrestrial structures, and direct aplication to stagnant water etc.

    Types of Formulations: Baits, dusts, granules, wettable powders, flowables, impregnated plastics and pressurized liquids.

    Types of Methods of Application: Ground and aerial, sprays and dust applications

    Application Rates: Range from 0.5 lbs. a.i./A to 3 lbs. a.i./A and crack and crevice treatment to broadcast treatment for indoor uses.

Usual Carriers: Synthetic clays, talc, various solvents

3. SCIENCE FINDINGS

   Summary Science Statement

   Chlorpyrifos has moderate mammalian toxicity. It is not considered to be oncogenic, mutagenic or teratogenic. However, the oncogenicity and mutagenicity studies used to draw these conclusions are not up to current Agency standards. Additional information from these studies is required.

   The Agency is imposing a 24-hour reentry restriction for crop uses until appropriate reentry studies are submitted and evaluated and a decision is reached whether a different time interval is more appropriate. The 24-hour interval also coincides with the requirements of California.

   Data are insufficient to fully assess the environmental fate of chlorpyrifos. The Agency is requesting necessary data to make this assessment and also to specifically assess whether or not chlorpyrifos has a potential to leach into groundwater. Data are also insufficient to measure human exposure in outdoor and indoor applications.

   Chlorpyrifos is extremely toxic to fish, birds and other wildlife. It is highly toxic to honey bees. Use precautions and restrictions are being imposed to reduce potential hazards.

   A full tolerance reassessment cannot be completed. The previous ADI (at 94% of the TMRC) was established based on a 2-year rat feeding study. The present ADI (313% of the TMRC) was calculated using a human study. Chronic feeding studies are required as well as metabolism and residue data on numerous commodities.

   Chemical Characteristics

   Physical State: Crystalline solid

   Color: White to tan

   Odor: Mild mercaptan

   Melting point: 41.5-43.5 °C

   Vapor Pressure: 1.87×10^{-5} mm Hg at 20°C

   Flash Point: None

Toxicology Characteristics

Acute Oral: 163 mg/kg, Toxicity Category II

Acute Dermal: 1505 mg/kg, Toxicity Category II

Primary Dermal Irritation: No irritation, Toxicity Category III

Primary Eye Irritation: Conjunctival irritation at 24 hours. Cleared at 48 hours. Toxicity Category III

Acute Inhalation: Data gap

Neurotoxicity: Not an acute delayed neurotoxic agent at doses up to 100 mg/kg (highest dose tested).

Oncogenicity: Two studies submitted but neither meet Agency standards. Neither suggest oncogenicity potential.

Teratogenicity: Three studies have been evaluated to determine the teratogenic potential of chlorpyrifos. The Agency has determined that this chemical is not teratogenic at levels up to 25 mg/kg/day.

Reproduction-2 generation: Two studies adequately demonstrate that chlorpyrifos does not produce reproductive effects. No effects were demonstrated at dose levels up to 1.2 mg/kg/day.

Metabolism: The submitted studies suggest that chlorpyrifos is rapidly absorbed and metabolized to 3,5,6-trichloro-2-pyridinal (TCP). The parent compound and metabolite are rapidly excreated in the urine. The submitted studies do not meet Agency standards.

Mutagenicity: Data Gap

Physiological and Biochemical Behavioral Characteristics

Mechanism of Pesticidal Action: An insecticide which is active by contact, ingestion, and vapor action and almost irreversibly causes phosphorylation of the acetylcholinesterase enzyme of tissues, allowing accumulation of acetylchloline at cholinergic neuroeffector junctions (muscarinic effects), and at skeletal muscle myoneural junctions and autonomic ganglia. Poisoning also impairs the central nervous system function.

Symptoms of poisoning include: headache, dizziness, extreme weakness, ataxia, tiny pupils, twitching, tremor, nausea, slow heatbeat, pulmonary edema, and sweating. Continual absorbtion at intermediate dosages may cause influenza-like illness which includes symptoms like weakness, anorexia, and malaise.

Metabolism and Persistence in Plants and Animals:

The metabolism of chlorpyrifos in plants and animals is not adequately understood. The major metabolite is 3,5,6-trichloro-2-pyridinol (TCP). The Agency does not have adequate data on TCP to determine if this metabolite should continue to be a part of the tolerance expression.

Environmental Characteristics

Available data are insufficient to fully assess the environmental fate of chlorpyrifos. Data gaps exist on all required studies except for aerobic and anaerobic soil studies.

Adsorption and Leaching in Basic Soil Types: The Agency is requesting data to determine if chlorpyrifos will contaminate groundwater.

Microbial Breakdown: Depending on the soil type, microbial metabolism of chlorpyrifos may have a half-life of up to 279 days.

Ecological Characteristics

Avian oral:
    Mallard duck--76.6 mg/kg
    Ring necked pheasant--17.7 mg/kg

Avian dietary:
    Mallard duck--136 ppm
    Bobwhite quail--721 ppm

Freshwater Fish:
    Coldwater fish (rainbow trout)--3.0 ppm
    Warmwater fish (bluegill sunfish)--2.4 ppm

Acute Freshwater Invertebrates:
    Daphnia--0.176 ppb

Acute Estuarine and Marine Organisms:
    Oyster--0.27 ppm
    Grass shrimp--1.5 ppm
    Killifish--3.2 ppm

Precautionary language is being required for hazards to birds, fish, and aquatic organisms. Chronic effects to non-target aquatic invertebrate species are not adequately characterized and therefore appropriate studies are required.

Tolerance Assessment

The Agency is unable to complete a full tolerance reassessment because of certain residue chemistry and toxicology data gaps.

Tolerances:

Commodity	Parts Per Million
Alfalfa, green forage	4.0
Alfalfa, hay	15.5
Almonds	0.05
Almonds, hull	0.05
Apples	1.5
Bananas (whole)	0.25
Bananas, pulp with peel removed	0.05
Bean forage	1.0
Beans, lima	0.05
Beans, lima, forage	1
Beans, snap	0.05
Beans, snap, forage	1
Beets, sugar, roots	1.0
Beets, sugar, tops	8.0
Broccoli	2
Brussels sprouts	2
Cabbage	2
Cattle, fat	2.0
Cattle, meat by-products (mbyp)	2.0
Cattle, meat	2.0
Cauliflower	2
Cherries	2.0
Citrus fruits	1.0
Corn, field, grain	0.1
Corn, fresh (inc. sweet corn; kernel plus cob with husk removed)	0.1
Corn, fodder	10.0
Corn, forage	10.0
Cottonseed	0.5
Cranberries	1.0
Cucumbers	0.1
Eggs	0.1
Figs	0.1
Goats, fat	1.0
Goats, mbyp	1.0
Goats, meat	1.0
Grapes	0.5
Hogs, fat	0.5
Hogs, mbyp	0.5
Hogs, meat	0.5
Horses, fat	1.0
Horses, mbyp	1.0

Tolerances (con't)

Commodity	Parts Per Million
Horses, meat	1.0
Milk, fat (reflecting 0.02 ppm in whole milk)	0.5
Mint, hay	1.0
Nectarines	0.05
Onions (dry bulb)	0.5
Pea forage	1.0
Peaches	0.05
Peanuts	0.5
Peanut hulls	15
Pears	0.05
Peppers	1.0
Plums (fresh prunes)	0.05
Poultry, fat (inc turkeys)	0.5
Poultry, mbyp (inc turkeys)	0.5
Poultry, meat (inc turkeys)	0.5
Pumpkins	0.1
Radishes	3
Seed and pod vegetables	0.1
Sheep, fat	1.0
Sheep, mbyp	1.0
Sheep, meat	1.0
Sorghum, fodder	6
Sorghum, forage	1.5
Sorghum, grain	0.75
Soybeans	0.5
Soybeans, forage	8.0
Soybeans, straw	15.0
Strawberries	0.5
Sunflower, seeds	0.25
Sweet potatoes	0.1
Tomatoes	0.5
Turnips (roots)	3
Turnips (greens)	1

Based on established tolerances the theoretical maximum residue contribution (TMRC) for chlorpyrifos residues in the human diet is calculated to be 0.5637 mg/day. The acceptable daily intake (ADI) of chlorpyrifos is 0.003 mg/kg/day. The maximum permissible intake (MPI) is 0.18 mg/day. The percent utilized ADI is 313%. To provide for conformity between U.S. tolerances for chlorpyrifos and tolerances established by the Codex Alimentarius, Canada and Mexico, the expression of the U.S. tolerances for chlorpyrifos would have to exclude the major metabolite TCP, but the Agency is not recommending this now.

U.S. tolerances for most raw agricultural commodities are supported by current residue chemistry data. In some cases however, more data are required.

4. SUMMARY OF REGULATORY POSITION AND RATIONALE

The Agency has determined that it should continue to allow the registration of chlorpyrifos. Adequate studies are available to assess the acute toxicological effects of chlorpyrifos to humans. None of the criteria for unreasonable adverse effects listed in section 162.11(a) of Title 40 of the U.S. Code of Federal Regulations have been met or exceeded. However, because of certain gaps in the data base a full risk assessment of chlorpyrifos cannot be completed.

Also, a full tolerance reassessment cannot be completed because of certain residue chemistry and toxicology data gaps.

The Agency is concerned whether or not the potential total human exposure to chlorpyrifos and its metabolites, from its widespread use and its ADI being exceeded three-fold, poses any unacceptable hazards. To resolve this concern, additional residue, metabolism and exposure data are required, and until it is resolved no significant new tolerances or uses will be granted.

A federal 24-hour reentry interval is established for treated crop areas until reentry data are submitted, as required, and the Agency decides on the most appropriate time interval.

Available data are insufficient to fully assess the environmental fate of chlorpyrifos. The Agency is requesting data to determine if chlorpyrifos will contaminate groundwater.

5. SUMMARY OF MAJOR DATA GAPS

Additional residue data on various processed commodities are being required. Also, additional chronic toxicity, oncogenicity and mutagenicity testing is needed to better define the long term effects of this chemical. Plant, animal and exposure data are required to better qualify and quantify human exposure to residues from dietary and nondietary sources.

Other requirements:

Acute inhalation
General metabolism
Hydrolysis study
Photodegradation studies
Soil metabolism studies
Mobility studies
Dissipation studies

Other Requirements (con't)

    Accumulation studies
    Fish embryo-larvae study
    Large scale field testing
    Monitoring for crop runoff
    Phytotoxic effects on algae and other aquatic plants
    Indoor monitoring

6. CONTACT PERSON AT EPA

    Jay S. Ellenberger
    Product Manager (12)
    Insecticide-Rodenticide Branch
    Registration Division (TS-767C)
    Office of Pesticide Programs
    Environmental Protection Agency
    401 M Street, S. W.
    Washington, D. C. 20460

    Office location and telephone number:
    Room 202, Crystal Mall #2
    1921 Jefferson Davis Highway
    Arlington, VA 22202
    (703) 557-2386

DISCLAIMER: The information presented in this Chemical Information Fact Sheet is for informational purposes only and may not be used to fulfill data requirements for pesticide registration and reregistration.

CHLORPYRIFOS-METHYL

Date Issued: June 30, 1985
Fact Sheet Number: 57

1. DESCRIPTION OF CHEMICAL

    Generic Name: O,O-dimethyl O-(3,5,6-trichloro-2-pyridyl) phosphorothioate

    Common Name: Chlorpyrifos-methyl

    Trade Name: Reldan

    EPA Shaughnessy Code: 059102

    Chemical Abstracts Service (CAS) Number: 510.0

    Year of Initial Registration: 1985

    Pesticide Type: Insecticide

    Chemical Family: Organophosphate

    U.S. Producer: Dow Chemical U.S.A.

2. USE PATTERNS AND FORMULATIONS

    Application Sites: Grains of rice, barley, oats, wheat, and grain sorghum intended for storage, empty grain storage bins, and equipment.

    Types of Formulations: Emulsifiable concentrate and dusts

    Types of Methods of Application: Spray and dusts

    Application Rates: 1.4-5.0 oz. active ingredient/1000 bushels of grain
    1% spray or 2-3% dust to grain equipment and storage walls/floors

    Usual Carriers: petroleum solvents and usual clay carriers

3. SCIENCE FINDINGS OF TECHNICAL CHEMICAL

Summary Science Statement

Chlorpyrifos-methyl is an organophosphate compound with moderate acute toxicity to both humans and wildlife species. The registered use precludes exposure to wildlife.

Chemical Characteristics

Physical State: Granular crystalline solid

Color: White

Odor: Mild mercaptan

Molecular weight and formula: 322.6 - $C_7H_7Cl_3NO_3PS$

Melting point: 45.5 - 46.5° C

Vapor Pressure: 7.40×10^{-7} mm Hg at 0° C

Solubility in various solvents: Solubility: g/100g of solvent at 24° C

Water	-0.0004
Acetone	-640
Benzene	-520
Ethyl alcohol	-30
Chloroform	-350

Toxicology Characteristics

Acute Oral: 1530 mg/kg, Toxicity Category III

Acute Dermal: greater than 2000 mg/kg, Toxicity Category III

Primary Dermal Irritation: Slight irritant-Toxicity Category III

Primary Eye Irritation: Slightly irritant-Corneal opacity persisted for 7 days. Toxicity Category III

Skin Sensitization: Positive

Acute Inhalation: 0.67 mg/l, Toxicity Category II

Neurotoxicity: Not an acute delayed neurotoxic agent at doses up to 500 mg/kg/day

Oncogenicity: Not shown to be an oncogen in rat or mouse studies at dose levels up to 3.0 mg/kg/day and 9.0 mg/kg/day (highest dosages tested), respectively

Teratogenicity: The Agency has determined that this chemical is not teratogenic at levels up to 200 mg/kg/day in rats and >16 mg/kg/day in rabbits (highest dosages tested)

Reproduction-3 generation: Studies adequately demonstrate that chlorpyrifos-methyl does not produce reproductive effects. No effects were demonstrated at dose levels up to 3.0 mg/kg/day (HDT)

Metabolism: The studies suggest that chlorpyrifos-methyl is rapidly excreted and there is no bioaccumulation

Mutagenicity: Data gap- These data are to be submitted in June of 1985

Physiological and Biochemical Behavioral Characteristics

Mechanism of Pesticidal Action: An insecticide which is active by contact, ingestion, and vapor action and causes phosphorylation of the acetylcholinesterase enzyme of tissues, allowing accumulation of acetylchloline at cholinergic neuro-effector junctions (muscarinic effects), and at skeletal muscle myoneural junctions and autonomic ganglia. Poisoning also impairs the central nervous system function.

Symptoms of poisoning include: headache, dizziness, extreme weakness, ataxia, tiny pupils, twitching, tremor, nausea, slow heatbeat, pulmonary edema, and sweating. Continual absorbtion at intermediate dosages may cause influenza-like illness which includes symptoms like weakness, anorexia, and malaise.

Metabolism and Persistence in Plants and Animals:

The metabolism of chlorpyrifos-methyl is adequately defined. A tolerance for stored grain has been established.

Environmental Characteristics

This use pattern precludes exposure to the environment. Therefore, most environmental effects are uncharacterized.

Ecological Characteristics

Avian acute oral:
Mallard Duck--1590 mg/kg

Avian dietary:
Mallard duck--5620 ppm
Bobwhite quail--2010 ppm

Freshwater Fish:
Coldwater fish (rainbow trout)--0.014 ppm
Warmwater fish (bluegill sunfish)--0.88 ppm

Acute Freshwater Invertebrates:
Daphnia--1.08 ppb

Tolerance Assessment

The tolerance for chlorpyrifos-methyl has been assessed according to the new Tolerance Assessment System (TAS). Under this method of analysis, the TMRC value for stored grain and grain product is 0.327 mg/day and the occupied ADI is 54 percent for the average U.S. population. The traditional system for figuring the TMRC gives an occupied ADI of 190%.

5. SUMMARY OF MAJOR DATA GAPS

   No major data gaps, only mutagenicity studies

6. CONTACT PERSON AT EPA

   Jay S. Ellenberger
   Product Manager (12)
   Insecticide-Rodenticide Branch
   Registration Division (TS-767C)
   Office of Pesticide Programs
   Environmental Protection Agency
   401 M Street, S. W.
   Washington, D. C.   20460

   Office location and telephone number:
   Room 202, Crystal Mall #2
   1921 Jefferson Davis Highway
   Arlington, VA   22202
   (703) 557-2386

DISCLAIMER: The information presented in this Chemical Information Fact Sheet is for informational purposes only and may not be used to fulfill data requirements for pesticide registration and reregistration.

CLIPPER (PACLOBUTRAZOL)

Date Issued: August 14, 1985
Fact Sheet Number: 62

1. Description of chemical:

    Common name: Paclobutrazol

    Code name: PP333

    Trade name: Clipper 50 WP

    EPA Shaughnessy code: 125601

    Chemical abstracts service (CAS) number: 76738-62-0

    Year of initial registration: 1985

    Pesticide type: Plant Growth Regulator

    U.S. and foreign producers: ICI Americas Inc.

2. Use patterns and formulations:

    Application sites: Ornamental trees (deciduous and broadleaf evergreen)

    Types of formulations: Wettable powder (50% a.i.)

    Types and methods of application: Applied in a pressurized tree injection system.

    Application rates: Rates vary depending upon tree size; dosage for smallest treatable tree is 0.006 oz. active ingredient per tree.

    Usual carriers: water

3. Science findings:

    Summary science statement: The data base for paclobutrazol (non-food uses) is well developed. Results of acute inhalation (50% formulation) and eye irritation studies indicate toxicity category II. The chemical is no mutagenic and is rapidly cleared from body tissue (rat and dog studies). Except for sandy soils, paclobutrazol does not exhibit a tendency to leac The chemical does not photodegrade and is not expected to hydrolyze. Hazards to aquatic and terrestrial wildlife are not anticipated because o low toxicity and low risk of exposure (tree injection system).

Chemical characteristics:

Physical state: Solid

Color: White

Odor: Not significant

Melting point: 165-166°C

Density: 1.22g/cm^3

Vapor pressure: 1.5 x 10^{-4} Pa at 50°C
8 x 10^{-6} Pa at 30°C
1 x 10^{-6} Pa at 20°C (by extrapolation)

Solubility:
Water	35 ppm
Cyclohexanone	18%
Methanol	15%
Acetone	11%
Methylene dichoride	10%
Xylene	6%
Propylene glycol	5%
Hexane	< 1%

Octanol/water partition coefficient: Log P 3.2

Unusual handling characteristics: None

Toxicology characteristics:

Acute toxicology results:

Technical
Acute oral toxicity (rat): 1.95 g/kg (male)
1.33 g/kg (female)
Toxicity category III

Acute dermal toxicity (rat): greater than 2g/kg
Toxicity category III

Primary skin irritation (rabbit): Paclobutrazol caused mild skin irritation. Toxicity category III

Primary eye irritation (rabbit): Paclobutrazol caused reversible corneal opacities with irritation lasting 72 hours. Toxicity category II

Dermal sensitization (guinea pig): Paclobutrazol is not a skin sensitizer.

Acute toxicology results:

<u>50% Formulation</u>
Acute inhalation toxicity (rat): greater than 766 mg/m^3 (male)
                                            359-766 mg/m^3 (female)
                                            Toxicity category II

Chronic toxicology results:

21-day dermal (rabbit): NOEL is 10 mg/kg/day
                            LEL is 100 mg/kg/day

90-day feeding (rat): NOEL is 250 ppm
                         LEL is 1,250 ppm

One year feeding (dog): NOEL is 15 mg/kg/day
                             LEL is 75 mg/kg/day

Teratology (rat): NOEL (maternal toxicity) is greater than 100 mg/kg/day (highest dose tested). NOEL (fetal effects) is 10 mg/kg/day.

Teratology (rabbit): Within limitations of study (low fertility), NOEL (maternal toxicity) is 25 mg/kg/day. LEL is 75 mg/kg/day.

Mutagenicity: Paclobutrazol does not cause mutagenic effects.

Metabolism: Results from rat and dog studies indicate that paclobutrazol and its metabolites are rapidly eliminated.

Major routes of exposure: Mixers, loaders and applicators would receive the most exposure via skin/eye contact and inhalation.

Physiological and biochemical behavioral characteristics:

Mechanism of pesticidal action: Paclobutrazol acts as a plant growth regulator and reduces regrowth in ornamental trees following trimming.

Environmental characteristics:

Adsorption and leaching in basic soil types: Paclobutrazol could leach in sandy soils with low organic content. In other soil types, the chemical does not have a high propensity to leach.

Loss from photodegradation: Paclobutrazol does not photodegrade after exposed to 10 days of simulated sunlight.

Resultant average persistance: Paclobutrazol degrades aerobically in soil with half-lives of about 1-7 months depending upon soil type. Paclobutrazol is not expected to hydrolyze in the environment.

Ecological characteristics:

Avian acute oral toxicity (Mallard): greater than 7,913 mg/kg

Avian dietary toxicity (Bobwhite quail): greater than 5,000 ppm

Avian dietary toxicity (Mallard): greater than 20,000 ppm

Fish acute toxicity (Bluegill): 23.6 mg/l

Fish acute toxicity (Rainbow trout): 27.8 mg/l

Aquatic invertebrate toxicity (Daphnia magna): 33.2 mg/l

Potential problems related to endangered species: Minimal hazard to endangered species is expected because of the low toxicity of paclobutrazol and proposed use (tree injection).

4. Summary of regulatory position and rationale:

Use classification: General

Use, formulation, manufacturing process or geographical restrictions: Paclobutrazol is not to be injected: (1) into trees that do not appear healthy, (2) into fruit or nut trees that will be harvested within one year after application and (3) into sugar maple or any other trees that are or could be tapped for sugar.

Unique label warning statements:

End-Use Product:

"Wear protective clothing, rubber gloves and a mask or pesticide respirator jointly approved by the Mining Enforcement and Safety Administration and the National Institute for Occupational Safety and Health."

5. Summary of major data gaps: None

6. **Contact person at EPA:**

   Robert J. Taylor
   Product Manager (25), TS-767C
   Environmental Protection Agency
   401 M Street, S.W.
   Washington, D.C. 20460
   (703) 557-1800

DISCLAIMER: The information presented in this Chemical Information Fact Sheet is for informational purposes only and may not be used to fulfill data requirements for pesticide registration and reregistration.

COMMAND HERBICIDE

Date Issued: June 20, 1986
Fact Sheet Number: 90.1

BACKGROUND

o This fact sheet addresses reports of phytotoxicity resulting from the use of Command, a herbicide manufactured by FMC Corporation. Command was registered by the Agency as a new chemical in February, 1986 for use on soybeans to control a broad spectrum of broadleaf and grassy weeds. The herbicide is marketed as two end-use products; a 4EC (Reg. No. 2793053) and a 6EC (Reg. No. 279-3054).

o The Agency has a complete data base for the chemical which shows very little potential for acute or long term health effects. However, because the adjuvants (inert ingredients) used in the formulated product can cause irreversible eye damage, the 6EC label bears the signal word danger.

o Phytotoxicity data indicate that some desirable plants are sensitive to Command and some off-target injury could occur. Because of this potential for injury to non-target plants, statements regarding chlorosis or bleaching of sensitive plants growing in proximity of treated fields are declared on the label. The Command product label states: "Some desirable plants, including ornamentals (e.g., roses), trees (e.g. flowering and edible cherries), agronomic crops (e.g., small grains, alfalfa, sunflowers) and vegetables (e.g., lettuce, cole crops, radish) are sensitive to Command herbicide. Foliar contact with spray drift or vapors may cause visual symptoms of chlorosis or bleaching to sensitive plants growing in the proximity of treated fields. These symptoms are temporary. Extra caution is advised when spraying near desirable plants found in residential or other commercial agricultural plantings."

PHYTOTOXIC PROBLEM

o The Agency has received numerous reports from several midwestern states in Region V & VII of damage to non-target plants when Command is applied to soybean fields as either a surface or soil incorporated treatment. Some of this damage is beyond the proximity of the treated field. Apparently, the chemical quickly volatilizes and moves from the treatment site to nontarget plants where it causes chlorosis (whitening) in plant foliage. This effect, in most cases, appears to be temporary. Reports from state officials indicate that the product was used according to label directions.

DIETARY EXPOSURE PROFILE

o Command is not oncogenic or teratogenic. The Acceptable Daily Intake is based on a NOEL of 100 ppm (4.3 mg/kg/day) in a 2 year rat feeding study with a safety factor of 100.

o Information submitted by the registrant do not show any detectable residues in non-target crops such as leafy vegetables (lettuce, spinach, kale, rhubarb, chinese cabbage), fruiting vegetables (peas, beans, tomatoes), root crops and fruits. Residues have been detected at levels up to 1.4 ppm in alfalfa and pasture hay.

o The registered use and tolerance on soybeans at 0.05 ppm uses 0.3% of the ADI. Even if every food item in the total diet had residues of .05 ppm only approximately 3% of the ADI would be used.

o The Agency does not have adequate data on the nature or magnitude in meat or milk from residues in feed. However, metabolism studies in rats show that Command was excreted rapidly (90-99% within 72 hours) with no significant retention in rats tissues. This, together with the structure of the chemical and other information indicate that significant bioaccumulation in edible tissues or milk is unlikely.

EPA POSITION

o The Agency has worked with FMC, regional personnel and state officials to determine the scope of the problem. EPA will be reviewing additional information as it becomes available and will be assessing this situation as to what can be done to address the off-target problem and related issues prior to the 1987 growing season.

o Since residues were detected in alfalfa and hay, the Agency will be discussing the matter with the Food and Drug Administration. However, on the basis of the available information, the Agency concludes that direct human consumption of the affected crops, or the feeding of affected forage items to animals, will not pose an unreasonable risk to the public.

o Off-target movement of Command is not expected to cause any adverse health effects to persons in the vicinity of treated areas.

(Contact Frank Sanders 557-1650 or Robert Taylor 557-1800)

COPPER SULFATE

Date Issued: March 21, 1986
Fact Sheet Number: 87

1. Description of chemical:

    Generic Name: Copper Sulfate
    Common name: Copper Sulfate
    Trade names: Copper Sulfate, Bluestone
    EPA Shaughnessy Code: 024401 (pentahydrate), 024402 (monohydrate),
                     008101 (basic copper sulfate)
    Chemical Abstracts Service (CAS) Number: 1344-73-6
    Year of Initial Registration:
    Pesticide Type: Herbicide, fungicide
    Chemical family: Copper sulfate
    U.S. Producers: The Anaconda Co., Cities Service Co., Inc., C.P Chemicals,
        Inc., Engelhard Minerals & Chemical Corp., Liquid Chemical Corp.,
        Madison Industries, Inc., Mallinckrodt, Inc., Phelps Dodge Corp.,
        Southern California Chemical Co., Inc., Transvaal, Inc., Univar Corp.,
        Gulf Chemical & Metallurgical Co., Pesticide Service Consultants.

2. Use patterns and formulations:

    Application sites: Primary use is to control alga growths in impounded waters,
                          lakes, ponds, reservoirs, and irrigation and irrigation
                          drainage conveyance systems. Other sites include foliar
                          applications for control of foliar pathogens on fruit,
                          nut, vegetable, and field crops, ornamentals, and
                          agricultural and home garden uses. Copper sulfate is
                          also registered for use as a weed seed treatment, for
                          treatment of tree wounds, for control of fungi and
                          fungal/bacterial slimes occurring in wood, and in
                          water systems, including sewer pumps and force mains,
                          pulp and paper mills, cooling towers and spray ponds.
    Types of formulations: Basic copper sulfate is formulated as crystalline
                          solids, 12.75 and 53.0 percent wettable powders,
                          30.0 percent liquid concentrate, and 2.0 and 52.0 percent
                          dusts.
    Types and Methods of Application: In impounded waters, lakes, ponds, reservoirs
                          and irrigation systems, copper sulfate can be applied
                          by spraying or dusting the water surface, large crystals
                          can be placed in a burlap bag and towed behind a boat,
                          or dumped in directly, or finer crystals can be metered in
                          continuously.
    Application Rates: From 0.0013 to 10.0 ppm for aquatic uses and from 0.24 to
                          21.2 lb a.i./A for terrestrial uses.
    Usual carriers: Water, or no carrier.

Copper Sulfate 201

3. Science Findings:

Summary science statement: Copper sulfate is exempt from the requirements of a tolerance when it is applied to growing crops. There is also a specific exemption from the requirement of a tolerance for copper in

meat, milk, poultry, eggs, fish, shellfish, and irrigated crops when copper sulfate pentahydrate is used as an algicide or herbicide in ponds, reservoirs, or other bodies of water. These exemptions are well founded. No additional toxicology studies are required. Copper sulfate is in Toxicity Category I as an eye and dermal irritant. Copper sulfate is toxic to fish and may present a hazard to aquatic organisms, especially when used as an algicide or molluscicide.

Chemical characteristics:

Physical state: Solid, crystalline
Color: Blue-green
Odor: None
Boiling point: N/A
Melting point: N/A
Flash point: N/A
Unusual handling characteristics: None

Toxicological characteristics:

Acute Effects:

Acute Oral Toxicity - Toxicity Category II
Acute Dermal Toxicity - Toxicity Category II
Dermal Irritation - Toxicity Category I
Dermal Sensitization - Copper Sulfate is a dermal sensitizer
Acute Inhalation Toxicity - N/A
Primary Eye Irritation - Toxicity Category I

Chronic Effects:

Oncogenicity - N/A (Data not required)
Teratology - N/A
Reproductive Effects - N/A
Mutagenicity - N/A
Feeding Studies - N/A

Major Routes of Exposure: Dermal, ocular, and inhalation

Physiological and Biochemical Behavioral Characteristics:

Foliar absorption: N/A
Translocation: N/A
Mechanism of pesticidal action: Inactivation of enzyme systems in fungus and algae
Metabolism and persistence in plants and animals: No build-up of copper occurs in plants. Because of the exemption from tolerance requirements, no further data will be required for plant and animal metabolism and persistence.

Environmental Characteristics:

    Adsorption and leaching in basic soil types: Data gap
    Microbial breakdown: N/A
    Loss from photodecomposition and/or volatilization: N/A
    Bioaccumulation: N/A
    Resultant average persistance: As a naturally occurring substance, copper persists indefinitely.
    Half-life in Water: N/A

Ecological characteristics:

    Hazards to Birds: Minimal hazards
    Hazards to Aquatic Invertebrates: Toxic to aquatic invertebrates.
    Hazards to Fish: Toxic to fish. Field studies have been required.
    Potential Problems with Endangered Species: Biological opinions from OES have stated that there may be hazards to the slackwater darter, freshwater mussels, and Solano grass. Statements will be required on all EP labels requiring the applicator to consult a specific Agency Bulletin before applying copper sulfate in certain counties.

Tolerance Reassessment:

    List of crops and tolerances: Copper sulfate is exempted from all tolerances.
    List of food contact uses: Copper sulfate is exempted from all tolerances.
    Results of tolerance reassessment: Exemption was found to be valid.
    Problems known to have occurred from use: No specific problems requiring regulatory action.

4. Summary of Regulatory Position and Rationale:

    Use Classification: None
    Use, Formulation or Geographic Restrictions: In certain counties, the user must consult a specific Agency Bulletin on endangered species before applying copper sulfate.

    Unique Label warning statements:

a. Labels for manufacturing-use copper sulfate products must bear statements reflecting the compound's acute human toxicity. Copper sulfate is in Toxicity Category I by eye and dermal irritation routes of exposure.

The following human hazard statement, based on data reviewed by the Agency, must appear on all MP labels, and on the labels of all EPs in the 99% a.i. crystalline form:

    "DANGER - Causes severe eye and skin irritation. Harmful if absorbed through the skin or inhaled. May cause skin sensitization reactions in certain individuals. Avoid contact with the skin, eyes, or clothing. Avoid breathing

Copper Sulfate   203

dust. Protective clothing, including goggles, should
be worn. Wash thoroughly with soap and water after
handling. Remove contaminated clothing and wash before
reuse."

One of the following statements of practical treatment,
based on data reviewed by the Agency, must appear on all
MP labels, and on the labels of all EPs in the 99% a.i.
crystalline form, under a heading that reads either "Practical
Treatment" or "First Aid":

"IF IN EYES, flush with plenty of water. Call a physician.
IF ON SKIN, wash with plenty of soap and water. Get
medical attention. IF SWALLOWED, call a physician or
Poison Control Center. Drink 1 or 2 glasses of water and
induce vomiting by touching the back of throat with
finger. Do not induce vomiting or give anything by mouth
to an unconscious person."; or

"IF IN EYES, flush with plenty of water. Call a physician.
IF ON SKIN, wash with plenty of soap and water. Get
medical attention. IF SWALLOWED, call a physician or
Poison Control Center. Drink 1 or 2 glasses of water and
induce vomiting by touching back of throat with finger, or,
if available, by administering syrup of ipecac. Do not
induce vomiting or give anything by mouth to an unconscious
person."

b. The following revised environmental hazard statement must
appear on all MP labels:

"This pesticide is toxic to fish and aquatic organisms.
Do not discharge effluent containing this product into
lakes, streams, ponds, estuaries, oceans, or public
water unless this product is specifically identified
and addressed in an NPDES permit. Do not discharge
effluent containing this product into sewer systems
without previously notifying the sewage treatment plant
authority. For guidance contact your State Water Board
or Regional Office of the EPA."

c. The following statements must appear on all EP labels:

1. "May cause skin sensitization in certain individuals"; and

11. "ENDANGERED SPECIES RESTRICTIONS

It is a violation of Federal laws to use any pesticide in
a manner that results in the death of an endangered species
or adverse modification of their habitat.

The use of this product may pose a hazard to certain Federally designated endangered species known to occur in specific areas within the following counties:

STATE
    Species
        (Bulletin No.)
	COUNTY

CALIFORNIA
    Solano Grass
        (EPA/ES-85-13)
	Solano

TENNESSEE
    Slackwater Darter
        (EPA/ES-85-04)
	Lawrence
	Wayne
	Hancock

Freshwater Mussels (EPA/ES-85-07)	Claiborne Hawkins Sullivan

ALABAMA
    Slackwater Darter
        (EPA/ES-85-05)
	Lauderdale
	Limestone
	Madison

VIRGINIA
    Freshwater Mussels
        (EPA/ES-85-06)
	Grayson
	Smyth
	Scott
	Lee
	Washington

Before using this product in the above counties you must obtain the EPA Bulletin specific to your area. This Bulletin identifies areas within these counties where the use of this pesticide is prohibited, unless specified otherwise. The EPA Bulletin is available from either your County Agricultural Extension Agent, the Endangered Species Specialist in your State Wildlife Agency Headquarters, or the appropriate Regional Office of the U.S. Fish and Wildlife Service. THIS BULLETIN MUST BE REVIEWED PRIOR TO PESTICIDE USE."

e. All copper sulfate products intended for direct application to water must bear the following statements:

i. "This pesticide is toxic to fish. Direct application of copper sulfate to water may effect a significant reduction in populations of aquatic invertebrates, plants and fish."

ii. "Do not treat more than one-half of lake or pond at one time in order to avoid depletion of oxygen from decaying vegetation. Allow 1 to 2 weeks between treatment for oxygen levels to recover."

iii. "Trout and other species of fish may be killed at application rates recommended on this label, especially in soft or acid waters. However, fish toxicity generally decreases when the hardness of water increases."

iv. "Do not contaminate water by cleaning of equipment or disposal of wastes."

v. "Consult your State Fish and Game Agency before applying this product to public waters. Permits may be required before treating such waters."

f. All copper sulfate products intended for end use applications to water where there is likelihood of effluent reaching natural waters, i.e. cooling towers, paper or pulp mills, spray ponds, sewer pumps, force mains, algicide and molluscicide uses, must bear the following statement:

"This pesticide is toxic to fish and aquatic organisms. Do not discharge effluent containing this product into lakes, streams, ponds, estuaries, oceans, or public waters unless this product is specifically identified and addressed in an NPDES permit. Do not discharge effluent containing this product to sewer systems without previously notifying the sewage treatment plant authority. For guidance contact your State Water Board or Regional Office of the EPA."

g. All copper sulfate products intended for end use on terrestrial sites must bear the following statement:

"This pesticide is toxic to fish and aquatic organisms. Do not apply directly to water. Drift and runoff from treated areas may be hazardous to fish and aquatic organisms in adjacent aquatic sites. Do not contaminate water by cleaning of equipment or disposal of wastes."

5. Summary of major data gaps:

Data Requirement:	Data Due Date:
Product Identity and Disclosure of Ingredients	6 months
Discussion of Formation of Impurities	6 months
Preliminary Analysis	12 months
Certification of Ingredient Limits	12 months
Analytical Methods and Data for Enforcement of Limits	12 months
Aerobic Soil Metabolism	27 months
Anaerobic Soil Metabolism	27 months
Anaerobic Aquatic Metabolism	27 months
Aerobic Aquatic Metabolism	27 months
Leaching and Adsorption/Desorption	12 months
Soil Dissipation - Field	27 months
Aquatic Dissipation - Field	27 months
Acute Toxicity to Freshwater Fish (on typical EP for Aquatic food and non-food uses)	9 months
Acute Toxicity to Freshwater Invertebrates (on typical EP for Aquatic food and non-food uses)	9 months
Acute Toxicity to Estuarine and Marine Organisms (on typical EP for Aquatic food and non-food uses)	12 months
Fish Early Life Stage	15 months
Invertebrate Life-Cycle	15 months
Field Testing - Aquatic Organisms	24-48 months

6. Contact person at EPA: Richard Mountfort   TS-767-C
   401 M Street SW
   Washington DC 20460
   (703) 557-1700

DISCLAIMER: The information presented in this Chemical Information Fact Sheet is for informational purposes only and may not be used to fulfill data requirements for pesticide registration and reregistration.

CRYOLITE

Date Issued: June 30, 1983
Fact Sheet Number: 2

1. Description of Chemical

Generic name: Sodium fluoaluminate or sodium aluminofluoride
Common name: Cryolite (natural or synthetic)
Trade name: Kryocide
EPA Shaughnessy code: 075101
Chemical Abstracts Service (CAS) number: 15096-52-3 or 1344-75-8
Year of initial registration: about 1967
Pesticide type: insecticide
Chemical family: aluminofluoride salt
U.S. and foreign producer: Pennwalt Corporation

2. Use Patterns and Formulations

Application sites: domestic and non-domestic terrestrial food crops and ornamentals
Types of formulations: wettable powder, dusts
Types and methods of application: aerial or ground application as a spray or dust
Application rates: varies from 6 to 50 lbs./acre
Usual carriers: Confidential Business Information

3. Science Findings

Summary science statement:

Based on data reviewed for the Cryolite Standard, cryolite exhibits low to moderate toxicity to humans (toxicity category III-IV), fish and wildlife.

However, since a number of toxicology and residue data gaps exist, the Agency is unable to complete a risk assessment.

Chemical characteristics:

Technical grade cryolite may be the naturally occuring fluoride of sodium and aluminum called "natural cryolite" or it may be the manufactured material of similar composition called "synthetic cryolite". Both forms are a solid with melting points of 1000°C for the synthetic and 1009°C for the natural cryolite. Natural cryolite may be white, black, purple or violet. The synthetic cryolite is white. Cryolite is completely stable under normal storage conditions. The are no unusual handling characteristics.

Toxicology characteristics:

No toxicological hazards of concern have been identified based on the studies reviewed for the standard.

- Acute dermal LD_{50}, rabbit: >2.1 g/kg (Tox category III)
- Acute inhalation LD_{50}: <5.03 mg/L, 2.06 mg/L (Tox category III)

- Primary dermal irritation, rabbit: P.I. score = 0.0, not an irritant (Tox category IV)
- Primary eye irritation, rabbit: moderate conjunctival irritation that disappeared within 7 days (Tox category III)

A sequential testing approach is being applied to this compound because the chemical properties are unique. Therefore, after review of the studies required under the standard (a 90-day feeding study in the rat and the dog), additional subchronic and chronic studies may be required.

Physiological and Biochemical Behaviorial Characteristics:

Mechanism of pesticidal action: stomach poison
Mechanism and persistence in plants and animals: Not known

Environmental Characteristics:

There are no available data on cryolite. However, based on the chemical and physical characteristics, none of the usual environmental fate data normally required would yield useful information, with the exception of hydrolysis and leaching studies which are required under the standard.

Ecological Characteristics:

Based on current data, cryolite has been determined to be practically non-toxic to bobwhite quail and mallard ducks in subacute doses; however, hazards to birds, fish and aquatic invertebtrates may be greater that previously supposed. Additional data are required to address this concern.

- Dietary LC_{50} (mallard duck and bobwhite quail):
  >10,000 ppm (practically nontoxic)
- Freshwater invertebrates LC_{50} (<u>Daphnia pulex</u>):
  10 ppm (moderately toxic)
- Fish acute LC_{50} (rainbow trout): 47 ppm (slightly toxic)
- Fish acute LC_{50} (bluegill sunfish): >300 ppm (practically nontoxic)

Efficacy review results:
  No efficacy data required.

Tolerance Assessments:

Tolerances have been established (40 CFR 180.145) of 7 ppm of combined fluorine for residues of the insecticidal fluorine compounds cryolite and synthetic cryolite in or on each of the following raw agricultural commodities: Apples, apricots, beans, beets (with or without tops) or beet greens alone, blackberries, blueberries (huckleberries), boysenberries, broccoli, brussels sprouts, cabbage, carrots, cauliflower, citrus fruits, corn, collards, cranberries, eggplants, grapes, kale, kohlrabi, lettuce, loganberries, melons, mustard greens, nectarines, okra, peaches, peanuts, pears, peas, peppers, plums (fresh prunes), pumpkins, quinces, radishes (with or without tops) or radish tops, raspberries, rutabagas (with or without tops) or rutabaga tops, squash, strawberries, summer squash, tomatoes, turnips (with or without tops) or turnip greens, youngberries.

Specific residue data are being required on a number of crops. The Agency will assess all cryolite tolerances after the necessary toxicity and residue data are received. Residue data are required to ascertain the need for food additive tolerances for the following processed commodities: Apple pomace (wet and dry), apple juice, bean cannery waste, citrus (peel, oil, dried pulp and molasses), corn (oil and milled fractions), mustard seed, tomato pomace (wet and dry, juice, puree, catsup).

Livestock feeding studies are required to ascertain the extent of carryover (if any) into meat, milk, poultry and eggs, and the need to establish tolerances for these commodities.

4. Summary of Regulatory Position and Rationale

   Use classification: not classified
   Use restrictions: none
   Unique warning statements: none

No toxicological hazards of concern have been identified in the studies reviewed for this standard. However, since a number of toxicology and residue data gaps exist, the Agency is unable to complete a risk assessment of cryolite.

5. Summary of Major Data Gaps

Product chemistry: most of these data are lacking for the natural cryolite. Data to be submitted by December, 1983.

Residue chemistry: residue data are lacking on all crops for which tolerances are established with the exception of grapes. Further, there are no data available on the processing of commodities other than raisins. Livestock feeding studies are required to ascertain the extent of carryover (if any) into meat, milk, poultry and eggs. Data must be submitted by June, 1986.

Toxicology: oral LD_{50} (rat), dermal sensitization, 90-day feeding (rat and dog), mutagenicity studies. Chronic studies may be required following receipt and evaluation of acute, subchronic, and residue data. Data must be submitted by December, 1983.

Environmental fate: hydrolysis and leaching studies. Data must be submitted by June, 1986.

6. Contact Person at EPA

William H. Miller
Product Manager (16)
Insecticide-Rodenticide Branch
Registration Division (TS-767)
Environmental Protection Agency
Washington, DC 20460

Tel. No. (703) 557-2600

DISCLAIMER: The information presented in this Chemical Information Fact Sheet is for informational purposes only and may not be used to fulfill data requirements for pesticide registration and reregistration.

CYANAZINE

Date Issued: December 31, 1984
Fact Sheet Number: 41

1. Description of chemical:

    Generic Name: cyanazine
    Common name: cyanazine (WSSA, BSI, ISO); SD 15418 and WL 19805 (code numbers)
    Trade name: BLADEX® 80 WP or 80W, BLADEX® 4-WDS or 4L, BLADEX® 15G
    EPA Shaughnessy Code: 100101
    Chemical Abstracts Service (CAS) Number: 21725-46-2
    Year of Initial Registration: 1971
    Pesticide Type: Herbicide
    Chemical family: Triazine
    U.S. and Foreign Producers: Shell Chemicals (U.S.A.)

2. Use patterns and formulations:

    Application sites: To control annual grasses and broadleaf weeds in corn, grain sorghum, cotton, and wheat fallow. Application for soybean use is pending.
    Types of formulations: Wettable powder, flowable suspension, granular form.
    Types and Methods of Application: Aerial and ground sprays, application through irrigation systems.
    Application Rates:

Crop	Application Timing	lbs active ingredient (ai) per acre
Corn	Preemergence	1.25-4.75
	Postemergence	1.2-2.0

Usual carriers: Water or liquid fertilizers for preemergent use on corn. Water only on postemergence treatments, grain sorghum and cotton.

3. Science Findings:

Summary science statement: Cyanazine has been found in groundwater. It has potential to create teratogenic effects; this potential is being evaluated in the Special Review process. Acute toxicity ratings are generally low.

Chemical characteristics:

Physical state:	crystalline
Color:	white
Odor:	none
Vapor pressure:	1.6×10^{-9} mm Hg at 20°C
Melting point:	166.5 - 167° C
Flammability:	non-flammable
Octanol/water partition coefficient:	not available
Stability:	Stable at pH values of 5, 7, and 9 for >30 days
	Stable in sunlight, and at 75° for 100 hours
Solubility:	Water (23°C) - 160 ppm Benzene (20°C) - 1.5%
	Xylene - <10% Ethanol (20°C) - 4.5%
	Chlorobenzene - <10% Chloroform - 21.0%
	Methylcyclohexanone - 21.0%

Unusual handling characteristics: None reported

Toxicological characteristics:

Acute Effects:

Acute Oral LD_{50} - 334 mg/kg (male rats), 156 mg/kg (female rats) (Category I
Acute Dermal LD_{50} - >2000 mg/kg (rabbits) (Category III)
Acute Inhalation Toxicity - LC_{50} >2.28 mg/kg (Category III)
Primary Eye Irritation - Mild eye irritation (Category II)

Cotton	Preemergence	0.5-1.3
	Early postemergence directed	0.6-1.0
	Postemergence layby	0.8-1.6
Grain Sorghum	Preemergence	
	Tank mixed with propachlor	1.0-1.6
	Tank mixed with propazine	0.8-1.2

Major Routes of Exposure: Dermal, inhalation

Chronic Effects:

Oncogenicity - Results inconclusive.

Teratology - F-344 Rats: Increased incidence of anophthalmia and microphthalmia at 25 mg/kg/day, NOEL = 10 mg/kg/day. Increased incidence of diaphragmatic hernia in all treated groups. More data are required to ascertain the nature of this effect.

Sprague-Dawley rats: Slight decrease in maternal body weight at 30 mg/kg/day. NOEL = 3 mg/kg/day

Rabbits: Maternal toxicity and fetotoxicity at 2 mg/kg/day. NOEL = 1 mg/kg/day

Mutagenicity - data gap
Immunotoxicity - data gap

Physiological and Biochemical Behavioral Characteristics:

Translocation: When applied to soil, cyanazine is absorbed by the roots and translocated to the leaves.

Environmental Characteristics:

    Absorption and leaching characteristics: Cyanazine is reversibly adsorbed to soil particles. The degree of adsorption varies with soil texture, water content, and organic matter content. Leaching rate into the soil was measured on a sandy loam soil and found to be comparable to that of atrazine.
    Microbial breakdown: Cyanazine is degraded in the soil primarily by microbes.
    Loss from photodegradation and/or volatilization: Under field conditions there is only a minimal loss of cyanazine by either photodecomposition or volatilization.

    Resultant average persistance: Half-life about 2 weeks under conditions favorable for plant growth.
    Half-life in Water: Unknown.

Ecological characteristics:

    Hazards to Birds: Data are incomplete. Preliminary data show low toxicity, suggesting minimal hazards.
    Hazards to Fish and Aquatic Invertebrates: Data are incomplete. Preliminary data show low toxicity, suggesting minimal hazards.
    Potential Problems with Endangered Species: No hazards indicated.

Tolerance Reassessment:

    List of crops and tolerances: (CFR 180.307)

COMMODITY	(PPM)
Corn, fodder	0.2
Corn, forage	0.2
Corn, fresh (Inc. sweet)(K+CWHR)	0.05
Corn, grain	0.05
Cotton, seed	0.05
Sorghum, forage	0.05
Sorghum, fodder	0.05
Sorghum, grain	0.05

Wheat, forage, green	0.1
Wheat, grain	0.1
Wheat, straw	0.1

List of food contact uses: Corn, cotton (oil), sorghum, wheat.
Results of tolerance assessment: No ADI can be set at this time

4. **Summary of Regulatory Position and Rationale:**

   Use Classification: Reclassified (by the Registration Standard) as a Restricted Use chemical because of teratogenic effects, and because it is found in groundwater.

   Use, Formulation or Geographic Restrictions: Manufacturing-use products may only be formulated into end-use products intended for use as an herbicide on corn, cotton, sorghum, or fallow land, or winter wheat.

   Unique Label Warning Statements:

a) <u>Use Classification Statements</u>:

   Labels of all formulated products must bear the following statements:

   "RESTRICTED USE PESTICIDE: Because cyanazine can leach into groundwater, and has produced birth defects in laboratory animals, this product may be applied only by certified applicators or persons under their direct supervision."; <u>and</u>

   "Cyanazine is a pesticide which can travel (seep or leach) through soil and can contaminate groundwater which may be used as drinking water. Cyanazine has been found in groundwater as a result of agricultural use. Users are advised not to apply cyanazine where the water table (ground water) is close to the surface and where the soils are very permeable, i.e. well drained soils such as loamy sands. Your local agricultural agencies can provide further information on the type of soil in your area and the location of ground water."

b) Precautionary Statements

Labels of manufacturing-use products and end-use products (EUPs) must bear the statements:

a. Hazards to Humans Statements

"WARNING: May be fatal if swallowed. Harmful if inhaled or absorbed through the skin. Causes substantial but temporary eye injury. Avoid breathing dust (vapor or spray mist). Avoid contact with skin, eyes or clothing. Do not get in eyes or on clothing. Wear a face shield. Wash thoroughly with soap and water after handling and before eating or smoking. Remove contaminated clothing and wash before reuse."; and

"Use of this product may be hazardous to your family's health. This product has been determined to cause birth defects in laboratory animals. Exposure of women of child-bearing age to cyanazine should be avoided."

c) Statements of Practical Treatment

"If on skin: Wash with plenty of soap and water. Get medical attention."

"If in eyes: Flush with plenty of water. Call a physician."

"If swallowed: Call a physician or Poison Control Center. Drink 1 or 2 glasses of water and induce vomiting by touching back of throat with finger. Do not induce vomiting or give anything by mouth to an unconscious person."

d) Environmental Hazard Statement

The following specific statements must appear on the labels of all manufacturing use products:

"Do not discharge effluent containing this product into lakes, streams, ponds, estuaries, oceans, or public waters unless this product is specifically identified and addressed in a NPDES permit. Do not discharge effluent containing this product to sewer systems without previously notifying the sewage treatment plant authority. For guidance contact your State Water Board or Regional Office of the EPA."

The labels of EUPs intended for outdoor use must bear one of the following statements, depending on the formulation of the product:

Granular products must bear the statement:

"Do not apply directly to water or wetlands. In case of spills, collect for use or properly dispose of the granules. Do not contaminate water by cleaning of equipment or disposal of wastes."

Non-granular products must bear the statement:

"Do not apply directly to water or wetlands. Do not contaminate water by cleaning of equipment or disposal of wastes."

The label of all products (except those, if any, intended solely for household use) must bear the appropriate container disposal statement, as will be given in Appendix IV-5 of the guidance package.

The required statements listed in this standard must appear on the labels of all MUPs and EUPs released for shipment after June 30, 1985. After review of data to be submitted under this standard, the Agency may impose additional label requirements.

Summary of risk/benefit analysis: Cyanazine produces teratogenic effects in laboratory animals. Exposure to the public through the dietary route is not sufficiently large to exceed the risk criterion in 40 CFR 162.11. Margins of safety are adequate for that route. However, the dermal exposure rates of mixer/loaders and applicators are comparable to levels at which effects occur in experimental animals. This triggers the risk criterion in 40 CFR 162.11, sending cyanazine into the Special Review Process. The benefits of cyanazine are primarily from its effectiveness as an herbicide on corn, which accounts for 96% of total use. Available alternatives to cyanazine do not have as broad a spectrum of weed control. The most widely used alternative is atrazine, which may have more persistent residues and leaches through the ground at the same rate as cyanazine. On cotton (3% total use) the alternatives again do not provide as broad a spectrum of weed control.

218 Pesticide Fact Handbook

5. Summary of major data gaps

Dates when major data gaps are due to be filled.

Data Requested	Due date (after publication of the Standard)
Statement of Composition	six months [1]
Discussion of formation of unintentional ingredients	six months [1]
Preliminary analysis of samples	six months [1]
Certification of limits	six months [1,2]
Analytical methods and data for enforcement of limits	six months [1,2]
Density, Bulk Density, or specific Gravity	six months [1]
Dissociation Constant	six months [1]
Octanol/water partition coefficient	six months [1]
pH	six months [1,2]
Oxidizing/Reducing Action	six months [1,2]
Flammability	six months [1,2]
Explodability	six months [1,2]
Storage Stability	six months [1,2]
Livestock residues	six months
Plant residues	six months
Animal residues	six months
Storage stability data	six months
Magnitude of the residue for each food use	six months
Hydrolysis	six months
Photodegradation (water, soil)	six months
Metabolism studies in lab	six months
Mobility studies - leaching and absorption/desorption	six months
Dissipation studies in field	six months
Monitoring of surface and groundwater	18 months
Primary eye irritation	six months [2,3,4]
Primary dermal sensitization	six months [4]
90-day feeding (rodent, non-rodent)	four years
Oncogenicity	four years
Teratogenicity	one year
Reproduction (2-generation)	20 months
Chromosomal aberration	20 months

Data Requested	Data Due (After publication of the Standard)
Other genotoxic effects	20 months
Avian dietary toxicity	six months
Acute toxicity to freshwater invertebrates	six months

1/ Data are required on the technical material
2/ Data are required on the 94% technical, which is used as a manufacturing use product, and the 28.2% flowable intermediate
3/ Data are required on all wettable powder formulations
4/ Data are required on all liquid formulations

6. <u>Contact person at EPA</u>: Robert Taylor, U.S. Environmental Protection Agency, TS-767-C, 401 M Street SW, Washington, DC 20460
(703) 557-1650

DISCLAIMER: The information presented in this Chemical Information Fact Sheet is for informational purposes only and may not be used to fulfill data requirements for pesticide registration and reregistration.

CYHEXATIN

Date Issued: June 30, 1985
Fact Sheet Number: 56

1. DESCRIPTION OF CHEMICAL

   - Generic Name: Tricyclohexylhydroxystannane
   - Common Name: Cyhexatin
   - Trade Name: Plictran®, Acarstin® and Dowco® 213
   - EPA Shaughnessy Code: 101601
   - Chemical Abstract Service (CAS) Number: 13121-70-5
   - Year of Initial Registration: 1972
   - Pesticide Type: Miticide
   - Chemical Family: Organotins
   - U.S. Producer: Dow Chemical Company

2. USE PATTERNS AND FORMULATIONS

   - Application sites: apples, pears, citrus, peaches, plums, nectarines, strawberries, almonds, walnuts, hops, and ornamental plants (including greenhouse grown)
   - Types and methods of applications: aerial and ground application as a spray
   - Application rates: 0.5 lb active ingredient (ai)/A to 2.0 lbs ai/A
   - Usual Carriers: wettable powders

3. SCIENCE FINDINGS

   Chemical Characteristics

   - Technical cyhexatin is a white, crystalline powder, nearly odorless that has no true melting point, degrades to bistricyclohexyltin oxide at 121 to 131° C and decomposes at 228° C.
   - It is soluble in some organic solvents and is very insoluble in water.
   - Vapor pressure is negligible at 25° C, it is stable in aqueous suspensions in neutral and alkaline pH, reacts ionically in the presence of a strong acid to form salts, converts to dicyclohexyltin oxide and further to cyclohexylstannoic acid by exposure to ultraviolet radiation.

Toxicology Characteristics

- Acute oral: 196 mg/kg (rat), Toxicity Category II.
- Acute dermal: Data gap.
- Primary Eye Irritation: Causes eye irritation--corneal and iris irritation (rabbit), Toxicity Category II.
- Acute Inhalation: 6.35 mg/l (rat), Toxicity Category III.
- Primary Skin Irritation: Nonirritant, Toxicity Category IV.
- Skin Sensitization: Not a sensitizer.
- Major Routes of Exposure: Human exposure from cyhexatin applications is greatest from mixing and loading of pesticide formulation and applying it. Exposure can be reduced by the use of goggles or face shield and gloves and other protective clothing.
- Neurotoxicity: Cyhexatin is not expected to be a delayed neurotoxin because it is neither an organophosphate nor an analog of a neurotoxic compound.
- Oncogenicity: Data gap. Study submitted does not meet Agency standards.
- Metabolism: Available data suggest that cyhexatin is not readily absorbed in tissues and is excreted in the feces. The minor amount that is not excreted is metabolized to an organotin form and is accumulated in the liver and kidney with lesser levels found in the brain, heart, adrenal, and muscle.
- Teratology: Adequate data are unavailable. Data gap.
- Reproduction: Adequate data are unavailable. Data gap.
- Mutagenicity: Adequate data are unavailable. Data gap.

Physiological and Biochemical Characteristics

- Mechanism of pesticidal action: It is suspected that cyhexatin inhibits adenosine triphosphate (ATP) enzymes.
- Metabolism and persistance in plants and animals: Available data suggest that plant degradates of cyhexatin are translocated following root exposure, however these data are insufficient to adequately characterize plant metabolism. Known animal metabolism is summarized above.

Environmental Characteristics

- Available data are insufficient to fully assess the environmental fate of cyhexatin. Data gaps exist for all required studies.

- The available data do suggest that cyhexatin can leach slowly in certain soils. Data are required to assess cyhexatin's environmental fate and ability to leach through soils.

Ecological Characteristics

- Avian acute oral (LD_{50}) toxicity: approximately 250 mg to 400 mg technical cyhexatin/kg body weight for quail (moderately toxic).
- Avian dietary (LC_{50}) toxicity: 195 ppm for bobwhite quail (highly toxic).
- Freshwater fish acute (LC_{50}) toxicity: cold water species (rainbow trout)--6 ppb for technical; warm water species (bluegill)--4 ppb for technical.
- Aquatic freshwater invertebrates toxicity: Daphnia--0.2 ug/l.
- Additional data are required to fully characterize the ecological effects of cyhexatin.

Required Unique Labeling Summary

All manufacturing-use and end-use cyhexatin products must bear appropriate labeling as specified in 40 CFR 162.10. In addition to the above, the following information must appear on the labeling:

- Manufacturing-use products must state that they are intended for formulation into other manufacturing-use products or end-use products for uses which are accepted by the Agency.
- Current labels must be revised to reduce the recommended spray gallonage and active ingredient per acre for pears, peaches, plums (prunes), and nectarines.
- Labels must be revised to incorporate the use of additional protective clothing such as mask or respirators and chemically resistant gloves.

Tolerance Assessment

- The Agency is unable to complete a full tolerance assessment for the established tolerances because of certain residue chemistry and significant toxicology data gaps.
- Established tolerances are published in 40 CFR 180.144 and they are:

Commodity	Parts Per Million
Almonds	0.5
Almonds, hulls	60
Apples	2
Cattle, fat	0.2
Cattle, kidney	0.5
Cattle, liver	0.5
Cattle, meat byproducts (mbyp)	0.2
Cattle, meat	0.2
Citrus fruits	2
Goats, fat	0.2
Goats, kidney	0.5
Goats, liver	0.5
Goats, mbyp	0.2
Goats, meat	0.2
Hogs, fat	0.2
Hogs, liver	0.5
Hogs, mbyp	0.2
Hogs, meat	0.2
Hops	30
Horses, fat	0.2
Horses, kidney	0.5
Horses, liver	0.5
Horses, mbyp	0.2
Horses, meat	0.2
Macadamia nuts	0.5
Milk, fat	0.05
Nectarines	4
Peaches	4
Pears	2
Plums (fresh prunes)	1
Sheep, fat	0.2
Sheep, kidney	0.5
Sheep, liver	0.5
Sheep, mbyp	0.2
Sheep, meat	0.2
Strawberries	3
Walnuts	0.5

- The data for cyhexatin residues in or on the following agricultural commodities are adequate to support the residue data requirements: hops, macadamia nuts, and strawberries.
- Additional residue data are required for the following commodities: peaches, plums, nectarines, apples, pears, almonds, almond hulls, walnuts, citrus fruits, dried hops and meat, milk, poultry, and eggs.
- Based on the established tolerances the theoretical maximum residue contribution (TMRC) for cyhexatin residues in the human diet is 0.33 mg/day (for a 60 kg person with a 1.5 kg

diet). However, this was based on an acceptable daily intake (ADI) which has been invalidated due to the lack of a sufficient chronic toxicology data base.
- Compatability of U.S. tolerances with Codex Maximum Residue Limits will be assessed when data have been submitted and evaluated.

4. SUMMARY OF REGULATORY POSITION AND RATIONALE

   - The Agency has determined that it should continue to allow the registration of cyhexatin. None of the criteria for unreasonable adverse effects listed in the regulations (§162.11(a)) have been met or exceeded. However, because of gaps in the data base a full risk assessment cannot be completed.
   - Also, a full tolerance reassessment cannot be completed because of major residue chemistry and toxicology data gaps. Until these gaps are filled cyhexatin will not be registered for significant new uses.
   - Available data are insufficient to fully assess the environmental fate of and the ecological effects from cyhexatin. Data are required to determine if cyhexatin will contaminate ground water.

5. SUMMARY OF MAJOR DATA GAPS

   - Additional crop residue studies on various commodities and plant and animal metabolism studies are required to support existing tolerances. The full compliment of chronic toxicology requirements are data gaps: chronic feeding, oncogenicity, reproduction, and teratology and mutagenicity.
   - The full compliment of environmental fate data requirements are data gaps. Studies on degradation (hydrolysis and photolysis), soil metabolism, mobility, dissipation, and accumulation are needed to fully characterize cyhexatin's environmental fate.
   - Additional data are required on avian toxicology (acute and subacute oral and reproduction) and freshwater and marine organism acute toxicology.
   - Other data gaps are product chemistry of technical cyhexatin, storage stability of residues, and acute and subchronic dermal toxicology.

CONTACT PERSON AT EPA

Jay S. Ellenberger
Product Manager (12)
Insecticide-Rodenticide Branch
Registration Division (TS-767)
Office of Pesticide Programs
Environmental Protection Agency
401 M Street SW.
Washington, DC 20460

Office location and telephone number:
Room 202, Crystal Mall Building #2
1921 Jefferson Davis Highway
Arlington, VA 22202
(703) 557-2386

DISCLAIMER: THE INFORMATION PRESENTED IN THIS CHEMICAL INFORMATION FACT SHEET IS FOR INFORMATIONAL PURPOSES ONLY AND NOT TO BE USED TO FULFILL DATA REQUIREMENTS FOR PESTICIDE REGISTRATION AND REREGISTRATION.

CYROMAZINE

Date Issued: December 1986
Fact Sheet Number: 105

1. DESCRIPTION OF CHEMICAL

    Generic Name: N-cyclopropyl-1,3,5-triazine-2,4,6 triamine

    Common Name: Cyromazine

    Trade Name: Larvadex® Technical and Premix (poultry feed through)
                    Trigard® (celery and lettuce)

    EPA Shaugnessy Code: # 121301

    Chemical Abstracts Service (CAS) Number: 66215-27-8

    Year of Initial Registration: 1984

    Pesticide Type: Insect Growth Regulator

    Chemical Family: Triazine

    Producer (U.S. only): Ciba-Geigy Corporation

2. USE PATTERN AND FORMULATION

    A. Application Site: Feed through for poultry - caged layers only. The active ingredient is passed through the chicken, leaving a residue in the manure that controls the growth of the fly larvae developing there.

        Type of Formulation: Premix - a coursely ground powder

        Application Rates: Mix 1 lb of Larvadex (0.3% Premix) per ton of feed for control of housefly and soldier fly; for lesser housefly, mix 3.33 lbs of Larvadex (0.3% Premix) per ton of feed.

B. Application Sites: Celery and lettuce

   Type of Formulation: Wettable Powder - this product is only packaged in 2 lb. water soluble bags

   Application Rates: One 2 lb. package mixed in the appropriate amount of water will treat 12 acres - this is 1/6 lb. Trigard 75W per acre.

3. SCIENCE FINDINGS

   Based on calculations and environmental chemistry data, there is no expectation for unreasonable adverse effects to nontarget organisms from the proposed uses of cyromazine. Sufficient data are available to support conditional registration of cyromazine as an insect growth regulator in and around poultry houses and for use on celery and lettuce.

   Chemical Characteristics of the Technical Chemical

   Physical state: crystalline solid
   Color: white
   Odor: odorless
   Boiling point: Not applicable
   Melting point: 220° to 222°C
   Flammability: Not applicable
   Solubility in water: 1.1%

   There are no unusual handling characteristics.

   Toxicology Characteristics

   Acute oral LD (rat); LD = 3387 mg/kg
     Toxicology Category III

   Acute oral LD (mouse); LD = 2029 mg/kg
     Toxicology Category III

   Acute oral LD (rabbit); LD = 1467 mg/kg
     Toxicology Category III

   Acute dermal LD (rat); LD >3100 mg/kg
     Toxicology Category III

   Primary dermal irritation (rabbit); PIS = 1.1/8.0
     Toxicology Category IV

Primary eye irritation (rabbit); Negative
  Toxicology Category IV

Skin Sensitization (guinea pig); nonsensitizing
  Toxicology Category: N/A

Mutagenic (Salmonella); Negative for point mutations in
            TA1535, TA1537, TA98, TA100 with
            and without activation

90-Day feeding (dog); NOEL = 300 ppm; LEL = 1000 ppm;
  levels tested = 0, 30, 300, and 3000 ppm.

6 months feeding (dog); NOEL = 30 ppm; LEL = 300 ppm;
  levels tested = 30, 300 and 300 ppm.

Teratology (rat); Negative for teratogenic effects up to
  and including 1000 mg/kg/day.

Teratology (rabbit); NOEL for teratogenic effects at
  5 mg/kg/day; Maternal NOEL at 10 mg/kg/day.

2-Generation reproduction: Systemic NOEL = 30 ppm;
  systemic LEL = 1000 ppm; levels tested = 0, 30, 1000
  and 3000 ppm.

2-Year feeding/Oncogenicity (rat); NOEL = 30 ppm;
  LEL = 300 ppm; negative for oncogenic effects at
  3000 ppm (HDT).

Mutagenic (dominant and nucleus); negative

Physiological and Biochemical Behavioral Characteristics

Mechanism of pesticidal action: Cyromazine is not a
cholinesterase inhibitor. As an insect growth regulator,
cyromazine exerts its toxic action by affecting the
nervous system of the immature stages (larvae) of certain
insects.

Environmental Characteristics

Cyromazine is practically nontoxic (acutely) to mammals
and birds. Exposure estimates for these organisms are
<0.05 ppm. Acute toxicity for birds is 1785 ppm maximum.
Safety factor is 10^5–10^6 for birds. Acute toxicity for
mammals is 1000 ppm maximum. Safety factor is again
10^5 – 10^6. A summary of the environmental chemistry factors
to be considered in this hazard assessment indicates that:

- Cyromazine is stable to hydrolysis; photolysis is the dominant physico-chemical degradation mechanism.

- Cyromazine is not expected to build up in soils; at 5 ppm in chicken manure (5 tons/A) <0.05 ppm cyromazine is expected in soils. Half lives in soil-manure are 1 year (lab) and 12 weeks (field/greenhouse).

- Cyromazine is not expected to build up in plants or crops: <0.05 ppm is expected from typical chicken manure applications.

- Cyromazine is not expected to build up in animal tissues; fish bioconcentration factors are expected to be between 0.2-7.8X.

- Cyromazine has no detrimental effects on soil microorganisms.

- Cyromazine has a low to moderate mobility in sandy or loam sand soils.

Ecological Characteristics:

Avian Acute Oral LD_{50}

Species	LD_{50}
Mallard duck	>2510 mg/kg
Bobwhite quail	1785 (1444-2206) mg/kg
Peking duck	>6000 mg/kg

Avian Dietary LC_{50}

Species	LD_{50}
Mallard duck	>5620 ppm
Bobwhite quail	>5620 ppm

Fish Acute LC_{50} (96-hr)

Species	96 hr LC_{50}
Bluegill sunfish	>89.7 mg/l
Rainbow trout	>87.9 mg/l
Channel catfish	>92.4 mg/l

Aquatic Invertebrate LC_{50}

Species	48 hr LC_{50}
Daphnia magna	>97.8 mg/l

Summary of Tolerances   40 CFR §180.414

Commodities	Parts Per Million
Eggs	0.25
Meat of Poultry	0.05
Fat of Poultry	0.05
Meat by-products of Poultry	0.05
Poultry feed (see 21 CFR)	5.0
Celery	10.0
Lettuce	5.0

4. SUMMARY OF REGULATORY POSITION

   Use classification: N/A

   Use, formulation or geographical restrictions: Larvadex has been approved for use in poultry feed for caged layers as a a feed through pesticide to control flies in the manure. Trigard has been approved for use on celery and lettuce as a conventional spray on emulsion to control leafminers. Products containing cyromazine are to be stored in cool, dry places. There are no geographical restrictions or unique warning statments required on the label.

5. SUMMARY OF MAJOR DATA GAPS

   Toxicology: None

   Environmental Fate: Applicator Exposure Study

   Environmental Safety: None

   Residue Chemistry: None

6. CONTACT PERSON AT EPA

   Arturo E. Castillo, PM-17
   Insecticide-Rodenticide Branch (TS-767)
   401 M Street SW.
   Washington, DC 20460
   (703) 557-4414

   DISCLAIMER: The information presented in this Chemical Information Fact Sheet is for informational purposes only and may not be used to fulfill data requirements for pesticide registration and reregistration.

DAMINOZIDE

Date Issued: June 30, 1984
Fact Sheet Number: 26

Description of chemical

Common Name: Daminozide
Chemical Name: Butanedioic acid mono(2,2-dimethylhydrazine),
              Succinic acid 2,2-dimethylhydrazide
Trade Name: Alar, Kylar, SADH, B-nine, B-995, aminozide,
EPA Shaughnessy Code: 035101
Chemical Abstracts Service (CAS) Number: 1596-84-5
Pesticide Type: Plant Growth Regulator
Chemical Family: Amino Acid Derivative
U.S. and Foreign Producers: Uniroyal and Aceto Chemical

Use Patterns and Formulations

Daminozide is a plant growth regulator registered as a 5 percent water soluble liquid or 85 percent water soluble dry concentrate formulation. The amount of daminozide that is applied as a field spray ranges from 0.9 to 6.8 pounds active ingredient per acre per year, depending of the crop, time of application and desired effects. Daminozide controls the vegetative and reproductive growth of orchard crops such as apples, cherries, nectarines, peaches, prunes and pears. In addition, daminozide use enhances shorter and more erect peanut vines or modifies the stem length and shape of ornamental plants. Other minor uses of daminozide include: Brussels sprouts (California), cantaloupes (California and Arizona), grapes, and tomatoes.

Scientific Findings

o Summary Science Statement

Daminozide is a white, water soluble solid. Data indicate that daminozide has low acute toxicity, low dermal irritation potential and is neither teratogenic nor mutagenic. Daminozide and its UDMH contaminant cause oncogenic effects. A tolerance reassessment cannot be performed at this time. Daminozide leaches from soil, but is not persistent. Daminozide is not an acute toxicant to fish and wildlife.

Chemical Characteristics

Daminozide is a white, crystalline solid with slight to no odor. Daminozide is soluble in water, methanol and acetonitrile, but insoluble in xylene and aliphatic hydrocarbons. Daminozide has a melting point range from 154 to 156°C. Technical daminozide contains at least 99 percent active ingredient.

Toxicological Characteristics

The LD50 and Toxicity Categories for daminozide are: acute oral (8.4 g/kg, IV), acute dermal (>16 g/kg, III), acute inhalation (>147 mg/kg, IV), primary eye irritation (mild, none at this time), dermal irritation (mild, IV). Daminozide does not produce mutagenic or teratogenic effects. Data are insufficient to judge the effects of daminozide on reproduction. Daminozide causes oncogenic effects in laboratory animals.

Physiological and Biochemical Behavioral Characteristics

Data indicate that daminozide is rapidly absorbed through the leaves, roots and stems. Daminozide is translocated in plants and can accumulate in roots, fruit, etc. Adequate methods are available to detect daminozide. A method to detect the UDMH metabolite down to 1 ppb must be validated to confirm the presence of UDMH residues in plants. Components of the final residues have not been adequately identified or quantified. The majority of daminozide residues ingested by milk animals is rapidly excreted in the urine and feces.

Environmental Characteristics

Degrades in water to unsymmetrical 1,1-dimethylhydrazine (UDMH), a known oncogen. Daminozide appears to resist photodegradation, but is degraded by soil microorganisms. Daminozide appears to leach, but since it does not persist in soil, the potential for ground water contamination is small. Daminozide does not bioconcentrate in fish nor does it accumulate in rotational crops.

Ecological Characteristics

Daminozide has low acute toxicity to fish and terrestrial wildlife. No data are available to assess the ecological hazard from the UDMH hydrolysis product/contaminant. Problems with Endangered Species: None known at this time.

Tolerance Assessment

A final reassessment of all tolerances cannot be made at this time until the data gaps specified by the Standard are filled.

o Problems with Use

Extended storage of solutions of daminozide result in excessive hydrolysis of the active ingredient to UDMH.

Regulatory Position & Rationale

o Use Clasiification

Daminozide is classified as a General Use Pesticide.

o Use Restrictions

None.

o Unique Warning Statements

Solutions of daminozide must be used within 24 hours after preparation.

o Benefit Analysis

Approximately 825,000 pounds of daminozide are produced annually with apples and peanuts accounting for 600,000 pounds and 225,000 pounds of the annual usage, respectively. Without daminozide, short term revenue losses are projected to range up to $30 million annually for apples and from $4.3 to $10.7 million annually for peanuts.

o Risk Analysis

Significant exposure to daminozide and UDMH can occur via consumption of raw and processed agricultural commodities treated with daminozide. The Agency's preliminary estimate of oncogenic dietary risk for daminozide is high. There are insufficient data to quantify the oncogenic dietary risk of UDMH at this time. The oncogenic nondietary risk for daminozide and UDMH may not be significant.

o Special Review

Registrants of daminozide products are notified, via the Guidance Document, that daminozide meets the oncogenicity risk criterion in 40 CFR 162.11(a) and will undergo a Special Review. The Agency will not reregister any current products and it will not register any new products containing daminozide until Special Review is completed and the Agency has received commitment to fulfill data requirements.

Summary of Major Data Gaps

Data gaps and time [in months] allowed to perform studies: toxicology (chronic testing [48], teratology [12], reproduction [24], general metabolism [12] and mutagenicity [6]), product chemistry (product identity, analysis and certification of product ingredients, physical and chemical characteristics) [6], environmental fate (degradation [6], photodegradation [6], metabolism [24], mobility [6], dissipation [24], accumulation [24] and reentry [24], residue chemistry (metabolism in plants and animals [12], analytical methods and residue data [12], residue data [12]), and ecological effects (avian and mammalian testing and aquatic organism testing) [48].

Contact Person

Robert Taylor, PM 25
Registration Division (TS-767C)
Environmental Protection Agency
401 M Street, SW
Washington, D.C. 20460
(703)-557-1800

Disclaimer

The information presented in this Chemical Information Fact Sheet is for informational purposes only and may not be used to fulfill data requirements for pesticide registration and reregistration.

DANTOCHLOR

Date Issued: September 12, 1984
Fact Sheet Number: 33

1. Description of chemical

Generic Name: Chlorinated Hydantoin Derivative
Common name: 1,3-dichloro-5-ethyl-5-methylhydantoin
Trade name: Dantochlor
EPA Shaughnessy code: #128826
Chemical Abstracts Service (CAS) number·
Year of initial registration: 1984
Pesticide type: Disinfectant (Slimicide, Algaecide, and Bactericide)
Chemical family: Hydantoins
U. S. and foreign producers: Glyco Inc.

2. Use patterns and formulations

Application sites: recirculating cooling water (Cooling Towers and air washers)
Types of formulations: Tablets (Briquettes)
Types and Methods of application: Manual
Application rates: 0.1 to 0.125 lb. per 1000 gal per day=12.90 ppm
Usual carriers: usually marketed in concentrated form with other related chlorinated and brominated hydantoins

3. Science Findings

Chemical Characteristics: solid, white powder, granules or briquettes, slight chlorine odor, melting point 106-130°c, Strong oxidizing agent, corrosive to tissues, and metals, capable of spontaneous reaction with water, organic contaminants.

Toxicology characteristics: Major routes of exposure in order of toxicological significance.

> Toxicity Category and values for each acute hazard: acute Oral-Category 3, Eye Irritation Category 1, Acute Dermal-Category 3, Skin Irritation-Category 1.

Chronic toxicology results.

Mutagenicity	no adverse efect was
Cytogenicity	shown in these studies
Skin Absorption Study	for the proposed uses
Exposure Assesment to address risk	

Physiological and Biochemical Behavioral Characteristics:

    Foliar absorption: N/A
    Transolocation: N/A
    Mechanism of pesticidal action: Hydrolyses in water to yield hypochlorous acid, (which is the disinfecting agent) and hydantoin.

    Metabolism and persistence in plants and animals. N/A

Environmental Characteristics:

    Absorption and leaching in basic soil types: N/A
    Microbial breakdown. N/A
    Loss from photodecomposition and/or volatization: neglible
    Bioaccumulation: N/A
    Resultant average persistance: N/A

Ecological Characteristics:

    Hazards to fish and wildlife (avian oral and dietary LD_{50}, fish LC_{50} and aquatic invertebrate LC_{50}).
    Potential problems related to endangered species: There are potential risks associated with the use of this product to aquatic organisms. NPDES permits are required and may mitigate hazards.

Tolerance assessments: No crop applications

Problems which are known to have occurred with use of the chemical (e.g., P.I.M.S. data). None

Summary of Regulatory Position and Rationale:

    Use Classification: General Use
    formulation or geographical restrictions: not to be registered for use in Spas, Hot Tubs, and swimming pools until additional chronic studies are submitted and accepted.
    Unique warning statements required on labels: None

Regulatory decisions in cases where risk assessments were conducted, including consideration of benefits:

Benefits exceeds risks for the use of the product in air recirculation water washers and cooling towers.

Summary of major data gaps and dates when these major data gaps are due to be filled: There are no data gaps for the approved uses.

Contact person at EPA (A. E. Castillo, 401 M St., SW; Washington, DC. 20460 TS-767C, 703-557-3964.

DISCLAIMER: The information presented in this Chemical Information Fact Sheet is for informational purposes only and may not be used to fulfill data requirements for pesticide registration and re-registration.

DCNA

Date Issued: January 9, 1984
Fact Sheet Number: 13

1. Description of the Chemical:

    Generic name: 2,6-dichloro-4-nitroaniline
    Common name: dicloran
    Trade name(s): DCNA, botran, ditranil, allisan, and resisan
    EPA Shaughnessy code: 031301
    Chemical abstracts service (CAS) number: 99-30-9
    Year of initial registration: 1961
    Pesticide type: Fungicide
    Chemical family: Nitroanaline
    U.S. and foreign producer: Upjohn Company

2. Use patterns and formulations:

    Application sites: DCNA is used to control a variety of pre- and post-harvest diseases on fruit and vegetable crops. Current major use sites include peaches, grapes, lettuce and celery. It is a protectant to ornamentals and vegetable seeds. It is also registered for use on cotton, a number of ornamentals, a seed-piece dip for sweet potatoes, a peanut seed treatment, and in greenhouses on cucumbers, lettuce, rhubarb, and tomatoes.
    Types of formulations: Formulated products are 4, 5, 6, 8, 10, 12, 15, 20, 30, 35, or 50% dusts; 48.8 or 75% wettable powders; 9 or 30% flowable concentrates: as 0.5 or 3% ready-to-use liquids; a 3% formulation in fruit wax; and a 0.2% impregnated fruit wrap.
    DCNA may be formulated with other pesticides such as captan, benomyl, and parathion. It is not compatable with some oil-based pesticides.
    Types and Methods of application: DCNA may be applied as pre-harvest and post-harvest uses on fruit and vegetable crops; seed and ornamental protectant. In the field, DCNA can be applied by ground and aircraft. DCNA can be applied as a post-harvest dip, spray, or dust to some fruits and vegetables and as a protectant to ornamental crops prior to storage and shipment.
    Application rates: 1.5 to 30 lb/A.

    Usual Carriers: either water or wax as diluent

3. Science Findings:

   Summary science statement:
   There are extensive data gaps for DCNA. No human toxicological hazards of concern, other than inducement of skin photosensitivity, and possible ocular toxicity, have been identified in studies reviewed by the Agency for this standard. The Agency has no information that indicates continued use will result in any unreasonable adverse effects to man or his environment during the time required to develop the data.

   Chemical characteristics:
   DCNA is a yellow, crystalline powder. The solubility of DCNA in ethanol is 0.2% at 20°C and its melting point is 192-194°C. The chemical does not present any unusual handling hazards.

   Toxicology characteristics:
   Acute toxicology studies:
   Acute Oral LD_{50} in rats: >10,000 mg/kg, Toxicity category IV
   Acute Dermal LD_{50} in rabbits: >2.0 g/kg, Toxicity category III
   Acute Inhalation LC_{50} in rats: >2 mg/l, Toxicity category III
   DCNA does not induce skin or eye irritation - Toxicity category III
   DCNA was found to be a skin sensitizer and may induce phototoxicity due to the presence of aniline in its compostion.
   Chronic toxicology studies:
   There is insufficient data to assess the subchronic dermal or subchronic inhalation hazard
   Chronic toxicity in rats: NOEL = 100 ppm: LEL = 3000 ppm
   Oncogenicity study in rats: no tumors at 3000 ppm
   Teratogenicity in rabbits: no abnormalities at 1000 ppm
   Reproduction in rats: NOEL = 100 ppm

   Major routes of exposure: Subchronic dermal or respiratory contact

   Environmental characteristics:
   Data are insufficient to fully assess the environmental fate of DCNA
   Ecological characteristics:
   Hazards to fish and wildlife:
   Avian dietary LC_{50}: Mallard duck - 9500 ppm
   Bobwhite quail - 2120 ppm
   Fish LC_{50}: >1.08 ppm
   Aquatic invertebrates LC_{50}: 2.3 ppm
   Potential problems for endangered species: Will be reassessed after review of environmental fate data.
   Physiological and biochemical behavioral characteristics:
   Mechanism of pesticidal action: Thought to be a non-specific inhibitor of cell division and can effect nuclear stability.
   General metabolism in rats: 1.7 and 8 mg/kg body weight was absorbed, no body tissue accumulation was detected.

Tolerance assessment:
: List of crops and tolerances (in ppm): apricots (pre and post H) 20, snapbeans 20, blackberries 15, boysenberries 15, carrots (post H) 10, celery 15, sweet cherries (pre and post H) 20, cotton seed 0.1, cucumbers 5, endive 10, garlic 5, grapes 10, kiwifruit (post H) 20, lettuce 10, nectarines (pre and post H) 20, onions 5, peaches (pre and post H) 20, plums (fresh prunes) (pre and post H) 15, potatoes 0.25, raspberries 15, rhubarb 10, sweet potatoes (post H) 10, tomatoes 5.
: Tolerance reassessment: Due to the absence of pertinent data, the Agency is unable to complete its reassessment of DCNA tolerances.

4. Summary of Regulatory Position and Rationale:

   Use classification: general use
   Unique label warning statements: Manufacturing-use labels must contain the statements "Do not discharge into lakes, streams, ponds or public waters unless in accordance with NPDES permit. For guidance contact your regional office of EPA."

5. Summary of Major Data Gaps:

   Product chemistry: data due 7/84
   : Identity of ingredients
   : Statement of composition
   : Discussion on formation of contaminants
   : Preliminary analysis
   : Certification of limits
   : Analytical methods
   : Odor
   : Density, bulk density, or specific gravity
   : Solubility
   : Vapor pressure
   : Dissociation constant
   : Octanol/water partition coefficient

   Toxicology: studies due 1/88
   : Ocular toxicity
   : Photosensitization
   : 21-day subchronic dermal
   : Oncogenicity in mouse (in progress)
   : Teratogenicity (in progress)
   : Mutagenicity testing

   Wildlife and Aquatic organisms: studies due 1/88
   : Single-dose oral LD_{50} on one avian species
   : Acute 48 hour toxicity study on a freshwater invertebrate
   : Fish embryo-larvae and/or aquatic field studies may be required depending on environmental fate data

Environmental Fate: studies due 1/88
  Hydrolysis studies
  Photodegradation studies in water
  Photodegredation studies on soil
  Aerobic soil metabolism study
  Anaerobic soil metabolism study
  Leaching and adsorption/desorption studies
  Laboratory volatility studies
  Terrestrial field dissipation studies
  Longterm field dissipation studies
  Confined accumulation studies on rotational crops
  Laboratory studies of pesticide accumulation in fish

Rentry Protection:
  Data requirement pending on results of toxicological testing

Residue Chemistry: studies due 1/88
  Nature of residue for plants and animals
  Residue analytical method for animals
  Storage stability data
  Crop field trials for potatoes, sweet potatoes, onions,
    apricots, cherries, nectarines, peaches, plums, black-
    berries, boysenberries, raspberries, kiwi
  Processed food/feed for potatoes, tomatoes, cottonseed

6. <u>Contact person at EPA</u>

   Mr. Henry M. Jacoby
   Environmental Protection Agency
   Fungicide-Herbicide Branch (TS 767-C)
   401 M St., SW
   Washington, D.C. 20460
   (703) 557-1900

DISCLAIMER: The information presented in this Chemical Information Fact Sheet is for informational purposes only and may not be used to fulfill data requirements for pesticide registration and reregistration.

DEMETON

Date Issued: February 27, 1985
Fact Sheet Number: 45

1. Description of chemical

    Chemical Name: Mixture of 2 isomers consisting of:
    O,O-diethyl O-[2(ethylthio) ethyl] phosphorothioate and
    O,O-diethyl S-[2ethylthio)ethyl] phosphorothioate.
    Common Names: Demeton-O + Demeton-S, mercaptophos and
    mercaptophos teolevy (USSR)
    Trade Name: Systox ®
    EPA Shaughnessy No: 057601
    Chemical Abstracts Service (CAS) Number: 8065-48-3
    Year of inital registration: July 15, 1955
    Pesticide type: systemic acaricide/insecticide
    Chemical family: organophosphate
    U. S. Producer: Mobay Chemical Corporation

2. Use Patterns and formulations

    Application sites: vegetable, field, orchard, and ornamental
    (including greenhouse).
    Types of formulations: liquid and granular
    Types/methods of application: ground or air.

3. Science Findings

    Chemical characteristics: light brown liquid; odor characteristic
    of sulfur compounds; soluble in most organic solvents;
    subject to hydrolysis under alkaline conditions; molecular
    weight: 258.32.

    Toxicity characteristics: Toxicity Category I by dermal
    route of exposure (14mg/kg for male rats and 8.2 mg/kg for
    female rats; Toxicity Category I by the oral route of
    exposure (6.2 mg/kg for male rats and 2.5 mg/kg for female
    rats); positive in mutagenicity studies _in vitro_ in cells;
    data gaps exist in the area of neurotoxicity; chronic
    toxicity, oncogenicity, teratogenicity, and reproduction.
    A gene mutation assay in mammalian cells in culture and a
    chromosome aberration assay _in vivo_ are required to be performed
    to assess the mutagenic potential of demeton.

Physiological and Biochemical Behavioral Characteristics: readily absorbed and translocated by plants; cholinesterase inhibitor.

Environmental Characteristics: no data are available to assess the environmental fate of demeton; no data are available to assess demeton's potential for contaminating groundwater.

Ecological Characteristics: highly toxic to birds (7.19 mg/kg for mallard duck and 8.21 mg/kg for pheasant); highly toxic to fish (0.1 ppm for bluegill sunfish and 0.6 ppm for rainbow trout); very highly toxic to freshwater invertebrates (0.014 ppm for daphnia pulex); special tests to monitor the residues of demeton on avian feed items and aquatic sites are required; interim labeling to protect endangered species to be imposed in time for the 1986 growing season if the generic (cluster) analysis has not been completed.

Efficacy review results, where conducted: NA

Tolerance assessments: refer to the attached table for the list of current tolerances established for demeton; available data are not sufficient to conduct a full tolerance assessment.

4. Summary of Regulatory Position and Rationale

Use classification: all end-use products containing demeton shall continue to be classified for restricted use.
Formulation or geographical restrictions: none
Unique warning statements required on labels: end-use (EP) products require the use of protective clothing, rubber gloves, rubber overshoes and goggles; reentry of 48 hours, and a crop rotation restriction.

5. Summary of Major Data Gaps:

Date Due

Toxicology:

		Date Due
82-1	90-Day Feeding-Rodent, Non-rodent	June, 1986
	90-Day Feeding-Rat-Thiol Sulfoxide	January, 1986
82-2	21-Day Dermal Toxicity-Rabbit	August, 1985
81-7	Delayed Neurotoxicity-Hen	August, 1985
83-1	Chronic Toxicity	March, 1985
83-2	Oncogenicity	December, 1987
83-3	Teratogenicity	December, 1985

83-4	Reproduction	December, 1987
84-2	Gene Mutation Assay in Mammalian Cells in Culture	August, 1985
	Chromosome Aberration Assay *in vivo*	August, 1985
85-1	General Metabolism	March, 1986

Environmental Safety:

70-1	Special Test - Monitoring of residues on avian feed items	September, 1986
70-1	Special Test - Monitoring of residues in aquatic sites	September, 1986
70-3	Acute Toxicity to Estuarine and Marine Organisms	September, 1986

Environmental Fate:

161-1	Hydrolysis	March, 1985
161-2	Photodegradation in Water	March, 1985
161-3	Photodegradation in Soil	March, 1985
161-4	Photodegradation in Air	August, 1985
162-1	Aerobic Soil Metabolism	June, 1986
162-2	Anaerobic Soil Metabolism	June, 1986
162-3	Anaerobic Aquatic Metabolism	June, 1986
163-1	Leaching and Adsorption Desorption	March, 1985
163-2	Volatility (Lab)	August, 1985
164-1	Soil Dissipation	June, 1986
164-3	Forestry	March, 1987
165-1	Rotational Crops (Confined)	March, 1987
165-2	Rotational Crops (Field)	March, 1988
165-4	Accumulation in Fish	February, 1988
132-1	Reentry Data	March, 1988
201.1	Droplet Size Spectrum Testing and Drift Field Evaluation	February, 1987

Residue Chemistry:

171-4	Residues in Livestock	February, 1987
171-4	Residues Analytical Method	February, 1987
171-4	Storage Stability	February, 1987
171-4	Residue Data on Crops	February, 1987

Product Chemistry:

61-2	Descripton of Beginning Materials and Manufacturing Process	March, 1985/ August, 1985
61-3	Discussion of the Formation of Impurities	-do-

62-1	Preliminary Analysis	March, 1985/ August, 1985
62-2	Certification of Ingredient Limits	-do-
62-3	Analytical Methods to Verify Certified Limits	-do-
63-	Physical/Chemical Properties	-do-

6. <u>Contact Person at EPA:</u>

William H. Miller, (PM-16)
Insecticide-Rodenticide Branch (TS-767)
401 M Street, SW.
Washington, DC 20460
Tel. No. (703) 557-2600

DISCLAIMER: The information presented in this Chemical Information Fact Sheet is for informational purposes only and may not be used to fulfill data requirements for pesticide registration and reregistration.

DIAZINON

Date Issued: September, 1986
Fact Sheet Number: 96

1. Description of chemical

    Chemical name: O,O-Diethyl O-(2-isopropyl-6-methyl-4-pyrimidinyl) phosphorothioate

    Common name: Diazinon

    Trade name: Spectracide, AG500, Alfa-tox, Sarolex, D-Z-N Diazinon 14G, Geigy Spectracide Lawn and Garden Insect Control, etc.

    EPA Shaughnessy code: 057801

    Chemical abstracts service (CAS) number: 333-41-5

    Year of Initial Registration: 1952

    Pesticide type: Insecticide

    Chemical family: Organophosphate

2. Use patterns and formulations

    Application sites: Field, fruit, nut, vegetable (including seed treatment) and nonfood crops (ornamentals and tobacco); forestry (including Christmas tree plantations); greenhouse food crops (vegetable bedding plants and ornamentals); livestock; range, pasture, and grassland; animal premises; lawns and turf; domestic outdoor and indoor (household); commercial indoor (including food handling establishments and processing plants); commercial and industrial outdoor sites.

    Types of formulations: Wettable powder, wettable powder/dust, emulsifiable concentrate, dust, microencapsulate, soluble concentrate, granular, oil solution, aerosol spray, spray concentrate, impregnated materials, soluble concentrate, liquid ready-to-use, and pressurized liquid.

Types and methods of application: Ground, aerial, and those methods unique to the formulation, such as aerosols.

3. **Science Findings**

    Physical and Chemical Characteristics-

    Physical state: Liquid
    Color: Colorless (Technical is amber to brown)
    Odor: Typical of organophosphates
    Boiling point: 83-84° C
    Melting point: Not applicable
    Flash point: 82° F for AG500
    >105° F for 4E and 4S
    Human Toxicology Characteristics-

    Acute rat oral LD_{50}: 66-635 mg/kg for females and 96-967 mg/kg for males, Toxicity Category II
    Acute rabbit Dermal LD_{50}: >2,000 mg/kg, Toxicity Category III
    Acute rat inhalation LD_{50}: 3.5 mg/l, Toxicity Category III
    Dermal rabbit irritation: Toxicity Category IV
    Eye rabbit irritation: data gap

    Ecological characteristics-

    Hazards to aquatic invertebrates and wildlife:

    Avian subacute dietary LC_{50} (ppm)
    191 for Mallard Ducks
    245 for Bobwhite Quail

    Avian acute oral LD_{50} (mg/kg)
    3.5 for Mallard Ducks
    10 for Bobwhite Quail

    These values characterize diazinon as very highly toxic to birds.

    Aquatic invertebrate LC_{50}
    0.079 ppm for bluegill sunfish
    0.635 ppm for rainbow trout
    0.522 ppb for *Daphnia* sp.

    These values characterize diazinon as very highly toxic to fish and aquatic invertebrates.

4. Summary of regulatory position and rationale

    On January 15, 1986, a FEDERAL REGISTER Notice was published concerning the Special Review of all pesticide products containing diazinon registered for use on golf courses and sod farms. The Special Review was based on the hazard to non-target birds from diazinon's use on these two sites. The Notice announced (1) the initiation of the Special Review on these two sites, (2) the Preliminary Determination proposing to cancel registrations and deny applications for diazinon products used on these two sites, and (3) the availability of the Support Document. The Support Document contained a risk/benefit analysis, which was the basis for the Agency's action.

In evaluating the hazard to birds, the Agency considered (1) acute toxicity studies which indicated that diazinon is very highly toxic to birds, (2) residue level and dose estimates on grass and seed which indicated a potential hazard, and (3) bird kills reported to the Agency in which diazinon was either confirmed or implicated as the primary cause. These kills have involved 23 species of birds and occurred throughout the country and throughout the year.

The Agency's concern for the hazard to non-target birds included a concern for the impact on populations of species at risk. The Agency reviewed information which indicated that diazinon caused a reduction of a local population of Atlantic Brant Geese when applied according to label directions to a golf course in New York.

The Agency also reviewed information concerning the avian and human hazards of the five major alternatives. Based on a comparative avian hazard assessment, the Agency determined that the major alternatives are not likely to be of greater hazard to birds than diazinon. Based on the available data concerning the hazard to humans, the Agency determined that the alternatives do not appear to pose a greater human health hazard than diazinon.

In the Support Document the Agency also reviewed the benefits of diazinon on golf courses and sod farms. Estimates indicate that at least 512,000 pounds are used per year on golf courses and 60,000 pounds on sod farms. The impact of cancellation that results would be a cost increase of $937,200 for golf courses and $300,000 for sod farms. The Agency anticipates that these impacts would be minor when compared to the maintenance cost of $1,900,000,000 for golf courses and the gross revenue of $210,000,000 on sod farms.

In weighing risks and benefits, the Agency reviewed a number of options to reduce the risk to birds. However, the Agency determined that cancellation was the only option that would reduce the hazard to birds adequately. The Agency concluded that the risks outweigh the benefits and proposed cancellation of all products registered for use on sod farms and golf courses.

During the comment period that followed publication of the FEDERAL REGISTER Notice, the Agency received additional information on the hazard to birds, which included reports of 26 additional bird kills. These kills once again confirm that the hazard from diazinon is widespread throughout the country and throughout the year. In addition the Office of Endangered Species, U.S. Department of the Interior, commented that certain endangered species could be seriously affected by the use of diazinon on golf courses and sod farms.

The Agency also received data concerning diazinon residues on grass and effects on Canada Geese penned on turf. The data demonstrated that birds foraging on treated turf would be exposed to lethal diazinon residues within a very short period of time and that these residues would be high enough to cause death to foraging waterfowl.

During the comment period the Agency also received information regarding the benefits of diazinon use, which included efficacy data on diazinon and its alternatives. The results indicated that the efficacy of the major alternatives is about the same as diazinon.

The Office of Pesticide Programs (OPP) transmitted the Support Document to the Scientific Advisory Panel (SAP) and the U.S. Department of Agriculture (USDA). The SAP's comments supported the Agency's conclusions concerning the avian hazard from diazinon application to golf courses and sod farms. The USDA commented that the proposed cancellation action was premature and could be inappropriate. The FEDERAL REGISTER Notice responds point-by-point to each of their comments.

The Agency also received 96 comments during the public comment period, most of which supported the proposed action.

The Agency carefully reviewed all of the new information and all of the comments that were submitted. The Agency still concludes that the hazard to birds from diazinon use on golf courses and sod farms outweighs the minor benefits and that cancellation is the only appropriate action.

5. Summary of major data gaps

The Agency is concerned about the hazard to birds from diazinon use on other sites. The data base is inadequate to evaluate the hazard on all of the remaining sites at this time. Consequently, the Agency is requiring the data necessary to fully evaluate this problem through the reregistration process. A Registration Standard is scheduled to be completed by the end of this year.

6. Contact person at EPA: Ingrid M. Sunzenauer
EPA
Office of Pesticide Programs
Registration Division (TS-767C)
401 M Street, S.W.
Washington, D.C. 20460

DISCLAIMER: The information in this Chemical Information Sheet is for informational purposes only and may not be used to fulfill data requirements for pesticide registration or reregistration.

3,5-DIBROMO

Date Issued: September 30, 1985
Fact Sheet Number: 67

1. Description of Chemical

    Generic Name: 3,4',5 Tribromosalicylanilide
    Common Name: Tribromsalan, TBS
    Trade Name: Temasept IV
    EPA Shaughnessy Code No: 077404
    Chemical Abstracts Service No: 87-10-5
    Generic Name: 3,5 Dibromosalicylanilide
    Common Name: N/A
    Trade Name: N/A
    EPA Shaughnessy Code No: 077405
    Chemical Abstracts Service No: 2577-72-2
    Generic Name: 4',5 Dibromosalicylanilide
    Common Name: N/A
    Trade Name: N/A
    EPA Shaughnessy Code No: 077402
    Chemical Abstracts Service No: 87-12-7
    Year of Initial Registration: July 6, 1964
    Pesticide Type: Antimicrobial, Preservative
    Chemical Family: Brominated Salicylanilides
    U.S. Producers: Hexcel Corporation, Sherwin Williams Chemicals

2. Use Patterns and Formulations

    Application Sites: Hard Surfaces, Laundry, Textiles, Manufactured Products
    Types of Formulations: Solid, Solutions, Sprays
    Application Rates: N/A
    Usual Carriers: Soap

3. Science Findings

    Chemical Characteristics:

    Physical State: Powder
    Color: White
    Odor: None
    Melting Point: 227-228°
    Density: No Data Available
    Solubility: Soluble (0.5-4.0%) in methanol, ethanol, isopropanol, ethyl ether and benzene; sparingly soluble (0.01-0.5%) in chloroform and cyclohexane; insoluble in water and petroleum ether.

Toxicity Characteristics:

There are no available data to assess the potential toxicity of the brominated salicylanilides.

Physiological/Biochemical Behavioral Characteristics: N/A

Environmental Characteristics: Indoor uses only; limited exposure is possible based on the registered uses of these products as disinfectants, laundry additives, textile preservatives and manufactured products.

Ecological Characteristics: Indoor uses only; limited exposure is possible based on the registered uses of these products as disinfectants, laundry additives, textile preservatives and manufactured products.

Efficacy Data: Product specific microbiological efficacy data will be required for all health related uses.

4. **Summary of Regulatory Position and Rationale**

No data are available to establish whether or not the risk criteria listed in Section 162.11(a) of Title 40 of the U.S. Code of Federal Regulations have been met or exceeded.

Unique Warning Statements: None

Tolerance Assessments: N/A The registered uses of the brominated salicylanilides do not include direct application to a food or feed crop.

Problems which are known to have occurred with use of the chemical (e.g., PIMS): None.

5. **Summary of Major Data Gaps:**

Toxicology, environmental fate, fish and wildlife and product chemistry data are required to assess the hazard associated with the registered uses of these chemicals.

6. Contact person at EPA: A.E. Castillo
   Product Manager 32
   Disinfectants Branch
   Registration Division (TS-767C)
   Telephone: (703) 557-3964

DISCLAIMER: The information presented in this Chemical Information Fact Sheet is for informational purposes only and may not be used to fulfill data requirements for pesticide registration and reregistration.

DICAMBA

Date Issued: October 17, 1983
Fact Sheet Number: 8

1. Description of Chemical:

    Generic name: 3,6-dichloro-o-anisic acid
    Common name: Dicamba
    Trade Names: Banvel, Banex, Brush Buster, Mediben, Velsicol 58-CS-11
    EPA Shaughnnessy Number: 029802
    Chemical Abstracts Service (CAS) Number: 1918-00-9
    Year of initial registration: 1967
    Pesticide Type: Herbicide
    Chemical family: Benzoic acid
    U.S. and foreign producers: Velsicol Chemical Corporation

2. Use Patterns and Formulations:

    Application sites: corn, small grains, grain sorghum, asparagus, sugarcane, pastures, rangeland and agricultural seed crops, noncrop sites, forest lands, lawns and ornamental turf

    Types of formulations: diethanolamine, monoethanolamine, dimethylamine and sodium salts as soluble concentrates or granulars.

    Types and methods of application: Applied by aerial or ground spray, invert system, tree injection or granular equipment. Dicamba is applied preplant, preemergence or postemergence.

    Application rates:

    1/4 pound active per acre to grain sorghum
    1/8-1/2 pound active per acre to small grains, asparagus
    1/4-3 pounds active per acre to sugarcane
    1/2-8 pounds active per acre to pasture, range and noncropland
    1/4-1 pound active per acre to turf and grass seed crops

    Usual carriers: water, fluid and dry fertilizer, oil in water emulsions, clay or vermiculite

3. Science Findings:

    Summary Science Statement:

    Dicamba appears to pose little acute toxicity or environmental hazard. The major problem appears to be the potential for a dimethylnitrosamine (DMNA) contaminant in the dimethylamine formulations. The level of DMNA is expected to be below 1 ppm and the risk level for dicamba with DMNA is 10^{-7} to 10^{-8} range.

Dicamba 255

Chemical Characteristics:

    It is a light tan slightly phenolic crystalline solid. It is stable to oxidation and hydrolysis and melts at temperatures between 90-100°C. Dicamba is nonflammable and does not present any unusual handling hazards.

Toxicology Characteristics:

    Acute Toxicology Results:

        Oral LD_{50} in rats: 2.74 mg/kg body weight. Toxicity Category III

        Dermal LD_{50} in rats >2,000 mg/kg, Toxicity Category IV

        Inhalation LC_{50} in rats: >200 mg/l, Toxicity Category IV

        Eye irritation in rabbits: Induced corrosiveness of conjunctival tissues and corneal injury which was reversible in 72 hours. In a recent study eye damage was irreversible and pannus was observed. Toxicity Category I.

        Dermal Irritation: slight dermal irritation.

    Chronic Toxicology Results:

        Teratology in rabbits: NOEL of 3.0 mg/kg/day for maternal toxicity; not teratogenic.

        Teratology in rats: Teratology NOEL - 400 mg/kg: maternal toxicity NOEL - 160 mg/kg

        Three-generation reproduction study in rats: No evidence of toxicity among the rats from any of the generations in the study. No test article related effects were evident for any reproductive indices examined. NOEL of 25 mg/kg/day.

        90-day subchronic feeding study with rats: The NOEL is 250 mg/kg/day. LEL was 500 mg/kg/day (slight decrease in comparative body weight gains and food consumption and evidence of reduced glycogen storage).

Major Routes of Exposure:

    Dermal and inhalation exposure to humans may occur during application, particularly via splashing during dilution, mixing and loading. Application by aircraft increases the potential for exposure of humans, livestock, and wildlife due to spray drift and ventilation.

Risk Assessment and Contaminants:

> The manufacturing process for dicamba has potential of resulting in traces of 2,7-dichlorodibenzo-p-dioxin as a contaminant. It is present at levels up to 50 ppb (parts per billion). The more toxic dioxin isomer, 2,3,7,8-tetrachlorodibenzo-p-dioxin, has not been found at the limit of detection (2 ppb) of the method and is not expected as an impurity in dicamba.
>
> Dicamba products formulated with the dimethylamine salt have the potential of adding a dimethylnitrosamine (DMNA) contaminant. Nitrosamine levels in the dimethylamine formulations are expected to be less than 1 ppm. The risk levels for the dicamba products with the nitrosamine contaminant are in the 1×10^{-7} to 1×10^{-8} range.
>
> The benefits outweigh the risks associated with the nitrosamines. The performance of the dicamba-containing herbicides is such that they are viable alternatives to the suspended uses (home lawns, pastures, ditchbanks and forests) of silvex and 2,4,5-T.

Physiological and Biochemical Behavioral Characteristics:

> Foliar absorption: Readily absorbed by leaves.
>
> Translocation: Dicamba is absorbed by leaves and readily moved to other plant parts.
>
> Mechanism of pesticidal actions: Exhibits properties of an auxin-like plant growth regulator.
>
> Plant metabolism: Rapidly absorbed and metabolized almost entirely into soluble metabolites and insoluble plant products (celluloses).
>
> Animal metabolism: Some dicamba is demethylated to the metabolite, 3,6-dichloro-2-hydroxybenzoic acid. Most dicamba is excreted rapidly in urine as the free and/or conjugated form.

Environmental Characteristics:

> Adsorption and leaching in basic soil types:
>
>> Dicamba (free acid and dimethylamine salt) is adsorbed to peat, but not appreciably adsorbed to soils ranging from heavy clay to loamy sand.

Dicamba is readily mobile in soils ranging from clay to loamy sand.

Microbial breakdown:

Under aerobic conditions in soil dicamba degrades with half-lives ranging from 1 to 6 weeks depending on soil texture. Degradation rates are slowed by decreasing temperatures (<20°C) and decreasing soil moisture below field capacity.

Loss from Photodecomposition and/or volatilization:

Phytotoxic dicamba (free acid) residues are photodegraded in water to nonphytotoxic levels.

Dicamba is volatile with losses of 60% in glass flow tubes and 49% from thin films. Data from sterile and nonsterile soil samples indicate that larger losses of dicamba are due to metabolism rather than to volatilization.

Resultant average soil persistence:

Dicamba has a half-life of 1 to 6 weeks. It may be leached out of the zone of activity in humid regions in 3 to 12 weeks. Dicamba may persist longer under conditions of low soil moisture and rainfall.

Ecological Characteristics:

Avian oral LD_{50} >2,510 mg/kg (practically non-toxic)
Avian dietary LC_{50} >10,000 ppm (practically non-toxic)
Aquatic invertebrates LC_{50} >100 mg/l (practically non-toxic)
Cold water fish LC_{50} = 135.3 mg/l (slightly toxicity)
Warm water fish LC_{50} >1,000 mg/l (practically non-toxic)

Available data indicate that dicamba is practically non-toxic to fish and wildlife and unlikely to directly effect these organisms. Use patterns of the chemical do not present any problem to endangered species.

Tolerance Assessments:
Crops and tolerances:

0.1 ppm on sugarcane, sugarcane fodder and sugar cane forage.

0.2 ppm on meat, fat and meat byproducts (except liver and kidney) of cattle, goats, hogs, horses and sheep

0.3 ppm on milk

0.5 ppm on barley grain and barley straw; corn fodder, forage, and grain; oat grain and oat straw, and wheat grain and wheat straw.

1.5 on kidney and liver of cattle, goats, hogs, horses and sheep.
2.0 ppm on sugarcane molasses (food/feed additive tolerance)

3.0 ppm on asparagus, sorghum fodder, forage and grain

40.0 ppm on grasses, hay; grasses, pasture; grasses, rangeland.

Results of tolerance assessment:

The available residue data support the existing tolerances.

Tolerances on sorghum milling fractions, poultry and eggs may be required once requested residue data and poultry feeding are submitted.

Based on a NOEL of 600 ppm (rat subchronic study) and a 2,000-fold safety factor, the existing tolerance utilizes 37.58% of the PADI.

Problems which are known to have occurred with the use of the chemical:

Based on the Pesticide Incident Monitoring System (PIM's) report, most reported incidents with dicamba involve phytotoxicity to adjoining crops because of drift.

4. <u>Summary of Regulatory Position and Rationals:</u>

Use Classification: General Use

Summary of risk/benefit review:

The risk level for dicamba products containing DMNA is in the 10^{-7} to 10^{-8} range. The Agency considers that the benefits outweigh the risk associated with the nitrosamines. The product performance of dicamba-containing herbicides is such that they are viable alternatives of several of the suspended uses of silvex and 2, 4, 5-T, such as for home lawns, pastures, along ditchbanks and brush control in pastures.

Use Restrictions:

Dicamba may not be used in any way which contaminates irrigation ditches or water for domestic purposes.

Unique label warning statement:

Crops for which dicamba is not registered may not be planted in dicamba-treated fields.

5. Summary of Data Gaps and Dates When These Gaps Are to be Filled:

   Residue data on poultry, eggs, and sorghum   October 1987

   Milling fractions   October 1987
   Poultry feeding study   October 1987
   Hydrolysis   October 1987
   Photodegradation   October 1987
   Laboratory metabolism studies   October 1987
   Mobility   October 1987
   Field dissipation studies   October 1987
   Accumulation studies   October 1987
   90-day feeding (Nonrodent)   October 1987
   Chronic feeding/oncogenicity (2 species)   October 1987
   Mutagenicity test   October 1987

6. Contact person at EPA:

   Robert J. Taylor
   Environmental Protection Agency (TS-767C)
   401 M St. SW.
   Washington, DC 20460
   (703) 557-1800

DICHLOBENIL

Date Issued: March 23, 1987
Fact Sheet Number: 122

1. Description of Chemical

    Generic Name: 2,6-dichlorobenzonitrile

    Common Name: dichlobenil

    Trade Names: Casoron®, H-133, Decabane®, 2,6-DBN, Code 133®

    EPA Shaughnessy Code: 027401

    Chemical Abstracts Service (CAS) Number: 1194-65-6

    Year of Initial Registration: 1964

    Pesticide Type: Herbicide

    Pests Controlled: Broadleaf weeds and grasses (annual and perennial), and aquatic weeds.

    Chemical Family: Benzonitrile

    U. S. and Foreign Producers: Duphar B.V. (Netherlands); PBI/Gordon Corporation; Shell International Chemical Company, Ltd., London; and Uniroyal Chemical, Div. of Uniroyal, Inc.

2. **Use Patterns and Formulations**

    Application Sites: Terrestrial food and nonfood crops, aquatic nonfood, forestry, commercial and industrial sites.

    Types and Methods of Application: Ground or aerial application.

    Application Rates: Alfalfa and Ladino Clover 1.4 to 2.0 lb ai/A; 4 to 6 lb ai/A in cranberry bogs, granular mix 3.4 to 6.75 lb ai/A and up to 10 lb ai/A on very weedy bogs; 4 to 6 lb ai/A for

control of annual weeds in bearing, nonbearing,
nursery stock, noncitrus, nut crops; 4 to 7
lb ai/A in citrus nurseries; terrestrial nonfood crops
10 to 20 lb ai/A; aquatic weed control 7 to 15 lb. ai/A.

Types of Formulations: 2%, 4%, and 10% active ingredient (ai) granule (G); 50% ai wettable powder (WP); 1.73%, 1.77%, 2.1% and 3.12% ai soluble concentrate/ liquid (SC/L); and 6.75% ai liquid ready to use (RTU).

Usual Carriers: Water.

3. **Science Findings**

The current data base does not suggest any major toxicological problems. However, there are several toxicology data gaps: acute inhalation, dermal sensitization, primary eye irritation, primary dermal irritation, oncology,reproduction studies and a teratology study in a second species. Dichlobenil has a low acute ora and moderate acute dermal toxicity. Acute inhalation toxicity, primary eye and dermal irritation, and dermal sensitization have not been characterized.

There is a potential toxicity concern regarding the plant and soil metabolite, dichlorobenzamide (BAM). Subchronic ingestion of BAM produces a neuromuscular effect in rats, not observed with the parent compound. This metabolite is readily absorbed and translocated by plants. Because BAM has tentatively been implicated in adverse toxicological symptoms not attributable to dichlobenil per se, we are requiring additional residue data depicting residues of BAM.

Chemical Characteristics: P = Pure
                           T = Technical
   Physical state - (P) Crystalline solid
                 (T) Powder
   Color - (P) White
        (T) Pale yellowish
   Odor - Characteristic aromatic
   Melting point - (P) 145 to 146 °C
                (T) 140 to 144 °C
   Solubility - (at 20 °C)
            (T) 25 ppm in water
            (T) 10 g/100 g in methylene chloride
            (T) 4 g/100 g in toluene
   Vapor pressure - (P) 5.5×10^{-4} mm Hg at 20 °C

Octanol/water partition coefficient - (P) 3.06

Toxicology Characteristics:

Toxicity Category and Value(s) for Each Acute Hazard

    Acute Oral Toxicity   4250 (3510-5150) mg/kg
        (rat)   (males and females)
                                  Toxicity Category III

    Acute Dermal Toxicity   1350 + 158 mg/kg (males)
        (rabbit)   Toxicity Category II

Subchronic Oral Toxicity - Rodent (Rat): Compound-related effects included increased absolute and relative liver and kidney weights at 1000 ppm and above; hepatic degeneration and an absolute neutropenia and leukopenia at 3000 ppm and above; and mortality (5 out of 6) and hepatic necrosis at 10,000 ppm.

    No Observed Effect Level (NOEL): 100 ppm (50 mg/kg/day) (rodent)

    Lowest Effect Level (LEL): 1000 ppm (500 mg/kg/day) (rodent)

Nonrodent - (beagle dogs): This requirement is satisfied by the 2-year chronic feeding study.

A core-minimum subchronic oral rat study was submitted. Rats (10/sex/dose) were treated with BAM in the diet for 13 weeks at doses of 0, 50, 180, 600, and 2300 ppm. The NOEL (180 ppm [14 mg/kg/day]) and the LEL (600 ppm [49 mg/kg/day]) were based on decreased body weight gain and food efficiency, increased blood urea nitrogen (BUN), and reduced coagulation times. There was also a possible reduction in muscle tone.

Major Routes of Exposure: The major route of exposure is through the skin during mixing and loading.

Chronic Toxicology Results:

Oncogenicity — The oncogenic potential of dichlobenil cannot be determined from the available rat study. A second oncogenicity study (not rat) is needed to determine its oncogenic potential.

Chronic feeding — Nonrodent (beagle dogs): The NOEL and LEL were based on increased absolute and relative liver and thyroid weight in both sexes; leukocytic infiltration and fibrinoid degeneration around the central hepatic veins in both sexes; increases in serum alanine aminotransferase (SGPT) (females) and serum alkaline phosphatase (SAP) (males and females); and liver enzyme glucose-6-phosphatase and glucose-6-phosphatase dehydrogenase (G6Pase and G6PD) activity (males and females).

NOEL = 50 ppm (1.25 mg/kg/day) (nonrodent)

LEL = 350 ppm (8.75 mg/kg/day) (nonrodent)

Teratogenicity — Rodent (Rat):

Maternal Toxic NOEL = 20 mg/kg/day

Developmental NOEL = 60 mg/kg/day

Mutagenicity: Battery of mutagenicity tests for gene mutation, chromosomal aberration, direct DNA damage and transformation are negative.

Physiological and Behavioral Characteristics:

Translocation — 2,6-Dichlorobenzamide is absorbed by roots and translocated in the plant.

Mechanism of Pesticide Action — It stimulates oxygen utilization and inhibits esterification of phosphorus, resulting in reduced meristemic cell growth and inhibition of germination.

Metabolism and Persistence in Plants and Animals -
    The available data are inadequate to evaluate
    the persistence of dichlobenil in plants
    and animals.

Environmental Characteristics:

Available data are insufficient to fully
assess the environmental fate and the potential
exposure of humans and nontarget organisms
to dichlobenil. Additional data are needed
to characterize the potential for dichlobenil
to enter ground water.

Ecological Characteristics:

Based on available data, the aquatic use of dichlobenil
presents the only major ecological concern. The
requirements for acute testing are only partially
satisfied. Dichlobenil is slightly toxic to game
birds and moderately toxic to fish and aquatic
invertebrates.

Toxicity to fish	Moderately toxic to coldwater (LC_{50} = 6.3 mg/L) and warmwater fish (LC_{50} = 5.7 to 8.3 mg/L)
Toxicity to Aquatic Invertebrates	Moderately toxic (EC_{50} = 3.2 ppm
Toxicity to birds	Slightly toxic to pheasants at an LC_{50} = 1500 ppm.

Potential Problems Related to Endangered Species: Ther
    are sufficient data to adequately evaluate the
    hazard to endangered avian and aquatic species.

Tolerance Assessment:

List of Crops and Tolerances - A tolerance at
    0.15 part per million (ppm) has been established
    for the combined negligible residues of
    dichlobenil and its metabolite 2,6 dichloro-
    benzoic acid (2,6-DCBA) in or on the raw agricultr

commodities: almond hulls, apples, avocados, blackberries, blueberries, citrus, cranberries, figs, grapes, mangoes, nuts, pears, raspberries, and stone fruits.

Results of Tolerance Assessment - A Provisional Acceptable Daily Intake (PADI) of 0.00125 mg/kg/day has been established for dichlobenil based on a 2-year dog feeding study and a thousand-fold safety factor. Current tolerances result in a Theoretical Maximal Residue Contribution (TMRC) of 0.000325 mg/kg/day and utilizes 26 percent of the PADI.

4. Summary of Regulatory Position and Rationale

   Warning Statements Required on Labels:

   Manufacturing use - Do not discharge effluent containing this product into lakes, streams, ponds, estuaries, oceans, or public waters unless this product is specifically identified and addressed in an NPDES permit. Do not discharge effluent containing this product to sewer systems without previously notifying the sewage treatment plant authority. For guidance, contact your State Water Board or Regional Office of the EPA.

   End use, aquatic weed control (nonfood) - Do not contaminate untreated water by cleaning of equipment or disposal of wastes. Treatment of weed areas can result in oxygen loss from decomposition of dead weeds. This loss can cause fish suffocation. Therefore, treat only 1/3 of the weed areas at a time and wait 14 days between treatments. Consult your State Environmental Regulatory Agency concerning the need for a permit before applying this product to public waters.

   End use, aquatic food (cranberry) - Do not contaminate water by cleaning of equipment or disposal of wastes.

   End use, nonaquatic (nongranular) - Do not apply directly to water or wetlands (swamps, bogs, marshes, and potholes). Do not contaminate water by cleaning of equipment or disposal of wastes.

End use, nonaquatic (granular) - Cover, collect or incorporate granules spilled on the soil surface. Do not apply directly to water or wetlands (swamps, bogs, marshes and potholes). Do not contaminate water by cleaning of equipment or disposal of wastes.

Worker safety label statements

IMPORTANT! Always wash hands, face, and arms with soap and water before smoking, eating, drinking, or toileting. [For nongranular formulations: Before removing gloves, wash them with soap and water.]

Keep all unprotected persons, children, livestock, and pets away from treated area or where there is danger of drift.

Do not rub eyes or mouth with hands. If you feel sick in any way, STOP work and get help right away. See Statement of Practical Treatment.

5. Summary of Major Data Gaps

Data	Due
Product Chemistry	6 to 15 months
Residue Chemistry	
Nature of Residue (Metabolism)	18 Months
Residue Analytical Method	15 Months
Storage Stability	6 Months
Magnitude of Residues For Each Food Use	18 Months
Environmental Fate	
Hydrolysis	9 Months
Photodegradation (water/soil)	9 Months
Metabolism	27 Months
Leaching and Adsorption/ Desorption	12 Months
Dissipation	27 to 50 Months
Accumulation	12 to 50 Months
Toxicology	
Acute Inhalation	9 Months
Primary Eye Irritation	9 Months

Primary Dermal Irritation	9 Months
Dermal Sensitization	9 Months
Chronic Toxicity (rodent)	6 Months
Oncogenicity (mouse)	48 to 50 Months
Oncogenicity (rat)	6 Months
Teratogenicity (non-rodent)	12 Months
Reproduction (rat)	30 Months
General Metabolism	24 Months

Wildlife and Aquatic Organisms

Acute Avian Oral Toxicity	9 Months
Avian Subacute Dietary (waterfowl)	9 Months
Acute Toxicity to Freshwater Invertebrates	9 Months
Fish Early Life Stage	15 Months
Aquatic Invertebrate Life Cycle	15 Months
Aquatic Organism Accumulation	12 Months
Nontarget Area Phytotoxicity	9 Months

6. **Contact Person at EPA**

Robert J. Taylor
U.S. Environmental Protection Agency
Registration Division (TS-767C)
401 M Street SW.
Washington, D.C. 20460
(703) 557-1800

DISCLAIMER: The information presented in this Pesticide Fact Shee is for informational purposes only and may not be used to fulfill data requirements for pesticide registration and reregistration.

2,4-DICHLOROPHENOXYACETIC ACID

Date Issued: March 1987
Fact Sheet Number: 94.1

1. DESCRIPTION OF CHEMICAL

    Generic Name: (2,4-Dichlorophenoxy) acetic acid

    Common Name: 2,4-D (includes parent acid as well as the 35 derivatives (esters and salts))

    EPA Shaughnessy code: 030001

    Chemical Abstracts Service (CAS) number: 94-75-7

    Year of Initial Registration: 1948

    Pesticide type: Herbicide, plant growth regulator

    Chemical family: chlorinated phenoxys

2. USE PATTERNS AND FORMULATIONS

    2,4-D is a systemic herbicide widely used to control broadleaf weeds. There are approximately 1500 registered products containing 2,4-D active ingredient. Over 60 million pounds of 2,4-D active ingredients are annually applied domestically with the majority used for control of broadleaf weeds in wheat, field corn, grain sorghum, sugar cane, rice, barley, and range and pastureland.

USE PATTERNS

Agricultural: wheat, field corn, grain sorghum, sugar cane, rice, barley, soybeans, orchard crops.

Aquatic Management: water hyacinth control; Tennessee Valley Authority (TVA) Eurasian watermilfoil; lakes, ponds where treated water is not used for domestic or irrigation purposes.

Pasture and Rangeland   brush control

Home and Garden: lawns, ornamental, turf, parks, recreation areas.

Forest Management: brush control, conifer release, tree injection.

TYPES AND METHODS OF APPLICATION: aerial and ground equipment, knapsack sprayers, pressure and hose-end applicators, and lawn spreaders.

TYPES OF FORMULATIONS: granular, amine and ester liquids, dust aerosol spray (foam).

TOLERANCES:

Tolerances are established for residues of 2,4-D acid in a variety of raw agricultural commodities, meat, milk, eggs, poultry, fish, and shellfish. The tolerances include residues of 2,4-D as a result of the application of the 2,4-D acid, as well as the salts and esters. (These tolerances are listed in Attachment 1).

Food additive tolerances have been established for sugarcane, molasses, milled fractions from barley, oats, rye, and wheat, and potable water as a result of the use of 2,4-D in specific aquatic management programs.

3. AGENCY REVIEW

The Agency initiated a review of the information available on 2,4-D in 1980 in order to address questions concerning potential health effects associated with the use of 2,4-D. This review was conducted to determine if 2,4-D should be reviewed under the Special Review process or if another regulatory action was appropriate. After a review of much of the existing toxicology data supporting 2,4-D registrations, the EPA concluded that the available information did not indicate that the continued use of 2,4-D posed a significant health hazard when used in accordance with label directions and precautions. The Agency did conclude, however, that more information on 2,4-D toxicological properties was necessary to better assess the potential health hazards associated with the use of 2,4-D.

On August 29, 1980, after consulting with the Scientific Advisory Panel (SAP), the Agency issued a data call-in letter under authority of Section 3(c)(2)(B) of the Federal Insecticide, Fungicide and Rodenticide Act (FIFRA) to registrants of 2,4-D. The letter required submission of data on the following potential health effects: acute toxicity (oral and dermal), oncogenicity (tumor formation) in rat and mouse, reproduction, teratogenicity (birth defects), neurotoxicity, and metabolism.

Major registrants of technical 2,4-D products subject to the data requirements have formed the "Industry Task Force on 2,4-D Research Data" (Task Force) to jointly produce the required data.

The following data have been submitted by the Task Force to support the 2,4-D registrations:

- Acute Oral and Dermal Toxicity Studies (Rabbit, Rat) for all manufacturing use products

- Teratology Studies (Rat) for 2,4-D acid and 2,4-D Dichlorophenol

- Reproduction Study (Rat) for 2,4-D acid

- Pharmacokinetic (Metabolism) Study (Rat, Mouse) for 2,4-D acid and isooctyl ester of 2,4-D

- Neurotoxicity Study (Rat) for 2,4-D acid

- Combined Chronic Toxicity and Oncogenicity Study (Rat) for 2,4-D.

- Oncogenicity Study (Mouse) for 2,4-D acid

All required studies have been submitted; the results of the major studies submitted are summarized below.

<u>Teratology Study</u> (Rat) 2,4-D acid

Fischer 344 female rats were administered doses of 2,4-D acid (technical grade) suspended in corn oil by gavage (oral intubation) of 8, 25, or 75 mg/kg on days 6-15 of gestation (35 rats/dose group). The results of the study were:

Maternal toxicity No Observed Effect Level (NOEL) = 75 mg/kg/
Maternal toxicity Lowest Effect Level (LEL) = not found, since NOEL was at the highest dose tested.
Fetotoxic NOEL = 25 mg/kg/day
Fetotoxic LEL = 75 mg/kg/day (delayed ossification of bones)

The study does not indicate a teratogenic (birth defects) effect up to 75 mg/kg/day.

<u>Teratology Study (Rat)</u> 2,4-Dichlorophenol (ester)

Fischer 344 female rats were gavaged on days 6-15 of gestation with 0, 200, 375 or 750 mg/kg of 2,4-Dichlorophenol suspended in corn oil at 4 ml/kg. The results of the study were:

Maternal toxicity LEL = 200 mg/kg
Fetotoxic NOEL = 375 mg/kg
Fetotoxic LEL = 750 mg/kg (delayed ossification of bones)

The test material was not teratogenic under the conditions of the study.

Reproduction Study (Rat) 2,4-D acid

Fischer 344 rats were administered 5, 20, or 80 mg/kg/day of 2,4-D acid in the diet. The results indicate:

Maternal toxicity NOEL = 5 mg/kg/day
Maternal toxicity LEL = 20 mg/kg/day
Fetotoxic NOEL = 5 mg/kg/day
Fetotoxic LEL = 20 mg/kg/day

The results indicated that at the lowest dose tested of 5.0 mg/kg/day neither parent nor offspring were affected by the administration of the chemical. However, at the next higher dose tested (20 mg/kg/day) there was a decrease in maternal body weight and reduced pup weight (weight of offspring).

There were no effects seen on fertility in male or female rats.

Neurotoxicity (Rat) Dimethylamine salt of 2,4-D

Four groups of male and female Fischer CDF 344 rats (15 rats/group) were used in a study to determine whether repeated dermal exposure to 2,4-dimethylamine on the peripheral nervous system of rats would result in pharmacological and/or toxicological effects. The skin of the animals in the three treatment groups was painted with a 12% 2,4-D amine solution for 2 hrs/day 5 days per week for 3 weeks. Control animals were treated with tap water.

Dermal exposure to 2,4-D resulted in two systemic effects: treated rats weighed less than controls and the kidneys of treated rats weighed more than the controls. Even though the rats had clear systemic effects of exposure to 2,4-D there were no treatment-related changes in the function or structure of the nervous system.

Oncogenicity (Rat)

Male and female rats (strain-CDF(F344)/CRL-BR)) were administered doses of 1, 5, 15, and 45 mg/kg/day of 2,4-D acid in the diet. The results of the study indicate that the administration of 2,4-D appears to produce increased numbers of astrocytomas in brains of male rats at 45 mg/kg/day and is suggestive of a carcinogenic effect. The Agency is currently reviewing the study and will make a final determination on the significance of the study in the Spring of 1987.

Oncogenicity (Mouse)

The mouse oncogenicity data were submitted in January 1987 and are under review by the Agency. The review is scheduled for completion in the spring of 1987.

Epidemiologic Studies

Approximately 30 epidemiologic studies have been published regarding the carcinogenic risk of herbicides, especially the chlorinated phenoxy herbicides which include 2,4-D. The Agency is undertaking a comprehensive review of these epidemiologic studies to evaluate the weight of evidence pertaining to the carcinogenic potential of 2,4-D.

4. SUMMARY OF MAJOR DATA GAPS

Environmental fate, residue chemistry, product chemistry, and ecological effects data gaps are, to be identified as part of the Registration Standard development which is due to be issued in early 1988.

5. REVIEW OF DATA

The Agency is reviewing the laboratory oncogenicity data and the epidemiologic studies. The Agency will classify 2,4-D under the Guidelines for Carcinogen Risk Assessment in the Spring of 1987. That classification will be presented to the Office of Pesticide Programs' Science Advisory Panel (SAP) in a public forum the Summer of 1987. After SAP review of the Agency's classification; the Agency will publish its determination regarding 2,4-D oncogenicity in the Federal Register.

6. CONTACT PERSON AT EPA

Douglas McKinney
Registration Division
U.S. Environmental Protection Agency
401 M Street, SW.
Washington, D.C. 20460
703-557-5488

1,3-DICHLOROPROPENE

Date Issued: September 1986
Fact Sheet Number: 95

1. Description of the chemical

    Generic name: 1,3-dichloropropene ($C_3H_4Cl_2$)
    Common name: None
    Trade names: Telone II®, Dow Telone®,
    Chemical Abstracts Service (CAS) Registry Number: 542-75-6
    Office of Pesticides Program Number: 029001
    Year of initial registration: 1966
    Pesticide type: Broad spectrum soil fumigant
    Chemical family: Chlorinated hydrocarbon
    U.S. producer: Dow Chemical Company

2. Use patterns and formulations:

    Application sites: A soil fumigant with nematicidal, fungicidal, insecticidal and herbicidal properties, for use on cotton, potatoes, tobacco, sugar beets, vegetables, grains, citrus fruit tree planting sites, deciduous fruit and nut-tree planting sites, bush and vine planting sites, floral/turf and ornamental tree sites.

    Types of formulations: 94% liquid concentrate formulations.

    Types and methods of applications: Chisel injection into the soil, using row (banded) or overall (broadcast) treatment.

    Application rates: 43 to 968 lbs ai/acre. One application per year.

    Usual carriers: None

3. Science Findings

Science summary: The Agency has categorized 1,3-dichloropropene as a probable human carcinogen. Chronic toxicity data show that the chemical is oncogenic at multiple sites in both sexes of rats and mice. Other data supporting the oncogenic finding of 1,3-dichloropropene are: (1) a subcutaneous injection study in mice which showed increased incidence of fibrosarcomas at the site of injection, (2) studies showing that the compound is a direct acting mutagen, and (3) the structural similarity between 1,3-dichloropropene and known human oncogens (i.e. vinyl chloride). The chemical has not been shown to cause teratogenic effects in rats or rabbits. 1,3-dichloropropene is acutely toxic by the oral and inhalation routes of exposure.

1,3-dichloropropene has low to moderate toxicity to birds and moderate toxicity to fish and aquatic invertebrates. Limited data indicate that 1,3-dichloropropene has the potential to leach ground water in sensitive environments. Available data are insufficient to more completely assess the environmental fate of the pesticide. Available data are also insufficient to assess the residues in raw agricultural commodities.

1,2-Dichloropropane, an impurity in 1,3-dichloropropene, is oncogenic in rats and mice. However, its oncogenic potency is lower than for 1,3-dichloropropene. Data indicate that 1,2-dichloropropane has the potential to leach to ground water.

Chemical characteristics:

   Color: Pale yellow liquid at room temperature.
   Molecular weight: 110.98.
   Miscible in hydrocarbon solvents.

Toxicological Characteristics:

   Acute oral toxicity - 470 to 713 mg/kg, Toxicity Category II

   Acute oral toxicity - 640 mg/kg, Toxicity Category III,

   Skin irritation - Moderate irritant, Toxicity Category III

   Eye irritation - Corneal opacity reversible within seven days, Toxicity Category II.

Subchronic toxicity

   Subchronic inhalation (rat and mouse) - NOEL 30 ppm

Chronic toxicity. A two-year rat oral gavage study had a NOEL of 25 mg/kg/day. Effects noted were an increased incidence of basal hyperplasia in treated males and females, increased kidney nephropathy in treated females, and urinary bladder edema at high dosages in males and females.

In a two year mouse feeding study, the effects noted were reduced female survival rate at the high dose and increased incidences of urinary bladder epithelial cell hyperplasia in both males and females in the mid-and high-dose levels. A NOEL was not determined in this study.

Oncogenicity: 1,3-dichloropropene has been classified as a probable human carcinogen based on studies in rats and mice. Oral studies in rats showed that it produced squamous cell papillomas and carcinomas of the forestomach in males at the high dose. In the liver, there were increases of neoplastic nodules in treated males at both dosage levels compared to the controls. In female rats, squamous cell papillomas of the forestomach, mammary gland adenomas/fibroadenomas and thyroid follicular cell adenomas/carcinomas were observed.

In mice, 1,3-dichloropropene produced an increased incidence of urinary bladder transitional cell carcinomas, squamous cell papillomas of the forestomach, and lung adenomas and carcinomas (combined) in females at the mid and high dosage levels. There was an increase in hepatocellular adenomas and carcinomas in the females at the mid dosage level. In male mice, there was an increased incidence of hepatocellular adenomas in combination with carcinomas, lung adenomas in combination with carcinomas, and squamous cell papillomas of the forestomach at the mid and high dosage levels. Urinary bladder carcinomas were also observed in the males at the high dosage level.

Studies also indicate that 1,2-dichloropropane, an impurity in 1,3-dichloropropene, is an oncogen in rats and mice, but the Agency believes it is a less potent oncogen than 1,3-dichoropropene.

Mutagenicity: 1,3-dichloropropene has been determined to be a direct acting mutagen. 1,3-dichloropropene has produced positive gene mutation in microbial systems. In addition, 1,3-dichloropropene produced positive results for DNA damage/repair in microbial strains. Data on structural chromosomal aberration are not available. These studies are required.

Reproductive effects: No data are available on reproductive effects. A two-generation reproduction study is required.

Developmental effects: In two inhalation studies, 1,3-dichloropropene was found to cause developmental effects in rats and rabbits. In the rat study, no teratogenic effects were observed, but maternal toxicity occurred at all dose levels. Developmental toxicity (delayed ossification of vertebral centra) occurred at the highest dose (120 ppm). In the rabbit study maternal toxicity occurred at the 2 highest dose levels (60 ppm and 120 ppm), but no evidence of developmental toxicity was apparent.

Exposure: The average inhalation exposure of applicators engaged in various work activities is estimated to range from 0.15 to 23.99 mg/kg/year.

Dietary exposure to 1,3-dichloropropene has not been estimated. Residue data for this assessment are required.

Risks to workers and applicators: The upper bound inhalation oncogenic risks to workers handling 1,3-dichloropropene during distribution and storage are estimated to be in the range of 10^{-2} to 10^{-5}. Risks to applicators during and immediately post-treatment are estimated to be in the range of 10^{-3} to 10^{-5}.

Physiological and biochemical behavioral characteristics: Mechanism of pesticidal action: 1,3-dichloropropene acts as a sterilant on contact with the pest.

Metabolism in plants: Metabolism of 1,3-dichloropropene in plants is not adequately understood.

Environmental characteristics: Data indicate that 1,2-dichloropropane, an impurity of technical 1,3-dichloropropene has the potential to leach to ground water. Additional studies on the environmental fate of 1,3-dichloropropene, its impurities and metabolites are required. Data indicate that 1,3-dichloropropene itself leaches to ground water when it is present in the most sensitive environment (shallow ground water and sandy soils of low percentage organic matter in areas of high rainfall or irrigation).

Ecological characteristics:

Avian acute oral toxicity:	Bobwhite quail - 152 mg/kg
Avian dietary toxicity:	Mallard duck - >10,000 ppm Bobwhite quail - >10,000 ppm
Aquatic invertebrate toxicity:	_Daphnia magna_ - 6.2 ppm
Freshwater fish toxicity:	Bluegill sunfish - 7.09 ppm Rainbow trout - 3.94 ppm

Based on these data, the Agency characterizes 1,3-dichloropropene as low to moderately toxic to waterfowl and upland game birds. 1,3-dichloropropene is moderately toxic to coldwater fish, warmwater fish, and freshwater invertebrates.

Endangered species: 1,3-dichloropropene has low to moderate toxicity to birds and aquatic species. Available data do not indicate a potential hazard to endangered species.

Tolerance assessment: No tolerances or exemptions from the requirement of tolerances for residues of 1,3-dichloropropene in or on food/feed commodities have been established in the United States, Mexico, Canada or by the Codex Alimentarius.

Current data are inadequate to determine if residues of 1,3-dichloropropene, its metabolites, or manufacturing impurities will result in food or feed from use of 1,3-dichloropropene. Data on the metabolism of 1,3-dichloropropene in crops grown in treated soils are required. The additional data will be used to assess possible residues in food and feed crops and may lead to additional data requirements in residue chemistry. Tolerances in food crops may be necessary if residues are found in crops, and may be necessary in animals if residues are found in animal feed items.

Reported pesticide incidents: The Agency's Pesticide Incidence Monitoring System does not report any incidents during the period 1966-1981. However, information from the State of California indicates that physicians treated an average of 2.20 1,3-dichloropropene incidents annually from 1981 through 1985. An additional 3.80 occupational cases were reported annually as dermal (1.80) or eye (2.00) injuries.

4. Summary of regulatory positions and rationale:

--- The Agency has placed 1,3-dichloropropene in Special Review because of oncogenic effects. 1,3-dichloropropene has been categorized as a probable human carcinogen, with effects demonstrated at multiple sites in both sexes of rats and mice.

--- The Agency is classifying the 1,3-dichloropropene products for Restricted Use by certified applicators only, based upon acute toxicity and oncogenicity. Restricted use will help ensure that mixer/loaders and applicators observe use precautions intended to protect against inhalation exposure.

--- The Agency is continuing on an interim basis the current reentry interval of 72 hours for 1,3-dichloropropene products.

--- The Agency is continuing to require protective clothing and equipment during mixing, loading, application, and if spillage occurs. Protective clothing and equipment include coveralls, gloves, heavy-duty footwear, safety goggles, and a mask or respirator approved for use with 1,3-dichloropropene.

--- Available residue data are inadequate to determine if residues of 1,3-dichloropropene, its metabolites, or manufacturing impurities will occur in food or feed as a result of soil fumigation. Although such fumigation use has in the past been considered non-food, the Agency may require the establishment of tolerances in food, feed, or animals if residues are found.

--- The Agency is requiring environmental fate data including ground water monitoring studies on 1,3-dichloropropene, its impurities, and metabolites to determine whether there is a potential for groundwater contamination.

--- The Agency will require a cancer hazard warning statement on 1,3-dichloropropene products, to provide the opportunity for informed consent by users, and to encourage compliance with protective measures that will reduce exposure.

--- Endangered species labeling statements are not warranted at this time, based upon available data. If environmental fate or other data indicate a potential hazard to endangered species, the Agency will request a consultation from the Office of Endangered Species, Department of the Interior.

5. Labeling Statements.

   a. Manufacturing-Use Products

   Precautionary Statements: "DANGER. Causes severe eye damage. May be fatal if inhaled, absorbed through skin, or swallowed. Do not get in eyes, on skin or on clothing. Wear chemical worker goggles, face shield or safety glasses. Wash thoroughly with soap and water after handling, and before eating and smoking".

   Effluent discharge statement: "Do not discharge effluent containing this product directly into lakes, streams, ponds, estuaries, oceans or public waters unless this product is specifically identified and addressed in a National Pollutant Discharge Elimination System (NPDES) permit. Do not discharge effluent containing this product into sewer systems without previously notifying the sewage treatment plant authority. For guidance, contact your State Water Board or Regional Office of the Environmental Protection Agency".

   Cancer Hazard Warning Statement: " The use of this product may be hazardous to your health. This product contains 1,3-dichloropropene which has been determined to cause tumors in laboratory animals. Risks can be reduced by closely following the use directions and precautions, and by wearing protective clothing specified elsewhere on this label".

   b. End Use Products.

   Precautionary statements: "DANGER. Causes severe eye damage. May be fatal if inhaled, absorbed through skin, or swallowed. Do not get in eyes, on skin or on clothing. Wear chemical worker goggles, face shield or safety glasses. Wash thoroughly with soap and water after handling, and before eating and smoking."

   Restricted use statement: "For retail sale to and use only by Certified Applicators or persons under their direct supervision and only for those uses covered by the Certified Applicator's certification."

Cancer Hazard Warning Statement: " The use of this product may be
hazardous to your health. This product contains 1,3-dichloropropene
which has been determined to cause tumors in laboratory animals.
Risks can be reduced by closely following the use directions and
precautions, and by wearing protective clothing and equipment
specified elsewhere on this label."

Enviromental hazard statement: "Do not contaminate water by cleaning
of equipment or disposal of wastes. In case of spills, properly
dispose of contaminated materials."

Protective clothing statements: "Required clothing and equipment for
mixing/loading and applying 1,3-dichloropropene":

"One-piece coveralls which have long sleeves and long pants constructed
of laminated fabric as specified in the USDA/EPA Guide for Commercial
Applicators".

"Liquid-proof hat such as a plastic hard hat with a plastic sweat
band".

"Heavy-duty liquid proof (neoprene/synthetic) work gloves and
boots".

"Any article worn while handling 1,3-dichloropropene must be washed
before reusing. Immediately remove all clothing which has been
drenched or has otherwise absorbed 1,3-dichloropropene from any
spill. Dispose of contaminated clothing in a sanitary landfill,
or by incineration if allowed by state and local authorities. If
burned, stay out of smoke."

Reentry statement: "Workers entering the treated area for 72 hours
after application of 1,3-dichloropropene must wear protective
clothing."

Storage And Disposal Statements: "Pesticide wastes are hazardous.
Improper disposal of excess pesticide is a violation of Federal
Law. If these wastes cannot be disposed of by use according to
label instructions, contact your State Pesticide or Environmental
Control Agency, or the Hazardous Waste representative at the nearest
EPA Regional Office for Guidance."

5. Summary of major data gaps:

Residue chemistry data
Nature of residue (plant metabolism)   May 30, 1987
Storage stability   May 30, 1987
Crop field Trials   April 30, 1988.

Toxicology data
Acute dermal toxicity   June 30, 1987
Acute inhalation toxicity   June 30, 1987
Dermal sensitization   June 30, 1987

21 day dermal	October 30, 1987
Chronic toxicity - dog	December 30, 1990
Oncogenicity - rat and mouse	July 30, 1988
Reproduction	July 30 1987
Structural chromosomal aberration	October 30, 1987
General metabolism	October 30, 1988
Environmental fate data	
Photodegradation in air	June 30, 1987
Aerobic soil metabolism	January 30, 1989
Anaerobic aquatic metabolism	January 30, 1989
Aerobic aquatic metabolism	Janaury 30, 1989
Mobility of degradates (leaching, adsorption/desorption)	October 30, 1987
Soil dissipation	January 30, 1989
Aquatic sediment	January 30, 1989
Soil dissipation (long-term)	December 30, 1990
Accumulation in rotational crops (confined)	January 30, 1990
Monitoring ground water & well water (protocol)	April 30, 1988.
Reentry protection data	
Soil dissipation	January 30, 1989
Dermal exposure	January 30, 1989
Inhalation exposure	January 30, 1989

Ecological effects data are not required for this Standard.

Product chemistry data are required during 1987.

6. <u>Contact Person at EPA:</u>

Lois Rossi,
Office of Pesticide Programs, EPA,
Registration Division (TS-767C)
Fungicide-Herbicide Branch
401 M Street., S.W.
Washington, DC 20460.
Phone (703) 557-1900

DISCLAIMER: The information presented in this Chemical Information Fact Sheet is for informational purposes only and may not be used to fulfill data requirements for pesticide registration and reregistration.

DICOFOL

Date Issued: December 30, 1983
Fact Sheet Number: 16

1. DESCRIPTION OF THE CHEMICAL

   ° Generic Name: Dicofol

   ° Common Name: Dicofol

   ° Trade Names: Acarin®, Carbax®, Decofol®, Kelthane®, Kelthane A®, p,p' Kelthane®, Mibol®, and Mitigan®

   ° Chemical Abstracts Service (CAS) Number: 115-32-2

   ° EPA Shaughnessey Number: 010501

   ° Year of Initial Registration: 1957

   ° Pesticide Type: Acaricide

   ° U.S. and Foreign Producers: Rohm & Haas, Aceto, Makhteshim Beer-Sniva, Agan Chemical, Drexel Chemical, Tricon Chem International, and Ida, Inc.

2. USE PATTERNS AND FORMULATIONS

   ° Foliar spray on agricultural crops and ornamentals, and in or around agricultural and domestic buildings for mite control.

   ° Formulated as emulsifiable concentrates, wettable powders, dusts, ready-to-use liquids, and aerosol sprays.

3. SCIENCE FINDINGS

   Summary of Science Findings

   ° There are insufficient data available to characterize the environmental fate of dicofol, however, groundwater contamination is not expected due to dicofol's lack of mobility in soil and low leaching potential.

   ° There are insufficient data available to determine the oncogenicity, teratogenicity, mutagenicity, and chronic feeding effects of dicofol.

- The data show that dicofol is highly toxic to aquatic organisms and only slightly toxic to birds, mammals, and beneficial insects.

- Dicofol bioaccumulates in some rotational crops and aquatic organisms.

- The DDT analog contaminants in dicofol may cause unreasonable adverse effects in certain bird and fish species.

Chemical Characteristics

- Manufacturing-use dicofol products contain a number of DDT analogs as manufacturing impurities. These impurities include the o,p' and p,p' isomers of DDT, DDE, DDD, and a substance called "extra-chlorine DDT" or Cl-DDT.

- Dicofol is a nonflowable liquid (or waxy solid), ranging from dark to yellow-brown in color. It is stable under cool and dry conditions, is practically insoluble in water, but soluble in organic solvents. Its melting point ranges from 58°C to 78°C. The chemical does not pose any unusual handling hazards.

Toxicological Properties

- Dicofol has a moderate acute oral toxicity (Toxicity Category III).
- Dicofol has a relatively high degree of acute dermal toxicity (Toxicity Category II).
- Results from an eye irritation test using a formulation intermediate indicates that the manufacturing-use product is probably a severe eye irritant (Toxicity Category I).

- Toxicity studies on dicofol are as follows:

    - Oral LD_{50} in rats: 684 - 1495 mg/kg body weight
    - Oral LD_{50} in rabbits: 1810 mg/kg body weight
    - Dermal LD_{50} in rabbits: 2.1 gm/kg body weight
    - Eye Irritation: tests with a formulation intermediate showed corneal damage in some rabbits that persisted for seven days
    - Subchronic Oral Toxicity in Rats: NOEL is 20 ppm.
    - Oncogenicity in Rats: Available data from the National Cancer Institute suggests that dicofol is a possible oncogen. However, the study is unacceptable due to the reported decomposition of the test material during the test.
    - Reproduction Study in Rats and Mice: The NOEL for both rats and mice is 100 ppm.

Environmental Characteristics

- Dicofol was stable for 350 days at 10 ppm in field plots under aerobic conditions.

- Aged dicofol residues are negligibly mobile in sandy loam soil column leaching studies.

- Field studies show that dicofol persists in soils for at least 4 years after application.

- Dicofol residues may accumulate in fish as well as some rotational crops.

Ecological Characteristics

- The DDT analog contaminants in the dicofol products can cause reproductive impairment in various fish and flesh eating birds (eggshell thinning).

- Avian dietary LC_{50}: 1237 to 3100 ppm for upland game birds (slightly toxic).

- Prolonged exposure to low levels (5 to 10 ppm) had no significant effects on reproduction behavior of mallards.

- Acute 96-hour LC_{50} for warm water fish: 0.31 to 0.51 ppm.

- Acute 96-hour LC_{50} for cold water fish: 0.053 to 0.086 ppm.

- Acute LD_{50} for marine grass shrimp: >0.439 ppm.

- Findings show that dicofol is highly toxic to aquatic organisms.

- Dicofol impairs the reproductive physiology of fish and aquatic invertebrates.

- Dicofol does not seem to be phytotoxic to most plants for which it is registered.

- Dicofol has been shown to be relatively nontoxic to honey bees and alfalfa leafcutting bees.

Tolerance Assessment

- Current application rates on labeling exceed the rates associated with the established tolerances. Tolerances and/or labeling may need to be revised to correct this discrepancy.

- Residues of the dicofol impurities will have to be examined for the tolerance reassessment.

4. SUMMARY OF REGULATORY POSITION AND RATIONALE

- No use, formulation, or geographical restrictions are required. General use classification.

- A Special Review of all dicofol end-use products containing detectable amounts of DDT analogs will be initiated.

- No new registrations will be issued for new dicofol products intended for outdoor use which would further increase the amount of DDT in the environment.

- Registrants and applicants for registration of manufacturing-use dicofol products must agree to develop the data as specified in the Guidance Document. Manufacturing-use dicofol products may be registered and/or reregistered subject to the terms and conditions specified in detail within the Guidance Document.

- Registrants and applicants for registration must submit data regarding the composition of their products and in particular the concentration of DDT contaminants in their manufacturing-use products. Registrants must indicate the lowest levels that could be achieved, the timeframe needed, and the cost. The Agency is concerned about the presence of DDT contaminants at levels substantially below 1% of the technical product because a no-effect-level has not been determined for avian reproductive effects. Agency files show that currently registered manufacturing-use products contain from 9 to 15 percent DDT and DDT analog contamination.

- Studies must be conducted by the registrants to measure whether DDT analog residues will be present on food or feed crops due to the use of dicofol products. A complete tolerance reassessment will be conducted after these data and other required residue data are submitted.

5. SUMMARY OF MAJOR DATA GAPS

- Product Chemistry: product identity, certification of limits, physical and chemical properties, and special requirements as mentioned in Item 4, above. Due April, 1984.

- Residue Chemistry: metabolism in plants and animals, residue analytical methods, updated residue data (including residue data on DDT analogs) on most crops. Due December, 1986.

- Environmental Fate: degradation studies, soil metabolism studies, field volatility, field dissipation, accumulation in fish and rotational crops. Due December, 1986.

- Toxicology: inhalation studies, subchronic feeding, chronic toxicity, oncogenicity, teratogenicity, mutagenicity, metabolism. Due December, 1986.

- Ecological Effects: avian reproduction, aquatic organism testing. Due December, 1986.

- Special Testing For DDT Analog Contaminants: all environmental fate studies conducted on Cl-DDT, chronic testing with Cl-DDT on chronic toxicity on birds, fish, and aquatic invertebrates. Due December, 1986.

6. Contact Person at EPA: Bruce A. Kapner (TS-767C)
Special Review Branch
Office of Pesticide Programs
401 M Street, S.W.
Washington, D.C. 20460

Telephone: 703/557-7400

DISCLAIMER: The information presented in this Chemical Information Fact Sheet is for informational purposes only and may not be used to fulfill data requirements for pesticide registration and reregistration.

DIFLUBENZURON

Date Issued: March 11, 1987
Fact Sheet Number: 68.1

1. Description of Chemical

    Generic name: N-[[(4-chlorophenyl)amino]carbonyl]-
                   2,6-difluorobenzamide

    Common Name: diflubenzuron
    Trade Names: Dimilin, Micromite, Vigilante
    EPA Shaughnessy Code: 108201
    Chemical Abstracts Service (CAS) Number: 35367-38-5
    Year of Initial Registration: 1976
    Pesticide Type: Benzamide chitin inhibitor
    U.S. and Foreign Producers: Duphar B.V., Amsterdam, Holland

2. Use Patterns and Formulations

    Application sites:
        Agricultural crops:
            cotton (boll weevils ; cotton leaf perforator)
            soybeans (velvetbean caterpillar; green cloverworm)
        Forests:
            woodland trees and shrubs including Christmas trees
               (gypsy moth; Douglas fir tussock mouth; forest
                tent caterpillar; Nantucket pine tipmoth)
        Ornamentals:
            commercial nursery (gypsy moth)
        Pasture:
            flood irrigated areas in California only
            (floodwater mosquitoes)
        Greenhouse:
            mushrooms (mushroom flies)

    Type of Formulation: 25% Wettable Powder

3. Science Findings

Summary Science Statement:

Diflubenzuron has demonstrated a low acute toxicity to mammals through the oral, dermal, and inhalation routes. The Agency has determined that diflubenzuron is not an oncogen, a teratogen, a mutagen, or a neurotoxin.

The allowable daily intake (ADI) of diflubenzuron is 0.02 mg/kg/day. This is based on a no observable effect level of 40 ppm (2 mg/kg/day) on changes in the blood of rats. The tolerances established add up to 2.9% of the ADI.

Diflubenzuron disrupts the normal molting of insects and other invertebrates by interfering with the deposition of chitin preventing proper formation of the new exoskeleton and shedding of the old one.

Chemical Characteristics:

    Physical state: solid
    Color: white
    Odor: none
    Boiling point: N/A
    Melting point: 210 to 230 °C
    Flammability: N/A
    Solubility in water: at 20 to 25 °C 0.2 mg/l.
    There are no unusual handling characteristics.

Toxicology Characteristics:

    Acute Oral, Rat: > 4640 mg/kg
        Toxicity Category IV

    Acute Dermal, Rabbit: > 4000 mg/kg
        Toxicity Category III

    Acute Inhalation, Rat: > 2.88 mg/L
        Toxicity Category III

The Major Routes of Exposure in Order of Toxicological Significance:

    Chronic Toxicology Results:

        Rat Chronic Feeding: NOEL = 40 ppm (2 mg/kg) bwt/day for met- and sulphemoglobin formation.

        Rat Oncogenicity: Not oncogenic to male and female rats under the conditions of the study; highest dose tested 10,000 ppm (500 mg/kg bwt/day).

Mouse Oncogenicity: Not oncogenic to male and female mice under the conditions of the study; highest dose tested was 10,000 ppm (500 mg/kg bwt/day).

Rat Teratology: Not fetotoxic or teratogenic to rats at levels up to 4 mg/kg bwt/day, the highest dose tested.

Rabbit Teratology: Not fetotoxic or teratogenic to rabbits at levels up to 4 mg/kg bwt/day, the highest dose tested.

Three-Generation Rat Study: No adverse effects on reproductive performance at 160 ppm (8 mg/kg/day), the highest dose tested.

Gene Mutation: At rates up to 1000 ug/plate there was no evidence of changes in spontaneous revertant frequency or any mutagenic effect.

Sheep 13-Week Feeding Study: No treatment-related effects were observed on food consumption, body weight gain, hematological parameters or urinalysis at 10,000 mg diflubenzuron/kg in the diet.

Dog 13-Week Feeding Study: No histopathological changes were reported attributable to feeding diflubenzuron at rates up to 160 ppm in the diet.

Physiological and Biochemical Behavioral Characteristics:

A. Translocation:

The available plant metabolism data show that diflubenzuron, when foliarly applied in doses approximating those registered or proposed for use on citrus, soybeans, and cotton will undergo very little, if any, translocation from treated areas. Additional metabolism data are needed to support the established mushroom tolerance.

B. Mechanism of Pesticidal Action:

Diflubenzuron interferes with arthropod chitin formation, thus disrupting the development of a new cuticle in preparation for molting of the exoskeleton. Exposed insect larvae continue to feed and otherwise develop normally until growth makes it necessary to shed the old cuticle. The old exoskeleton is shed, after which the growing insect swells by taking air into the respiratory system, expanding the body, and stretching the formed, elastic new cuticle before it hardens. Insects exposed to diflubenzuron die while trying to molt since the new cuticle is not properly formed.

C. Metabolism and Persistence in Plants and Animals:

The metabolism of diflubenzuron in plants has been adequately described for higher plants following foliar treatments. No residues were found in the milk, fat, kidney, liver, or meat of cattle fed at 25ppm of the total diet. In an exaggerated rate study, residues were found in milk, liver, and kidneys from cattle fed diflubenzuron at 250ppm of the diet. Metabolites found included p-chlorophenylurea, p-chloroaniline, and 4-chloroacetanilide.

Environmental Characteristics:

Due to rapid binding with soil particles and organic matter and breakdown by soil biota, ground water contamination problems are unlikely. In aerobic soil, radiolabeled diflubenzuron with a particle size of approximatly 2 microns had a half-life of less than 2 weeks at 20 °C.

Ecological Characteristics:

Avian Oral Acute Toxicity:
> 5000 mg/kg - Bobwhite quail and mallard duck

Avian Dietary Toxicity:
> 20,000 ppm - Bobwhite quail and mallard duck

Fish Acute Toxicity:

> 25 ppm - Yellow perch
> 50 ppm - Brook trout
> 100 ppm - Rainbow trout, channel catfish and bluegill sunfish
> 500 ppm - Fathead minnow

Freshwater Invertebrate Acute Toxicity:

560 ppb - Chironomus sp
16 ppb - Daphnia magna
30 ppb - Gammarus pseudolimnaeus

Estuarine/Marine Organism Acute Toxicity:

2.06 ppb - Mysid shrimp
0.64 ppm - Grass shrimp
> 130.00 ppm - Oyster larvae
255.00 ppm - Mummichug

> 1000.00 ppm   - Uca pugilator,
                    Carcinus maenus, Anodonta sp.
                    and Mercenaria mercenaria

Honeybee Acute Toxicity:

> 114.8 micrograms per bee. (rel. non-toxic)

Based on these studies, diflubenzuron is of low toxicity to birds, finfish, and honeybees. However, based on current studies, it is extremely toxic to aquatic invertebrates. Therefore, additional studies are required to complete a hazard assessment for aquatic invertebrates.

Tolerance Assessment:

Sufficient data are available to determine that the established tolerances for residues of diflubenzuron (DFB) in or on the following commodities are adequate: cottonseed, pasture grass, soybeans, soybean hulls, soybean soapstock, milk, eggs, and the meat, fat, and meat byproducts of cattle, goats, hogs, horses, sheep, and poultry. However, additional data are required for complete elucidation of the nature of the residue in mushrooms to support the established tolerance in mushrooms.

The Theoretical Maximum Residue Concentrate (TMRC) for diflubenzuron is 0.0352 mg/day based on a 1.5 kg diet. The ADI is 0.011 mg/kg/day. based on a 1.5 kg/day diet. The tolerances established add up to 2.9% of the ADI.

4. Summary of Regulatory Position and Rationale

The Agency has determined that it should continue to allow the registration of diflubenzuron. However, because of gaps in the data base, additional data are required as specified in the registration standard whitch was published on September 10, 198 and is available from the National Technical Information Servic in Springfield Virginia. Additional tolerances and label chang will be considered as applications are submitted.

Because of toxicity to crab, shrimp, and other aquatic inve tebrate animals, diflubenzuron is classified as a restricted pesticide for use on forests and field crops. Cautionary statements are required on the label warning of hazards to aquatic invertebrates.

The only geographic limitation for use of products containing diflubenzuron is for control of mosquitoes in temporarily flooded areas of pastures in Central California. The primary concern with these mosquitoes breeding in wastewater from irrigation projects is their potential for carrying diseases affec ing humans.

5. Summary of Major Data Gaps

   Toxicology:

   None.

   Environmental Safety: (Data due in 1987)

   Avian Reproduction Studies
   Freshwater Invertebrate Acute LC_{50}
   Estuarine/Marine Organism Acute EC_{50}
   Freshwater Invertebrate Life-Cycle
   Estuarine Invertebrate Life-Cycle.

   Residue Chemistry: (Data due in 1987)

   Metabolism in mushrooms
   Metabolism in cattle, poultry, and swine

6. Contact person at EPA:

   Arturo E. Castillo, (PM-17)
   Insecticide-Rodenticide Branch (TS-767)
   401 M Street, SW
   Washington, DC 20460
   Telephone: (703) 557-2690

DISCLAIMER: The information presented in this Pesticide Fact Sheet is for informational purposes only and may not be used to fulfill data requirements for pesticide registration and reregistration.

DINOCAP

Date Issued: October 1986
Fact Sheet Number: 65.1

1. Description of the Chemical

    Chemical name:
    a mixture of 2, 4-dinitro-6-octylphenol crotonate and
    2, 6-dinitro-4-octylphenol crotonate

    Common name: Dinocap

    Trade names: Karathane®, Mildex®, Dikar®

    EPA Shaughnessy number: 036001

    Chemical abstracts service (CAS) number: 131-72-6

    Year of initial registration: mid-1950's

    Pesticide type: Fungicide

    Chemical Family: Dinitrophenolic

2. Use Patterns and Formulations

    Application sites: apples, pears, grapes, apricots, melons, cantaloupes, cucumbers, peaches, pumpkins, raspberries, squash, nursery stock, ornamentals, and home gardens.

    Types of formulations: liquid, wettable powder, dust, aerosol

    Types and methods of application: ground boom, mist blower, airblast, handheld sprayers

    Application rates: rate and frequency vary according to site application

3. Science Findings:

   Physical and Chemical Properties

   Color -- Dark reddish-brown

   Boiling Point -- 138 to 140°C

   Solubility -- Soluble in most organic solvents, but almost insoluble in water

   Appearance at room temperature -- White, crystalline solid at 23°C

   Tolerance Assessment: Tolerances were established in 40 CFR 180.341 for residues of dinocap

   Toxicology Summary:

   Acute toxicity - For technical dinocap, the acute oral LD_{50} in rats is 980 mg/kg.

   Chronic Toxicity - Dinocap produces developmentally toxic effects in rabbits. For oral exposure, the NOEL for these effects is .5 mg/kg/day; for dermal exposure, the NOEL is 50 mg/kg/day.

4. Summary of Proposed Regulatory Position

   The Agency is proposing that current registrations for the use of dinocap on agricultural and nonfood crops will be allowed to continue provided the following conditions and label modifications are met to reduce applicator exposure:

   1. require ground boom, airblast and mist blower equipment to be drawn by enclosed vehicles
   2. reclassify as a restricted use pesticide requiring the physical presence of the certified applicator
   3. require use of a developmental toxicity warning on the label:

      "Use of this product may be hazardous to your health. This product has been determined to cause birth defects in laboratory animals."

   4. require use of water-soluble packaging for wettable powder formulations

5. Contact person:

   Paul Parsons
   Review Manager
   Special Review Branch
   Registration Division
   Office of Pesticide Programs
   (703)-557-1632

DINOSEB

Date Issued: October, 1986
Fact Sheet Number: 130

1. Description of Chemical

    Common Name: Dinoseb (includes parent chemical as well as four salts)

    Chemical Name: 2-(sec-butyl)-4,6-dinitrophenol

    Trade Names: DNBP, DNOSBP, "dinitro", dinoseb (F-ISO), Caldon, Sinox, Vertac General and Selective Weed Killer, Basanite, Chemox General & PE, Chemsect, Dinitrex, Dinitro-3, Dinitro General, Drexel Dynamite 3, Dynamite, Elgetol 318, Gebutox, Hel - Fire, Kiloseb, Nitropone C, Subitex, Unicrop DNBP, Vertac Dinitro Weed Killer 5, Dynanap, Premerge Plus with Dinitro, and Klean Krop.

    EPA Shaughnessy Code: 037505

    Chemical Abstracts Service (CAS) Number: 88-85-7

    Year of Initial Registration: 1948

    Pesticide Type: Fungicide, Herbicide, Insecticide, Desiccant

    Chemical Family: Dinitrophenol

    U.S. and Foreign Producers: Baird and McGuire, Uniroyal Chemical Company, Cedar Chemical Corp., Hoechst AG, S.H. Marks Co. Ltd., Universal Crop Protection Ltd., S.N.P.E., Combinatal Chimic Fararar.

2. Use Patterns and Formulations

   Dinoseb is a contact herbicide widely used to control broadleaf weeds. Approximately 180 registered products contain dinoseb (or its 4 salts) as an active ingredient. The four salt formulations are alkanolamine, triethanolamine, sodium and ammonium. Between 7 and 11 million pounds of dinoseb active ingredient are sprayed annually as a liquid from airplanes, tractor-drawn equipment and hand held equipment. The major use sites by volumne include soybeans (40%), cotton (15%), potatoes (16%), peanuts (9%), alfalfa (4%), snap beans (2%), peas (2%), grapes (2%) and almonds (1%).

   * Uses of Dinoseb

     Agricultural Uses: Food Crops- alfalfa, almonds, clovers, apples, apricots, barley, beans, blackberries, blueberries, boysenberries, cherries, citrus, corn, cotton, cucurbits,

     currants, dates, figs, garlic, gooseberries, grapes, hops, loganberries, nectarines, oats, olives, onions, peaches, peanuts, pears, peas, pecans, plums, potatoes, raspberries, rye, soybeans, strawberries, walnuts, and wheat. Terrestrial Non-Food Crops- clovers, birdsfoot trefoil, dichondra, flax, and timothy.

     Aquatic Management: drainage ditches

     Ornamentals: bulbous iris, daffodil, gladiolus, ligustrum, lilac, narcissus, roses, spirea, tulip, yew.

     Forest Management: brush control, conifer release.

   * Methods of Application: aerial, tractor-drawn ground boom, hand-held sprayer.

   * Types of Formulations: The majority of products are soluble and emulsifiable concentrates. Other formulations include granular and flowable concentrates.

3. Scientific Findings

   * Chemical Characteristics

     Dinoseb is a dark brown/reddish brown solid or viscous liquid with a pungent odor. Depending on the technical formulation, dinoseb is soluble in toluene, petroleum oil, ether, ethyl alcohol, ethanol, n-heptane, slightly soluble in water, and miscible in ethyl ether and xylene. Dinoseb has a melting point range from 30-34°C. Technical dinoseb contains at least 90 percent active ingredient.

- Toxicological Characteristics

    Dinoseb causes developmental toxicity in laboratory animals and may pose a risk of birth defects in pregnant women. Biologically and statistically significant increases in malformations and/or anomalies were observed in a recently submitted rabbit teratology study. The primary malformations or anomalies observed were in the neurological and skeletal systems. Based upon a NOEL of 3/mg/kg/day, the Agency has calculated Margins of Safety (MOS). At all use sites the MOS is less than 100, and in many cases less than 1. The risk of birth defects is greatest for those women exposed at the dinoseb application site.

    Dinoseb is acutely toxic to humans and has been classified into Toxicity Category I (acute oral LD_{50} rat = 40-60 mg/kg). Several human fatalities have been attributed to dinoseb piosonings. The EPA Toxicity Peer Review Committee has tentatively concluded that dinoseb be classified as a Category C carcinogen (limited evidence of carcinogenicity in animals)

    due to statistically significant elevations of liver adenomas in the mid and high dose groups of test animals.

    Studies in laboratory animals indicate that dinoseb has the potential to cause male sterility, cataracts, and damage to the immune system. Also, several formulations are contaminated with nitrosamines (cancer causing substances).

- Physiological and biochemical behavorial characteristics

    Dinoseb interferes with the basic energy metabolism of cells. Specifically, it "uncouples" oxidative phosphorylation by preventing the conversion of adenosine diphosphate (ADP) to adenosine triphosphate (ATP). This is a basic energy conserving step in cell biochemistry and, when disrupted, results in other cellular changes such as increased oxygen uptake and increased permeability of mitochondria to hydrogen ions.

    The above effect on energy metabolism may be a common denominator for all toxic events observed in man and laboratory animals when exposed to dinoseb.

- Environmental Characteristics: Available data indicate that dinoseb has the potential to leach to groundwater. Residues in the range of 1-5 ppb have been found in groundwater in potato growing regions of New York, Maine, and Massachusetts.

- Ecological Characteristics: Dinoseb is highly toxic to birds, mammals, and invertebrates and may pose a risk to non target organisms including endangered or threatened species.

* Tolerance Assessment: Tolerances were established in 40 CFR 180.281 for residues of dinoseb and its hydrolyzable salts.

4. Summary of Agency's Regulatory Position and Rationale

After considering a range of regulatory options to mitigate exposure to dinoseb, the Agency has decided to emergency suspend all registrations. Data indicating that the use of dinoseb may pose a risk of inducing developmental toxic effects in humans is the basis for this action. This risk is presented not only to mixers, loaders and applicators of dinoseb, but also to persons entering treated fields or exposed through spray drift. There is no hazard to persons consuming food that has been treated with dinoseb. In conjuncture with this action, a Notice of Intent to Cancel all dinoseb registrations has been issued. The Administrator has determined that continued registration of dinoseb poses an imminent hazard during the period in which administrative hearings could delay the effectiveness of the cancellation of these registrations.

* Benefits Analysis: Suspension of all herbicide, fungicide and desiccant uses of dinoseb would be expected to result in first year losses primarily at the farm level of $80 to 90 million.

5. Contact Person

Michael McDavit
Special Review Branch, Registration Division
Office of Pesticide Programs (TS-767C)
401 M Street, S.W.
Washington, D.C. 20460
(703) 557-1787

DIPHENAMID

Date Issued: June 24, 1987
Fact Sheet Number: 136

1. Description of Chemical

    The following chemical is covered by this pesticide fact sheet:

    Common name: Diphenamid
    Chemical name: N,N-dimethyldiphenylacetamide
    CAS Number: 957-51-7
    OPP (Shaughnessy) Number: 036601
    Empirical Formula: $C_{16}H_{17}NO$
    Trade Names: Difenamid, Dymid, Enide®, A-831010, L-34-314
    Chemical Family: Acetamide
    Pesticide Type: Herbicide
    Year of Initial Registration: 1964
    Registrants of Technical Products: Nor-AM Chemical Co.

2. Use Patterns And Formulations

    Application Sites: Field, bush, vine, vegetable crops, fruit, citrus fruit, cotton, ornamental plants, forestry and turf.

    Methods of Application: Primarily applied as broadcast or banded spray by ground or aerial equipment.

    Application Rates: Terrestrial food crops: 2.0 to 9.0 lb., active ingredient (ai)/acre, Terrestrial non-food crops: 2.0 to 6.0 lb., ai/acre, Ornamentals and forest trees 0.50 to 1.25 lb., ai/acre.

    Types of Formulations: 90%, 80% & 50% (ai) wettable powder; 5% & 1.42% (ai) granular; 42.5% & 15.4% (ai) flowable concentrates.

    Usual Carrier: Water

3. Science Findings

Diphenamid related effects include increased liver weights and slight histological changes in a chronic toxicity study in dogs. Except for mild liver reactions in weanlings, this chemical does not induce reproductive effects in rats. This pesticide has shown low acute oral and dermal toxicity in test animals and it is not an eye irritant. It is low in avian toxicity and slightly to moderately toxic to fish and aquatic invertebrates.

Although the present data base for diphenamid does not indicate major toxicological concerns, there are many toxicology data gaps: including acute inhalation toxicity, primary dermal irritation, dermal sensitization, subchronic dermal toxicity (21-day), chronic toxicity (rat), oncogenicity in two species, teratology in two species, mutagenicity and general metabolism studies.

In addition, the environmental fate, product chemistry, and the metabolism in food crops, in ruminants and poultry for this compound have not been characterized.

Chemical Characteristics:

Diphenamid is a white to off-white granular solid with a sweet aromatic odor at room temperature. Its melting point is 133.8-135.4° C, and its molecular weight is 239.3. Its water solubility is 0.26 g/liter in water at 25°C.

Toxicological Characteristics:

Acute Oral: Toxicity Category III - 1,373 mg/kg (rats, both sexes)
Acute Dermal: Toxicity Category III - > 6,320 mg/kg (rabbits, both sexes)
Primary Eye Irritation: Toxicity Category IV - No irritation with 0.1 ml solution (rabbits)

Chronic Toxicity:

In a 2-year dog feeding study, males and females were offered diets containing 0, 120, 400, and 1200 ppm technical diphenamid in the diet for 103 weeks. The only compound-related effects included slightly increased liver weights and a slightly increased incidence of portal macrophages and/or fibroblasts in the livers of the dogs in the 400 and 1200 ppm dose groups. The no-observed-effect level (NOEL) was determined to be 120 ppm (3 mg/kg/day) in the diet.

Human Exposure

Major routes of exposure: Applicators and mixer/loaders handling this pesticide would be exposed primarily through skin contact and inhalation.

Physiological & Biochemical Chracteristics:

Translocation: Diphenamid is absorbed by the roots and translocated through the roots, stem, and leaves.

Mechanism of pesticidal action: It inhibits growth in terminal leaves and roots.

Metabolism & Persistence in Plants & Animals: The available data are inadequate to evaluate the persistence of diphenamid in plants and animals.

Environmental Chracteristics:

Available data are insufficient to fully assess the environmental fate and potential exposure of humans and non-target organisms to diphenamid. Additional data are required to characterize the potential for diphenamid to reach ground water supplies.

Ecological Characteristics:

There is sufficient information to characterize technical diphenamid as "slightly toxic" to freshwater invertebrates: Daphnia magna (LC_{50} of 58.0 ppm). Formulations containing 50% technical were classified as very low in toxicity to avian species: Bobwhite quail (LC_{50} of 18,000 ppm) and Mallard duck (LC_{50} of 30,000 ppm), slightly toxic to warmwater fish: Bluegill (LC_{50} of 32.0 ppm) and moderately toxic to coldwater fish: Rainbow trout (LC_{50} of 1.2 ppm). In an acute contact study, diphenamid was shown to be very low in toxicity to honey bees.

Endangered species

Since the registrant has voluntarily deleted the forestry use pattern for diphenamid products, endangered species labeling which would have been required under forestry uses will not be required in this Standard.

Tolerance Assessment

According to 40 CFR 180.230, tolerances for diphenamid have been approved for the raw agricultural commodities (RAC's) listed below.

Crop	Tolerance(ppm)
Apples	0.10
Cattle, fat	0.05
Cattle, meat	0.05
Cattle, mbyp	0.05
Cottonseed	0.10
Cotton forage	0.20
Fruiting vegetables	0.10
Goat, fat	0.05
Goat, meat	0.05
Goat, mbyp	0.05
Hogs, fat	0.05
Hogs, meat	0.05
Hogs, mbyp	0.05
Horses, fat	0.05
Horses, meat	0.05
Horses, mbyp	0.05
Milk	0.01
Okra	0.10
Peaches	0.10
Peanuts	0.10
Peanut, hay & forage	2.00
Peanut hulls	0.50
Potatoes	1.00
Raspberries	1.00
Sheep, fat	0.05
Sheep, meat	0.05
Sheep, mbyp	0.05
Strawberries	1.00
Soybeans	0.10
Soybean, hay & forage	0.50
Sweet potatoes	0.10

The Provisional Acceptable Daily Intake (PADI) for diphenamid is based on the 103 week dog feeding study with a NOEL of 120 ppm (3 mg/kg/day). Other toxicology data considered in suport of these tolerances include a 3-generation rat reproduction study indicating that reproductive performance was not affected at dietary intake up to and including 30 mg/kg/day. Utilizing a safety factor of 100, the PADI was calculated to be .03 mg/kg/day. A Tolerance Assessment System (TAS) printout has recently been completed which compares the PADI to the Theoretical Maximum Residue Contribution (TMRC). The theoretical dietary exposure expressed as a percentage of PADI utilized by the Theoretical Maximum Residue Contribution (TMRC) ranged from 3.8% to 11.8% of the PADI for various subgroups with 5.6% being the U.S. population average.

Additional data are required for plant and animal metabolism, analytical methods, and storage stability. Processing studies are required for potatoes, soybeans, tomatoes, apples and peanuts. Since the data required for individual commodities are dependent on metabolism data, the Agency recommends that metabolism data be obtained and submitted prior to any required residue data.

There is a Canadian tolerance of 1.0 ppm for diphenamid residues in or on strawberries but no tolerances for residues of diphenamid on the other raw agicultural commodities listed in 40 CFR 180.230. There are no Mexican tolerances or Codex Maximum Residue Levels for residues of diphenamid on the raw agicultural commodities listed in 40 CFR 180.230.

Existing tolerances for soybean hay and forage; and peanut hay are adequately supported by data. Therefore, the registrant may propose removal of the grazing and feeding restrictions currently in effect for diphenamid on soybeans and peanuts.

Reported Pesticide Incidences

The Pesticide Incident Monitoring System (PIMS) does not have any incident involving diphenamid at this time. Based upon data from California, five incidences indicating skin or eye irritation occurred from 1981 through 1985. Although four of the incidences involved applicators and one involved a non-applicator, none of these cases required hospitalization.

4. Summary of Regulatory Positions & Rationale:

-- The Agency will not place diphenamid in Special Review because diphenamid does not exceed any of the risk criteria for adverse effects in 40 CFR, Section 154.7. It does not pose a risk of serious acute injury to humans, domestic animals or avian species.

-- The Agency has determined that certain toxicological studies are required to support the reregistration of diphenamid products: acute inhalation, primary dermal irritation, dermal sensitization, subchronic dermal toxicity (21-day), chronic toxicity in rats, oncogenicity in two species, teratology in two species, mutagenicity and metabolism studies.

-- The Agency has determined that present precautionary statements for persons handling or applying diphenamid products are sufficient for the labels of manufacturing-use and end-use products. Available data indicate that diphenamid causes low oral (Category IV) and dermal (Category III) toxicities in test animals and it is not an eye irritant(Category IV). Therefore, the labeling of these products contain statements that caution persons applying or handling this compound, provide first aid instructions, and require the use of precautionary measures to ensure safe handling of the pesticide products.

-- The Agency will require the following environmental fate data necessary to support the reregistration of diphenamid: photodegradation in water and soil, mobility (leaching & adsorption/desorption), mobility [volatility (lab)], soil dissipation, rotational crops (confined) and accumulation in fish studies.

-- The Agency has determined that certain groundwater data are required. The necessary environmental fate data (aerobic and anaerobic metabolism studies) are being repeated at the present time. The Agency expects to receive these data by July 1987. Until these data are received and found acceptable by the Agency, data gaps for ground water exist.

-- The Agency has determined that reentry intervals for workers are not required for diphenamid products. The low acute toxicity of this chemical does not warrant significant concern about exposure of workers reentering treated areas, according to the criteria in 40 CFR Part 158.140.

--The Agency will require the following residue chemistry data necessary to support the reregistration of diphenamid: plant and animal metabolism, analytical methods, storage stability, and processing studies for potatoes, soybeans, tomatoes, apples, cottonseed and peanuts.

-- The Agency has determined that grazing and feeding restrictions for soybean hay and forage, and peanut hay are not required for diphenamid products because existing tolerances for these commodities are adequately supported by residue chemistry data.

--Since the registrant has voluntarily deleted the forestry use patterns for diphenamid products, endangered species labeling which would have been required under forestry uses will not be required in this standard.

--This pesticide, when applied at recommended rates, does not present unreasonable hazards to birds. Existing studies on diphenamid indicate that its dietary toxicity to avian species is very low.

5. Precautionary Statements

   a. Manufacturing-Use Product Statements

   All diphenamid products intended for formulation into end-use products must bear the following statements:

   "Do not discharge effluent containing this product into lakes, streams, ponds, estuaries, oceans, or public water unless this product is specifically identified and addressed in an NPDES permit. Do not discharge effluent containing this product into sewer systems without previously notifying the sewage treatment plant authority. For guidance contact your State Water Board or Regional Office of the EPA."

   b. End-Use Product Statements

   The following precautionary statements must appear on the following diphenamid EP labels:

   1. (Non-granular)

   "Do not apply directly to water or wetlands (swamps, bogs, marshes, and potholes). Do not contaminate water by cleaning of equipment or disposal of wastes."

   2. (Granular)

   "Cover, incorporate or collect granules spilled on the soil surface. Do not contaminate water by cleaning of equipment or disposal of wastes."

6. Summary of Major Data Gaps

The following data are required for this standard.

158.120 Product Chemistry data are required during 1988.

158.125 Residue Chemistry:
- 171-4 Nature of Residue (Plant & Animal Metabolism)
- 171-4 Residue Analytical Methods
- 171-4 Storage Stability
- 171-4 Residue Studies on Crops
- 171-4 Residue Studies on Processed Food/Feed Commodities

158.135 Toxicology:
- 81-3 Acute Inhalation Toxicity (Rat)
- 81-5 Primary Dermal Irritation (Rabbit)
- 81-6 Dermal Sensitization (Guinea pig)
- 82-2 Subchronic Dermal (21-day)
- 83-1 Chronic Toxicity (Rat)
- 83-2 Oncogencity (Two species) (Mouse) Ongoing Study* (Rat)
- 83-3 Teratogenicity (Two species)
- 84-2 Mutagenicity
- 85-1 Metabolism

158.130 Environmental Fate
- 161-2 Photodegradation in Water
- 161-3 Photodegradation on Soil
- 162-1 Aerobic Soil Metabolism
- 162-2 Anaerobic Soil Metabolism
- 163-1 Mobility Studies (Leaching & Adsorption/Desorption)
- 163-2 Mobility Studies (Volatility/lab)
- 164-1 Soil Dissipation
- 165-1 Rotational Crops (Confined)
- 165-4 Accumulation in Fish

158.145 Wildlife and Aquatic Organisms
- 71-1 Avian Oral Toxicity
- 71-2 Avian Dietary Toxicity (Two species)
- 72-1 Freshwater Fish Acute Toxicity (Two psecies)
- 122-2 Aquatic Plant Growth

7. Contact Person at EPA

    Robert J. Taylor, PM-25
    Office of Pesticide Programs, EPA
    Registration Division (TS-767C)
    401 M Street, S.W.
    Washington, DC 20460
    Phone (703) 557-1800

DISCLAIMER: The information presented in this Pesticide Fact Sheet is for informational purposes only, and may not be used to fulfill data requirements for pesticide registration or reregistration.

DIPROPETRYN

Date Issued: June 30, 1985
Fact Sheet Number: 55

1. Description of the chemical:

    Generic name: 2(ethylthio)-4,6-bis(isopropylamino)-s-triazine,
    Empirical formula: $C_{11}H_{21}N_5S$
    Common name: Dipropetryn
    Trade name: Sancap®, Cototar® and GS-16068.
    Chemical Abstracts Service (CAS) Registry number: 4147-51-7
    Office of Pesticides Program's EPA Chemical Code Number: 104401
    Year of initial registration: 1973
    Pesticide type: Herbicide
    Chemical family: S-triazine
    U.S. producer: Ciba-Geigy Corporation

2. Use patterns and formulations:

    Application sites: Dipropetryn is registered for preemergence control of pigweed and Russian thistle on cotton. In addition, dipropetryn is registered for use only on cotton grown on the sandy soils in Oklahoma, Texas, Arizona and New Mexico.

    Type of formulation: Dipropetryn is available in a wettable powder formulation.

    Types and methods of applications: Dipropetryn can be banded or broadcast applied to the soil surface with ground equipment or broadcast applied with aerial equipment as a preemergence spray.

    Application rates: 1.2 to 2.0 lbs a.i./A on crop sites.

    Usual carriers: Water.

3. Science Findings:

    Summary science statements:

Dipropetryn is not acutely toxic by the oral, dermal, and eye irritation routes of exposure. The available data are insufficient to show that any of the risk criteria listed in § 162.11(a) of Title 40 of the U.S. Code of Federal Regulations have been met or exceeded for the uses of dipropetryn at the present time. There are no valid chronic toxicity and mutagenicity studies for dipropetryn. There are also extensive residue chemistry and environmental fate data gaps.

Available data indicate that dipropetryn is slightly toxic to birds and moderately toxic to fish and freshwater invertebrates. A detailed ecological hazard assessment cannot be made until certain environmental chemistry data requirements are fulfilled.

Chemical characteristics:

Dipropetryn is a solid at room temperature. Its molecular weight is 255.40. The melting point is 104-106°C. Dipropetryn is soluble in water (at 20°C) to 16.0 ppm and soluble in aromatic and chlorinated hydrocarbon solvents.

Toxicological characteristics:

Acute toxicity effects of dipropetryn are as follows:

Acute Oral Toxicity in rats: >2,197 mg/kg body weight, Toxicity Category III
Acute Dermal Toxicity in rabbits: >10,000 mg/kg body weight, Toxicity Category IV
Acute Inhalation Toxicity in rats: >320 mg/l (4 hour exposure) Toxicity Category IV
Skin Irritation in rabbits: Not an irritant, Toxicity Category IV
Eye Irritation in rabbits: Not an irritant, Toxicity Category IV.

Subacute toxicity effects on dipropetryn are as follows:

A 19-week rat feeding study and a 14-week dog feeding study indicated effects on various organs at 1200 ppm dosage (the Lowest Effect Level). The No Observable Effect Level is 400 ppm.

Chronic toxicological effects of dipropetryn have not been evaluated because there are no valid chronic toxicity studies in both the rat and dog, oncogenicity studies in both the rat and mouse, teratogenicity studies in both the rat and rabbit, a two-generation reproduction study in the rat; a general rat metabolism study; and no mutagenicity tests (gene mutation in bacteria, gene mutation in mammalian cells in culture, chromosomal aberration analysis in mammalian cells in culture, or DNA damage in mammalian cells in culture).

Major routes of human exposure:

Non-dietary exposure to dipropetryn by a farmer as an applicator during mixing, loading, spraying and flagging is possible.

Physiological and biochemical behavioral characteristics:

Absorption characteristics: Dipropetryn appears to penetrate foliage rapidly, minimizing removal from foliage by rain.
Translocation: Readily translocated through the xylem from roots and foliage, accumulating in the apical meristems and leaf tips.
Mechanism of pesticidal action: Dipropetryn inhibits photolysis of water in the photosynthetic process.
Metabolism in plants: In cotton, dipropetryn's total residues are evenly distributed in the seed with crude oil extracts containing less than the intact seed.

Environmental characteristics:

Adsorption and leaching in basic soil types: Dipropetryn is more readily adsorbed on soils with high clay and organic matter content. Dipropetryn and other alkylthio-s-triazines are adsorbed to a greater extent than most other commercial triazines.
Microbial breakdown: Soil microorganisms do appear to play a significant role in degradation of dipropetryn.
Loss from photodecomposition and/or volatilization: Slight losses.
Average persistence at recommended rates: When used at recommended rates under normal environmental conditions, dipropetryn's residual activity is from 1 to 3 months.

Ecological characteristics:

Avian acute oral toxicity: >1,000 mg/kg.
96-hour fish toxicity): 3.130 ppm for bluegill sunfish (moderately toxic), and 2.430 ppm for rainbow trout (moderately toxic).

Potential problem for endangered species:

The Agency has evaluated dipropetryn under the cotton cluster/use patterns. Available data show a low order of dipropetryn toxicity suggesting that the potential hazard to endangered species is low.

Currently, there are no identifiable endangered plants that would be affected by dipropetryn.

The Agency believes that the conventional labeling for dipropetryn is adequate to properly inform the users on how to protect any endangered species.

Tolerance assessments:

Data are not available for calculating the Acceptable Daily Intake (ADI) for dipropetryn.

The Agency is unable to complete a full tolerance reassessment because the available dipropetryn toxicology and residue data do not fully support the established tolerance listed below. The metabolism of dipropetryn in animals and plants is not fully understood. Therefore, the Agency is requiring data on the metabolism of dipropetryn and related metabolite(s) in crops and animals. Additional long term rodent and nonrodent toxicological studies are also required. The additional data will be used to assess dietary exposure to dipropetryn and may lead to revisions in the existing tolerance. The Agency will not grant any significant pending or new tolerances for dipropetryn until the data are submitted.

Commodities	Parts per million
Cottonseeds	0.1

International Tolerances

Presently, there are no tolerances for residues of dipropetryn in Canada, Mexico, or in the Codex Alimentarius.

Problems known to have occurred with use:

The Pesticide Incident Monitoring System (PIMS) does not indicate any incident involving agricultural uses of dipropetryn.

4. Summary of regulatory position and rationale:

Based on the review and evaluation of all available data and other relevant information on dipropetryn the Agency has made the following determinations:

The available data are insufficient to show that any of the risk criteria listed in § 162.11(a) of Title 40 of the U.S.Code of Federal Regulations have been met or exceeded for the uses of dipropetryn at the present time.

The Agency has concerns about dipropetryn's lack of chronic toxicity data and its use on a food and feed crop when it is structurally related to certain other s-triazine pesticides which are known to be potential ground water contaminants and to cyanazine, a s-triazine pesticide that may be regulated as a teratogen.

The Agency will not allow any significant new uses to be established for dipropetryn until the toxicological, residue chemistry, and ground water data deficiencies identified in the registration standard have been satisfied.

The Agency is imposing restrictions on rotational crops. The extent of the restrictions will be reconsidered when additional data are received.

Specific label precautionary statements:

Hazard Information

The human precautionary statements must appear on all manufacturing-use products (MPs) labels as presribed in 40 CFR 162.10.

Environmental Hazard Statements

All MPs intended for formulation into end-use products (EPs) must bear the following statements:

"This pesticide is toxic to fish. Do not discharge effluent containing this product directly into lakes, streams, ponds, estuaries, oceans or public waters unless this product is specifically identified and addressed in a National Pollutant Discharge Elimination System (NPDES) permit. Do not discharge effluent containing this product into sewer systems without previously notifying the sewage treatment plant authority. For guidance, contact your State Water Board or Regional Office of the Environmental Protection Agency".

"Do not apply directly to water. Do not contaminate water by cleaning of equipment or disposal of wastes. In case of spills, cover or incorporate spills."

Restrictions on Rotational Crops

"Do not plant food and feed crops in dipropetryn-treated fields unless dipropetryn is authorized for use on those crops".

5. Summary of major data gaps and due dates:

The following toxicological studies are required:

A 90-day inhalation study (October 30, 1986),
Chronic toxicity studies and oncogenicity studies (September 30, 1989),
Teratology studies (October 30, 1986),
A two generation reproduction study (October 30, 1988),
Mutagenicity data (April-July 1986), and
A general metabolism study (July 30, 1987).

The following environmental fate data are required:

A hydrolysis study (April 30, 1986),
Photodegradation studies in water and on soil (April 30, 1986),
Metabolism tests in aerobic soil and in anaerobic soil (October 30, 1986),
A mobility test involving leaching and adsorption/desorption (July 30, 1986),
Accumulation studies on rotational crops (confined: October 30, 1988 and field: September 30, 1989), and
An accumulation study in fish (July 30, 1986).

The following ecological effects data are required:

Two subacute dietary studies (April 30, 1986), and
An acute freshwater invertebrate toxicity study (April 30, 1986).

Product chemistry data are required during 1986.

The following residue chemistry data are required:

Additional plant metabolism data (July 30, 1987), and
Metabolism studies utilizing ruminants and chickens (January 30, 1987).

6. Contact Person at EPA:

   Robert J. Taylor   (703) 557-1800
   Office of Pesticide Programs, EPA
   Registration Division (TS-767C)
   Fungicide-Herbicide Branch
   401 M Street, S.W.
   Washington, DC 20460
   Telephone (703) 557-1800

DISCLAIMER: The information presented in this Chemical Information Fact Sheet is for informational purposes only and may not be used to fulfill data requirements for pesticide registration and reregistration.

DISULFOTON

Date Issued: December 31, 1984
Fact Sheet Number: 43

1. DESCRIPTION OF CHEMICAL

    Generic Name: 0,0-diethyl S-[2-(ethylthio)ethyl] phosphorodithioate

    Common Name: Disulfoton

    Trade Name: Di-Syston

    EPA Shaughnessy Code: 032501

    Chemical Abstracts Service (CAS) Number: 298-04-4

    Year of Initial Registration: 1958

    Pesticide Type: Insecticide/Acaricide

    Chemical Family: Organophosphate

    U.S. and Foreign Producers: Mobay Chemical Corp.

2. USE PATTERNS AND FORMULATIONS

    Application Sites: Grain crops, nut crops, cole crops, root crops, pome, strawberry and pineapple fruits, forage, field and vegetable crops, sugarcane, seed crops, forest plantings, ornamentals, and potted plants (including houseplants)

    Types of Formulations: Emulsifiable Concentrates, granulars, pelleted/tableted, and ready to use liquids

    Types of Methods of Application: Soil incorporation of granulars, ground and aerial spray and granular broadcast applications

Application Rates: Range from 0.25 lbs. a.i./A to 8 lbs. a.i./A for broadcast applications and .25 oz ai.i./1000 ft. row to 10 oz. a.i./1000 foot row for band treatment; also individual potted plant soil treatment uses at various rates depending on plant and pot size.

Usual Carriers: Synthetic clays, various solvents, fertilizers

3. SCIENCE FINDINGS

Summary Science Statement

Disulfoton is very highly toxic to all mammalian systems by all routes of exposure and is assigned to Toxicity Category I, on the basis of acute toxicity requiring the most stringent labeling precautions and use restrictions. It is not considered to be oncogenic, mutagenic, or teratogenic based upon existing data. However, additional studies in a second species are being requested to fully assess oncogenic and teratogenic potential. Additional mutagenicity studies are also being required. Reproduction data is lacking and is required.

Due to the high acute toxicity and cholinesterase inhibition of disulfoton the Agency is imposing a 24-hour reentry restriction for crop uses until appropriate reentry studies and dermal absorption data are submitted and evaluated and a decision is reached whether a different time interval is more appropriate.

Data are insufficient to assess the environmental fate of disulfoton. The Agency is requesting necessary data to make this assessment and also to specifically assess whether or not disulfoton will leach into groundwater. Considering the high acute toxicity of disulfoton spray drift data are being required to measure human and non-target organism exposure resulting from spray applications and dermal and inhalation exposure data are being required to measure worker exposure in outdoor applications.

Disulfoton is very highly toxic to fish, mammals, highly toxic to birds, and moderately toxic to honey bees. Full field monitoring studies are required for the terrestrial uses to assess the exposure potential. Based on these results and on the results of the outstanding environmental fate data, chronic studies for both aquatic and terrestrial species may be required as well as full field monitoring studies for the aquatic uses. Use precautions and restriction are being imposed in the interim to reduce potential hazards.

A number of terrestrial and aquatic endangered species have been identified as at risk from the use of a number of chemicals, including disulfoton on certain crops. This issue is currently being addressed as part of a cluster approach. Interim labeling to protect these species may be necessary if the cluster analysis is not completed by 1986.

A full tolerance reassessment cannot be completed. The previous ADI was established using a rat chronic feeding study which was found to be unacceptable. The present Provisional ADI was based on a dog chronic feeding study. The percent of the PADI utilized is 169%. A second rat chronic feeding study is required, as well as animal metabolism data to quantify and qualify disulfoton oxidation metabolites in meat, milk, poultry and eggs and residue data on numerous commodities. The Agency is requiring that when the tolerance reassessment is made, after receipt of the requested data, all tolerances are to be calculated and expressed in terms of disulfoton sulfone, the major metabolite, rather than as demeton (which is how the Agency previously expressed tolerances for disulfoton).

Chemical Characteristics

Physical State: Liquid

Color: Pale yellow

Odor: Unknown

Boiling Point: 62° C at 0.01 mm/Hg

Vapor Pressure: 1.8×10^{-4} millibars at 20° C

Flash Point: >180° F(TOC)

Toxicology Characteristics

Acute Oral: 1.9 - 6.2 mg/kg, Toxicity Category I

Acute Dermal: 3.6 - 15.9 mg/kg. Toxicity Category I

Primary Dermal Irritation: NA since chemicals toxicity would preclude testing for this requirement

Acute Inhalation: One study, which did not meet Agency standards, indicated toxicity at 0.2 mg/l, which would place it in Toxicity Category I.

Neurotoxicity: One study was submitted which did not meet Agency standards. The study did not indicate delayed neurotoxic effects.

Oncogenicity: Two studies have been evaluated; one was acceptable and did not suggest oncogenic potential.

Teratogenicity: Two studies were evaluated; one was acceptable and the other did not fully meet the Agency standards. The chemical is not teratogenic at 0.3 mg/kg/day.

Reproduction - 2 generation: Data gap

Metabolism: The available studies suggest that disulfoton is rapidly absorbed and may undergo sequential oxidation steps that enhance anti-cholinesterase properties. Excretion is complete and rapid via urine. Major metabolites include the 0-analog of disulfoton, and the sulfoxide and sulfone derivatives of both disulfoton and its 0-analog. Data to further describe the nature and dynamics of this process are necessary.

Mutagenicity: Contradictory reports are available on the mutagenic potential of disulfoton. The Agency has concluded that the mutagenic potential is not adequately defined and further testing is necessary.

Physiological and Biochemical Behavioral Characteristics

Mechanism of Pesticidal Action: A plant systemic insecticide which is active by contact, ingestion, and vapor action. Disulfoton and its major metabolites are potent cholinesterase inhibitors primarily attacking acetylcholinesterase. Poisoning and death results from excessive stimulation of both the parasympathetic and central nervous systems, and the consequent myoneural junction effect as a result of acetylcholinesterase accumulation.

Symptoms of Poisoning: headache, dizziness, extreme weakness, ataxia, tiny pupils, twitching, tremor, nausea, slow heartbeat, pulmonary edema, and excessive sweating. Continual daily absorption at intermediate doses may cause influenza-like illness characterized by weakness, anorexia, and malaise.

Metabolism and Persistence in Plants and Animals:

The metabolism of disulfoton in plants is adequately understood. The major plant metabolite appears to be disulfoton sulfone. Consequently the Agency believes that the tolerances for disulfoton residues should be expressed as disulfoton sulfone.

The metabolism in animals is not well understood. More data are required to quantify and qualify animal metabolites, and to quantify plant metabolites.

Environmental Characteristics

Available data are insufficient to assess the environmental fate of disulfoton. Data gaps exist for virtually all required studies. In order to characterize the potential of the chemical to contaminate groundwater adsorption and leaching studies are being requested by the Agency.

Droplet Size Spectrum Testing and Drift Field Evaluation studies are being requested in order to determine the magnitude of exposure to non-target organisms.

Ecological Characteristics

Avian Oral:
  Mallard duck — 6.54 mg/kg
  Bobwhite Quail — 12-31 mg/kg

Avian dietary:
  Mallard duck — 510-692 ppm
  Bobwhite quail — 541-715 ppm
  ring-necked pheasant — 634 ppm

Freshwater fish:
  coldwater fish (rainbow trout) — 3.0 ppm
  warmwater fish (bluegill sunfish) — 0.039 ppm

Acute Freshwater Invertebrates: (All studies listed were not conducted according to Agency standards)

Acute Estuarine and Marine Organisms: Data gaps

Precautionary language is being required to mitigate hazards to birds, fish, and aquatic organisms. Additional labeling to protect identified endangered species may be required at a later date. Because of the lack of environmental fate and field monitoring data to quantify exposure of disulfoton to these organisms, the Agency can not quantify the hazard potential. Additional chronic toxicity studies may be required depending on the results of the environmental fate and field monitoring data.

Tolerance Assessment

The Agency is unable to complete a tolerance reassessment because of certain residue chemistry and toxicology data gaps.

Tolerances:

Commodity	Parts Per Million
alfalfa (fresh)	5.0
alfalfa (hay)	12.0
asparagus	0.1
barley (fodder, green)	5.0
barley (grain)	0.75
barley (straw)	5.0
beans (dry)	0.75
beans (lima)	0.75
beans (snap)	0.75
beans (vines)	5.0
beets, sugar (roots)	0.5
beets, sugar (tops)	2.0
broccoli	0.75
brussels sprouts	0.75
cabbage	0.75
caulifower	0.75
clover (fresh)	5.0
clover (hay)	12.0
coffee beans	0.3
corn, field (fodder)	5.0
corn, field (forage)	5.0
corn, grain	0.3
corn, pop	0.3
corn, pop (fodder)	5.0
corn, pop (forage)	5.0

Tolerances (con't):

Commodity	Parts Per Million
corn, sweet (fodder)	5.0
corn, sweet (forage)	5.0
corn, sweet, grain (kernels plus cob with husks removed)	0.3
cottonseed	0.75
hops	0.5
lettuce	0.75
oats (fodder, green)	5.0
oats (grain)	0.75
oats (straw)	5.0
peanuts	0.75
peanuts (hay)	5.0
peanuts (hull)	0.3
peas	0.75
peas (vines)	5.0
pecans	0.75
peppers	0.1
pineapples (forage)	5.0
potatoes	0.75
rice	0.75
rice (straw)	5.0
sorghum (fodder)	5.0
sorghum (forage)	5.0
sorghum (grain)	0.75
soybeans	0.1
soybeans (forage)	0.25
soybeans (hay)	0.25
spinach	0.75
sugarcane	0.3
tomatoes	0.75
wheat (fodder, green)	5.0
wheat (grain)	0.3
wheat (straw)	5.0

Based on established tolerances the theoretical maximum residue contribution (TRMC) for disulfoton residue in the human diet is calculated to be 0.2544 mg/day. The provisional acceptable daily intake (PADI) of disulfoton is 0.0025 mg/kg/day. The maximum permissable intake (MPI) for a 60 kg person is 0.15 mg/day. The percent of the ADI utilized is 169%. However a reassessment of the current tolerances based on actual constituents of the plant residues (metabolites) is necessary as well as toxicity data on the most toxic metabolite. Conformity of U.S. tolerances with Canada and Codex Alimentaruis tolerances is withheld pending receipt and evaluation of appropriate data referred to above.

U.S. tolerances for most raw agricultural commodities are not supported by current residue data. More data are required.

SUMMARY OF REGULATORY POSITION AND RATIONALE

The Agency has determined that it should continue to allow the registration of disulfoton. Adequate studies are available to assess the acute toxicological effects of disulfoton to humans. None of the criteria for unreasonable adverse effects listed in section 162.11(a) of Title 40 of the U.S. Code of Federal Regulations have been met or exceeded. However, because of certain gaps in the data base a full risk assessment of disulfoton cannot be completed.

Also, a full tolerance reassessment cannot be completed because of certain residue chemistry and toxicology data gaps.

The Agency is concerned about whether or not the potential total human exposure to disulfoton and its metabolites, both from direct and indirect human contact and the exceeded ADI, poses any unacceptable hazards. To resolve this concern, additional residue, metabolism and exposure data are required, and until it is resolved no new uses will be granted.

All end-use products formulated at greater than 2% are classified for Restricted Use, pending receipt and evaluation of appropriate acute toxicity data. Acute toxicity data on products 2% and less are being required in order to determine the appropriateness of a Restricted Use classification. These steps are being taken due to the extreme toxicity of disulfoton and the lack of product specific acute toxicity data.

A federal 24-hour reentry interval is established for treated crop areas until reentry and dermal absorption data are submitted, as required, and the Agency decides on the most appropriate time interval.

Available data are insufficient to fully assess the environmental fate of disulfoton. The Agency is requesting data to determine if disulfoton will contaminate groundwater.

Toxicity data available for disulfoton indicates that it is highly toxic to aquatic, terrestrial and avian species. Data to assess the extent of the potential exposure is currently lacking and is required to complete the hazard assessment.

5. SUMMARY OF MAJOR DATA GAPS

Additional residue data on various raw agricultural and processed commodities are being required. Also additional chronic toxicity, oncogenicity, and mutagenicity data are needed to better define the long term effects of this chemical. Plant and animal metabolism, exposure, spray drift, reentry and subchronic toxicity data are required to better qualify and quantify human exposure to residues of disulfoton and its metabolites, both from dietary and non-dietary sources.

Other requirements

Acute Inhalation
Acute oral, dermal and inhalation studies on formulating intermediates and end-use products
Acute delayed neurotoxicity
Dermal absorption study
Product Integrity study
Hydrolysis study
Photodegradation studies
Soil and Water Metabolism studies
Mobility studies
Volatility studies
Dissipation studies
Accumulation studies
Large Scale Field Monitoring studies
Acute freshwater invertebrates testing
Acute estuaring and marine organisms testing
Honey bee toxicity of residues on foliage study

6. COMPLIANCE DATES FOR REVISED LABELING

-For addition of RESTRICTED USE classification to product formulations containing greater than 2% disulfoton. All such products released for shipment after September 1, 1985 must bear RESTRICTED USE labeling. All such products in the channels of trade after September 1, 1986 must be labeled for RESTRICTED USE.

-For intrastate products the Agency is requiring submission of applications for full registration of all intrastate products containing disulfoton by December 31, 1985. Holders of such intrastate products who request withdrawal or who fail to respond to the notification, may not distribute or sell the intrastate product after December 31, 1985. Products already in the channels of trade as of that date may continue to be distributed and sold by dealers and retailers until June 30, 1986. Any product found in the channels of trade after June 30, 1986 will be considered to be in violation of FIFRA sec. 12(a)(1)(A).

CONTACT PERSON AT EPA

George T. LaRocca
Product Manager (15)
Insecticide-Rodenticide Branch
Registration Division (TS-767C)
Office of Pesticide Programs
Environmental Protection Agency
401 M Street, S.W.
Washington, D.C. 20460

Office location and telephone number:
Room 204, Crystal Mall #2
1921 Jefferson Davis Highway
Arlington, VA 22202
(703) 557-2400

DISCLAIMER: The information presented in this Chemical Information Fact Sheet is for informational purposes only and may not be used to fulfill data requirements for pesticide registration and reregistration.

DIURON

Date Issued: September 30, 1983
Fact Sheet Number: 9

1. Description of the chemical:

   Generic name: 3-(3,4-Dichlorophenyl) 1,1-dimethylurea ($C_9H_{10}Cl_2N_2O$)
   Common name: Diuron
   Trade name: Cekiuron®, Dailon®, Diater®, Di-on®, Diurox®, Diurol®, Drexel Diuron 4L®, Dynex®, Karmex®, Unidron®, Urox® and Vonduron®
   EPA Shaughnessy Code: 035505
   Chemical Abstracts Service (CAS) Registry number: 150-68-5
   Year of initial registration: 1966
   Pesticide Type: Herbicide
   Chemical family: Substituted urea
   U.S. and foreign producers: E.I. duPont de Nemours and Company, Vertac Chemical Corp., Bayer AG, Makhteshim-Agan, Pennwalt Holland B.V., Rhone-Poulenc, Staveley Chemicals Ltd., Universal Crop Protection Ltd.

2. Use patterns and formulations:

   Application sites: Diuron is a substituted urea compound registered for use as a herbicide to control a wide variety of annual and perennial broadleaf and grassy weeds on both crop and noncrop sites. Diuron is registered for use on numerous crop sites such as forage crops, field crops, fruits, vegetables, nuts, and ornamental crops. In noncrop applications, diuron is used on industrial sites, on rights-of-way, around farm buildings, and on irrigation and drainage ditches.

   Types of formulations: Diuron is available in wettable powder, granular, flowable, pelleted/ tableted, liquid suspension, and soluble concentrate formulations.

   Types and methods of applications: Diuron is applied as follows: broadcast or banded on soil surface using ground or aerial equipment.

   Application rates: 0.6 lbs. a.i./A to 8.0 lbs. a.i./A on crop sites; and 15.0 lbs. a.i./A to 48.0 lbs. a.i./A on non-crop sites.

   Usual carriers: Water, oil and clay.

3. **Science Findings:**

   Summary science statement:

   Diuron has low acute toxicity and it's uses are not expected to affect avian wildlife. But diuron is structurally related to linuron, whose studies have exhibited testicular adenomas in rats and liver cell adenomas in female mice. Therefore, the protocols for related chronic studies of diuron should reflect the oncogenic concerns raised by the linuron data.

   Chemical characteristics:

   Technical diuron is a white, crystalline, odorless solid. It is stable towards oxidation and moisture under conventional conditions and decomposes at 180-190°C. The chemical does not exhibit any unusual handling hazards.

   Toxicological characteristics:

   Acute toxicology studies on diuron are as follows:

   Oral LD_{50} in rats: 3,400 mg/kg body weight, Toxicity Category III
   Dermal LD_{50} in rabbits: > 20,000 mg/kg body weight, Toxicity Category IV
   Skin irritation in rabbits: mild irritant, Toxicity Category IV
   Eye irritation in rabbits: mild conjunctival irritant, Toxicity Category IV.

   Chronic toxicology studies on diuron are as follows:

   The requirement for a subchronic inhalation study is being deferred until an acute inhalation study has been completed.

   In a two-year chronic feeding study, the no-observed-effect-level (NOEL) was 25 ppm in male and female rats. No evidence of tumorigenicity was found.
   In a two-year feeding study, the no-observed-effect-level in dogs was 25 ppm. No evidence of tumorigenicity was found.

   Major routes of human exposure:

   Current data does not indicate that the U.S. population is exposed to diuron through the dietary or non-dietary routes.

   Physiological and Biochemical Behavioral Characteristics:

   Foliar absorption characteristics: Diuron is most readily absorbed through the root system.
   Translocation: Diuron is translocated upward primarily in the xylem.
   Mechanism of pesticidal action: It is a strong inhibitor of photosynthesis (Hill reaction).

   Environmental characteristics:

   Adsorption and leaching characteristics in basic soil types: Diuron's adsorption increases as clay content and/or organic matter content of soil increases.
   Microbial breakdown: Microbes are the primary factor in the breakdown of diuron in soils and the aquatic environment.
   Loss from photodecomposition and/or volatilization: Diuron's loss from photodecomposition is minimal.

Ecological characteristics:

Avian oral LD_{50}: >2,000 ppm
Avian dietary LC_{50}: >1,730 ppm
Fish LC_{50}: 3 to 60 ppm.

Potential problems for endangered species: Additional ecological effects data must be submitted before a complete hazard assessment can be made.

Tolerance assessments:

Tolerances are currently established for residues of the herbicide 3-(3,4-Dichlorophenyl) 1,1-dimethylurea in or on the following raw agricultural commodities:

0.1 ppm (Negligible residues) in Bananas, Nuts, and Peaches;

0.5 ppm in Papayas;

1.0 ppm in Apples; Artichokes; Barley grain; Blackberries; Blueberries; Boysenberries; Fat of cattle, goats, hogs, horses, and sheep; Meat of cattle, goats, hogs, horses, and sheep; Meat Byproducts of cattle, goats, hogs, horses, and sheep; Citrus fruits; Field corn, ear, and grain; Popcorn, ear; Sweetcorn, ear; Cotton, seed; Currants; Dewberries; Gooseberries; Grapes; Huckleberries; Loganberries; Oats grain; Olives; Pears; Peas; Pineapple; Potatoes; Raspberries; Rye grain; Sorghum grain; Sugarcane; Vetch, seed; and Wheat grain;

2.0 ppm in Alfalfa; Barley forage, hay, and straw; Clover forage and hay, Corn fodder and forage; Popcorn fodder and forage; Sweetcorn fodder and forage; Grass crops and grass hay (except Bermuda grass and Bermudagrass hay); Rye forage, hay, and straw; Pea forage and hay; Peppermint hay; Sorghum forage and fodder; Oats forage, hay, and straw; Trefoil, birdsfoot forage and hay; Vetch forage and hay; Wheat forage, hay, and straw;

4.0 ppm (food additive) in Dried citrus pulp;

7.0 ppm in Asparagus; and Bermudagrass and Bermudagrass hay.

A reassessment of the diuron tolerances indicates that those originally set for certain commodities in 40 CFR, § 180.106 were too high. The Agency will propose the reduction of certain tolerances during the next year.

Problems known to have occurred with use:

The Pesticide Incident Monitoring System (PIMS) indicated several incidents involving diuron alone from 1971 to 1980. Two fish kills were reported after aquatic areas were sprayed for weed control and the dying weeds depleted the water of oxygen. Three instances of crop injury were reported involving an accidental aerial application, the rotation of a sensitive crop onto previously treated land, and injury to wheat resulting from wet weather. Two applicators received medical attention after exposure from spraying. Symptoms included vomiting, dizziness, and diarrhea. No fatalities were reported. PIMS is a voluntary reporting system and does not include detailed followup or validation of reported incidents.

4. **Summary of regulatory position and rationale:**

   Use classification:

   General use classification:

   Use, formulation, or geographical restrictions:

   No use, formulation, or geographical restrictions are required.

   Unique label warning statement:

   Reserved pending filling data gaps.

   Summary of risk/benefit review:
   No risk/benefit assessment was conducted.

5. **Summary of major data gaps:**

   The following toxicology data are required within 48 months after receipt of this guidance package unless otherwise noted:

   Two oncogenicity tests are required, one in rat and one in another species,
   Two teratogenicity studies are required, one in rat and one in another species (rabbit),

   The following mutagenicity data are required:
   A test for gene mutations in bacterial (Salmonella typhimurium) plate test,
   A test for gene mutation in mammalian cells in culture,
   A test for DNA repair induction: in vivo mammalian sister chromatid exchange test,
   A test for chromosome effects (either in vivo or in vitro mammalian chromosome aberration analysis.

   An acute inhalation study is required within 6 months after receipt of this guidance package.

The following environmental fate data are required within 48 months after receipt of this guidance package:

Hydrolysis test,
Photodegradation test in water,
Photodegradation test in soil,
Metabolism test in anaerobic soil,
Metabolism test in aerobic aquatic site,
Mobility (volatility) test in the lab,
Mobility (volatility) test in the field,
Dissipation study in soil,
Dissipation study in an aquatic site.

The following ecological effects data are required: within 48 months after receipt of this guidance package.

Acute and chronic tests of diuron on estuarine fish, shrimp, and oysters.

An aquatic field study may be needed for the aquatic uses pending the outcome of the environmental fate studies.

6. Contact Person at EPA:

Robert J. Taylor
Product Manager (25)
Environmental Protection Agency (TS-767C)
401 M Street, S.W.
Washington, D.C. 20460
(703) 557-1800

DISCLAIMER: The information presented in this Chemical Information Fact Sheet is for informational purposes only, and may not be used to fulfill data requirements for pesticide registration and reregistration.

DODINE

Date Issued: February 1987
Fact Sheet Number: 135

1. Description of the Chemicals

Chemical Name:	Dodecylguanidine acetate	Dodecylguanidine hydrochloride (DGH)	Dodecylguanidine terephthalate (DGT)
Common Name:	Dodine		
Brand Names:	Doguadine, Tsitrex, Dodine acetate, CL 7521, Dodecylguanidine monoacetate, AC 5223, Cyprex, Melprex, Carpene, Curitan, Syllit, Venturol, and Vondodine	Cytox 2013	Durotex 7487-A
CAS Registry Number:	2439-10-3	13590-97-1	19727-17-4
EPA/OPP Pesticide Chemical Code:	044301	044303	044302
Empirical Formula:	$C_{15}H_{33}N_3O_2$	$C_{13}H_{30}N_3Cl$	Not Available
Pesticide Type:	Fungicide and industrial biocide/preservative	Industrial biocide/preservative	Industrial preservative
Major U.S. Producers:	American Cyanamid, Onyx Chemical, Aceto Chemical	Betz	Ventron Chemical

First registered: 1956

2. Use Patterns and Formulations

	Dodecylguanidine acetate	Dodecylguanidine hydrochloride (DGH)	Dodecylguanidine terephthalate (DGT)
Registered Sites:	Fruit and nut trees, ornamentals, industrial aquatic sites, pulp and paper products	Industrial aquatic sites	Industrial preservative
Predominant Use(s):	Apple trees	Industrial biocide/ preservative	Non-clothing textiles
Formulation Types Registered:	Wettable powder/ dust, soluble concentrate/ liquid	Soluble concentrate/ liquid	Soluble concentrate/ liquid
Method(s) of Application:	Foliar spray, Air blast spray, Dusting equipment drawn by tractor or truck, Boom sprayer, Aerial spray, Automatic pump, Drip Feed, Pour	Automatic pump, Drip Feed, pour	Manual pour, Conventional padding equipment

3. Science Findings

Summary: EPA has only limited data pertaining to dodecylguanidine acetate (dodine), dodecylguanidine hydrochloride (DGH), and dodecylguanidine terephthalate (DGT). The available data describing DGH and DGT are limited to some product identity and disclosure of ingredients. There are no data available describing environmental fate, ecological effects, or toxicology for DGH and DGT.

Chemical Characteristics (dodine, DGH, and DGT):

	Dodine	DGH	DGT
Color:	White	Not Available	Not Available
Physical State:	Crystals, Slightly waxy	Not Available	Not Available
Melting Point:	136C	Not Available	Not Available
Solubility:	Soluble at 7-23% in low molecular weight alcohols at room temperature. Soluble in acids and 0.06% soluble in water at 25C. Insoluble in most other solvents. Soluble in ethanol.	Not Available	Not Available
Molecular Weight:	287.4	263.9	Not Available

Environmental Fate (dodine): There are no available data allowing EPA to assess the environmental fate of dodine.

Ecological Effects (dodine): EPA does not have data describing toxicity of technical grade dodine to birds or aquatic organisms. However, data on a multiple active ingredient formulation containing 95% dodine indicate that dodine is slightly toxic to birds and highly toxic to freshwater fish. A laboratory acute contact study indicates that dodine is relatively nontoxic to honey bees.

Although the absence of data on technical dodine limits conclusions, available data suggest that there would be no immediate hazard to avian species from terrestial food crop residues. Data show that the acute LD_{50} for avian species for a formulated product are in the range of 700 to 2000 ppm. Estimation of residues to which birds will be exposed are approximately 70 - 170 ppm on leaves and 3-7 ppm on fruit. The estimated environmental concentration (EEC) in aquatic systems, 15 ppb, is well below the toxic level to fish (650 - 870 ppb). Based on the incomplete data base, terrestrial dodine food residues pose no apparent acute hazards to any avian or aquatic species that are Federally designated threatened or endangered.

Toxicology (dodine): The available data allow EPA to adequately characterize the acute effects of dodine. The findings of acute oral LD_{50} of 1.46 g/kg, a one hour LC_{50} of 1.05-1.79 mg/l, and no deaths in test animals after acute dermal exposure to 2 g/kg indicate moderate lethal potency (Category III). Severe irritation in both washed and unwashed eyes and severe dermal irritation including erythema and edema were seen in rabbits (Category I). Based on a 21-day skin sensitization study, there is no evidence of sensitization in humans.

Although many of the available studies of dodine do not satisfy EPA's data requirements, various effects of dodine can be identified based on existing studies. Subchronic dietary exposure of dogs caused changes in thyroid cells indicative of stimulation (No Observed Effects Level (NOEL) = 1.25 mg/kg). In another study, chronic dietary exposure in the rat caused reduced weight gain in both sexes, accompanied by a comparable reduction in food consumption in males (NOEL = 10 mg/kg). EPA classifies this rat study as supplementary because the histopathology analysis was inadequate. In a reproductive study, dietary exposure of parent mice caused decreases in the number of pups per litter surviving until day 5 and weaning (Lowest Observed Effect Level (LOEL) = 74-89 mg/kg. The results of an Ames assay for mutagenicity of 5 strains of bacteria with and without metabolic activation were negative.

Tolerance Reassessment (dodine): The initial Acceptable Daily Intake (ADI) and Provisional ADI (PADI) for dodine are both based on a chronic 12 month dog study. Treated dogs exhibited histological changes in the thyroid described as increased vascularity and changes in the shape of follicular epithelial cells from squamous to cuboidal. These changes are considered to be indicative of thyroid stimulation. The No Observed Effect Level (NOEL) of 50 ppm (1.25 mg/kg) reported in the study was confirmed in the reassessment for this standard.

EPA calculated a PADI of 0.0013 mg/kg for dodine. This value is based on a safety factor of 1000 for interspecific and intraspecific extrapolations and the uncertainty of risk due to gaps in the chronic data base. The PADI is equivalent to a Maximum Permissible Intake (MPI) of 0.078 mg/day. The Theoretical Maximum Residue Concentration (TMRC) of dodine in the daily diet is 0.005117 mg/kg/day (0.307 mg/day for a 60 kg person). Under these assumptions, 393.6% of the PADI is used.

4. Summary of Regulatory Positions and Rationales

— EPA will not, at this time place dodine, DGH, or DGT into Special Review because, based on available data, none of these chemicals meets or exceeds the criteria for conducting a Special Review (40 CFR Part 154.7).

— EPA will not require restricted use classification for end-use products containing dodine, DGH, or DGT. Dodine is not acutely toxic via the inhalation or oral routes. There are data demonstrating that exposure to dodine can cause severe ocular and dermal irritation. However, labeling provisions for protective clothing will minimize acute risks from use of products containing dodine.

- EPA will not require specific label statements pertaining to the protection of Federally designated threatened or endangered species because the available data for dodine do not suggest that terrestrial food residues pose acute hazards to threatened or endangered species.

- EPA is not proposing a label advisory statement regarding groundwater because there are, at present, no data suggesting cause for groundwater concerns.

- EPA is adopting the following positions regarding dodine residues and tolerances: (1) EPA will revoke, within one year of issuance of this standard, the spinach tolerance if no use for this crop is proposed. (2) EPA will revoke or replace the zero tolerances for meat and milk. (3) EPA will delete the restriction against feeding apple pomace to livestock after registrants submit data demonstrating either no residues in pomace or the need for a tolerance.

- EPA is not imposing a reentry internal for dodine at this time. However, EPA is requiring data needed to establish a reentry interval.

- Labels of products containing dodine, must bear language requiring the use of either protective clothing or closed mixing - loading systems to reduce the potential risk of skin and eye exposure.

- EPA will not allow any significant new uses of dodine, DGH, or DGT until data gaps identified in this Standard are filled because the available data are insufficient for EPA to evaluate risks associated with the uses of these chemcials.

- EPA will review immediately upon receipt of the data:

    40 CFR Part 158.125 Residue chemistry

        171-4   Nature of metabolism in plants - dodine

    40 CFR Part 158.135 Toxicology

82-1	90-day feeding - DGT
82-2	21-day dermal - dodine
83-1	Chronic toxicity, rodent and non-rodent - dodine
83-2	Oncogenicity, 2 species - dodine
83-3	Teratogenicity, 2 species - dodine
83-4	Reproduction, 2 generation - dodine

    40 CFR Part 158.145 Ecological effects

72-3	Oyster LC_{50} - dodine, DGH, and DGT
72-7	Aquatic residue monitoring - dodine, DGH, and DGT

5. Summary of Major Data Gaps

Study	Due Date		
	Dodine	DGH	DGT
Product Chemistry			
Product identity and composition	Jan. 1988	Nov. 1987	Nov. 1987
Analysis and certification of product ingredients	July 1988	May 1988	May 1988
Physical and chemical characteristics	Jan. 1987	Nov. 1987	Nov. 1987
Residue Chemistry (only dodine)			
Metabolism	July 1989		
Analytical methods	Jan. 1989		
Storage stability	July 1989		
Residue in plants	July 1988 July 1989		
Residue in animal products	Jan. 1989		
Environmental Fate (dodine and DGH)			
Degradation (laboratory)	Apr. 1988	Feb. 1988	–
Photogradation	Apr. 1988	Feb. 1988	–
Metabolism (laboratory)	Oct. 1989	Aug. 1989	–
Mobility	July 1988		
Dissipation (field)	Oct. 1989 – Sept 1991	Aug. 1989	–
Accumulation	July 1988 – Sept 1990	May 1988 July 1990	– –
Toxicology			
Acute studies	–	Apr. 1988	Feb. 1989
90 day feeding (rodent)	–	–	Sept 1988
90 day feeding (non-rodent)	–	Jan. 1989	Reserved
21 day dermal	July 1988	May 1988	Reserved
90 day dermal	–	–	Reserved
90 day inhalation	–	–	Reserved
Chronic toxicity (rodent & non-rodent)	Sept 1991	–	Reserved
Oral oncogenicity (rat and mouse)	Sept 1991	–	Reserved
Teratogenicity (2 species)	Oct. 1988	–	Reserved
Reproduction	Jan. 1990	–	Reserved
Mutagenicity	July 1988	May 1988	Reserved
General metabolism	July 1989	May 1989	Reserved
Wildlife and Aquatic Organisms			
Avian and mammalian studies	Apr. 1988 – July 1989	Feb. 1988	Feb. 1988
Aquatic organisms	Apr. 1988	Feb. 1988	Feb. 1988
Estuarine and marine organism	July 1988	May 1988	May 1988
Aquatic monitoring	July 1989	May 1989	May 1989

	Due Date		
Study	Dodine	DGH	DGT

Reentry

Foliar dissipation	Sept 1989	-	-
Dermal exposure	Sept 1989	-	-

6. Contact Person at EPA

   John Lee (PM -31)
   Registration Division (TS-767C)
   Disinfectants Branch
   401 M Street, S.W.
   Washington, D.C. 20460
   Tele: (703) 557-3675

DISCLAIMER: The information presented in this Pesticide Fact Sheet is for informational purposes only and may not be used to fulfill data requirements for pesticide registration and reregistration.

EPN

Date Issued: April 30, 1987
Fact Sheet Number: 127

1. DESCRIPTION OF CHEMICAL

    Generic Name: O-ethyl O-p-nitrophenyl phenylphosphonothioate

    Common Name: EPN

    Trade and Other Names: None

    EPA Shaughnessy Code: 041801

    Chemical Abstracts Service (CAS) Number: 2104-64-5

    Year of Initial Registration: 1949

    Pesticide Type: Insecticide

    Chemical Family: Organophosphate

    U.S. and Foreign Producers: Nissan Chemical Works, Ltd. of Japan

    There are no U.S. producers and there are no U.S. registrations for technical EPN. There were 32 end use EPN products registered in the U.S. as of April 29, 1987.

2. USE PATTERNS AND FORMULATIONS

    Application Sites: Cotton, soybeans, field corn, pecans, almonds, apples, apricots, beans (green beans, lima beans, navy beans red kidney beans, snap beans), black-eyed peas, cherries (sweet and sour), citrus (citron, grapefruit, lemons, limes, oranges, tangelos, tangerines), corn (sweet), cowpeas, grapes, kumquats, nectarines, olives, peaches, pears, pecans, plums, prunes, sugar beets, tomatoes, walnuts, and earthworm farms.

Types and Methods of Application: Foliar broadcast using aerial or ground equipment.

Application Rates: Recommended application rates range from 0.125 to 5.0 pounds of active ingredient per acre.

Types of Formulations: Emulsifiable concentrates; granular; wettable powder.

3. SCIENCE FINDINGS

Summary Science Statement

EPN is a non-halogenated, aromatic, phosphonothioate organophosphorus compound with high acute toxicity. A single oral dose of the chemical has been shown to produce organophosphate type delayed neurotoxicity in the domestic hen. EPN has been shown to be non-teratogenic. Based on all mutagenicity tests reviewed, EPN is considered to be non-mutagenic although EPN technical was weakly mutagenic in a single replicate of one of the tests utilized. Data gaps exist for chronic feeding, oncogenicity, and reproduction. EPN is highly toxic to aquatic organisms and birds. Tolerances have been established for a number of raw agricultural commodities, however additional data are required to support many of them. The theoretical maximum residue contribution (TMRC) for EPN is 0.9859 mg/kg/day. A provisional acceptable daily intake (PADI) of 0.00001 mg/kg/day has been calculated for EPN based on the most sensitive study for delayed neurotoxicity, a 90-day oral dosing study in the hen with a NOEL of 0.01 mg/kg/day for irreversible histopathological changes in the spinal cord. This effect has been identified as the most sensitive indicator of EPN toxicity. The maximum permissible intake (MPI) for EPN is calculated as 0.0006 mg/day for a 60 kg person. Based on these figures and actual residue data for 5 representative crops, the TMRC occupies 120% of the PADI (using the percent of the crop treated in the calculation). Applicators, mixer/loaders and field workers (making early reentry into treated areas) are primarily acutely exposed to EPN during their work activities. Based on average exposure values from surrogate pesticide studies, the Agency calculated the daily margin of safety for applicators to be 0.05 for cotton, 0.01 for soybeans, and 1.4 for field corn. Margins of safety for field workers, based on dislodgeable residue dissipation data for EPN and no protective clothing, were calculated to be 30 at 7 days after application to cotton; 30 at 2 days after application to soybeans, corn, and pecans; and 30 at 35 days after application to citrus. Certain uses of EPN also have sufficient exposure to pose a potential hazard to certain endangered and/or threatened species of mammals, birds, aquatic organisms, crustaceans, reptiles and insects. EPN is slightly persistent; however, insufficient data are available for the Agency to fully assess the environmental fate and transport of the compound.

Chemical Characteristics of the Technical Material

Physical State: Oily liquid (technical); crystalline powder (pure).

Color: Reddish-yellow (technical); light-yellow (pure).

Odor: Aromatic odor.
Molecular weight and formula: 323.3 - $C_{14}H_{14}NO_4P_5$.

Melting Point: 34.5°C.

Vapor Pressure: 0.03 mmHg at 100°C (technical).

Specific Gravity: 1.27 at 20°C.

Solubility in various solvents: Miscible with benzene, toluene xylene, acetone, isopropyl alcohol, and methanol; slightly soluble in in water.

Toxicology Characteristics

Acute Oral: High acute oral toxicity to mammals with toxicity values of approximately 52.8 mg/kg/body weight in male rats and 13.2 mg/kg/body weight in female rats.
Toxicity Category I.

Acute Dermal: 354 mg/kg/body weight in male rabbits and 500 mg/kg/body weight in female rabbits.
Toxicity Category II.

Primary Dermal Irritation: Technical EPN does not produce dermal irritation.
Toxicity Category IV.

Primary Eye Irritation: Technical EPN does not produce eye irritation
Toxicity Category IV

Skin Sensitization: Data gap.

Acute Inhalation: Extremely acute inhalation toxicity with values of 0.076 mg/L in male rats and 0.024 mg/L in female rats
Toxicity Category I

Major routes of exposure: Dermal exposure, with some inhalation exposure, to applicators.

Delayed neurotoxicity: EPN causes organophosphate-type delayed neurotoxic effects in test animals. Histopathological changes were seen at 0.1 mg/kg/day; clinical effects (ataxia were seen at 2.5 mg/kg/day.

Oncogenicity: Data gap.

Chronic Feeding: Data gap.

Metabolism: Partial data gap. Data presently available to the Agency show the existence of several possible metabolites, however, these metabolites were not identified. Further work must be performed to identify the metabolites of EPN.

Teratogenicity: EPN is not teratogenic.
Reproduction: Data Gap.

Mutagenicity: EPN is not considered to be mutagenic based on all of the mutagenicity tests reviewed, although EPN technical was weakly mutagenic in a single replicate of one of the tests utilized.

Physiological and Biochemical Characteristics

Mechanism of Pesticidal Action: Cholinesterase inhibition following contact with, or ingestion of, EPN treated surfaces.

Metabolism and Persistence in Plants and Animals: Available data indicate that EPN is slightly persistent, however, these data are insufficient to fully assess the transport of EPN. Although the metabolism of EPN in plants is not adequately understood, detailed characterization of residues in whole, 10-week old cotton plants has revealed the presence of EPN and two metabolites, O-ethyl phenylphosphonic acid and phenylphosphonic acid. This study also revealed that ^{14}C-residues of [^{14}C] EPN are absorbed into plant tissues following foliar application. Submitted data pertaining to the metabolism of EPN in animals indicate that EPN residues will transfer to tissues of poultry, but otherwise are inadequate to show the nature of those residues in poultry or other animals.

Environmental Characteristics

Available data are insufficient to fully assess the environmental fate and transport of EPN and the potential exposure of humans and nontarget organisms to EPN. Data gaps exist for nearly all applicable studies. However, available preliminary information indicate general trends of EPN behavior in the environment. EPN is slightly persistent and degrades in aerobic sandy loam soil with a half-life of 4 to 8 weeks. Phenyl phosphonic acid, O-ethyl phenyl phosphonic acid, and O-ethyl phosphonothioic acid are expected to be the main degradates under aerobic conditions. Data currently available are insufficient to characterize EPN's leaching potential for contamination of ground water. Data to characterize the potential of EPN to contaminate groundwater are being required. Treated areas should not be re-entered for at least 7 days for corn or cotton crops, 35 days for citrus, and 2 days for all other crops, unless protective clothing is worn

Ecological Characteristics

Avian acute toxicity: LD_{50} values of 7.09 to 27 mg/kg for mallard duck; 53.4 mg/kg for ring-necked pheasants, and 5.25 mg/kg for coturnix.

Avian dietary toxicity: 8-day Dietary LC_{50} values of 168 ppm in mallard duck and 349 ppm in bobwhite quail.

Freshwater fish acute toxicity: 96-hour LC_{50} values ranged from 80 ug/L for rainbow trout to 190 ug/L for bluegill.

Marine fish acute toxicity: 96-hour LC_{50} values ranged from 37 ug/L for Spot to 140 ug/L for Sheepshead.

Freshwater invertebrate toxicity: 48-hour LC_{50} values ranged from 0.32 ug/L for Daphnia magna to 36.0 ug/L for Gammarus lacustris.

Marine invertebrate toxicity: 96-hour LC_{50} values ranged from 4.6 ug/L for Penaeus Stylirostris to 13.0 ug/L for Mysidopsis bahia.

TOLERANCE REASSESSMENT

Tolerances have been established for residues of EPN in a variety of raw agricultural commodities (40 CFR 180.119). The Agency has evaluated the residue and toxicology data supporting these tolerances and has determined that a full tolerance reassessment for EPN cannot be made at this time because of extensive residue chemistry and toxicology data gaps. Because of these extensive data gaps, no significant new uses, including group tolerances, will be granted until the Agency has received data sufficient to thoroughly evaluate the dietary exposure to EPN. (The Agency has actual residue data for five crops: soybeans, drybeans, tomatoes, corn and cotton. Taking the percent of crop treated into account, the Agency believes that it is unlikely that EPN residues on crops will be present at the tolerance levels listed.

The present United States tolerances and Mexican tolerances are listed below. No Canadian tolerances or Codex MRLs have been established for EPN.

Summary of Present EPN Tolerances

Commodity	Tolerance (ppm) United States	Mexico
Apples	3.0	3.0
Apricots	3.0	---
Beans	3.0	3.0
Beets	3.0	---
Beet greens	3.0	---
Blackberries	3.0	---
Boysenberries	3.0	---
Cherries	3.0	---
Citrus fruits	3.0	3.0
Corn	3.0	3.0
Dewberries	3.0	---
Grapes	3.0	3.0
Lettuce	3.0	3.0
Loganberries	3.0	---
Nectarines	3.0	---
Olives	3.0	---
Peaches	3.0	3.0
Pears	3.0	3.0
Pineapples	3.0	3.0
Plums (fresh prunes)	3.0	---
Quinces	3.0	---
Raspberries	3.0	---
Rutabagas	3.0	---
Rutabagas tops	3.0	---
Spinach	3.0	3.0
Strawberries	3.0	3.0
Sugarbeet tops (but not sugar beet tops)	3.0	---
Tomatoes	3.0	3.0
Turnips	3.0	---
Turnip greens	3.0	---
Youngberries	3.0	---
Almonds	0.5	---
Cottonseed	0.5	0.5
Pecans	0.5	---
Walnuts	0.5	---
Soybeans	0.05	0.05
Nuts	---	0.5

The data for EPN residues in or on beans (dried), cottonseed, tomatoes, and corn are adequate to support the respective established tolerances. However, the Agency requires that the tolerance for residues in or on corn be changed to two tolerances, each at 3 ppm, for residues in or on field corn grain and sweet corn (kernels plus cob with husks removed) and that the tolerance for residues in or on beans be changed to three tolerances for residues in or on dried beans at 3 ppm, and lima and snap beans for which additional data are required.

Data are not adequate to support the established tolerance for residues in or on almonds, apples, apricots, beans (snap and lima only), cherries, citrus, grapes, lettuce, nectarines, olives, peaches, pears, pecans, plums, soybeans, sugar beets (without tops), and walnuts.

There are currently no registered use for the following crops for which tolerances are established: beets and beet greens, blackberries, boysenberries, dewberries, loganberries, pineapples, quinces, raspberries, rutabaga spinach, strawberries, turnips and turnip greens, and youngberries. The Agency will revoke the currently established tolerances for these raw agricultural commodities, unless the registrant submits usage proposals and the required data to support the tolerance.

The Agency will also move to revoke the tolerance for residues of EPN in or on rutabaga tops since rutabaga tops are not presently considered a raw agricultural commodity of rutabagas and no registered use of EPN on rutabagas exists.

The theoretical maximum residue contribution (TMRC) for EPN is 0.9859 mg/kg/day. A provisional acceptable daily intake (PADI) of 0.00001 mg/kg/ day has been calculated for EPN based on the most sensitive study for delayed neurotoxicity, a 90-day oral dosing study in the hen with an LEL of 0.1 mg/kg/day for histopathological evidence of toxicity in the spinal cord, a NOEL of 0.0 mg/kg/day and a safety factor or uncertainty factor of 1000. Histopathologi changes in the spinal cord have been identified as the most sensitive indica of EPN toxicity. The maximum permissible intake (MPI) for EPN is calculated as 0.0006 mg/day for a 60 kg individual. Based on these figures the TMRC is equivalent to 164,000% of the PADI. However, the Agency believes that this figure is actually much lower. For five crops for which the Agency has actu field residue data, the TMRC occupies 120% of the PADI, assuming the appropr percent of crop treated for each crop.

4. Required Unique Labeling and Regulatory Position Summary

The Agency is initiating a Special Review for all registered uses of EPN based on the results of the delayed neurotoxicity studies, and the risks to the public from consumption of food commodities containing EPN residues and risks to workers involved with EPN application and working in fields treated with EPN. The use of EPN meets or exceeds the criteria for adverse effects (40 CFR 154.7 (a)(2)).

The Agency previously reviewed EPN in the special review process in 1979 when a Notice of Rebuttable Presumption Against Registration was issued (44 FR 54384) based upon studies showing that EPN caused delayed neurotoxic effects in test animals, and was acutely toxic to aquatic organisms. The Agency concluded that special review with publication of its final notice of determination in the Federal Register on August 31, 1983 (48 FR 39494) announcing:

- ° the cancellation of the mosquito larvicide use of EPN;
- ° the prohibition of the use of human flaggers during aerial application of EPN unless the flaggers were in totally enclosed vehicles;

- ° the requirement for standardized labeling statements
for the use of protective clothing;
- ° the requirement for labeling statements regarding
drift of EPN from treated areas; and
- ° the requirement for labeling statements warning of
the hazard of EPN application to crops visited
by bees.

The previous special review of EPN relied primarily on the NOEL of 0.1 mg/kg/day for EPN-induced depression of plasma and red blood cell cholinesterase levels in humans as the NOEL for delayed neurotoxicity. On that basis, the Agency calculated dietary and applicator risks posed by continued uses of EPN. The Agency concluded at that time that adequate margins of safety existed for human dietary exposure and for applicators with the exception of flaggers.

Information received since that time on various cholinesterase-inhibiting compounds indicates an effect on blood cholinesterase is not the most sensitive indicator of toxicity for organophosphates. New information on recovery after a single large dose of EPN in hens indicates that the spinal histopathological changes are a more sensitive indicator of toxicity and that these changes are irreversible. This finding has led the Agency to conclude that the most appropriate NOEL to use for risk assessment purposes and the most sensitive indicator of potential human toxicity for this histopathological effect is 0.01 mg/kg/day from a 90-day subchornic feeding study in hens.

No significant new uses, including group tolerances, will be granted until the Agency has received data sufficient to thoroughly evaluate the dietary exposure to EPN.

The Agency is continuing the restricted use classification of all liquid formulations and any formulation greater than 4 % EPN. Application must be made by, or under the direct supervision of, Certified Applicators. Direct supervision for EPN products is defined as the Certified Applicator being physically present during mixing, loading, equipment repair, and equipment cleaning. Applicators must ensure that all persons involved in these activities under their direct supervision are informed of the precautionary statements.

The Agency is requiring a label statement concerning the histopathological changes in the spinal cord to be used in conjunction with the restricted use statement.

Endangered species labeling will be required at a later date for certain use patterns of EPN. Specific labeling requirements will be imposed through a Pesticide Registration (PR) Notice.

Preliminary evaluation of recently submitted toxicity and dissipation data indicate that the 24-hour reentry interval established in 1974 for EPN under 40 CFR 170.3 (b)(2) does not provide adequate protection for field-workers. Therefore, until the Agency receives acute delayed neurotoxicity data on which to base the risk assessment to determine the most effective reentry interval for EPN, the Agency is requiring the following reentry intervals: 7 days for corn and cotton; 35 days for citrus; and 2 days for all other crops.

Work safety rules, precautionary statements, and protective clothing statements for mixer/loaders and applicators are required to be included on the label of EPN products.

All EPN end-use product labeling is required to contain work safety rules, precautionary statements, and protective clothing statements.

The Agency is requiring processing data for the following raw agricultural commodities: sugar beets, soybeans, tomatoes, citrus, prunes, grapes, apples, cottonseed, corn, and olives.

The Agency is requiring the addition of the telephone number of the National Pesticide Telecommunications Network to all end-use EPN products.

While data gaps are being filled, currently registered manufacturing use products and end use products containg EPN may be sold, distributed, formulated, and used, subject to the terms and conditions specified in the Registration Standard for EPN. Registrants must provide or agree to develop additional data in order to maintain existing registrations.

5. <u>Summary of Major Data Gaps</u>

   <u>Toxicology</u>
   Acute delayed neurotoxicity - single dose NOEL
   Dermal sensitization
   Chronic toxicity   (two species - rodent and non-rodent)
   Oncogenicity study   (two species - rat and mouse preferred)
   Reproductive study
   Metabolism

   <u>Environmental Fate</u>

   Hydrolysis
   Photodegradation, water
   Photodegradation, soil
   Aerobic metabolism
   Anaerobic metabolism
   Leaching and adsorption/desorption
   Soil dissipation
   Long-term soil dissipation
   Rotational crop (confined)
   Rotational crop (field)
   Fish accumulation
   Droplet size spectrum
   Spray drift field evaluation

EPN   345

Ecological Effects

Avian reproduction study
Fish early life-stage study
Aquatic invertebrate life-cycle study
Aquatic monitoring or mesocosm study

Residue Chemistry

Storage stability study
Plant metabolism study
Animal metabolism study
Residue data for almonds, apples, apricots, beans (snap and lima), cherries, citrus, grapes, lettuce, nectarines, olives, peaches, pears, pecans, plums, soybeans, sugar beets (without tops), and walnuts.
Residue data and usage proposal for beets and beet greens, blackberries, boysenberries, dewberries, loganberries, pineapples, quinces, raspberries, rutabagas, spinach, strawberries, turnips and turnip greens, and youngberries. (tolerances for these items will be revoked if residue data and usage proposal are not submitted)
Processing data for residues of EPN in sugar beets, soybeans, tomatoes, citrus, prunes, grapes, apples, cottonseed, corn, and olives.
Residue data and tolerance proposals for bean vines and hay, sugar beet tops.

Special Processing Studies to provide:

>   Residue data for cooked (microwaving and boiling) sweet corn.
>   Residue data on fresh unwashed tomatoes
>   A washing study to provide residue data on lettuce

7. CONTACT PERSON AT EPA

    Dennis Edwards
    Acting Product Manager (12)
    Insecticide-Rodenticide Branch
    Registration Division (TS-767C)
    Office of Pesticide Programs
    Environmental Protection Agency
    401 M Street, S. W.
    Washington, D. C.   20460

    Office location and telephone number:

    Room 202, Crystal Mall #2
    1921 Jefferson Davis Highway
    Arlington, VA   22202
    (703) 557-2386

DISCLAIMER: The information presented in this Chemical Information Fact Sheet is for informational purposes only and may not be used to fulfill data requirements for pesticide registration and reregistration.

EPTC

Date Issued: October 3, 1983
Fact Sheet Number: 6

1. Description of the chemical:

   Generic name: S-ethyl dipropylthiocarbamate ($C_9H_{19}NOS$)
   Common name: EPTC
   Trade name: Chemolimpex®, Eptam®, and Eradicane®
   EPA Shaughnessy Code: 041401
   Chemical Abstracts Service (CAS) Registry number: 759-94-4
   Year of initial registration: 1969
   Pesticide Type: Herbicide
   Chemical family: Thiocarbamate
   U. S. and Foreign Producer: Stauffer Chemical Company

2. Use patterns and formulations:

   Application sites: EPTC is a selective "thiocarbamate" herbicide which is registered for use in preemergent control of certain annual grasses, broadleaf weeds and perennial weeds, such as quackgrass, bermudagrass, and nutsedge on field, vegetable, orchard, ornamental and noncrop sites.
   Types of formulations: EPTC is available in granular and emulsifiable liquid concentrate formulations.
   Types and methods of application: EPTC must be incorporated in the soil by discing, applied with subsurface injection equipment, or metered into irrigation water to obtain proper weed control. The specific method of application and type of equipment are determined by site, formulation, and availability of equipment.
   Application rates: 2.0 lbs. a.i./A to 7.0 lbs. a.i./A
   Usual carriers: Water and clay.

3. Science Findings:

   Summary science statement:

   EPTC is potentially a strong eye irritant but it has low acute toxicities. Since EPTC is incorporated into the soil immediately or just after planting, it will not present a hazard to birds and aquatic organisms. At least two chronic toxicology studies are required to complete the data base.

   Chemical characteristics:

   Technical EPTC is a light yellow liquid. At room temperature, EPTC is a liquid and it has an amine odor. EPTC boils at ca 235° C. The chemical does not present any unusual handling hazards.

Toxicological characteristics:

    Technical EPTC is potentially a strong eye irritant (Toxicity Category II). The available data also indicate that technical EPTC has relatively low acute oral, dermal, and inhalation toxicities, and low potential for primary dermal irritation.

    Studies on butylate (a close structural relative of EPTC), as well as studies on EPTC itself, have indicated effects of impaired clotting function in test animals. Therefore, any additional studies on EPTC must include a test for clotting function and hematology in the protocol design to better define the effects on clotting and clearly establish a no-observed-effect-level (NOEL).

Acute toxicology studies on EPTC are as follows:

    Oral LD_{50} in rats: 1,652 mg/kg body weight, Toxicity Category III
    Oral LD_{50} in mice: 3,160 mg/kg body weight, Toxicity Category III
    Inhalation LC_{50} in rats: 4.3 mg/l, Toxicity Category III
    Dermal LD_{50} in rabbits: 2,750 mg/kg bodyweight, Toxicity Category III
    Primary Eye Irritation in rabbits: severe eye irritant, Toxicity Category II
    Primary Dermal Irritation in rabbits: mild skin irritant, Toxicity Category IV.

Chronic toxicology studies on EPTC are as follows:

    In a 54-week feeding study of EPTC in rats: the NOEL was 20 mg/kg/day; the LEL was 80 mg/kg/day.
    In a two year chronic feeding and oncogenicity study of EPTC in mice: the NOEL was 20 mg/kg/day; the LEL was 80 mg/kg/day and no evidence of tumorigenicity was found.
    General metabolism: EPTC studies in male and female rats were reviewed and satisfied the current data requirements for animal metabolism.

Major routes of human exposure:

    Current data does not indicate that the U.S. population is exposed to EPTC through the dietary or non-dietary routes.

Physiological and Biochemical Behavioral Characteristics:

    Foliar absorption: EPTC is absorbed mostly through the plant roots with little or no foliar penetration.
    Translocation: EPTC is readily absorbed by roots and translocated upward to the leaves and stems.
    Mechanism of pesticidal action: EPTC disrupts the growth of meristematic regions of the leaves and protein synthesis.
    Metabolism and Persistence in Plants: EPTC is rapidly metabolized by plants to CO_2 and common plant constituents (amino acids and fructose).

Environmental characteristics:

    Adsorption and leaching characteristics in basic soil types: EPTC is adsorbed onto dry soil. The amount of leaching decreases as clay and organic matter increases.
    Microbial breakdown: Microbes are the primary factor in the breakdown of EPTC in soils.
    Loss from photodecomposition and/or volatilization: EPTC is readily lost from volatilization unless it is immediately incorporated in the soil at time of application.
    Average persistence at recommended rates: The half life of EPTC in moist loam soil at 21 to 27C° is approximately one week.

Ecological characteristics:

    Avian oral LD_{50}: > 26,000ppm,
    Avian dietary LC_{50}: > 20,000 ppm,
    Fish LC_{50}: 17.0 ppm to 27.0 ppm,
    Problems with aquatic organisms are not anticipated because an estimated environmental concentration of 0.141 ppm in EPTC runoff is well below the LC_{50} for aquatic organisma.
    The low toxicity and placement of EPTC into the soil immediately prior to or just after planting should reduce exposure and provide an adequate safety margin.

Tolerance assessment:

    Tolerances are established for negligible residues (N) of the herbicide S-ethyl dipropylthiocarbamate in or on the following raw agricultural commodities:

Commodity	Parts per million
Almonds, hulls	0.1(N)
Asparagus	0.1(N)
Beans, castor	0.1(N)
Cotton, forage	0.1(N)
Cottonseed	0.1(N)
Flaxseed	0.1(N)
Fruits, citrus	0.1(N)
Fruits, small	0.1(N)
Grain crops	0.1(N)
Grasses, forage	0.1(N)
Legumes, forage	0.1(N)
Nuts	0.1(N)
Pineapples	0.1(N)
Safflower, seed	0.1(N)
Strawberries	0.1(N)
Sunflower, seed	0.1(N)
Vegetables, fruiting	0.1(N)
Vegetables, leafy	0.1(N)
Vegetables, roof crop	0.1(N)
Vegetables, seed and pod	0.1(N)

Problems known to have occurred with use:

> The Pesticide Incident Monitoring System (PIMS) indicated at least four cases involving EPTC alone from 1966 to 1979. In these reports, agricultural workers such as mixers, loaders and applicators received medical treatment after the pesticide contacted their eyes. The exposures were caused by equipment failure, or splashing and resulted in irritation, swelling, redness, and inflammation of the eyes. No fatalities were reported. PIMS is a voluntary reporting system and does not include detailed followup or validation of reported incidents.

4. Summary of regulatory position and rationale:

   Use classification:

   General use classification.

   Use, formulation or geographical restrictions:

   No use, formulation, or geographical restrictions are required.

   Unique label warning statement:

   None

   Summary of risk/benefit review:
   No risk assessments were conducted.

5. Summary of major data gaps:

   The following toxicological studies are required within four years after the receipt of the guidance package.

   A 1-year or longer dog feeding study,
   A rat oncogenicity study,
   A teratology study in two species*,
   A 2-generation reproduction study**.

   * A teratology study (CDL:247780) submitted by Stauffer Chemical Company is pending review by the Agency.

   ** A reproduction study (CDL:249077) submitted by Stauffer Chemical Company is pending review by the Agency.

   The requirement for a subchronic neurotoxicity study is being deferred until an acute neurotoxicity study has been completed in order to determine whether the subchronic study is necessary. This study must be submitted within six months after receipt of the guidance package.

The following environmental fate data are required within four years after the receipt of the guidance package.

Hydrolysis test,
Photodegradation test in water,
Metabolism test in anaerobic soil,
Mobility (volatility) test in the lab,
Dissipation study in soil,
Accumulation study in combined rotational crops.

6. Contact Person at EPA:

Robert J. Taylor
Product Manager (25)
Registration Division (TS-767C)
Fungicide-Herbicide Branch
401 M Street, S. W.
Washington, D. C. 20460
(703) 557-1800

DISCLAIMER: The information presented in this Chemical Information Fact Sheet is for informational purposes only and may not be used to fulfill data requirements for pesticide registration and reregistration.

ETHALFLURALIN

Date Issued: June 30, 1985
Fact Sheet Number: 58

1) Description of Chemical

   Generic Name: N-ethyl-N-(2-methyl-2-propenyl)-2,6-dinitro-4-(trifluoromethyl) benzenamine

   Common Name: ethalfluralin

   Chemical Abstracts Service (CAS) number:

   Year of initial registration: 1983

   Pesticide type: herbicide

   U.S. producer: Elanco Products Co., Division of Eli Lilly & Co.

2) Use Patterns and Formulations

   Emulsifiable concentrate: pre-plant incorporated (PPI) to dry peas, dry beans, soybeans

   surface applied, cucurbits direct seeded: cantaloupe, cucumber, pumpkin, watermelon

3) Summary of Science Findings

Chemical Characteristics (From Elanco General Summary 161/ZGSUm1/AM/1)

Ethalfluralin is a yellow crystalline solid with a faint amine odor. It has a molecular weight of 333.3, a specific gravity of 1.32/ml, a melting range of 57 to 59°C, an n-octanol-to-water partition coefficient of 5-11, a vapor pressure of 8.2×10^{-5} mmHg at 25°C, and is susceptible to decomposition by ultraviolet light. Ethalfluralin is readily soluble in organic solvents; its solubility in water is 0.3 ppm at 25°C.

Toxicological characterisitics:

Ethalfluralin is considered a skin sensitizer.
E.C. formulation is irritating to the eyes and skin.
E.C. formulation is toxic to fish.
Toxicology studies on ethalfluralin are as follows:

Oral LD_{50}, rat: greater than 10g/kg
Oral LD_{50}, mouse: " " 10g/kg
Oral LD_{50}, dog: greater than 200 mg/kg
Oral LD_{50}, cat: " " 200 mg/kg

Acute dermal LD_{50}, rabbit: greater than 2 gm/kg
Primary eye irritation, rabbit: conjunctivitis; no corneal irritation
Acute inhalation LC_{50}, rat: greater than 0.028 mg/L/hour

Toxicology studies on E.C. formulation are as follows:
Oral LD_{50}, rat: greater than 2mL/kg
Acute dermal LD_{50}, rabbit: greater than 2mL/kg
Primary dermal irritation, rabbit: slight to moderate irritation
Primary eye irritation, rabbit: slight iritis & conjunctivitis; corneal dulling reversible in 21 days
Acute inhalation LC_{50}, rat: greater than 74.4 uL/L/hour

Ethalfluralin is oncogenic in the rat: mammary gland fibroadenomas (benign) in females at mid (250) and high (750) ppm dose levels. Using a "one-hit" model, the worst-case dietary oncogenic risk is calculated to be 3.77 incidences in one million.

Ethalfluralin is considered a teratogen in the rabbit: teratogenic No-observed-effect level (NOEL) is 75mg/kg

Environmental Characteristics

Ethalfluralin binds readily to soil particles and is, thus, not prone to leaching. It is volatile and readily photodegraded. Ethalfluralin concentrations in water would be expected to decline rapidly with a half-life of 1-2 days. Ethalfluralin residues may transport to aquatic environments via soil erosion. Label warnings include: Do not apply directly to any body of water or wetlands. Runoff or erosion from treated areas may be hazardous to fish in neighboring areas.

Ecological Characteristics:

Avian acute oral LD_{50}: Bobwhite quail > 2000 mg/kg
Avian dietary LC_{50}: Bobwhite quail > 5000 ppm
Avian dietary LC_{50}: Mallard duck > 5000 ppm
Fish acute LC_{50}: Bluegills: 0.032 ppm (highly toxic)
             Rainbow trout: 0.037 ppm (highly toxic)
Aquatic invertebrate acute LC_{50}: Daphnia: 0.365 ppm (highly toxic)
mallard and bobwhite one-generation reproduction: supplementary dietary levels up to 1000 ppm did not affect reproduction.

Tolerance assessment:

Tolerance levels of 0.05 ppm established for parent compound ethalfluralin in commodities: dry beans, dry peas, soybeans and cucurbit vegetable group. Real residues of ethalfluralin detectable in forage items, thus label restriction: Do not graze or forage crop grown in treated soil or cut for hay or silage.

Summary of Regulatory Position and Rationale

Ethalfluralin is regulated as an oncogen and teratogen. Benefits associated with registered crop uses are considered to outweigh oncogenic risks. There are adequate margins of safety for teratogenic effects. Analysis of benefits and risks is provided in an Agency position document accompanying the tolerance regulation published December 21, 1983, in the FR for this pesticide.

Summary of major data gaps

Chronic non-rodent feeding (dog)
Teratology: second species

ETHOPROP

Date Issued: February 1, 1985
Fact Sheet Number: 3.1

1. Description of chemical:

    Generic name: O-ethyl S,S-dipropyl phosphorodithioate
    Common name: ethoprop
    Trade name: Mocap®
    EPA Shaughnessy code: 041101
    Chemical Abstracts Service (CAS) number: 13194-84-4
    Year of initial registration: 1967
    Pesticide type: soil insecticide-nematicide
    Chemical family: organophosphate
    U.S. and foreign producers: Rhone-Poulenc, Inc.

2. Use patterns and formulations:

    Application sites: registered for use on a variety of tropical fruits, vegetables, ornamentals, field crops, commercial turf, and home lawns.
    Types of formulations: granulars, emulsifiable concentrates.
    Types and methods of application: applied as a soil incorporated, broadcast, or band treatment, as a root dip treatment for citrus and certain ornamentals. Ethoprop formulations are commonly applied by using ground equipment and are incorporated into the soil immediately after application with the use of cultivating equipment and/or by irrigation. Also, spray equipment (i.e., backpack sprayers) and watering cans may be used for application.
    Application rates: vary according to formulation and crop.
    Usual carriers: Confidential Business Information

3. Science findings:

    Summary science statement:

    Adequate studies are available to assess the acute toxicological hazards of technical ethoprop. No toxicological hazards of concern were identified. Available studies indicate that ethoprop is very toxic to birds, marine/estuarine crustaceans and fish species. It is moderately toxic to coldwater fish species and moderately toxic to honeybees.

    Although a full tolerance and risk assessment cannot be completed due to a number of data gaps, there is no evidence

to suggest that current tolerances are likely to expose
the public to unreasonable adverse effects.

Chemical characteristics:

Ethoprop is a clear yellow-tinted liquid with a strong
mercaptan odor. The empirical formula is $C_8H_{19}O_2PS_2$
and the molecular weight is 242.307. The boiling point
is 86-91°C at 0.2 mmHg. Ethoprop is soluble in water to
750 ppm and soluble in most organic solvents.

Toxicological characteristics:

Currently available toxicology studies on ethoprop are as
follows:

- Oral LD_{50} in rats: 56.2 mg/kg for males and 30.2 mk/kg
    for females.
- Dermal LD_{50} in rabbits: 23.7 ul/kg
- Primary dermal irritation: Unknown since death prevented
    manifestation of a skin irritation response.
- Primary eye irritation: Unknown since all rabbits
    died within 1 hr of administration. Substance is
    too toxic to determine an eye irritation potential.
- Inhalation LC_{50}: 0.123 mg/liter
- Acute delayed neurotoxicity: Hens treated with ethoprop
    were not shown to produce signs of delayed neurotoxicity.
- Teratology in rats: Maternal NOEL=1.6 mg/kg; terata
    NOEL=1.6 mk/kg
- 90-day dog feeding study: Systemic NOEL=100 ppm at
    the highest dose tested; cholinesterase NOEL=1 ppm.

Adequate studies are available to assess the acute toxicological effects of technical ethoprop. No toxicological
hazards of concern have been identified in the studies reviewed for the Standard.

Physiological and biochemical behavioral characteristics:

Metabolism and persistence in plants and animals: ethoprop
  in bean and corn plants is metabolized by hydrolysis
  following uptake from the soil. The metabolites identified are ethyl propyl sulfide, propyl disulfide, ethyl
  propyl sulfoxide and water soluble high-boiling phosphoric
  acids and/or their salts. The only residue of toxicological concern is the parent compound.

  There are no data to indicate how ethoprop is metabolized
  or excreted by animals. In one feeding study in which
  dogs were fed unlabeled ethoprop for 21 days at rates

up to 2.3 ppm of their diet, no detectable residues
(i.e. >0.01 ppm) were found in their tissues. Because
feed items with established negligible residue tolerances
(<0.02 ppm) are expected to have little, if any, ethoprop
residues, it is concluded that the metabolism in animals,
although not completely known, is at this time sufficiently
detailed.

Environmental characteristics:

Ethoprop is not expected to contaminate drinking water
supplies. It degrades fairly rapidly with half-lives of
3-56 days in soil. Increase in soil temperature tends to
increase the dissipation rate. However, ethoprop is very
mobile in sandy soil and has a potential for contaminating
groundwater in areas of sandy soil with a high water table.
Because of this, soil dissipation study monitoring may,
depending on the leaching data, have to be conducted at a
depth greater than 6 inches.

Ecological characteristics:

Based on studies available to assess hazards to wildlife
and aquatic organisms, ethoprop is characterized as very
highly toxic to birds, marine/estuarine crustaceans, and
marine/estuarine fish species. It is moderately to highly
toxic to coldwater fish species and moderately toxic to
honey bees.

Results of currently available studies are as follows:

- avian oral LD_{50}: ranges from 4.21 - 61 mg/kg.
- avian dietary LC_{50}: ranges from 33 - 118 ppm in
  upland gamebirds and 287 - 550 ppm in waterfowl.
- fish LC_{50}: ranges from 1.02 mg/l - 1.85 mg/l in
  rainbow trout.
- aquatic invertebrate LC_{50}: ranges from 13 ug/l -
  25.3 ug/l.

Tolerance assessment:

Tolerances are established for negligible residues of the
insecticide nematicide ethoprop in or on the following raw
agricultural commodities: bananas, cabbage, corn-grain,
corn fodder and forage, cucumbers, fresh corn including
sweet corn (kernels plus cob with husk removed), lima
beans, lima bean forage, peanuts, peanut hay, pineapples,
pineapple fodder and forage, soybeans, soybean forage and
hay, sugarcane, sugarcane fodder and forage, and sweet
potatoes at 0.02 parts per million. These tolerances are
published in 40 CFR 180.262. There are no international

tolerances nor Codex Maximum Residue Limits (MRLs) for residues of ethoprop.

Tolerances for most raw agricultural commodities (with the exception of potatoes and tobacco) are supported by currently available residue data. However, the Agency is unable to complete a tolerance reassessment because of residue chemistry and toxicology data gaps.

A provisional acceptable daily intake (PADI) is calculated to be 0.000075 mg/kg/day, the provisional maximum permissible daily intake (PMDI) is 0.0045 mg/day and the theoretical maximum residue concentration (TMRC) is 0.0052 mg/day for a 1.5 kg diet. The percent of the PADI utilized is 115.02%. Although the PADI has been exceeded, it is concluded that exceeding the PADI does not necessarily in and of itself represent an immediate hazard to the public. The actual levels of residues to which the public is exposed are likely to be considerable less than this theoretical maximum would indicate for some of the following reasons:

- Not every crop for which a tolerance has been established has been treated with the pesticide.
- Most treated crops have residue levels which are below the established tolerance level.
- Processing or time to market often result in further residue reductions.
- Preparing food for consumption often results in residue reductions.
- Not all crops contributing to the TMRC (PADI) are likely to be consumed by an individual.
- Market basket surveys conducted by FDA indicate that little if any real pesticide residues of organophosphates actually remain in/on finished foods.

After review of the requested toxicology data, the ADI will be reevaluated.

4. Summary of regulatory position and rationale:

The "Restricted Use" classification for all ethoprop emulsifiable concentrate formulations containing 40% and greater will continue. Granular end-use products containing 10% or more ethoprop with disulfoton 5% and greater are classified for "Restricted Use". All other granular and/or fertilizer end-use products containing ethoprop are also classified "Restricted-Use", however, registrants of these products have the option of accepting the restricted-use classification, or submitting data to show that the the product does not warrant this restriction. All such products released for shipment on September 1, 1985, or thereafter, must be labeled for restricted use. Similarly, all such products which are in channels of trade on or after September 1, 1986 must bear restricted use labeling.

Although the Agency is unable to complete a tolerance reassessment because of certain residue chemistry and toxicology data gaps, the Agency concludes that no change in present tolerances are indicated at this time. Further, although the provisional acceptable daily intake has been exceeded, future requests for tolerances will not be automatically rejected on this basis.

No federal or state reentry intervals have been established for ethoprop. There is no need to establish any reentry intervals, since the practice of soil incorporation is expected to minimize exposure during reentry operations. This is also true for turf areas where the pesticide is watered-in.

The Agency has identified a potential avian adverse effect concern. Based on acceptable subacute dietary studies, it is calculated that the expected residues on avian food-stuffs following a single application of ethoprop at a rate of 6 pounds a.i. per acre (maximum registered corn rate) exceed the subacute dietary LC_{50} level in avian species. In order to determine whether significant evidence relating to this adverse effect would raise prudent concerns of unreasonable adverse risk to man or the environment, the Agency is requiring additional avian field testing data.

The following additional human hazard statement must appear on all manufacturing-use product labels:

> "Poisonous if absorbed through the eye.
> Do not get in eyes."

Although a full tolerance reassessment and risk assessment cannot be completed, there is no evidence to suggest that current tolerances are likely to expose the public to unreasonable adverse effects.

5. Summary of major data gaps:

- 90-day feeding; rodent, non-rodent
- 21-day dermal
- chronic toxicity; 2 species
- oncogenicity study; 2 species
- teratogenicity; 1 species
- reproduction; 2 species
- mutagenicity testing; gene mutation, chromosomal aberration, other mechanisms of mutagenicity
- general metabolism
- hydrolysis study
- photodegradation study; in water
- aerobic and anaerobic soil study
- leaching and adsorption/desorption study
- field soil dissipation study
- accumulation studies; rotational crops and fish
- data on residues in potatoes and tobacco
- 2 acute field studies,
- freshwater fish LC_{50}
- acute LC_{50} freshwater invertebrates

All studies are to be submitted to the Agency by June, 1986.

6. Contact Person at EPA

William H. Miller
Product Manager (16)
Registration Division (TS-767C)
Insecticide-Rodenticide Branch
Environmental Protection Agency
Washington, DC 20460

Tel. No. (703) 557-2600

Disclaimer: The information presented in this Chemical Information Fact Sheet is for informational purposes only and may not be used to fulfill data requirements for pesticide registration and reregistration.

ETHYLENETHIOUREA (ETU)

Date Issued: June 1987
Fact Sheet Number: 139

DESCRIPTION OF CHEMICAL:

Generic Name: NA

Common Name: Ethylenethiourea

Trade Names: NA

Chemical Abstracts Service (CAS) Number: 96-45-7

Year of Initial Registration: NA

Pesticide Type: Contaminant, degradation product, and metabolite of the EBDC pesticides.

U.S. and Foreign Producers: The principal producers of various of the EBDC pesticides include FMC Corporation, Pennwalt Corporation, and Rohm & Haas Company.

USE PATTERNS AND FORMULATIONS:

ETU is a contaminant, metabolite, and degradation product of the family of pesticides known as the EBDCs. There are five such pesticides currently registered: mancozeb, maneb, metiram, nabam, and zineb. All registrations of amobam products are now cancelled. In their agricultural uses, the EBDC pesticides are broad spectrum fungicides used to prevent crop damage by fungi and to protect harvested crops from deterioration. Some EBDC pesticides products are also used as industrial biocides in some applications.

Although these pesticides are used on a wide variety of crops, the principal crops on which they are used are apples, potatoes, and tomatoes. All types of formulation types are registered. Application is done by various methods, including airblast, ground boom, and sprinkler irrigation.

SCIENCE FINDINGS

Summary Science Statement: ETU has been classified by the Agency as a Group B2 oncogen (probable human carcinogen). ETU has induced an increased incidence of thyroid tumors in rats, and liver tumors in mice. In addition, ETU has been shown to be a teratogen in studies with rats and hamsters, and has shown thyroid effects in rat feeding studies.

Toxicology Characteristics

Subchronic Toxicity:

    Oral (rats): NOEL for thyroid effects (Sprague-Dawley rats) - 5 ppm (0.25 mg/kg/day) (Freudenthal, 1977).

    Oral (mouse): NOEL for liver effects - 10 ppm

        NOEL for thyroid effects - 10 ppm (1.72 mg/kg/day, males; 2.38 mg/kg/day, female) (O'Hara & DiDonata, 1985).

    Oral (monkey): NOEL for thyroid effects - < 50 ppm.

NOEL not determined for thyroid effects in the dog.

Chronic Toxicity:

    Chronic Toxicity: No chronic NOEL determined for thyroid effects in the rat. Study required in Sprague-Dawley rats.

    Oncogenicity: Potency estimate of 0.1428 mg/kg/day based on liver tumors in mice (Innes 1969).

        Produced thyroid tumors in rats (Ulland 1972; Graham 1975).

    Teratology: Teratogenic effects: NOEL (rats) - 5 mg/kg (Khera 1973). The dose resulting in abnormalities such as hydrocephalus in the hamster was 27 times that resulting in hydrocephalus in the rat.

    Reproduction: Study required.

Mutagenicity: Mixed results - weakly genotoxic.

Tolerance Assessment: Tolerances have been set for maneb and zineb at 40 CFR 180.110 and 40 CFR 180.115, respectively. Tolerances for mancozeb, expressed as zinc ethylene bisdithiocarbamate, have been established for raw agricultural commodities at 40 CFR 180.176, for meat byproducts at 40 CFR 180.319, and in processed food and feed at 21 CFR 193.460 and 21 CFR 561.410, respectively.

SUMMARY OF REGULATORY POSITION & RATIONALE

The Agency is initiating a Special Review of the EBDC pesticides because of concern about the oncogenic risk to consumers from dietary exposure to ETU from foods treated with these pesticides, and the risks of teratogenicity and adverse thyroid effects to applicators and mixer/loaders from exposure to ETU. ETU is present as part of the residue of the EBDC pesticides on treated agricultural commodities. In addition, a portion of the EBDC pesticide residues convert into ETU in the body after ingestion. The Agency estimates that the lifetime dietary oncogenic risk to consumers from these two sources of exposure to ETU is 2.2×10^{-5}. This estimate is based on exposure to ETU from the residues of only one of the EBDC pesticides. Consequently, the overall dietary risk may be higher.

Applicators and mixer/loaders are exposed to ETU during application and mixing and loading of the EBDC pesticides. Because ETU is a contaminant, degradation product, and metabolite of these pesticides, it is present in tank mixes and in the formulations of the EBDC pesticides.

The Agency calculated margins of safety for teratogenic and thyroid effects from exposure to ETU in mancozeb, one of the EBDC pesticides. Several margins of safety were below the threshold of concern. (A safety factor of 100 is used to measure the acceptable margin between actual exposure and the levels at which chronic effects may occur. A margin of safety for chronic effects of less than 100 is unacceptable). For example, airblast applicators applying mancozeb to apples had margins of safety for exposure to ETU of 87 for teratogenic effects, and 11 for thyroid effects. Ground boom applicators applying mancozeb to potatoes, onions and tomatoes had margins of safety for exposure to ETU of less than 100 for thyroid effects. The Agency is not initiating a Special Review on oncogenic risks to applicators and mixer/loaders because the relative risks and the number of persons potentially exposed are low.

This elevated level of potential risk warrants a careful examination during the Special Review process of the risks and benefits of using the EBDC pesticides, and a determination whether such uses should be cancelled or otherwise regulated.

Contact Person:

David Giamporcaro
Review Manager
Special Review Branch
Registration Division
Office of Pesticide Programs
(703)557-0481

FENAMINOSULF

Date Issued: October 14, 1983
Fact Sheet Number: 10

1. Description of the Chemical:

    Generic name: p-(dimethylamino)benzenediazo sodium sulfonate
    Common name: Fenaminosulf
    Trade name: Lesan, Dexon, Bayer 22555, Bayer 5072, diazoben
    EPA Shaughnessy Code: 034201
    Chemical abstracts service (CAS) number: 140-56-7
    Year of initial registration: 1959
    Pesticide type: Fungicide
    Chemical Family: sulfonated phenyl diazo
    U.S. and foreign producer: Mobay Chemical Corporation

2. Use Patterns and Formulations:

    Application sites: vegetables, ornamentals, lawns, and turf
    Types and methods of application: dry or as a slurry, soil drench, foliar spray or protectant seed treatment
    Application rates: 0.005 to 17 lb ai/A on crop sites
    Types of formulations: Available as a wettable powder, dust and granular formulations (80% technical product; 70,35,32, and 5% WP; 2,5% D; 5 and 0.4% G)
    Usual carrier: water

3. Science Findings:

    Summary science statement:
    There are extensive data gaps for Fenaminosulf. No human toxicological hazards of concern, other than mutagenicity, have been identified in studies reviewed by the Agency for the standard. The Agency has no information that indicates continued use will result in any unreasonable adverse effects to man or his environment during the time required to develop the data.

    Chemical characteristics:
    Fenaminosulf is a yellowish-brown, odorless powder. It is soluble in water (2-3% at 25°C). The chemical does not present any unusual handling hazards.

    Toxicologic characteristics:
    Chronic toxicology studies:
      Mutagenicity: positive results
      Fenaminosulf is structurally related to the azo dyes that are known carcinogens

    Major routes of exposure: dermal and inhalation from field uses

Physiological and biochemical behavioral characteristics:
  Mechanism of pesticidal action: appears to inhibit mitochondrial function
  Metabolism and persistence in plants and animals: There is no data on plant and animal metabolism or bioaccumulation potential.

Environmental characteristics:
  Absorption and leaching in basic soil types: appears to degrade with a half-life of 2 days in soil
  Studies are inadequate to reach a final conclusion about the fate of Fenaminosulf in the environment or the exposure potential of the chemical to humans and wildlife

Ecological characteristics:
  Hazards to fish and wildlife:
    Avian acute oral LD_{50}: 13 mg/kg
    Avian subacute dietary LC_{50}: 250 ppm
    Fish LC_{50}: >60 ppm
    Aquatic invertebrates LC_{50}: Scuds 3.7 ppm; Stoneflies 24 ppm
  Potential problems for endangered species: The use patterns do not present any problem to endangered species.

Tolerance Assessments:
  There are no tolerances established for Fenaminosulf residues in or on any crop, or animal product. When the chemical was first registered, seed treatments were considered non-food uses. The Agency's current position is that seed treatments are considered food uses until data is submitted showing no uptake of the chemical or its' degradates. Therefore metabolic studies are needed to determine the residues of concern and if tolerances must be established. If tolerances need to be established, additional data requirements may have to be met.

Problems known to have accurred from use: (PIMS) There have been 21 cases of exposure reported from mixing, loading, or field reentry. No exposures have resulted in fatality.

4. Summary of Regulatory Position and Rationale:

   Use classification: general

5. Summary of Major Data Gaps:

   Product Chemistry: studies due 5/84
     Discussion of formation of ingredients
     Preliminary analysis
     Analytical methods
     Certification of limits
     Physical and chemical characteristics
   Residue Chemistry: studies due 11/87
     Plant residue
     Animal residue
     Meat/milk/poultry/eggs residue

Environmental Fate: studies due 11/87
  Hydrolysis
  Photodegradation
  Aerobic metabolism
  Leaching and absorption/desorption
  Dissipation
  Fish accumulation
Toxicology: studies due 11/87
  Primary eye irritation - rabbit
  Primary dermal irritation
  Dermal sensitization
  Acute Oral LD_{50} - rat
  Acute Dermal LD_{50} - rat
  Acute Inhalation LC_{50}
  21 day dermal
  Chronic toxicity - 2 species
  Oncogenicity - 2 species
  Teratogenicity - 2 species
  Reproduction, 2 generation
  Gene mutation
  Chromosomal aberration
  General metabolism
Wildlife and Aquatics: studies due 11/87
  Avian dietary - upland game
  Acute LC_{50} - invertebrates

6. Contact person at EPA:

   Mr. Henry M. Jacoby (PM-21)
   Environmental Protection Agency
   Fungicide-Herbicide Branch (TS 767-C)
   401 M St., S.W.
   Washington, D.C. 20460
   (703) 557-1900

DISCLAIMER: The information presented in this Chemical Information Fact Sheet is for informational purposes only and may not be used to fulfill data requirements for pesticide registration and reregistration.

FENBUTATIN-OXIDE

Date Issued: March 31, 1987
Fact Sheet Number: 119

1. Description of Chemical

-Generic Name: bis[tris(2-methyl-2-phenylpropyl)tin] oxide or hexakis(2-methyl-2-phenylpropyl)-distannoxane

-Common Name: Fenbutatin-oxide

-Trade Name: Vendex, Fenbutatin oxyde, SD 14114, Torque, Neostanox, and Osadan

-EPA Shaughnessy Number: 104601 Fenbutatin-oxide

-Chemical Abstracts Service (CAS) Number: 13356-08-6

-Year of Initial Registration :   1974

-Pesticide Type: Acaricide

-Chemical Family: Organotin

-U.S. Producer: E.I. duPont de Nemours and Company

2. Use Patterns and Formulations

-Application sites: fruit, field crops, and ornamentals

-Types and method of applications: Foliar application by ground equipment only

-Application rates: 0.5 to 2.0 lb a.i. per acre

-Types of formulations: technical, wettable powder and flowable concentrate

-Usual carriers: Confidential Business Information

3. Science Findings

Chemical Characteristics:

Technical fenbutatin-oxide is a white crystalline solid with a melting point at around 145°C, that is insoluble in water and slightly soluble in aromatic solvents. The empirical formula is $C_{60}H_{78}OSn_2$ and the molecular weight is 1053.

Toxicological Characteristics:

- Acute Oral: Data Gap
- Acute Dermal: >2000 mg/kg (rabbit)
- Primary Eye Irritation: Severe eye irritant (rabbit), Toxicity Category I
- Acute Inhalation: Data Gap
- Primary Skin Irritation: Mild skin irritant (rabbit)
- Dermal Sensitization: Not a dermal sensitizer (guinea pig)
- Major routes of exposure: Dermal followed by inhalation- Human exposure occurs from mixing, loading and application Exposure can be reduced by the use of goggles or face shield and protective clothing.
- Oncogenicity: A rat chronic/oncogenicity study showed no oncogenic effects at the highest dose tested (600 ppm). Mouse oncogenicity study: Data Gap
- Metabolism: Data Gap
- Teratology: Not teratogenic in rats (No Effect Level of 60 mg/kg with highest dose tested) or rabbits (No Effect Level of 5 mg/kg).
- Reproduction: Data Gap
- Mutagenicity: Data Gap

Physiological and Biochemical Characteristics

- Mechanism of Pesticide Action: It is suspected that fonbutatin-oxide inhibits adenosine triphosphate (ATP) enzymes. That is, it paralyzes the insects cardiovascular and respiratory systems by inhibiting oxidative phosphorylation.
- Metabolism and perisistence in plants and animals: Available data demonstrate that fenbutatin-oxide is relatively persistent on leaf surfaces and fruit and is gradually degraded to dihydroxy-bis-(2-methyl-2-phenylpropyl) stannane, 2-methyl-2-phenylpropyl stannonic acid and inorganic tin. Although submitted metabolism studies are not adequate to assess the nature of fenbutatin-oxide in animals, metabolites of fenbutatin-oxide were identified in feces of dairy cattle.

Environmental Characteristics

- Available data are insufficient to fully assess the environmental fate of fenbutatin-oxide. Data gaps exist for all required studies. The data required in this Standard will also allow us to assess the potential of this pesticide to contaminate groundwater.

Ecological Characteristics

- Acute avian oral toxicity: $LD_{50} > 2510$ mg/kg for Bobwhite Quail (nontoxic).
- Avian dietary toxicity: $LC_{50} > 5620$ ppm for Mallard Duck (nontoxic). Supplemental data indicate an LC_{50} of 5065 ppm for Bobwhite Quail (nontoxic).
- Freshwater fish acute toxicity: $LC_{50} = 0.0017$ ppm for Rainbow trout (very highly toxic). $LC_{50} = 0.0048$ ppm for Bluegill sunfish (very highly toxic).
  Freshwater aquatic invertebrate toxicity: Supplemental data indicate that an LD_{50} value for freshwater invertebrates is 0.040 ppm (very highly toxic).

Tolerance Assessment

- Tolerances have been established for residues of fenbutatin-oxide in a variety of raw agricultural commodities including meat, fat and meat by products (refer to 40 CFR 180.362 for listing of tolerances), and in processed food (21 CRF 193.236) and feed (21 CFR 561.255).
- The Agency is unable to complete a full tolerance assessment for the established tolerances due to residue chemistry and toxicology data gaps.

Commodity	Tolerances (ppm) U.S.	Canadian	Mexican	(MRL) Codex
Almonds	0.5	-	-	-
Almonds, hulls	80.0	-	-	-
Apples	15.0	3.0	-	5.0
Cattle, fat	0.5	-	-	-
Cattle, meat by product	0.5	-	-	-
Cattle, meat	0.5	-	-	0.02
Cherries, sour	6.0	-	-	5.0
Cherries, sweet	6.0	-	-	5.0
Citrus fruits	2.0	2.0	-	5.0
Cucumber	4.0	0.5	-	1.0
Eggplant	6.0	-	-	1.0
Eggs	0.1	-	-	-
Goats, fat	0.5	-	-	-
Goats, meat by product	0.5	-	-	-
Goats, meat	0.5	-	-	-
Grapes	5.0	-	-	5.0
Hogs, fat	0.5	-	-	-
Hogs, meat by products	0.5	-	-	-
Hogs, meat	0.5	-	-	-
Horses, fat	0.5	-	-	-
Horses, meat by product	0.5	-	-	-
Horses, meat	0.5	-	-	-
milk fat	0.1	-	-	-

Commodity	U.S.	Tolerances (ppm) Canadian	Mexican	(MRL) Codex
Papayas	2.0	-	-	-
Pecans	0.5	-	-	-
Peaches	10.0	-	-	7.0
Pears	15.0			
Plums	4.0	-	-	3.0
Poultry, fat	0.1	-	-	-
Poultry, meat by products	0.1	-	-	-
Poultry, meat	0.1	-	-	-
Prunes	4.0	-	-	-
Sheep, fat	0.5	-	-	-
Sheep, meat by products	0.5	-	-	-
Sheep, meat	0.5	-	-	-
Strawberries	10.0	-	-	-
Walnuts	0.5	-	-	-
Apple pomace, dried	20.0	-	-	-
Citrus pulp, dried	7.0	-	-	-
Grape pomace, dried	100	-	-	-
Raisin waste	20.0	-	-	-
Prunes, dried	8.0	-	-	-
Raisins	20.0	-	-	-

—The data for the combined residues of fenbutatin-oxide and its organotin metabolites in or on almonds, almond hulls, apples, cherries(sweet and sour), citrus fruits, eggplant, grapes, papayas, peaches, pears, pecans, plums, raspberries, strawberries and walnuts are adequate to support the established tolerances.

—Data are not adequate to support the established tolerance for residues in or on cucumbers. Processing studies are required for apples, citrus fruits, grapes and plums.

—An acceptable Daily Intake (ADI) of 0.05 mg/kg/day, based a 2 year rat study has been established using a 100 fold safety factor. The Maximum Permissible Intake (MPI) is 3.0 mg/kg/day for a 60 kg person. The Theoretical Maximum Residue Contribution (TMRC) to the human diet from the existing tolerances is 0.0357 mg/kg/day for a 1.5 kg diet which is 71.4% of the MPI.

4. Summary of Regulatory Position

A. The Agency is requiring the submission of acute aquatic toxicity data and aquatic field monitoring data on the end-use formulations and aquatic life stage data and avian reproduction data on the technical formulation in order to complete the wildlife risk assessment.

B. Because of California incidence reports and the fact that this chemical is acutely toxic due to adverse eye effects, the Agency is requiring a 24 hour reentry interval, as established by California and Texas, until the Agency has received and evaluated the required reentry data.

C. Because of acute toxicity (eye effects), the Agency is requiring that all fenbutatin-oxide product labels contain protective clothing statements for early reentry.

D. No new tolerances or new food uses will be granted until the Agency has received data to evaluate dietary exposure to fenbutatin-oxide.

E. Based on fenbutatin-oxide use pattern and toxicity data, the Agency, in cooperation with the Fish and Wildlife Service, has determined that fenbutatin-oxide does not trigger endangered species concerns at this time. No endangered species labeling is required. If the required monitoring data indicate a hazard, then a re-evalution will be conducted.

F. The Agency has determined that the following data are essential to the Agency's assessment and should receive a priority review as soon as they are received by the Agency:

158.140 Reentry Protection

132-1 Foliar Dissipation (Re-entry)

158.145 Ecological Effects

72-4 Fish Early Life Stage and Aquatic Invertebrate Life Cycle
72-3 Acute Toxicity- Aquatic Estuarine and Marine Organism
72-7 Field Testing for Aquatic Organism

G. The Agency is requiring that the fenbutatin-oxide organotin metabolites be listed separately in the tolerance regulation so that established tolerances can be made compatible with the Maximum Residue Limits established by the Codex Alimentarius Commission.

5. Summary of Major Data Gaps

-The following studies are required to assess the toxicological characteristics of technical fenbutatin-oxide:
Acute Oral, Acute Inhalation, 90 Day Feeding (non-rodent), 21 Day Dermal Toxicity, Chronic Toxicity (non-rodent), Oncogenicity (mouse), 2 Generation Reproduction, Mutagenicity and General Metabolism Testing.

- The following data are required to fully characterize fenbutatin-oxide's environmental fate: Re-entry (foliar dissipation), Hydrolysis, Photodegradation (in water and on soil) Aerobic and Anaerobic Soil Metabolism, Leaching, Volatility (Lab) Soil Dissipation, Rotational Crops (Confined), and Accumulation in fish.
- Additional residue studies on commodities, animal metabolism studies, and storage stability studies are required to support existing tolerances.
- The following data are required to complete a wildlife risk assessment: Avian Subacute Dietary Toxicity, Avian Reproduction, Freshwater Fish Toxicity, Acute Toxicity to Freshwater Invertebrates, Acute Toxicity to Estuarine and Marine Organisms, Fish Early Life Stage and Aquatic Invertebrate Life Cycle, Aquatic Organism Accumulation Testing, and Simulated or Actual Field testing for Aquatic Organisms.
- Product Chemistry and Acute toxicity data are required.

Contact person at EPA:

Dennis Edwards
Product Manager (12)
Insecticide-Rodenticide Branch
Registration Division (TS-767C)
Office of Pesticide Programs
Environmental Protection Agency
401 M. Street, S.W.
Washington,D.C. 20460

Office location and telephone number:
Room 202, Crystal Mall Building #2
1921 Jefferson Davis Highway
Arlington,Va. 22202
703-557-2386

Disclaimer: The information presented in this pesticide Fact Sheet is for informational purposes only and may not be used to fulfill data requirements for pesticide registration and reregistration.

FENOXYCARB

February 1, 1986
Fact Sheet Number: 78

1. Description of Chemical

    Generic Name: (Ethyl[2-(4-phenoxyphenoxy)ethyl] carbamate)

    Common Name: fenoxycarb

    Trade Names: Logic, Ro 13-5223

    EPA/OPP Chemical Code (Shaughnessy Number: 128801

    Chemical Abstracts Service (CAS) Number: 72490-01-8

    Year of Initial Registration: 1985

    Pesticide Type: Insecticide/Miticide

    U.S. and Foreign Producers: MAAG Agrochemicals

2. Use Patterns and Formulations

    Fenoxycarb is registered for use as a bait to control fire ants in or on turf, lawns, and nonagricultural land such as airport turfs, parks and golf courses. Applications are made as single mound (1 to 3 level tablespoons/ mound) and broadcasting (apply uniformly with ground equipment calibrated to give correct dosage [1 to 1.5 lb/A]).

    Fenoxycarb is formulated as a 1.0 percent bait.

3. Science Findings

    Summary Science Statement:

    Fenoxycarb is an Insect Growth Regulator (IGR) which is moderately acutely toxic to human and other nontarget terrestrial organisms. Fenoxycarb is highly toxic to aquatic invertebrates. The current No Observable Effect Level (NOEL) is based on a rat 52-week interim report, which satisfies the requirement for a rat 90-day feeding study. Available data are sufficient to fully assess the toxicology of fenoxycarb for the subject use pattern, its fate in the environment, and the exposure of humans and nontarget organisms to the chemical or its degradates. Environmental fate data requirements has been satisfeid with the exception of soil photolysis (data gap).

Chemical Characteristics:

Technical fenoxycarb is a light brown lumpy powder, stable under normal conditions. Slight hydrolysis occurs in aqueous solution at pH 3,7, and 9 at 35°C and 50°C for period of time up to 70 days. It is very soluble (250 g/l solvent) in most organic solvents(e.g., acetone, chloroform, diethyl ether, diethyl formamide, ethyl acetate, methanol, toluene) hexane. It is slightly soluble in hexane(5 g/l solvent).

Toxicological Characteristics:

Fenoxycarb is moderately toxic (Tox. Category III from acute oral and dermal routes of exposure).

Toxicology studies on fenoxycarb are as follows:

- Oral LD_{50} in rats: LD_{50} > 16,800 mg/kg
- Dermal LD_{50} in rats: LD_{50} > 5000 mg/kg; effects included dyspnea, curved body position, ruffled fur, sedation, diarrhea. No deaths occurred.
- Metabolism study in rats: 90-92% excreted in 96 hours. Organs did not show persistent residues.
- Rat 52-week Interim Report: NOEL = 200ppm
- Teratogenicity: Teratogenic effects were not observed at dose levels up to 300 mg/kg/day which was the highest dose tested.
- Mutagenicity: Nonmutagenic
- Dermal sensitization: Not sensitizing
- 21-Day dermal: NOEL = 200mg/kg/day(slight erythema; elevated liver weight at top dose)

Physiological and Biological Characteristics:

Fenoxycarb is an insecticide that disrupts the development of the pest. Fenoxycarb has an insect specific mode of action exhibiting strong juvenile hormone activity. It inhibits metamorphosis to the adult stage, induces interference with the molting of early instar larvae, and produces certain ovicide and delayed larvicide/ adulticide effects in various insect species.

Environmental Characteristics:

Fenoxycarb is stable to hydrolysis at pH 3-9 and temperatures up to 50 °C. However, fenoxycarb is expected to photodegrade in pure and natural water with a half-life of 5 hours.

Field studies indicate that under exaggerated and normal use conditions dissipation of fenoxycarb in soil is rapid; residues are no longer detectable 3 days after application.

Soil column studies using fresh and aged soils indicate a low potential for leaching. Adsorption/desorption studies indicate moderate to strong soil binding. Fish exposed to fenoxycarb in water will bioaccumulate fenoxycarb to concentration 300x greater than the concentration in the water. However, the fish will release 99% of the residues within 2 weekss when placed in water containing no fenoxycarb.

Ecological Characteristic:

Avian Oral LD_{50} = > 3000 mg/kg (mallard ducks)

Avian Dietary LC_{50} = 11574 ppm (bobwhite quail). Data show that fenoxycarb is practically nontoxic to birds.

Fish LC_{50} = 1.6 ppm for rainbow trout and 1.86 ppm for bluegill. Data show that fenoxycarb is moderately toxic to fish.

Data submitted indicate that fenoxycarb is low to moderate in toxicity to honeybees. Bee hazard is decreased because the pesticide will be formulated as a grit or corncob bait, which will result in little or no bee exposure. Thus, currently registered uses of the formulated product should present no hazard to bees.

It is unlikely that the use of fenoxycarb would adversely affect endangered aquatic species because of the low acute toxicity of fenoxycarb to this group and the low use rates. The expected concentrations in 6 inches of water are less than 1/20 the fish LC_{50} and 1/20 the daphnia LC_{50}.

It is unlikely that the use of fenoxycarb would affect avian species because of its low acute toxicity and low use rate.

Tolerance Assessment:

There are no tolerances established for fenoxycarb.

4. Summary of Regulatory Position and Rationale

Fenoxycarb qualifies for a Section 3(c)(7)(C) registration based on submission of an almost complete data base, a finding of no unreasonable adverse effects, and a finding that registration is in the public interest. The public interest finding is based on the chemical's relatively low toxicity and its effectiveness (as attested to by USDA) in fire ant control. The only study still to be submitted is a soil photolysis study which is not typically required of a surface applied formulation. The study is being required in this instance because vegetation may shield the product from sunlight. Because this data gap is a minor one, and because the registrant could not have been expected to know of the data requirement, the chemical is being conditionally registered at this time.

- ° The following Environmental Hazards text is required on manufacturing-use products because of the hazards posed to aquatic invertebrates.

    This pesticide is toxic to aquatic invertebrates. Do not discharge effluent containing this product into lakes, streams, ponds, estuaries, oceans, or public water unless this product is specifically identified and addressed in an NPDES permit. Do not discharge effluent containing this product into sewer systems without previously notifying the sewage treatment plant authority. For guidance contact your State Water Board or Regional Office of the EPA.

- ° The following environmental statements are required for end-use products.

    This pesticide is toxic to aquatic invertebrates. Drift and runoff from treated areas may be hazardous to aquatic invertebrates in adjacent aquatic sites. Do not apply directly to water. Do not contaminate water by cleaning the equipment or disposal of wastes.

Due to the use pattern and the low toxicological characteristic to mammals, a reentry precautionary statement is not required.

5. **Summary of Major Data Gaps**

   ○ Environmental Fate   Due Date

     - Soil Photolysis   2/86

6. **Contact Person at EPA**: Timothy A. Gardner
   Product Manager 17
   Insecticide-Rodenticide Branch (TS-767C)
   401 M Street SW.
   Washington, DC 20460
   (703) 557-2690

DISCLAIMER: The information presented in this Chemical Information Fact Sheet is for informational purposes only and may not be used to fulfill data requirements for pesticide registration and reregistration.

FENSULFOTHION

Date Issued: February 28, 1985
Fact Sheet Number: 14.1

1. Description of chemical

    Generic name: 0,0-diethyl 0-[p-(methylsulfinyl) phenyl] phosphorothioate
    Common name: fensulfothion
    Trade names: Dasanit, BAY 25141, S-767, TERRACUR P
    EPA Shaughnessy code: 032701-5
    Chemical abstracts service (CAS) number: 115-90-2
    Year of initial registration: 1965
    Pesticide type: insecticide-nematicide
    U.S. and foreign producer: MOBAY Chemical Corporation

2. Use Patterns and Formulations

    Fensulfothion is registered for use as preplant or at-planting soil application to tobacco, and various fruits and vegetables. Postplant topical applications are permitted in addition to the at-planting application on corn, peanuts, and rutabagas. Topical application is also permitted on commercial and ornamental turf.

    Fensulfothion is formulated into a 63% (6lb/gallon) E.C.(restricted use), and 10% and 15% granulars, and at various percentages with disulfoton, thiram, or pebulate.

3. Science Findings

    Summary science statement: Fensulfothion is an organophosphate insecticide-nematicide of high toxicity to man and other non-target terrestrial and aquatic organisms. Tier III data (field studies) have been requested in response to the three RPAR triggers (mammalian, avian, and aquatic) that have been exceeded. The current NOEL, ADI, and MPI are now only partially supported by data, but will be used in the interim. Available data are insufficient to fully assess the toxicology of fensulfothion, its fate in the environment, or the exposure of humans and non-target organisms to the chemical or its degradates.

    Chemical Characteristics: Technical fensulfothion is a brown liquid organophosphate, stable under normal use conditions, with a boiling point of 138-141°C at 0.01 mm Hg. It is soluble in most organic solvents except aliphatics. The chemical is acutely toxic and extreme caution is necessary in handling of contaminated articles and during mixing, loading, and application. Respirator and protective clothing are required during these operations.

Toxicoloxical Characteristics:

Fensulfothion is highly toxic (Tox. Category I) from acute oral and dermal routes of exposure.
Many of the toxicology studies do not meet <u>present</u> guideline requirements and need to be replaced, however no significant risks have been identified from the existing data base.
Toxicology studies on fensulfothion are as follows:

- Oral LD_{50} in rats: 2.2 mg/kg (female), 10.5 mg/kg (male) [Acceptable]
- Dermal LD_{50} in rats: 3.5 mg/kg (female), 30.0 mg/kg (male) [Acceptable]
- Acute delayed neurotoxicity in chickens: none observed [Supplementary]
- Metabolism study in rats: [Acceptable]
- 90 Day feeding-rodent; non-rodent: [Supplementary; Invalid - data gap]
- Chronic toxicity, non-rodent: [Invalid - data gap]
- Chronic toxicity, rodent: No NOEL found [Supplementary]
- Oncogenicty, rat: [Supplementary]
- Teratogenicity, rabbit: [Supplementary]
- Reproduction study: NOEL 1 or 4 ppm [Supplementary]
- Mutagenicity, gene mutation: Negative [Partially satisfied study]

Physiological and Biological Characteristics

Fensulfothion is an organophosphate insecticide-nematicide that kills primarily by contact action but also provides some systemic control of insects attacking the foliage of treated plants.

The mode of action is by phosphorylating the acetylcholinesterase enzyme of tissues, allowing accumulation of acetylcholine at nerve junctions with subsequent blocking effects upon the central nervous system.

The metabolism of fensulfothion is basically similar in both plants and animals. By the processes of hydrolysis, oxidation and reduction the parent compound may be broken down to 13 known metabolites, 5 of which are themselves cholinesterase inhibitors. On the basis of this knowledge, all presently established fensulfothion tolerances are expressed in terms of the combined residues of the parent compound and these five cholinesterase inhibiting metabolites. Fensulfothion is metabolized fairly rapidly by both plants and animals. In animals, hydrolytic degradation in liver and other tissues results in excretion of low toxicity degradation products, with half the pesticide eliminated within 24 hours and almost total elimination of the pesticide and its metabolites within a week.

Environmental Characteristics:

Fensulfothion is degraded in soils under aerobic conditions with half-lives of 3-28 days and is due to microbial degradation. Half-life is rapid in silty clay loam and organic soil (3-7 days) and fairly rapid in sandy loam, silt loam and loam soils (around 28 days). Fensulfothion degrades rapidly in the water and silt of a simulated pond with half-lives of 10 and 12 days, respectively.

The mobility of fensulfothion and aged residues is low to moderate in a wide range of soils. Dissipation of fensulfothion is fairly rapid from field soils with half-lives ranging from <30 days to >182 days.

3. **Science Findings** (continued)

   Environmental Characteristics (continued):

   Fensulfothion residues are taken up by rotational crops grown in the greenhouse but are not taken up by field rotational crops.

   Available data are insufficient to fully assess the fate of fensulfothion in the environment; however, ground water contamination does not appear to be a problem with this chemical.

   Fensulfothion has a low potential to bioaccumulate in bluegill sunfish.

   Data are insufficient to fully assess the exposure of humans and non-target organisms to the chemical or its degradates, however, human exposure should be minimal by use of current restricted use classification and labeling precautions requiring approved respirators and protective clothing.

   Exposure during reentry operations should be minimal, however, data are not available to fully assess such exposures. A 7 day reentry period is being required for unprotected workers following soil application if the soil is wet. A 24 hour reentry period is being required for applications of fensulfothion where agricultural practice will involve hand labor with prolonged, intimate foliar contact, or if the soil is dry.

   Ecological Characteristics:

   Avian oral LD_{50}: 0.749 ppm (very high toxicity)
   Avian dietary LC_{50}: 22 ppm (high toxicity)
   Fish LC_{50}: 0.07 ppm (very high toxicity)

   The toxicity of fensulfothion to terrestrial and aquatic non-target organisms is very high. Residue calculations indicate that 3 RPAR triggers (mammalian, avian, and aquatic) may be exceeded. In all cases the Standard has asked for Tier III data (field studies) to gather qualitative and quantitative data to support the registration and/or need for special review.

   Because of fensulfothion's extensive number of use patterns and its high toxicity to wildlife, numerous endangered species have been identified that could be impacted. The Agency is currently considering various approaches to address the problem for this and other chemicals, and the Standard may be amended to incorporate the results of this additional review.

Science Findings (continued)

Tolerance assessment:

The following tolerances (in parts per million) have been established for fensulfothion:

Commodity	United States	Canada	Mexico	International (Codex)
Bananas	0.02	-	-	0.02
Beets, sugar	0.05	-	-	-
Beets, sugar, tops	0.05	-	-	-
Cattle, fat	0.02	-	-	0.02
Cattle, MBYP	0.02	-	-	0.02
Cattle, meat	0.02	-	-	0.02
Corn, field, fodder	1.0	-	-	-
Corn, field, forage	1.0	-	-	-
Corn, fresh (inc. sweet) (K+CWHR)	0.1	0.1	-	0.1
Corn, grain	0.1	0.1	-	0.1
Corn, pop, fodder	1.0	-	-	-
Corn, pop, forage	1.0	-	-	-
Corn pop, grain	0.1	0.1	-	0.1
Corn, sweet, fodder	1.0	-	-	-
Corn, sweet, forage	1.0	-	-	-
Cotton, seed	0.02	-	-	-
Goats, fat	0.02	-	-	0.02
Goats, MBYP	0.02	-	-	0.02
Goats, meat	0.02	-	-	0.02
Hogs, fat	0.02	-	-	-
Hogs, MBYP	0.02	-	-	-
Hogs, meat	0.02	-	-	-
Horses, fat	0.02	-	-	-
Horses, MBYP	0.02	-	-	-
Horses, meat	0.02	-	-	-
Onions, dry bulb	0.1	0.1	-	0.1
Peanuts	0.05	-	-	0.05
Peanuts, hulls	5.0	-	-	-
Pineapples	0.05	-	-	0.05
Pineapples, forage	0.05	-	-	-
Plantains	0.02	-	-	-
Potatoes	0.1	0.1	-	0.1
Rutabagas, roots	0.1	0.1	-	0.1
Sheep, fat	0.02	-	-	0.02
Sheep, MBYP	0.02	-	-	0.02
Sheep, meat	0.02	-	-	0.02
Sorghum, fodder	1.0	-	-	-
Sorghum, forage	1.0	-	-	-
Sorghum, grain	0.1	-	-	-
Soybeans	0.02	-	-	-
Soybeans, forage	0.1	-	-	-
Sugarcane	0.02	-	-	-
Sweet potatoes	0.05	-	-	-
Tomatoes	0.1	-	-	0.1

3. Science Findings (continued)

   Tolerance assessment (continued):

   Most tolerances for residues are supported with data, however, additional data must be submitted to support tolerances for residues in or on the following commodities: bananas, peanuts, peanut hulls, plantain, and potatoes (processing data only).

   There is no reasonable expectation of finite residues in milk, eggs, poultry meat, fat, or meat by-products and no tolerances are required.

   Reported pesticide incidents involving fensulfothion alone between 1966 and 1983 include 25 involving human injury and 4 involving animals. Most of the human incidents resulted from failure to use safety equipment while applying fensulfothion. Other incidents were the result of improper disposal, handling, or storage. Because the incidents involve occasions of misuse, no additional precautionary statements are necessary at this time to minimize the risk of injury.

4. Summary of Regulatory Position and Rationale

° The previous "Restricted Use" classifications required in 40 CFR §162.31 will be continued. In addition, granular formulations are now being restricted. All granular formulation products released fro shipment after September 1, 198 must be labeled for restricted use. Also, all products still in channels of trade after September 1, 1986 must be labeled for restricted use.

° The following Environmental Hazards text will be required on manufacturing use products because of the hazards posed to non-target terrestrial and aquatic wildlife:

   "This product is toxic to fish and extremely toxic to wildlife. Do not discharge into lakes, streams, ponds, or public waters unless in accordance with an NPDES permit. For guidance, contact your Regional Office of the EPA."

° The following environmental statements are required for end-use products:

   "This product is toxic to fish and extremely toxic to wildlife. Use with care when applying in areas frequented by wildlife. Birds feeding on treated areas may be killed. Cover, disc, or incorporate spill areas. Drift and runoff from treated areas may be hazardous to aquatic organisms in neighboring areas. Do not apply directly to water or wetlands. Do not contaminate water by cleaning of equipment or disposal of wastes. This product is highly toxic to bees exposed to direct treatment on blooming crops or weeds. Do not apply this product or allow it to drift to blooming crops or weeds while bees are actively visiting the treatment area."

° The following reentry precautions are required on end-use products in the interim until requested reentry data has been received and reviewed by the Agency:

   "Unprotected workers should not re-enter treated fields until 24 hours after application. Unprotected workers should not re-enter fields where the soil is wet until 7 days after soil application."

Summary of Major Data Gaps

	Due Date
° Product Chemistry	
– Description of manufacturing process	6/84
– Description of formation of impurities	6/84
– Preliminary analysis	6/84
– Certification of limits	6/84
– Odor	6/84
– Solubility	6/84
– Vapor pressure	6/84
– Dissociation constant	6/84
– Octanol/water partition coefficient	6/84
– pH	6/84
– Analytical method for enforcement of limits	6/84
– Oxidizing or reducing action	6/84
– Flamability	6/84
– Explodability	6/84
– Viscosity	6/84
– Miscibility	6/84
° Residue Chemistry	
– Storage stability	1/87
– Processed food/feed studies on potatoes	1/87
– Crop field trials on bananas, peanuts, and plantain	1/87
° Toxicology	
– Inhalation LC_{50} – rat	6/84
– Acute delayed neurotoxicity – hen	6/84
– 90-Day feeding – rodent, non-rodent	1/87
– 90-Day inhalation – rat	1/87
– Chronic toxicity – 2 species	1/87
– Oncogenic study – 2 species	1/87
– Teratogenicity – 2 species	1/87
– Reproduction – 2 generation	1/87
– Gene mutation	1/87
– Chromosomal aberration	1/87
° Wildlife and Aquatic Organisms	
– Avian reproduction	1/87
– Simulated and actual field testing (mammals and birds)	1/87
– Acute LC_{50} freshwater invertebrates	1/87
– Acute LC_{50} estuarine and marine organisms (shrimp, marine fish, and oyster)	1/87
– Fish early life stage and aquatic invertebrate life-cycle	1/87
° Environmental Fate	
– Hydrolysis	1/87
– Photodegradation in water	1/87
– Anaerobic soil metabolism	1/87
– Volatility	1/87
– Soil dissipation	1/87
– Accumulation in rotational crops	1/87
– Accumulation in fish	1/87

° Reentry Protection
  - Foliar dissipation   1/87
  - Soil dissipation   1/87
  - Dermal exposure   1/87
  - Inhalation exposure   1/87

6. Contact person at EPA: George T. LaRocca
                             Product Manager 15
                             Insecticide-Rodenticide Branch (TS-767C)
                             401 M Street
                             Washington, DC 20460

DISCLAIMER: The information presented in this Chemical Information Fact Sheet is for informational purposes only and may not be used to fulfill data requirements for pesticide registration and reregistration.

FLUCHLORALIN

Date Issued: June 30, 1985
Fact Sheet Number: 52

1. Description of chemical:

    Generic Name: Fluchloralin
    Common name: Fluchloralin
    Trade name: Basalin®
    EPA Shaughnessy Code: 108701
    Chemical Abstracts Service (CAS) Number: 33245-39-5
    Year of Initial Registration: 1970
    Pesticide Type: Herbicide
    Chemical family: Chloroaniline
    U.S. and Foreign Producers: BASF Wyandotte, Inc.

2. Use patterns and formulations:

    Application sites: Dry and succulent peas and beans, cotton, okra, peanuts, soybeans, and sunflowers.
    Types of formulations: Emulsifiable concentrate (4 lbs ai per gallon).
    Types and Methods of Application: Pre-plant broadcast or banded spray, using ground equipment. Soil incorporation recommended.
    Application Rates: 0.5 - 1.5 lbs ai/A on beans (including soybeans), okra, peas, peanuts, and sunflowers.
    Usual carriers: Water

3. Science Findings:

    Summary science statement: No valid acute or chronic toxicity data are available. One metabolite has shown potential for leaching through soil, but the toxicological properties of this metabolite are unknown. Toxicity to fish is very high.

Chemical characteristics:

Physical state: Crystalline solid
Color: Orange-yellow
Odor: Faint, unusual
Melting point: 42-43°C
Solubility (at 20°C):

Solvent	Solubility
ethyl acetate	>100 g/100 g
benzene	100 g/100 g
cyclohexane	25.1 g/100 g
ether	>100 g/100 g
acetone	>100 g/100 g
chloroform	>100 g/100 g
ethanol	17.7 g/100 g
water	<7 g/100 g

Vapor pressure: 6×10^{-6} mm Hg at 20°C, 2.5×10^{-5} mm Hg at 30°C
Stability: Sensitive to ultraviolet light. Stable in aqueous solution over range of pH 5 to 9.

Toxicological characteristics:

Acute Effects:

Acute Oral LD_{50} - data gap
Acute Dermal LD_{50} - data gap
Dermal Irritation - data gap
Acute Inhalation Toxicity - data gap
Primary Eye Irritation - data gap

Chronic Effects:

Oncogenicity - data gap
Teratology - data gap
Reproductive Effects - data gap
Mutagenicity - data gap
Feeding Studies - data gap

Major Routes of Exposure: Dermal, ocular, and ingestion

Physiological and Biochemical Behavioral Characteristics:

Foliar absorption: Not applicable - fluchloralin is soil-incorporated.
Translocation: Residues are taken up and translocated through the roots and shoots of cotton and soybean plants. Parent compound and some metabolites have been identified in cotton and soybean foliage following exposure of the roots to fluchloralin.
Mechanism of pesticidal action: Believed to affect seed germination and other physiological growth processes.
Metabolism and persistence in plants and animals: Plant metabolism is not adequately understood. Fluchloralin residues did not transfer to ruminant tissues at exaggerated rates (67X the expected intake by beef cattle). Degradation in animals is step-wise through N-dealkylation and N-hydrolysis.

Environmental Characteristics:

Adsorption and leaching in basic soil types: Fluchloralin (unaged) and fluchloralin residues (aged 30 days) are relatively immobile to slightly mobile in loamy sand and sandy soils, but the degradate 2,6-dinitro-4-trfluoromethylphenol is highly mobile in loamy sand soil and mobile in sandly clay loam soil. Fluchloralin is slightly mobile in runoff from plots (5-12% slope) of silt loam soil.
Microbial breakdown: No data

Loss from photodecomposition and/or volatilization: Fluchloralin photodegrades rapidly (half-life 27 minutes) in water (pH 5.6) when exposed to artificial sunlight. Photodegradation of solid fluchloralin film is slower (half-life 48 hours in artificial sunlight). No valid volatilization data are available.

Bioaccumulation: No valid data

Resultant average persistance: Half-life ranges from <32 to 120 days, depending on soil type.

Half-life in Water: Stable in water over pH range from 5.0 to 9.0, if not exposed to light.

Ecological characteristics:

Hazards to Birds: Data gap

Hazards to Aquatic Invertebrates: Data gap

Hazards to Fish: High toxicity poses potential threat to fish populations. Hazard cannot be evaluated until receipt of certain environmental fate data.

Potential Problems with Endangered Species: USDI has made a jeopardy assessment, finding threats to slackwater darter and eleven freshwater mussels from use of fluchloralin on soybeans.

Tolerance Reassessment:

List of crops and tolerances:

(N=negligible)

CROP	TOLERANCE (ppm)
Cotton, Seed	0.05N
Peanuts	0.05
Peanuts, forage	0.05
Peanuts, hay	0.05
Peanuts, hulls	0.1
Soybeans	0.05N
Sunflower, seeds	0.05
Vegetables, seed & pod (dry/succulent)	0.05
Vegetables, seed & pod, forage	0.1
Vegetables, seed & pod, hay	0.1

List of food contact uses: Beans (dry and succulent), okra, peas (dry and succulent), peanuts, soybeans, sunflower seeds.

Results of tolerance assessment: Current PADI is 0.0026 mg/kg/day, based on a NOEL of 5.250 mg/kg/day (210 ppm) and an LEL = 15.75 mg/kg/day (hemisiderosis in the liver) using a safety factor of 2000. The portion of the PADI currently occupied is <3%. However, the feeding study on which the PADI was based has been declared invalid, and no other toxicological data are available.

Problems known to have occured from use: No PIMS data available.

4. Summary of Regulatory Position and Rationale:

Use Classification: Not classified.

Use, Formulation or Geographic Restrictions: Manufacturing-use products may only be formulated into end-use products intended for use as an herbicide on dry and succulent peas and beans, cotton, okra, peanuts, soybeans, and sunflowers.

Unique Label warning statements:

a) Use Pattern Statements:

Labels of manufacturing-use products must bear the statement:

"For formulation into end-use herbicide products intended only for use on kidney, lima, navy, green, pinto, Great Northern or edible soy beans, blackeyed, cow, field, or garden peas, cotton, okra, peanuts, or sunflowers."

b) Precautionary Statements:

Labels of manufacturing-use products must bear the statement:

"Do not discharge effluent containing this product into lakes, streams, ponds, estuaries, oceans, or public waters unless this product is specifically identified and addressed in a NPDES permit. Do not discharge effluent containing this product to sewer systems without notifying the sewage treatment plant authority. For guidance contact your State Water Board or Regional Office of the EPA."

The labels of all end-use products must bear the statement:

"Do not apply directly to water. Do not contaminate water by cleaning of equipment or disposal of wastes."

5. Summary of major data gaps

DATA REQUESTED	DUE DATE (# of months after issuance of the Standard)
Statement of Composition	six months
Discussion of Formation of Impurities	six months
Preliminary Analysis	twelve months
Certification of Limits	twelve months
Analytical Methods for Enforcement of Limits	twelve months
Density, Bulk Density, or Specific Gravity	six months
Dissociation constant	six months
Octanol/water Partition Coefficient	six months
pH	six months
Stability	six months
Metabolism in Plants	twenty-four months
Residue - Dill - crop field trials	twenty-four months
Residue - Okra - crop field trials	twenty-four months
Hydrolysis	nine months
Photodegradation - on soil	nine months
Anaerobic Soil	twenty-seven months
Leaching and Adsorption/Desorption	twelve months
Field Dissipation - Soil	twenty-seven months
Accumulation - confined rotational crops	thirty-nine months
Accumulation - field rotational crops	fifty months

DATA REQUESTED	DUE DATE (# of months after issuance of the Standard)
Accumulation - in fish	twelve months
Acute Oral Toxicity	nine months
Acute Dermal Toxicity	nine months
Acute Inhalation Toxicity	nine months
Primary Eye Irritation	nine months
Primary Dermal Irritation	nine months
Dermal Sensitization	nine months
21-Day Dermal Toxicity	twelve months
Chronic Toxicity, Rodent and Non-rodent	fifty months
Oncogenicity - two species	fifty months
Teratogenicity	fifteen months*
Reproduction - 2-generation	thirty-nine months
Gene Mutation	nine months
Chromosomal Aberration	twelve months
Other Mechanisms of Mutagenicity	twelve months
General Metabolism	twenty-four months

* A study has been submitted, but not yet reviewed.

6. <u>Contact person at EPA</u>: <u>Robert Taylor,</u> TS-767-C, 401 M Street SW, Washington DC 20460
   (703) 557-1800

DISCLAIMER: The information presented in this Chemical Information Fact Sheet is for informational purposes only and may not be used to fulfill data requirements for pesticide registration and reregistration.

FLUOMETURON

Date Issued: December 1985
Fact Sheet Number: 88

1. Description of chemical

    Generic name: FLUOMETURON (ANSI, WSSA, BSI, ISO)

    Trade name: Cotoran®, Lanex®

    EPA Shaughnessy code: 035503

    Chemical Abstracts Service (CAS) number: 2164-17-2

    Year of initial registration: 1960

    Pesticide type: Herbicide

    Chemical family: Substituted urea

    U.S. and foreign producers: Ciba-Geigy, FBC Chemicals

2. Use patterns and formulations

    Application sites: To selectively control certain annual grasses and broadleaf weeds in cotton and sugarcane.

    Formulations: Flowable liquid concentrate and wettable powder.

    Types and methods of application: Postemergent directed or over-the-top spray, preplant soil incorporated, preemergent broadcast or banded spray, a directed layby treatment, and aerial application.

    Application rates: Cotton 1.0 to 2.0 pounds of active ingredient per acre with one to three applications per season, sugarcane 2.0 to 4.0 pounds of active ingredient per acre with one to two applications per season.

3. Science findings

    Summary science statement: Fluometuron has been found to have many data gaps and to be a potential ground water contaminant.

Fluometuron is not acutely toxic in general, however, it does produce some skin irritation and does induce corneal opacity.

Chemical Characteristics:

Physical State: Powder or crystals

Color: White to tan

Odor: Amine-like

Melting Point: 163 to 164.5

Specific Gravity or Density: 1.40 g/cm^3 @ 20 °C

Solubility

Solvent	20 °C
Water	105 ppm
Acetone	15%
Chloroform	2%
Methanol	14%
Hexane	< 4%

Vapor Pressure: 5 x 10 mmHg @ 20 °C

Octanol/Water Partition Coefficient: PK = 3.2 @ 20 °C

Stability: Hydrolyzed by acid or base. Stable at room temperature.

Toxicological Characteristics:

Acute Effects:

Acute Oral Toxicity: 8.9 gm/kg male rats
7.8 gm/kg female rats
Toxicity Category IV

Acute Dermal Toxicity: 3.03 gm/kg rabbits
Toxicity Category III

Acute Inhalation: > 2.07 mg/L
Toxicity Category III

Primary Eye Irritation: Induces corneal opacity.
Toxicity Category: I

Major Routes of Exposure: Dermal, inhalation

Subchronic Effects:

The NOEL for rats is 7.5 mg/kg

The NOEL for dogs is 400 ppm

Chronic Effects:

No adequate feeding studies are available to establish the chronic toxicity potential of technical Fluometuron.

Oncogenicity: Data gap

Teratogenicity: The compound is not teratogenic up to 500 mg/kg. No LeL can be established based on current data.

2-Generation Reproduction: Data Gap

Mutagenicity: Data Gap

Physiological and Biochemical Behavioral Characteristics:

Metabolism: Data Gap

Environmental Characteristics:

Preliminary Adsorption and Leaching Characteristics:

Fluometuron is moderately to very mobile in a variety of soils ranging in texture from sand to silty clay loam. These characteristics are not adequately understood.

Resultant Average Persistence:

Fluometuron dissipates in sandy clay loam with a half-life of < 30 days (Europe).

Ecological Characteristics:

Hazards to Birds:

Data are sufficient to characterize fluometuron as slightly to practically nontoxic to birds.

Hazards to Fish: Data Gap

Tolerance Assessment:

List of Crops and Tolerances: (CFR 40 180.229)

Commodity	ppm
Cottonseed	0.1
Sugarcane	0.1

Results of Tolerance Assessment:

   Tolerances cannot be adequately assessed until data gaps have been filled.

4. **Summary of Regulatory Position and Rationale:**

   Based on review and evaluation of all available data and other relevant information on fluometuron, the agency has made the following determinations:

   The available data are insufficient to indicate that any of the risk criteria listed in § 162.11(a) of title 40 of the U.S. Code of Federal Regulations have been met or exceeded for the uses of fluometuron at the present time.

   Fluometuron is a toxicity category I chemical regarding corneal opacity. Skin irritation places fluometuron in toxicity category II. However the agency has determined that the protective clothing required by this standard will sufficantly reduce the risk associated with its use.

   The absence of other toxicological data prevents the agency from determining the subacute and chronic effects of fluometuron. Given the lack of data, the most appropriate action is to move quickly to fill the data gaps. When data are submitted and reviewed, the agency will determine the registerability of the affected use pattern.

   Fluometuron tolerances for cotton and sugarcane will be reevaluated when the requested data in the standard are received and reviewed. The data base supporting these tolerances is inadequate. The data required to fill these data gaps are listed under number 6, Summary of Major Data Gaps, on this fact sheet.

   Tolerances for fluometuron on meat, milk, poultry, and eggs have not been established therefore the label restrictions on feeding crop material to livestock will remain on product labeling.

   Fluometuron has been placed on the list of possible ground water contaminants and the Agency has requested ground water data on an accelerated basis under the data-call-in program.

Use, Formulation or Geographic Restrictions:

>Manufacturing-use products may only be formulated into end-use products intended for use as an herbicide on cotton and sugarcane.

5. <u>Unique Label Warning Statements</u>:

Labels of all MPs must bear the statement:

Use Pattern Statements

>For formulation into end-use herbicide products intended for use on cotton and sugarcane.

Precautionary Statements

>Do not contaminate water by cleaning of equipment or disposal of wastes.

>Do not discharge effluent containing this product into lakes, streams, ponds, estuaries, oceans, or public waters unless this product is specifically identified and addressed in an NPDES permit. Do not discharge effluent containing this product into sewer systems without previously notifying the sewage treatment plant authority. For guidance contact your State Water Board or regional office of EPA.

Labels of EPs must bear the statements:

Hazards to Humans Statements:

>Mixer/loader/applicators must wear protective clothing, consisting of coveralls, a long-sleeved shirt, shoes, impermeable gloves, and eye protection when handling this product.

Environmental Hazard Statements:

> The labels of EPs intended for outdoor use must bear one of the following statements, depending on the formulation of the product.

Granular products must bear the statement:

> Do not apply directly to water or wetlands. In case of spills, collect for use or properly dispose of the granules. Do not contaminate water by cleaning of equipment or disposal of wastes.

Nongranular products must bear the statement:

> Do not apply directly to water or wetlands. Do not contaminate water by cleaning of equipment or disposal of wastes.

All products registered for use on:

Cotton:

> Do not feed foliage from treated cotton plants or gin trash to livestock.

> Do not apply within 60 days of cotton harvest.

Sugarcane:

> Do not apply within 180 days of harvest.

> Do not graze treated fields with livestock.

All Products Phytotoxicity Warnings:

> Do not plant crops other than sugarcane or cotton within one year after the last application or crop injury may result.

6. **Summary of Major Data Gaps:**

Dates when major data gaps are due to be filled.

Data Requested	Date Due
	(After issuance of the Standard)
Product Chemistry	7 Months
Magnitude of the Residues Sugarcane	19 Months
Magnitude of the Residues Cotton	19 Months
Nature of Residue (Metabolism)	25 Months
Hydrolysis	Study under review
Photodegradation Water	Study under review
Soil	Study under review
Metabolism-lab Aerobic Soil	Study under review
Anaerobic Soil	Study under review
Leaching and Adsorption/Desorption	Study under review
Soil Dissipation Field	Study under review
Accumulation, Rotational Crops	Study under review
21-Day Dermal	13 Months
Chronic Toxicity	50 Months
Oncogenicity Study	50 Months
Teratogenicity	15 Months
Reproduction	39 Months
Gene Mutation	9 Months

Data Requested (cont.)	Date Due (cont.)
Chromosal Aberration	12 Months
Other Mechanisms of Mutagenicity	12 Months
General Metabolism	24 Months
Freshwater Fish Warmwater	9 Months
Foliar Dissipation	25 Months
Soil Dissipation	25 Months
Dermal Exposure	25 Months
Inhalation Exposure	25 Months
Analytical Methods to Verify Certified limits	13 Months
Submittal of Samples	7 Months
Dermal Sensitization and Photosensitization	9 Months

7. Contact Person at EPA:

Robert J. Taylor
U.S. Environmental Protection Agency
TS-767-C
401 M Street SW.
Washington, DC 20460
(703) 557-1800

DISCLAIMER:

The information presented in this Chemical Information Fact Sheet is for informational purposes only and may not be used to fulfill data requirements for pesticide registration and reregistration.

FLURIDONE

Date Issued: March 1986
Fact Sheet Number: 81

1. Description of chemical

    Generic name: 1-methyl-3-phenyl-5-[3-(trifluoromethyl)phenyl]-4[1H]-pyridinone
    Common name: Fluridone
    Trade name: Sonar
    EPA Shaughnessy code: 112900-6
    Year of initial registration: 1986
    Pesticide type: Aquatic herbicide
    U.S. and foreign producers: Elanco Products Company

2. Use patterns and formulations

    Application sites: freshwater ponds, lakes, reservoirs, drainage canals, irrigation canals, rivers
    Types of formulations: aqueous suspension; pellet
    Types and methods of application: surface spray, weighted-hose dragged near bottom; broadcast (pellet)
    Application rates: 0.5 lb ai/surface acre--4.0 lb ai/surface acre
    Usual carrier: water

3. Science findings

    Summary science statement: Supporting data base for fluridone registration and tolerance proposals supporting aquatic use is complete and acceptable except for a second species (rat) teratology study. Original study submitted did not produce teratogenic response at any level tested. The study, however, is not adequate for regulatory purposes because the highest dose tested did not produce frank maternal toxicity or fetotoxicity. The study is presently being repeated.

    Chemical characteristics: Fluridone is a white (to offwhite) crystalline solid with no odor. The melting point is 154-155° C. The flashpoint for the aqueous suspension formulation is greater than 200° C. Fluridone is not corrosive to application equipment.

    Toxicological characteristics:

    Acute toxicology: Technical fluridone is in Category IV for acute oral effects in the rat and is moderately toxic through acute inhalation exposure. Eye irritation potential has been demonstrated as moderate to severe (Cat. III and Cat. II). The aqueous suspension and pellet formulations are in Cat. III for oral, dermal, skin, and eye irritation effects.

Chronic toxicology: A complete, acceptable chronic toxicity data base is available, except for a rat teratology study (second species). A valid rabbit teratology study indicates no teratogenic response up to a dose level of 300 mg/kg/day. Fluridone is not considered to have produced an oncogenic response in the mouse or rat. Mutagenicity assays submitted do not indicate genotoxic potential, gene mutation, or structural chromosomal aberration.

Physiological and biochemical behavioral characteristics: Fluridone is a systemic herbicide; it is absorbed from water by plant shoots and from hydrosoil by roots. Inhibits carotenoid synthesis which enhances degradation of chlorophyll, producing white (chlorotic) growing points in susceptible plants.

Environmental characteristics:

Degradation: Fluridone is stable to hydrolysis. It will photodegrade (half-life 34 hours in natural pond water).

Persistence: Under anaerobic aquatic conditions, fluridone has a half-life of 9 months. Half-life for fluridone in water is estimated to be 20 days; for hydrosoil, 90 days.

Bioaccumulation: Fluridone has a low potential for bioaccumulation in fish.

Ecological characteristics:

Avian studies: Acute oral (bobwhite quail) > 2000 mg/kg (slightly toxic). Avian dietary (bobwhite quail and mallard duck) > 5000 ppm. No impairment on reproduction for above species up to 1000 ppm dietary exposure.

Aquatic species studies:

Daphnia magna 48-hour acute is 6.3 mg/L (moderately toxic)
Bluegill sunfish 96-hour acute is 12 mg/L (moderately toxic)
Rainbow trout 96-hour acute is 11.7 mg/L (moderately toxic)
Sheepshead minnow 96-hour acute is 10.91 mg/L (moderately toxic)
Oyster embryo-larvae 48-hour acute is 16.51 mg/L (moderately toxic)
Maximum acceptable theoretical concentration (MATC) value for fathead minnow (second generation fry) was calculated to be > 0.48 < 0.96 mg/L. No treatment related effects were observed at or below 0.48 mg/L. Total length of 3 day-old fry was reduced at 2 mg/L fluridone.

Potential problems for endangered species:

Acute and MATC values indicate a potential hazard for aquatic organisms in shallow areas at higher treatment rates described on the label. Formal consultation with Office of Endangered Species (OES) has been initiated. To minimize hazard, label directions provide for use of lowest listed rates for shallow areas and consultation with Fish and Game agency or U.S. Fish and Wildlife Service if questions arise concerning aquatic resources in the area to be treated.

Tolerances proposed:

A tolerance is proposed for residues of the herbicide fluridone (1-methyl-3-phenyl5-[3-(trifluoromethyl)phenyl]-4(1H)-pyridinone) and its metabolite (1-methyl-3-(4-hydroxyphenyl)-5-[3-(trifluoromethyl)phenyl]-4(1H)-pyridinone) in fish at 0.5 ppm.

A tolerance is proposed for residues of the herbicide fluridone in the following raw agricultural commodities:

Commodity	Parts Per Million
Cattle, fat	0.05
Cattle, kidney	0.1
Cattle, liver	0.1
Cattle, meat byproducts	0.05
Cattle, meat (except liver and kidney)	0.05
Eggs	0.05
Goats, fat	0.05
Goats, kidney	0.1
Goats, liver	0.1
Goats, meat byproducts	0.05
Goats, meat (except liver and kidney)	0.05
Hogs, fat	0.05
Hogs, kidney	0.1
Hogs, liver	0.1
Hogs, meat byproducts	0.05
Hogs, meat (except liver and kidney)	0.05

Horses, fat	0.05
Horses, kidney	0.1
Horses, liver	0.1
Horses, meat byproducts	0.05
Horses, meat (except liver and kidney)	0.05
Milk	0.05
Poultry, fat	0.05
Poultry, kidney	0.1
Poultry, liver	0.1
Poultry, meat byproducts	0.05
Poultry, meat (except liver and kidney)	0.05
Sheep, fat	0.05
Sheep, kidney	0.1
Sheep, liver	0.1
Sheep, meat byproducts	0.05
Sheep, meat (except liver and kidney)	0.05

Tolerances are proposed in the following irrigated crops and crop groupings for residues of the herbicide fluridone resulting from use of irrigation water containing residues of 0.15 ppm following applications on or around aquatic sites. Where tolerances are established at higher levels from other uses of fluridone on the following crops, the higher tolerance also applies to residues in the irrigated commodity. The tolerances follow:

Commodity	Parts Per Million
Avocados	0.1
Citrus	0.1
Cottonseed	0.1
Cucurbits	0.1
Forage Grasses	0.15
Forage Legumes	0.15
Fruiting vegetables	0.1
Grain crop	0.1
Hops	0.1
Leafy vegetables	0.1
Nuts	0.1
Pome fruit	0.1
Root crops - vegetables	0.1
Seed and pod vegetables	0.1
Small fruit	0.1
Stone fruit	0.1

Based on the NOEL of 8 m/kg/day in the chronic rat feeding study and a 100-fold safety factor, the acceptable daily intake (ADI) has been set at 0.08 mg/kg/day with a maximum permissible intake (MPI) of 4.8 mg/day for a 60-kg person. There are no previously established tolerances for this herbicide.

The Agency is designating an acceptable residue level for fluridone in potable water at 0.15 ppm. This concentration reflects the maximum application rate for the herbicide registration(s) issued pursuant to FIFRA. Consumption of water is estimated at 2.0 liters per day for a 60-kg adult. These tolerances and the acceptable residue level in potable water result in a theoretical maximum residue contribution of 0.4112 mg/day in a 1.5-kg diet (including 2 liters of water) and use 8.57 percent of the ADI.

No Mexican, Canadian, or Codex maximum residue levels have been established. Residue studies are adequate to support the proposed tolerances. Plant and animal metabolism is adequately understood and adequate analytical methods are available to enforce the tolerance levels. The residue of concern in drinking water is parent compound, i.e., fluridone.

4. Summary of Regulatory Position and Rationale

Risk/benefit review: None of the risk criteria set forth in Title 40 Code of Federal Regulations §162.11 have been exceeded for fluridone.

Fluridone has been proposed only for direct application to aquatic sites. No ground water contamination issue is associated with this use.

5. Summary of Major Data Gaps

An additional rat (second species) teratology is underway. Schedule for submission: July 1, 1986.

6. Contact Person at EPA

Richard F. Mountfort
Product Manager (23)
Environmental Protection Agency
401 M St. SW.
Washington, DC 20460
(703) 557-1830

DISCLAIMER: The information presented in this Chemical Information Fact Sheet is for informational purposes only and may not be used to fulfill data requirements for pesticide registration and reregistration.

FLUVALINATE

Date Issued: March 31, 1986
Fact Sheet Number: 86

1. Description of Chemical

    Generic Name: N-[2-chloro-4-(trifluoromethyl)phenyl]-
    DL-valine(±)-cyano(3-phenoxyphenyl)
    methyl ester.

    Common Name: Fluvalinate.

    Other Proposed Names: None.

    Trade Names: MAVRIK®

    EPA/OPP Chemical Code (Shaughnessy) Number: 109302

    Chemical Abstracts Service (CAS) Number: 69409-94-5

    Year of Initial Registration: 1983

    Pesticide Type: Insecticide/Acaricide.

    Chemical Family: Synthetic pyrethroid.

    U.S. and Foreign Producers: Zoecon Corporation.

2. Use Patterns and Formulations

    Application Site: Nonbearing fruit and nut trees and vines; turf; certain crops grown for planting seed; and cotton.

    Pests Controlled: Tobacco budworm; cotton bollworm; tarnished plant bug; cotton leafperforator; lygus bug; cabbage looper; pea looper; alfalfa looper; mint looper; pink bollworm; whitflies; fleahoppers; flea beetles; salt marsh caterpillar; boll weevil; cotton leafworm; garden webworm; grasshoppers; aphids; twospotted spider mite; pacific spider mite; strawberry spider mite; thrips; leafhoppers; beet armywrom; fall armyworm; midge; diamondback moth; corn earworm (and other Heliothis spp.); imported cabbageworm; cabbage pod weevils; Japanese Beetle; Mexican bean beetle; blister beetle;

Pests Controlled (continued): cabbage flea leaf beetle; bean leaf
                              beetle; black pollen beetle; greenbug; fall webworm;
                              leaftiers; onion maggot; velvetbean caterpillar;
                              green cloverworm; cutworm (climbing and surface
                              feeding); Mormon cricket; chinch bugs; earwigs;
                              ticks; ants; lawn moths (sod webworms); Bermuda
                              mites; crickets; Brown Dog Ticks; chiggers; clover
                              mites; sowbugs; millipedes; springtails (Collembola);
                              scales (crawlers); leafrollers; Stink bug; borers
                              (Lesser peach tree); Gypsy moth; Tussock moth; Red
                              humped caterpillar; tufted apple budmoth; cankerworms;
                              rose chafer; curculios; tent caterpillars.

Types and Methods of Application: Aerial and ground (excluding ULV).

Application Rates: 0.025 to 0.15 pounds of active ingredient per acre per
                   application. Maximum of 0.5 pounds of active ingredient
                   per acre per year to cotton.

Restrictions: 30-day pre-harvest interval following last application to
              cotton.

Types of Formulations: Emulsifiable concentrate (2 pounds per gallon)
                       Flowable concentrate (2 pounds per gallon)

3. <u>Science Findings</u>

   <u>Summary Science Statement:</u>

   Fluvalinate, a synthetic pyrethroid insecticide/acaricide, has moderate
   to low acute mammalian toxicity except that the formulated products are
   highly corrosive to the eye. The Agency has a complete battery of
   subchronic and chronic toxicity data that demonstrate no significant
   adverse effects.

   Fluvalinate is slightly toxic to birds and highly toxic to some aquatic
   organisms. Additional studies on ecological effects of fluvalinate are
   required.

   A. <u>Chemical Characteristics:</u>

      1. Physical State: Viscous oil.

      2. Color: Yellow to brown.

      3. Odor: Undetectable.

      4. Boiling Point: >450°C.

      5. Melting Point: N/A

      6. Vapor Pressure: <1 X 10^{-7} Torr at 25°C.

Chemical Characteristics (continued):

7. Density: 1.29 g/cm^3.

8. Solubility:
   Water: 2 ppb.
   Organic solvents: Very soluble
   Aromatic hydrocarbons: Very soluble
   Hexane: Slightly soluble

9. Dissociation Constant: N/A

10. Octanol/Water Partition Coefficient: >7000 (log K >3.8)

11. pH: N/A

12. Oxidizing or Reducing Agent: N/A

13. Storage Stability: In glass 24 months at 24°C and 42°C.

B. Toxicological Characteristics:

1. Acute Oral: 261 mg/kg (rat).

2. Acute Dermal: LD_{50} >20.1 g/kg (rat).

3. Primary Eye Irritation: Moderately irritating.

4. Primary Skin Irritation: Mildly irritating- erythema and edema.

5. Skin Sensitization: Non-sensitizing.

6. Dermal Repeated Application: Systemic NOEL is 100 mg/kg.

7. 90-day Rat Feeding Study: No effects at 3.0 mg/kg/day

8. 6-months Dog Feeding Study: No effects at 5.0 mg/kg/day

9. Teratology: (rat) No effects at 50.0 mg/kg for terata (HDT). Fetotoxicity noted at 50.0 mg/kg.

   (rabbit) No effects at 125.0 mg/kg for terata. Fetotoxicity noted at 50.0 mg/kg.

10. Reproduction: (rat) No effects at 20 ppm for reproductive effects. Fetotoxicity noted at 250 and 500 ppm (500 ppm was the highest dose level tested).

Toxicological Characteristics (continued):

11. Oncogenicity: (mouse) No oncogenic effects were noted at dosage levels of 2, 10, and 20 mg/kg/day (20 mg/kg/day was the highest dosage level tested).

   (rat) No oncogenic effects were noted at dosage levels of 0.25, 0.5, 1.0 and 2.5 mg/kg/day (2.5 mg/kg/day was the highest dosage level tested).

12. Mutagenic Effects: Non mutagen.

13. 21-day Delayed Neurotoxicity: (hen) No effects at 20,000 mg/kg/day.

C. Physiological and Biological Characteristics:

   1. Foliar absorption: N/A.
   2. Translocation: N/A.
   3. Mechanism of Pesticide Action: Neuropathy characteristic of pyrethroid insecticides.

D. Environmental Characterisitics:

   1. Absorption and leaching in soil: Fluvalinate has little potential to leach. However, major soil metabolites have the potential to leach in some soils.

   2. Hydrolysis: Fluvalinate is stable to hydrolysis at environmental pH's and temperature.

   3. Photolysis: Fluvalinate photodegrades in aqueous solutions with a half-life of 0.6 - 1.0 day yielding the haloaniline, anilino acid and 3-phenoxybenzoic acid. Photodegrada-on soil surface does not appear to occur.

   4. Aerobic Soil Metabolism: Fluvalinate degrades in soil under aerobic conditions with half-lives of 4 - 8 days in sandy loam, sandy clay and clay soils.

   5. Anaerobic Soil Metabolism: Fluvalinate will anaerobically degrade in soil with a half-life of 15 days in sandy loam soil.

Fluvalinate 409

Environmental Characterisitics (continued):

6. Enviromental Fate and Surface and Ground Water
    Contamination Concerns: Data submitted indicate that parent fluvalinate has little potential to leach. Major metabolites do have potential to leach in some soils. Low application rates, usually below 0.1 lb. a.i./acre will tend to decrease the potential for ground and surface water contamination.

7. Exposure of Humans and Nontarget Organisms to
    Chemical or Degradates: Fluvalinate products are classified for restricted use. Minimal exposure to humans is anticipated from use. Additional data are needed to fully assess the enviromental fate and ecological effects from fluvalinate.

8. Exposure During Reentry Operations: Not applicable for cotton use.

E. Ecological Characteristics:

1. Hazards to Fish and Wildlife:

    Birds – Dietary and Acute Toxicity:

    Mallard ducks: Avian dietary LC_{50} >5620 ppm.

    Bobwhite quail: Avain acute oral LD_{50} >2510 mg/kg; Avian dietary LC_{50} >5620 ppm.

    Fish:

    Bluegill sunfish 96 hr. LC_{50} 0.09 (0.7 – 1.1) ug/l.

    Rainbow trout 96 hr. LC_{50} 2.9 (2.3 – 3.6) ug/l.

    Freshwater Invertebrates:

    Daphnia magna: 48 hr. LC_{50} 74 (61 – 89) ug/l.

    Mysid shrimp: 48 hr. LC_{50} 2.9 (2.3 – 3.6) ug/l.

F. **Tolerance Assessment:**

The established tolerances for residues of fluvalinate in or on raw agricultural commodities are published in 40 CFR 180.427. A summary of these tolerances follows:

U. S. Tolerances

Commodities	Maximum Residue Limits in Parts Per Million (PPM)
Cottonseed	0.1
Cattle, fat	0.01
Cattle, mbyp	0.01
Cattle, meat	0.01
Eggs	0.01
Goat, fat	0.01
Goat, mbyp	0.01
Goat, meat	0.01
Hogs, fat	0.01
Hogs, mbyp	0.01
Hogs, meat	0.01
Horses, fat	0.01
Horses, mbyp	0.01
Horses, meat	0.01
Milk	0.01
Poultry, fat	0.01
Poultry, mbyp	0.01
Poultry, meat	0.01
Sheep, fat	0.01
Sheep, mbyp	0.01
Sheep, meat	0.01

Tolerance Assessment (continued):

The established food additive tolerances for residues of fluvalinate are published in 21 CFR 561.437. A summary of these tolerances follows:

U. S. Tolerances

Commodities	Maximum Residue Limits in Parts Per Million (PPM)
Cottonseed hulls	0.3
Cottonseed oil	1.0

The established Acceptabale Daily Intake (ADI) for fluvalinate is 0.01 mg/kg/day based on a 2-year rat feeding study and a 100 fold safety factor. The Maximum Permissible Intake (MPI) has been calculated to 0.6 mg/kg/day for a 60 kg person. Based on the established tolerances for residues of fluvalnate as cited under 40 CFR 180.427 and 21 CFR 561.437, the Theoretical Maximum Residue Concentration (TMRC) is 0.0504 mg/day for a 1.5 kg food diet for a 60 kg person. The TMRC is 8.40 % of the ADI.

G. Reported Pesticide Incidents: Several.

Caused general complaints in numerous workers, such as:
1. transient coughing, sneezing and throat irritation;
2. itching, burning sensation of the arms and/or face with or without a rash, blisters and desquamation;
3. ocular irritation; and
4. headache and/or nausea.

4. Summary of Regulatory Position and Rationale

Fluvalinate, formulated as a flowable concentrate and an emulsifiable concentrate, is registered for use on nonbearing fruit and nut trees and vines, turf, certain crops grown for planting seed, and cotton. Because of toxicity to fish and aquatic invertebrates, precautionary labeling (including endangered species identification) is required to warn against contamination of bodies of water with products containing fluvalinate and restricted use classification is required. The condititonal registration for fluvalinate is issued for a period ending August 31, 1989.

5. **Summary of Data Gaps**

>Acute LC_{50} Estuarine and Marine Organisms Study (96 hour LC_{50} for marine fish plus 48-hour EC_{50} for oyster embryo larvae) - Guideline Reference No. (GRN 72-3) - Due Date - June 30, 1986
>
>Fish Early life stage study (GRN 72-4) -
>Due Date - June 30, 1986
>Two Field Dissipation Studies (GRN 164-1) -
>Due Date - August 31, 1988
>
>Rotational Crop Studies (GRN 165-1) -
>Due Date - August 31, 1989
>
>Aquatic Organism Accumulation (GRN §72-6) -
>Due Date - March 31, 1987
>
>Fish - Life Cycle Test (GRN §72-5) -
>Due Date - June 30, 1988
>
>Simulated Field Testing - Aquatic Organisms
>(GRN 72-7) - Due Date December 31, 1988

6. **Contact Person at EPA:** George LaRocca
Product Manager 15
Insecticide-Rodenticide Branch (TS-676C)
401 M Street SW.
Washington, DC 20460
(703) 557-2400

DISCLAIMER: The information presented in this Chemical Information Fact Sheet is for informational purposes only and may not be used to fillfill data requirements for pesticide registration and reregistration.

FONOFOS

Date Issued: February 1, 1985
Fact Sheet Number: 22.1

1. Description of chemical

    Generic name: 0-ethyl S-phenyl ethylphosphonodithioate
    Common name: Fonofos
    Trade name: Dyfonate
    EPA Shaughnessy code: 041701
    Chemical abstracts service (CAS) number: 944-22-9
    Year of initial registration: 1967
    Pesticide type: Insecticide
    Chemical family: organophosphate
    U.S. and foreign producers: Stauffer Chemical Co.

2. Use patterns and formulations

    Fonofos is a soil applied insecticide used primarily on corn (95%). It is used also on various vegetable crops, ornamentals, home lawns and home vegetable gardens and commercial turf. Fonofos is applied mainly with ground equipment. Aerial applications are made to hybrid seed corn. Application rates vary from 1-4 lbs./acre. The usual carrier is water.

3. Science Findings

    Fonofos is a yellow liquid with a mercaptan-like odor. The boiling point is 212°F (100°C) and the melting point is -32°C at 0.3mm H Fonofos is almost insoluble in water and miscible in common organic solvents.

    Toxicology characteristics:

    Technical fonofos is highly toxic based on acute oral, dermal, eye and inhalation effects.

    Results of toxicological studies on fonofos are as follows:

    -Oral LD_{50}, ranges from 3.16-18.5 mg/kg
    -Dermal LD_{50}, ranges from 121-359 mg/kg
    -Primary Eye Irritation, negative to 0.01 ml; 0/6 dead
    -Inhalation LC_{50}, 0.9 mg/L (male and female combined)
    -3 generation reproduction rat, reproductive and fetotoxic NOEL= 31.6 ppm (highest dose tested)
    -2 year dog feeding study - NOEL, ChE and non-cholinergic= 8 ppm; LEL, ChE and non-cholinergic= 60 ppm

Available data are insufficient to fully assess the toxicological properties of fonofos. Data gaps must be filled in areas of neurotoxicity, subchronic and chronic toxicity, oncogenicity and mutagenicity before a total risk assessment can be made.

Physiological and Biochemical Behavioral Characteristics:

Fonofos is not absorbed by foliage and is not translocated in the plant body. It is a cholinesterase inhibitor and accumulates in carrots.

Envirionmental Characteristics:

Fonofos is immobile in sandy loam and silt loam soils. It is mobile in quartz sand. It decomposes in aerobic soils by microbes i 4-8 weeks. Fonofos is non-volitile from soil but volatile from water. It degrades in aerobic soils with a half like of 3-16 weeks. Fonofos is moderately persistent.

Ecological Characteristics:

Fonofos is moderately to highly toxic to birds and highly toxic to freshwater fish and salt water organisms.

Simulated avian field studies indicate granular treatments of fonofos may result in some mortality, as well as brain AChE inhibition, but that effects are not likely to diminish wildlife resources.

See under Data Gaps for additional data requirements.

Tolerance assessments:

Tolerances are established for residues of the insecticide O-ethyl S-phenyl ethylphosphonodithioate, including its oxygen analog O-ethyl S-phenyl ethylphosphonothioate, in or on raw agricultural commodities as follows (40 CFR 180.221):

- 0.5 part per million in or on asparagus.
- 0.1 part per million (negligible residue) in or on bean forage, bean vine hay, fresh corn including sweet corn (kernels plus cob with husk removed), corn grain (including popcorn), corn forage or fodder (including sweet corn, field corn, and popcorn), fruiting vegetables, leafy vegetables, mint (peppermint, spearmint, peppermint hay, and spearmint hay), pea forage, pea vine hay, peanuts, peanut forage, peanut hay, peanut

hulls, root crop vegetables, seed and pod vegetables, sorghum (grain, fodder, and forage), soybean forage, soybean hay, strawberries, sugar beet tops, and sugarcane.

4. Summary of Regulatory Position and Rationale:

   The Agency has determined that certain formulations of fonofos warrant classification as restricted use pesticides. These include all emulsifiable concentrates 44% or greater and the 20% granular formulation. All products of these types which are released for shipment after September 1, 1985 must be labeled for restricted use. All products of these types which are in channels of trade after September 1, 1986 must be labeled for restricted use.

   A 24 hour interim reentry interval has been established for all uses of fonofos including the home lawn and home vegetable garden use.

   Gloves and shoes must be worn when applying fonofos.

5. Summary of Major Data Gaps *

   - Delayed neurotoxicity - hen
   - 90 day rodent feeding study
   - 90 day neurotoxicity study - hen/mammal
   - Chronic toxicity study-rodent
   - Oncogenicity study
   - Teratogencity study - 1 species
   - Gene mutation study
   - Chromosomal aberration study
   - Reentry Data
   - Acute LC_{50} - freshwater invertebrates
   - Fish early life cycle stage and aquatic invertebrate lifecycle studies
   - Residue data in:
     - root and tuber vegetables
     - leaves of root and tuber vegetables
     - fruiting vegetables (except curcubits)
     - cereal grains
     - forage, fodder and straw of cereal grains
     - miscellaneous crops (asparagus, peanuts, sugarcane and tobacco)
   - Poultry feeding study and ruminant feeding study
   - Photodegradation in water, soil and air
   - Hydrolysis study
   - Metabolism study in anaerobic soil

- Mobility studies (leaching and adsorption/desorption, volatilty lab, and volatility field)
- Soil dissipation study
- Accumulation studies - rotational crops and fish

*All major data gaps are to be filled by March 31, 1987.

6. <u>Contact person at EPA</u>   (Name, address, and telephone number)

   <u>Contact Person</u>

   William H. Miller, PM 16
   Registration Division (TS-767)
   Office of Pesticide Programs
   Environmental Protection Agency
   401 M Street SW
   Washington, DC 20460
   (703) 557-2600

DISCLAIMER: The information presented in this Chemical Information Fact Sheet is for informational purposes only and may not be used to fulfill data requirements for pesticide registration and reregistration.

FORMETANATE HYDROCHLORIDE

Date Issued: September 30, 1983
Fact Sheet Number: 11

1. DESCRIPTION OF CHEMICAL

    Generic Name: N,N-dimethyl-N'-[3-[[(methylamino)carbonyl]oxy]phenyl] methanimidamide monohydrochloride

    Common Name: Formetanate hydrochloride(HCl)

    Trade Name: Carzol

    EPA Shaughnessy Code: 097301

    Chemical Abstracts Service (CAS) Number: 23422-53-9

    Year of Initial Registration: 1969

    Pesticide Type: Acaricide/Insecticide

    Chemical Family: methylcarbamate hydrochloride

    U.S. and Foreign Producers: Imported into the U.S. from West Germany where it is manufactured by Schering A.G. Chemical Company. In the U.S., the sole importer and distributor of the chemical is Nor-Am Chemical Company.

2. USE PATTERNS AND FORMULATIONS

    Application Sites: Citrus, pome and stone fruits and alfalfa grown-for-seed.

    Types of Formulations: 92% soluble powder

    Types of Methods of Application: Ground and aerial

    Application Rates: Range from 0.0575 lbs. a.i./A to 1.38 lbs. a.i./A

    Usual Carriers: Water

3. SCIENCE FINDINGS

Summary Science Statement

Formetanate HCl is highly toxic from an oral route of exposure. It demonstrates low toxicity from the dermal route of exposure and moderate toxicity from the inhalation route of exposure. It is considered to be an eye irritant. The chronic studies are data gaps.

Formetanate HCl appears to leach in soil and therefore has a potential for contamination of groundwater. However, available data are insufficient to fully assess the environmental fate of this chemical. A 24-hour reentry interval has been established.

This chemical is highly toxic to birds. It is slightly toxic to fish and moderately toxic to estuarine and marine organisms.

A tolerance reassessment cannot be made until the studies required to fill the toxicology data gaps have been submitted and validated. A petition for a tolerance for residues in or on alfalfa hay and alfalfa forage must be submitted. A food additive petition must be submitted for apple pomace.

Chemical Characteristics

Physical State: Crystalline solid

Color: White

Odor: Essentially odorless

Melting point: 200-202° C

Toxicology Characteristics

Acute Oral LD_{50}: 26.4 mg/kg (rat), Toxicity Category I

Acute Dermal LD_{50}: > 10,000 mg/kg (rabbit), Toxicity Category II

Primary Dermal Irritation: No irritation (rabbit), Toxicity Category IV

Primary Eye Irritation: One-fifth of the animals (rabbit) showed irritation at a 7-day observation period. The effect was considered to be presumptively not reversible in 21 days.

Acute Inhalation LC_{50}: 0.29 mg/liter (rat), Toxicity Category II

The chronic studies are data gaps. The original studies were conducted by IBT and have been determined to be invalid. No replacement studies have been submitted. A commitment has been made by Nor-Am Chemical Company to replace the IBT studies.

A risk assessment cannot be completed at this time due to gaps in the data base.

Physiological and Biochemical Behavioral Characteristics

Mechanism of Pesticidal Action: A contact insecticide which causes reversible carbamylation of the acetylcholinesterase enzyme. Poisoning also impairs the central nervous system function.

Metabolism and Persistence in Plants and Animals: In plants, formetanate HCl is absorbed into the leaves but is not translocated to untreated areas. On the fruit, weathering and growth dilution appear to be the primary cause of the dissipation of formetanate HCl. In animals, formetanate is eliminated in the urine and feces. The major metabolic pathway involves hydrolysis to metabolites and subsequent formation of the glucuronide and ethereal sulfate conjugates of m-acetamidophenol.

Environmental Characteristics

Available data are insufficient to fully assess the environmental fate of formetanate HCl. It appears to rapidly degrade in soil under aerobic conditions. This chemical appears to leach in soil and, therefore has a potential for groundwater contamination. A 24-hour reentry interval has been established for this chemical on orchard crops.

Ecological Characteristics

Avian oral LD_{50} -
Mallard duck: 11.7 mg/kg
Bobwhite quail: 43.1 mg/kg

The avian dietary LC_{50} studies, freshwater fish LC_{50} studies and acute LC_{50} freshwater invertebrate studies are data gaps.

Formetanate HCl is characterized as highly toxic to birds. It is slightly toxic to warmwater fish and moderately toxic to estuarine and marine organisms. It is moderately toxic

to honey bees with direct contact spray, but it has very
low toxicity to bees when they are exposed to residues on
plants.

Tolerance Assessments

A full tolerance reassessment cannot be completed because of
certain residue chemistry and toxicology data gaps. The
current values for the acceptable daily intake intake(ADI)
of 0.025 mg/kg/day and the maximum permissible intake(MPI)
of 1.5 mg/kg/day are provisional. This is because a number
of multiple dose toxicity studies submitted to establish the
ADI/MPI were conducted by IBT and have been determined to be
invalid. Another reassessment and recalculation will be
made when the studies required to fill the toxicology data
gaps are submitted and reviewed.

A petition for tolerance for residues in or on alfalfa
forage and alfalfa hay must be submitted to support the
alfalfa grown-for-seed use. No residue data are available.
A petition for a food additive tolerace for apple pomace
must be submitted.

Tolerances:

Commodity	Parts Per Million
Peaches	5
Grapefruit	4
Lemons	4
Limes	4
Nectarines	4
Oranges	4
Tangerines	4
Apples	3
Pears	3
Plums(fresh prunes)	2

4. SUMMARY OF REGULATORY POSITION AND RATIONALE

The Agency has determined that it should continue to allow
the registration of formetanate HCl. Adequate studies are
available to assess the acute toxicological effects of
formetanate HCl to humans. None of the criteria for unreasonable adverse effects listed in 40 CFR §162.11(a) have
been met or exceeded. However, because of gaps in the data

base a full risk assessment of formetanate HCl cannot be made at this time.

A full tolerance reassessment cannot be completed because of residue chemistry and toxicology data gaps.

Available data are insufficient to fully assess the environmental fate of carbaryl. A 24-hour reentry interval has been established for formetanate HCl. This chemical appears to have the potential to contaminate groundwater.

5. SUMMARY OF MAJOR DATA GAPS

Residue data on alfalfa
Chronic feeding studies - 2 species
Oncogenicity studies - 2 species
Teratogenicity studies - 2 species
Reproduction studies - 2 species
Hydrolysis study
Mobility studies
Photodegradation studies
Dissipation studies
Avian dietary LC_{50} studies
Aquatic organism testing

6. CONTACT PERSON AT EPA

Jay S. Ellenberger
Product Manager (12)
Insecticide-Rodenticide Branch
Registration Division (TS-767C)
Office of Pesticide Programs
Environmental Protection Agency
401 M Street, S. W.
Washington, D. C. 20460

Office location and telephone number:
Room 202, Crystal Mall #2
1921 Jefferson Davis Highway
Arlington, VA 22202
(703) 557-2386

DISCLAIMER: The information presented in this Chemical Information Fact Sheet is for informational purposes only and may not be used to fulfill data requirements for pesticide registration and reregistration.

GLYCOSERVE

Date Issued: March 8, 1985
Fact Sheet Number: 47

1. Description of chemical

    Generic Name: Substituted Dimethyl Hydantoin
    Common name: 1,3-Bis(Hydroxymethyl)-5,5-dimethylhydantoin
    Trade name: Glycoserve
    EPA Shaughnessy code: 115501
    Chemical abstracts service (CAS) number: 273AB
    Year of initial registration: 1985
    Pesticide type: Disinfectant
    Chemical family (e.g., carbamates) Hydantoins
    U.S. and Foreign producers: Glyco

2. Use patterns and formulations

    Application sites: End-use products as a preservative in paints, room deodorizers, and dishwashing detergents as a Manufacturing Use Pesticide for reformulating use as a preservative in paints, room deodorizers and detergents.
    Types of formulations: Liquids
    Types of Methods of application: manual
    Application rates: 0.2% to 1% in paints, 0.25 to 1% in room deodorizer and detergents.
    Usual carriers: water

3. Science Findings

    Summary science statement: No mutagenic, or oncogenic effects are anticipated at the use dilutions recommended for the proposed uses.
    Chemical Characteristics: physical state-liquid, color-water white, odor-slight formaldehyde odor, boiling point-101.2°C
    Unusual handling characteristics: Protect eyes and skin when handling.

    Toxicology characteristics:

       Toxicology Category and value for each acute hazard.

    Testing Laboratory Warf Institute

    Acute Oral LD_{50}, Toxicity Category III

    * LD_{50} = 2.00 to 3.65 g/kg (M)

          3.00 to 5.00 g/kg (F) no signs of toxicity

Testing Laboratory: IBT

3.72 mg/kg hypoactivity, lacrimation, weakness, labored respiration.

b. Testing Laboratory: Warf Institute

   *Acute inhalation toxicity category IV
   $LC_{50} > 204$ mg/L/hr gasping

c. Testing Laboratory: Food and Drug Research Lab.

   $LC_{50} > 377.8$ ug/L/4 hrs slight irritation of mucous membranes envanol nose and eyes

d. Testing Laboratory: Bio/dynamics

   *Primary eye irritation Toxicity Category IV
   Negative using 0.1 ml of 0.1% solution

e. Testing Laboratory: Warf Institute
   *Acute Dermal

   $LD_{50} = >20$ gm/kg

f. Testing Laboratory: Warf Institute

   *Acute dermal irritation
   PIS = 1.5

** Acute oral LD_{50} - rat; $LD_{50} = 2.0$ to 3.65 g/kg(male)
   Warf Institute; #6051229 $LD_{50} = 3.0$ to 5. g/kg(female)
   9/30/76   (highest dose tested 10g/kg)

** Acute oral LD_{50}- rat; $LD_{50} = 3.720$ g/kg
   IBT; #853009420; 9/22/76   (highest dose tested 15,380 g/kg)

Key * Test conducted using 1,3-hydroxymethyl)-5,5-dimethyl hydrontoin 44% and hydroxymethyl-5,5 dimethyl hydantoin
   ** = Test conducted using the product as formulated glyco

   Major routes of exposure in order of toxicological significance.

** Dermal patch test - human; IBT; #8537-09670; 12/3/76	IBT - invalid Dynamac Corp. 2/25/83
** Dermal absorption - rat; EG & G Mason Research Institute; Report MRI-206-GC-83-35; 9/9/83	Body burden <1% < than 3% was absorbed and the main portion was recovered in the urine. (Note 0.1 ml = 3.15 mg) <u>Addendum</u> Maximum absorption in 10 hrs. 10.01528 ml = 0.4813 mg 1/2 hr. (minimal absorption) 0.02406 mg
** Patch test- human; Food and Drug Research Lab,; #18;10/11/81	10 repeated insult path test in 31 males and 171 females with 0.21 ml of test material produced negative results for sensitization
** Acute oral LD_{50} - rat; Warf Institute; #6051229; 9/30/76	LD_{50} = 2.00 to 3.65 g/kg (M) 3.00 to 5.00 g/kg (F) (no signs of toxicity)
** Mutagenic; unscheduled DNA synthesis; Lab not stated; no. M0017;5/7/82	Positive induced a dose-related increase in the number of grains per nucleus
** Mutagenic- Ames; Litton Bionetics; # 20838; 3/78	Negative with and without activation
** Mutagenic- Ames; Microbiological Associates T 1804.502; 9/3/82	Positive for TA 100 and TA 98 with and without metabolic activation
** Mutagenic- Ames; Microbiological Associates; T 1804.502; 9/3/82	Negative for TA 1535, TA 1538 and TA 98
** Mutagenic- chinese hamster ovary <u>in Vitro</u>; Microbiological Assoc.; #003-560-723-2; 2/23/82	Positive; significant increase in chromosomal aberrations with and without S-9 activation
** Mutagenic- induction in mammalian cells- mice; Microbiological Assoc.; #003-561-724-7; 4/30/82	Positive with and without activation

Key * Test conducted using 1,3-hydroxymethyl)-5,5-dimethyl hydrontoin 44% and hydroxymethyl-5,5 dimethyl hydantoin
  ** = Test conducted using the product as formulated glyco

** Mutagenic- DNA synthesis -rat; Microbiological Assoc.; #M0017; 5/7/82; Report # 25

Positive for induction of unscheduled DNA synthesis

Note: Contains up to 2% Free Formaldehyde

Chronic toxicology results, include size of data base; conclusions in each major toxicology area:

Metabolism - rat; EG & E Mason Research Institute; @206-GC-35; 9/9/83

97% recovered from treatment site
2% recovered in urine
1% recovered in tissue and feces (body burden less than 1%)

* Mutagenic Ames salmonella; Litton Bionetics; #20838; 3/78

Negative with and without activation

* 90 Day oral - rat; Food and Drug Research Lab.; #7059; 3/16/82

NOEL > 400 mg/kg (HDT). Levels tested by gavage in Sprague-Dawley strain - 0,100, 200 and 400 mg/kg/day for weeks 1-8, than 0, 100, 300 and 600 mg/kg/day for weeks 9-13.

* Mutagenic Ames salmonella mammalian microsome; Microbiological Assoc; #T1804-502; 9/3/81

Positive on tester strains TA100 and TA98 with or without metabolic activation

* Mutagenic Ames salmonella mammalian microsome; Microbiological Assoc; #003-520-723-1; 2/23/82

Negative with without metabolic activation

* Mutagenic cytogenicity - CHO cell in vitro; Microbiological Assoc; #T1721.337; 5/25/82

Positive with or without metabolic activation. Produced chromosomal aberrations

Key * Test conducted using 1,3-hydroxymethyl)-5,5-dimethyl hydrontoin 44% and hydroxymethyl-5,5 dimethyl hydantoin
** = Test conducted using the product as formulated glyco

Physiological and Biochemical Behavioral Characteristics:

   Foliar absorption (if applicable). N/A
   Translocation (if applicable) N/A
   Mechanism of pesticidal action. Hydrolysis in water to
      yield free formaldehyde (which is the disinfecting agent)
   Metabolism and persistence in plants and animals. N/A

Environmental Characteristics: (for indoor use only)

   Adsorption and leaching in basic soil types. N/A
   Microbial breakdown. N/A
   Loss from photodecomposition and/or volatization. N/A
   Bioaccumulation N/A
   Resultant average persistence. N/A

Ecological Characteristics:

   As proposed, the product provides for minimal hazard to
   non-target organisms and due to the low toxicity of the
   product no F & W precautionary toxicity statements are
   required.

Tolerance assessments: N/A. Registered use is not for use
   on food or feed stuffs.

Problems which are known to have occurred with the use of the
   chemical (e.g., PIMS data). None

4. <u>Summary of Regulatory Position and Rationale:</u>

   Use, formulation, manufacturing process or geographical
      restrictions (including classification for restricted
      use). No restrictions
   Unique warning statements required on labels (e.g.,
      protective clothing requirements, restrictions
      regarding use around water, reentry intervals). None
   Summary of risk/benefit review in cases where risk
      assessments were conducted. Benefits exceed risks
      for use of the product as a preservative in paints,
      room deodorizers and soft detergents.

5. <u>Summary of major data gaps</u>

   There are no data gaps for the approved uses.

6. Contact person at EPA:

>A. E. Castillo (PM 32)
>Disinfectants Branch
>Registration Division (TS-767C)
>Environmental Protection Agency
>401 M St., SW.
>Washington, DC 20460
>Phone: (703) 557-3964

DISCLAIMER: The information presented in this Chemical Information Fact Sheet is for informational purposes only and may not be used to fulfill the data requirements for pesticide registration and reregistration.

HELIOTHIS NPV

Date Issued: June 30, 1984
Fact Sheet Number: 27

1. Description of Chemical

    Generic Name: Heliothis zea (NPV)

    Common Name: Heliothis

    Trade Names: Elcar, Biotrol VHZ and Viron/H

    EPA Shaughnessy Code: 107301

    Chemical Abstract Service (CAS) Number: 456H

    Year of Initial Registration: 1975

    Pesticide Type: Viral Insecticide

2. Use Patterns and Formations

    Elcar® (Heliothis zea NPV) is registered for use on a variety of crops. There is also an exemption from the requirements of a tolerance established for the residues of this microbial insecticide nuclear polyhedrosis virus of Heliothis zea, in or on all raw agricultural commodities including:

    Application Sites and Rates:

Terrestrial Food Use	Rates (lbs. active ingredient)
Beans Corn Lettuce Peanuts Sorghum Soybeans Strawberry Tomato Cotton	Application rates may vary from 0.001 to 0.25 pounds per acre

    Terrestrial Non Food

    Tobacco

3. Science Findings

   Summary Science Statements:

   Heliothis zea NPV does not appear to be toxic to man or animal; it does not appear to have any deleterious effects on the environment. Heliothis zea NPV has not been shown to be neurotoxic, oncogenic, teratogenic or mutagenic in the studies that have been reviewed to date.

   Additional data are needed to determine the toxicity of Heliothis zea NPV to wildlife. From the available data, limited exposure to large animals, estuarine/marine organisms and freshwater organisms is expected as a result of the current label uses.

   Chemical/Microbial Characteristics:

   The nuclear polyhedrosis virus (NPV) of cotton bollworm Heliothis zea is a member of the occluded virus group of baculoviruses which have unusually complex structure. It contains double-stranded deoxyribonucleic acid genome enclosed within a nucleocapsid in the polyhedral inclusion body. The viral DNA is heterogenous in size with contour lengths ranging from 15 to 45 millimicrons, and has an estimated molecular weight of 30×10^6 daltons.

   They are rods or cylinders and ellipsoids or oval structures. The size of the enveloped virions ranges from 60 to 70 millimicrons in width by 260 to 300 millimicrons in length. The polyhedral inclusion bodies of Heliothis zea NPV which also vary in both shape and size, ranging from 0.65 to 1.93 microns in diameter, can be observed easily under the light microscope.

   The polyhedra which are made of large crystals of protein, are insoluble in water, withstand freezing and a wide range of environmental temperatures, and are resistant to bacterial putrefaction and to chemical treatments. However, the polyhedral inclusion bodies of Heliothis zea can be readily inactivated by short exposure to natural and artificial sunlight, and are susceptible to the treatment at high pH with solutions such as 0.1 M sodium carbonate. The treatment of polyhedra at high pH can be used to free the virions.

Toxicology Characteristics:

The means by which the Heliothis NPV infect the insect hosts are not fully understood at present. Virus may be aquired either by mouth, transovarially, or as a result of injury, for example, through parasitism; oral infection is the most common by far of these three modes of acquisition. It is reasonable to believe that the high pH (pH 9.5 to 10) of the anterior midgut with some enzymic virus degradation activity is adequate to initiate polynedral dissolution and virion release. Once released, virions are directly exposed to the chemical conditions within the midgut lumen and enter into midgut cells. Virion occlusion begins by the disposition of polyhedral material around single virion and later the polyhedron enlarges to incorporate additional virions. It is uncertain how the polyhedral enlargement occurs. However, when a large dose of Heliothis NPV was fed to animals, the infective Heliothis NPV was only detected in the alimentary tract, but greater than 99.9% of the total virus infectivity was either destroyed or excreted within 48 hours. All animals remained healthy and no adverse effects of the treatment were reported. Similarly, infective Heliothis NPV was completedly destroyed after two hours incubation in human gastric juice.

Ecological Characteristics:

There are no data on the acute oral toxicity of **Heliothis zea** to birds. This data requirement remains unfulfilled. No data were received relating to the acute pathogenicity of Heliothis NPV to birds. This guideline requirement remains unfilled. No data were received relating to toxicity/pathogencity of Heliothis to freshwate. Invertebrates, thus this guideline requirement remains unfulfilled.

Six studies on a formulated product of Heliothis were found to be acceptable for use in a hazard assessment. They are: LC_{50} values for rainbow trout at 10% virus were reported at approximately 96 hours, as being 2.06 ppm; white sucker at 10% test product, no effect at 100 ppm after 96 hours; black bullhead at 10% virus product showed no effect at 10 ppm; rainbow trout at 10% virus product resulted in 100% mortality in test solutions containing greater than 10 ppm concentrations after 96 hours; white sucker and black bullhead showed no effect at 70 ppm after 96 hours.

To address the ecological hazards, the Agency is requiring that the test results from all of the Tier I nontarget organism tests, as noted in the current guidelines, be submitted in support of registration or reregistration of all products containing the active ingredient **Heliothis zea**.

Tolerance Exemption Assessments as per 40 CFR §180.1027:

Exemptions from the requirement of a tolerance are established for the residues of the microbial insecticide nuclear polyhedrosis virus of **Heliothis zea**, in or on all raw agricultural commodities including: corn, cottonseed, beans, lettuce, okra, peppers, sorghum, soybeans, tobacco, and tomatoes.

4. **Summary of Regulatory Position and Rationale:**

   Because all Heliothis zea products were registered under the conventional pesticides data requirement guideline before the promulgation of the current biorational guidelines, the Agency is requiring that all product analysis data be submitted as in Tier I of the biorational data guidelines, as well as most of Tier I data for Ecological Effects.

5. **Summary of Major Data Gaps**

   The data requirements represent major data gaps for Heliothis NPV. These data are required to be submitted to the Agency within 2 years from the date of the issuance of the registration document.

   Tier I §158.165 Nontarget Organisms

   - Primary dermal study - Avian oral toxicity study
   - Wild mammal study
   - Freshwater fish study
   - Freshwater aquatic invertebrate study
   - Estuarine/marine animal testing study
   - Plant studies
   - Non-target insect study
   - Honeybee testing study
   - Product analysis data

6. **Contact Person at EPA**

   Timothy A. Gardner
   Product Manger (PM) 17,
   Registration Division (TS-767C),
   Office of Pesticide Programs,
   Environmental Protection Agency,
   401 M St., SW.,
   Washington, DC 20460

Office location and telephone number:
Rm. 202, CM #2,
1921 Jefferson Davis Highway,
Arlington, VA 22202,
(703)-557-2690.

DISCLAIMER: The information presented in this Chemical Information Fact Sheet is for informational purposes only and may not be used to fulfill data requirements for pesticide registration or reregistration.

HEPTACHLOR

Date Issued: December 1986
Fact Sheet Number: 107.1

1. DESCRIPTION OF CHEMICAL

    Generic Name: 1,4,5,6,7,8,8-heptachloro-3a,4,7,7a-tetra-
    (Chemical)   hydro-4,7-methano-1H indene
    Common Name: Heptachlor
    Trade and Other Names: 1,4,5,6,7,8,8-heptachloro-3a,4,7,7a-
      tetrahydro-4,7-methanoindene; E-3314; Velsicol 104; Hep-
      tagran; Heptalube; heptachlore; Drinox H-34; Gold Crest
      H-60; Heptamul; and Heptox
    EPA Shaughnessy Code: 044801
    Chemical Abstracts Service (CAS) Number: 76-44-8
    Year of Initial Registration: 1952
    Pesticide Type: Insecticide
    Chemical Family: Chlorinated cyclodiene
    U.S. and Foreign Producers: Velsicol Chemical Corporation

2. USE PATTERNS AND FORMULATIONS

    Application Sites: Soil surrounding wooden structures for
      termite control; control of fire ants in buried cable
      closures; above-ground structural application for control
      of termites and other wood-destroying insects.
    Types of Formulations: Emulsifiable concentrates; granular.
    Types and Methods of Application: Trenching, rodding, subslab
      injection, and low-pressure spray for subsurface termite
      control; caulking gun, trowel, or brush for applying to
      structural wood.
    Application Rates: 0.06 to 1.0% emulsion for termite control;
      0.2 oz/buried cable closure size 1 sq ft.

3. SCIENCE FINDINGS

    Summary Science Statement: Heptachlor is a chlorinated
    cyclodiene with moderate acute toxicity. The chemical has
    demonstrated adverse chronic effects in mice (causing liver
    tumors). Heptachlor may pose a significant health risk of

chronic liver effects to occupants of structures treated with heptachlor for termite control. This risk may be determined to be of regulatory concern, pending further evaluation. Heptachlor is extremely toxic to aquatic organisms and birds. Heptachlor is persistent and bioaccumulates. Heptachlor may have a potential for contaminating surface water; thus, a special study is required to delineate this potential. Applicator exposure studies are required to determine whether exposure to applicators may be posing health risks. Special product-specific subacute inhalation testing is required to evaluate the short-term respiratory hazards to humans in structures treated with heptachlor. An inhalation study of one (1) year duration using rats is required to assess potential hazards to humans in treated residences from this route of exposure. The Agency has been apprised of reported cases of optic neuritis associated with termiticide treatment of homes with a related cyclodiene, chlordane. To determine whether this is a significant health effect, and whether heptachlor plays a role, the registrant must have eye tissue from the required 2-year rat oncogenicity study analyzed by neuropathologists specializing in optic tissue pathology. Data available to the Agency show an occurrence of misuse and misapplication of heptachlor. The Agency is requiring restricted use classification of all end-use products (EPs) containing heptachlor. Application must be made either in the actual physical presence of a Certified Applicator or if the Certified Applicator is not physically present at the site, each Uncertified Applicator must have completed a State-approved training course for termiticide application meeting minimal EPA training requirements and be registered in the State in which the Uncertified Applicator is working.

Chemical/Physical Characteristics of the Technical Material

Physical State: Crystalline solid
Color: White
Odor: Mild camphor-like odor
Molecular Weight and Formula: 373.3 - $C_{10}H_5Cl_7$
Melting Point: 95 to 96 °C
Boiling Point: 135-145 °C at 1-1.5 mmHg
Density: 1.65-1.67 g/mL at 65° C
Vapor Pressure: 0.0003 mmHg at 25 °C
Solubility in Various Solvents: Soluble in ethanol, xylene, carbon tetrachloride, acetone, and benzene; practically insoluble in water.
Stability: Stable in daylight, air, moisture, and moderate heat.

Toxicology Characteristics

Acute Oral: Data gap (except for a 74% technical formulation which showed the oral LD_{50} value for male and female rats to be 208 mg/kg and 158 mg/kg, respectively. This places the 74% technical into Toxicity Category II).

Acute Dermal: Data gap

Primary Dermal Irritation: Data gap

Primary Eye Irritation: Data gap

Skin Sensitization: Not a sensitizer

Acute Inhalation: Data gap

Subacute Inhalation (2-week product-specific test using rats or guinea pigs): Data gap

Chronic Inhalation (1 year using rats): Data gap

Major Routes of Exposure: Inhalation exposure to occupants of treated structures; dermal and respiratory exposure to termiticide applicators.

Delayed Neurotoxicity: Does not cause delayed neurotoxic effects.

Oncogenicity: This chemical is classified as a Group B_2 oncogen (probable human oncogen).

There are three long-term carcinogenesis bioassays of heptachlor in mice, which were independently conducted by investigators affiliated with the National Cancer Institute, the International Research and Development Corporation, and the Food and Drug Administration. Reported in these studies were significant tumor responses in three different strains of mice (C_3H, CF_1, and $B6C3F_1$) in males and females with a dose-related increase in the proportion of tumors that were malignant. Available data from five existing carcinogenicity bioassays in rats are inadequate and inconclusive and a well-designed study in rats for heptachlor epoxide is needed to determine the carcinogenic potential of heptachlor in this species.

Chronic Feeding: Based on a dog chronic feeding study with heptachlor epoxide, a lowest effect level (LEL) of 0.0125 mg/kg/day for liver effects has been calculated.

Data gaps exist for rodents and nonrodents for heptachlor epoxide and for heptachlor in nonrodents.

Metabolism: In biological systems, heptachlor is readily epoxidized to heptachlor epoxide.

Teratogenicity: Data gap

Reproduction: A NOEL of 1.0 ppm has been set for reproductive effects to the young; the liver is the target organ of effect.

Mutagenicity: Sufficient evidence exists to conclude that neither heptachlor nor heptachlor epoxide possesses mutagenic activity in bacteria. Further testing is required to fulfill mutagenicity testing requirements in all three categories (gene mutation, structural chromosome aberrations, and other genotoxic effects).

Physiological and Biochemical Characteristics

The precise mode of action in biological systems is not known. In humans, signs of acute intoxication are primarily related to the central nervous system, including hyperexcitabilty, convulsions, depression, and death.

Environmental Characteristics

Data gaps exist for all applicable studies. However, available supplementary data indicate general trends of heptachlor behavior in the environment. Heptachlor is persistent and bioaccumulates. Heptachlor is not expected to leach, since it is insoluble in water and should adsorb to the soil surface. Thus, it should not reach underground aquifers. However, additional data are necessary to fully assess the potential for ground water contamination as a result of heptachlor's termiticide use.

Ecological Characteristics (technical grade)

Avian Oral Toxicity: Data gap

Avian Dietary Toxicity: 92, 224, and 480 ppm in bobwhite
(8 days)   quail, pheasant, and mallard duck, respectively.

Freshwater Fish Acute Toxicity: 13 ug/L for bluegill;
(96-hr LC_{50}) 7.4 ug/L for rainbow trout.

Freshwater Invertebrate Toxicity: 42 ug/L for <u>Daphnia</u> <u>pulex</u>;
(48-hr or 96-hr EC_{50}) 1.1 ug/L for <u>Pteronarcys</u>.

4. <u>Required Unique Labeling and Regulatory Position Summary</u>

   o EPA is currently evaluating the potential human health risks of 1) nononcogenic chronic liver effects, and 2) oncogenic effects to determine whether additional action on heptachlor may be warranted.

   o In order to meet the statutory standard for continued registration, retail sale and use of all EPs containing heptachlor must be restricted to Certified Applicators or persons under their direct supervision. For purposes of heptachlor use, direct supervision by a Certified Applicator means 1) the actual physical presence of a Certified Applicator at the application site during application, or 2) if the Certified Applicator is not physically present at the site, each Uncertified Applicator must have completed a State-approved training course in termiticide application meeting minimal EPA training requirements and be registered in the State in which the Uncertified Applicator is working. The Certified Applicator must be available if and when needed.

   o In order to meet the statutory standard for continued registration, heptachlor product labels must be revised to provide specific disposal procedures and provide fish and wildlife toxicity warnings.

   o The Agency is requiring a special monitoring study to evaluate whether and to what extent surface water contamination may be resulting from the use of heptachlor as a termiticide.

   o A new 2-year rat oncogenicity study is needed to determine the carcinogenic potential of heptachlor epoxide.

   o Special product-specific subacute inhalation testing is required to evaluate the respiratory hazards to humans in structures treated with termiticide products containing heptachlor.

   o Evaluation of eye tissue from the required 2-year rat oncogenicity study is required to determine whether heptachlor's termiticide use may be causing optic neuritis in humans.

- o The Agency is requiring the submission of applicator exposure data from dermal and respiratory routes of exposure.

- o While data gaps are being filled, currently registered manufacturing-use products and EPs containing heptachlor may be sold, distributed, formulated, and used, subject to the terms and conditions specified in the Registration Standard for heptachlor, and any additional regulatory action taken by the Agency. Registrants must provide or agree to develop additional data in order to maintain existing registrations.

5. TOLERANCE REASSESSMENT

No tolerance reassessment for heptachlor is necessary, since there are no longer any food or feed uses. EPA is proceeding to revoke existing heptachlor tolerances and replace them with action levels. A Final Rule is scheduled for publication in the FEDERAL REGISTER in early 1987.

6. SUMMARY OF MAJOR DATA GAPS

- o Hydrolysis
- o Photodegradation in water
- o Aerobic soil metabolism
- o Anaerobic soil metabolism
- o Leaching and adsorption/desorption
- o Aerobic aquatic metabolism
- o Soil dissipation
- o Chronic feeding - nonrodents and rats (heptachlor epoxide) non-rodents (heptachlor)
- o Oncogenicity - rats (heptachlor epoxide)
- o Teratogenicity
- o Rat oncogenicity study
- o Mutagenicity studies
- o Acute toxicity studies
- o Optic tissue pathology

- o Special surface water monitoring studies
- o Applicator exposure studies
- o Indoor air exposure studies
- o Special product-specific subchronic inhalation study (2-week duration using rats or guinea pigs)
- o Subchronic inhalation study (1-year duration using rats)
- o Avian acute oral toxicity
- o All product chemistry studies

7. <u>CONTACT PERSON AT EPA</u>

   George LaRocca
   Product Manager (15)
   Insecticide-Rodenticide Branch
   Registration Division (TS-767C)
   Office of Pesticide Programs
   Environmental Protection Agency
   401 M Street SW.
   Washington, DC 20460

   Office location and telephone number:
   Room 204, Crystal Mall #2
   1921 Jefferson Davis Highway
   Arlington, VA 22202
   (703) 557-2386

DISCLAIMER: The information presented in this Pesticide Fact Sheet is for informational purposes only and may not be used to fulfill data requirements for pesticide registration and reregistration.

HYBREX

Date Issued: April 1986
Fact Sheet Number: 85

1. Description of chemical

    Generic name: Potassium 1-(4-chlorophenyl)-1,4-dihydro-6-methyl-4-oxopyridazine-3-carboxylate

    Common name: (ISO) Fenridazone-potassium

    Trade name: Hybrex

    EPA Shaughnessy code: 119001

    Year of initial registration: 1986

    Pesticide type: Hybridizing agent

    Chemical family: dihydro-oxopridizine

    U.S. and foreign producers: Rohm and Haas

2. Use patterns and formulations

    Application sites: wheat hybridizing agent (a chemical which allows production of hybrid plants; in this particular case, through allowing easier cross-pollination of normally self-pollinating species of plant such as wheat).

    Hybrex® detrimentally affects the pollen produced by treated wheat plants so that they cannot be self-pollinated, but must rely instead on a different untreated variety of wheat to supply the pollen. The offspring is a hybrid wheat variety.

Method of Application: Apply between the initiation of jointing and flag leaf tip emergence stages of growth. Apply no later than 50 days before harvest with a ground rig, as a foliar spray in enough water to cover foliage.

Application Rates: Use 2 to 10 pints (0.5 to 2.5 lb active ingredient) per acre with up to 4 quarts of surfactant per 100 gallons of spray mixture.

Types of Formulations: Liquid concentrate end use product 21.97% active ingredient.

Usual Carrier: Water, with surfactant.

3. SCIENCE FINDINGS

Summary Science statement: Hybrex has been found to be acceptable for the proposed use and all data requirements have been met. It is relatively non-toxic by the oral, dermal an inhalation routes, mildly irritating to the skin and causes corneal opacity at 24 hours when exposed to the eyes. Hazard to nontarget organisms is considered to be minimal.

Chemical Characteristics:

Physical State:	Aqueous solution
Color:	Deep red to red/brown
Odor:	Slight xylene odor
Boiling Point:	100.2 °C
Molecular Weight:	302.76

Solubility:

Solvent	Solubility [a]
Distilled water, pH 5.0	1.97 ± 0.10 g/100 ml
Distilled water, pH 6.9[b]	1.444 ± 0.4 g/100 ml
Methanol	7.02 ± 0.05 g/100 ml
Absolute ethanol	0.247 ± 0.010 g/100 ml
Ethyl cellosolve	1.892 ± .002 g/100 ml
Dimethyl sulfoxide	0.670 ± .039 g/100 ml
Acetonitrile	200 ± 90 mg/kg
Mixed xylenes	< 60 mg/kg

[a] Mean value ± standard deviation of duplicate determinations

[b] This mixture was quite viscous; the solubility would be expected to be still higher at pH 9.

Unusual Handling Characteristics:

No special handling needed.

Toxicology Characteristics:

Acute effects: Hybrex is relatively non-toxic by oral, dermal and inhalation routes, mildly irritating to the skin and causes corneal opacity at 24 hours when exposed to the eyes. Acute test results indicate Toxicity Categories II, III, and IV[1]/ as follows:

Acute oral toxicity (mouse)
    Greater than 500 ppm (75 mg/kg/day)
    Toxicity Category IV

Acute oral toxicity (rat)
    Greater than 500 ppm (25 mg/kg/day)
    Toxicity Category IV

Acute oral toxicity (mouse)
    Greater than 2000 ppm (300 mg/kg/day)
    Toxicity Category IV

Acute dermal toxicity (rabbit)
    Greater than 5 g/kg
    Toxicity Category III

Acute inhalation toxicity (rat)
    Greater than 20.2 mg/l
    Toxicity Category IV

Primary eye irritation (rabbit)
    Corneal opacity at 24 hours
    Toxicity Category II

Primary dermal irritation (rabbit)
    Mildly irritating
    Toxicity Category IV

---

[1]/ Precautionary labeling for: 1/ Toxicity Category II- Causes eye irritation. Do not get in eyes, on skin, or clothing. Harmful if swallowed. (appropriate first aid statement required); Toxicity Category III- Harmful if swallowed. Avoid contact with skin eyes or clothing. In case of contact immediately flush eyes or skin with plenty of water. Get medical attention if irritation persists; and Toxicity Category IV- No precautions are required.

Subchronic effects: Tests indicate systemic effects at the lowest dose tested (LDT).

14-Day dose finding study: (rat)	No-observed-effect level (NOEL) 6250 ppm (312.5 mg/kg/day)
90-Day feeding study (rat)	NOEL less than 500 ppm (LDT) (25 mg/kg/day)

Chronic effects: Tests indicate no oncogenic or teratogenic potential and no reproductive toxicity, and only weak mutagenicity in mouse lymphoma cells with three negative mutagenicity studies reported.

1-year dietary toxicity study (beagle dogs)	NOEL (62.5 mg/kg/day)
2-year rat feeding study	NOEL 300 ppm (15 mg/kg/day) No oncogenic effects observed.
18-month rat feeding study:	NOEL 2,000 ppm (300 mg/kg/day) No oncogenic effects observed.
3-generation reproduction study (rat)	NOEL 100 ppm, (5 mg/kg/day) for systemic effects, (45 mg/kg/day) for reproductive effects.
Teratology (rabbit)	NOEL (1,000 mg/kg/day), fetal toxicity NOEL (100 mg/kg day), maternal toxicity NOEL (100 mg/kg/day).
Teratology (rat)	NOEL 10,000 ppm (500 mg/kg/day), fetal toxicity NOEL 30,000 ppm (1,500 mg/kg/day), teratogenic NOEL 30,000 ppm (1,500 mg/kg/day).

Mutagenicity	Ames test	: Negative
Mutagenicity	Cytogenic	: Negative
Mutagenicity	Mouse Lymphoma cells:	Weakly mutagenic at 1250 g/ml under both activated and non-activated test conditions
Mutagenicity	in vitro transformation:	Negative

The Acceptable Daily Intake (ADI)
NOEL of (15.0 mg/kg/day)

Major Routes of Exposure: Dermal, inhalation

Physiological and Biochemical Behavoiral Characteristics:

    Foliar Absorption: Absorption occurs through the foliage.

    Translocation: The material is translocated to the fruiting parts from the foliage.

    Mechanism of Action: Hybrex® detrimentally affects the pollen produced by treated wheat plants so that they cannot be self-pollinated but must rely instead on a different untreated variety of wheat to supply the pollen. The offspring is a hybrid wheat variety.

    Persistence: Hybrex is not persistent in plants or animals.

    Metabolism in plants and animals: The nature of the residues in plants and animals is adequately understood for the use of hybrex on wheat. The parent compound and its decarboxylated metabolite (2% plants, 1% or less in animals) are the residues of concern. The residue analytical method is adequate for the determination of residues in wheat and for enforcement purposes.

Environmental Characteristics: Hybrex is stable to hydrolysis at pH 5.0 and 6.9. It is soluble in distilled water at 1.44 to 1.97 g/100 ml.

    Absorption and leaching in basic soil types: Hybrex is strongly adsorbed, and not very mobile in soil.

    Microbial breakdown: Hybrex is decarboxilated slightly, however the parent compound stays relatively intact and is strongly adsorbed to soil particles.

    Loss from photodecomposition and volatilization: Photodecomposition is very slight and volatilization does not occur.

    Resultant average persistence: Hybrex should not persist beyond 52 days.

Ecological Characteristics: The following test results indicate that Hybrex is practically nontoxic to avian species, finfish, aquatic invertebrates and honeybees.

Hazards to Birds:

Avian acute oral toxicity: Greater than 5,000 ppm. (mallard duck and bobwhite quail)

Avian dietary toxicity: Greater than 5,390 ppm. (mallard duck and bobwhite quail)

Hazards to Fish:

Fish acute toxicity:

Bluegill sunfish	141.3 ppm
Rainbow trout	266.5 ppm
Fathead minnow	43.7 ppm

(fathead minnows were very young)

Aquatic invertebrate toxicity:

Daphnia magna	188.0 ppm

Honeybees:

Nontoxic to honeybees.

Tolerance Assessment:

List of Crops and Tolerances

Commodity	Parts per million
Wheat grain	40.0
Wheat straw	25.0
Meat and meat byproducts (except kidney and liver) of cattle, goats, hogs, horses, and sheep.	0.05
Fat of cattle, goats, hogs, horses, and sheep	0.05
Kidney and liver of cattle, goats, hogs, horses, and sheep.	1.00

Fat, meat, and meat byproducts of poultry.	0.30
Eggs	0.05
Milk	0.05

4. Summary of Regulatory Position and Rationale:

   Position: The agency has accepted the use of this pesticide for the production of hybrid wheat seed.

   Rationale: The agency has reviewed the data submitted and found these data to be adequate for the proposed use of this pesticide. The proposed use of this pesticide is not expected to result in any adverse effects to humans or the environment and will fill a need for producing large quantities of hybrid wheat seed.

   Use Restrictions: For use by Rohm and Haas company or or under the direct supervision of their representative only.

   Unique label statements: See use restrictions above.

5. Summary of major data gaps:

   None.

6. Contact Person at EPA

   Robert J. Taylor (PM-25)
   U.S. Environmental Protection Agency (TS-767C)
   401 M Street, SW.
   Washington, DC 20460
   (703) 557-1800

DISCLAIMER:

The information presented in this Chemical Information Fact Sheet is for informational purposes only and may not be used to fulfill data requirements for pesticide registration and reregistration.

IMAZAQUIN

Date Issued: March 20, 1986
Fact Sheet Number: 83

1. DESCRIPTION OF CHEMICAL

    Generic Name: 2-[4,5-dihydro-4-methyl-4-(1-methylethyl)-5-oxo-1H-imidazol-2-yl]-3-quinoline carboxylic acid

    Common Name: Imazaquin

    Trade Name: Scepter

    EPA Shaughnessy Code: 128821

    Chemical Abstracts Service (CAS) Number: None

    Year of Initial Registration: 1986

    Pesticide Type: Herbicide

    U.S. and Foreign Producers: American Cyanamid

2. USE PATTERNS AND FORMULATIONS

    Application site: Soybeans

    Method of application: Scepter may be applied preplant incorporated, preemergence, or postemergence. In southern locations only, two applications are possible: an initial preplant incorporated or preemergence treatment followed by a postemergence treatment.

    Application rates: Ground or aerial application rates are 2 oz active ingredient per acre (a.i./A). Ground applications are in 10 or more gallons of water per acre and aerial applications are in 5 or more gallons of water per acre. Where two applications are permitted, no more than 0.25 lb/a.i./A per season may be applied.

    Type of formulation: 95% technical grade and 17.3% active ingredient liquid concentrate end-use product.

    Usual carrier: Water

3. SCIENCE FINDINGS

Summary science statement: Scepter has been found to be acceptable for the proposed use. It is relatively non-toxic by the oral, dermal and inhalation routes, non-irritating to slightly irritating to the eye and skin; it is not a dermal sensitizer. Hazard to nontarget organisms is considered to be minimal.

Chemical characteristics:

Physical state	: Solid
Color	: Light tan
Odor	: None
Melting Point	: 219-224°C
Solubility	: 60 ppm at 25°C
Octanol/water partition coefficient	: 2.2
pH	: 3.8[1]

Toxicology characteristics:

Acute effects. Imazaquin is relatively nontoxic by oral, dermal and inhalation routes, nonirritating to slightly irritating to the eye and skin and not a dermal sensitizer. Acute test results indicate Toxicity Categories III and IV[2] as follow:

Acute oral toxicity (rat):	Greater than 5,000 mg/kg bwt Toxicity Category IV
Acute inhalation (rat)	: Greater than 5.7 mg/L Toxicity Category III
Acute dermal (rabbit)	: Greater than 2,000 mg/kg bwt Toxicity Category III
Acute dermal sensitization (guinea pigs)	: Not a sensitizer

[1] Slurried into 100 ml of de-ionized water at 23°C; pH is for this slurry at 23°C.
[2] Toxicity Category III = Harmful if swallowed, inhaled or absorbed through the skin. Contact with skin, eyes or clothing requires immediate first aid and may require medical attention. Toxicity Category IV = No precautions are required.

Primary dermal irritation: Mildly irritating
(rabbit) Toxicity Category IV

Primary eye irritation : Nonirritating
(rabbit) Toxicity Category IV

Subchronic effects. Tests indicate no systemic toxicity at the highest dose tested (HDT):

21-day dermal (rabbit) : No-observed-effect level (NOEL) 1,000 mg/kg bwt/day (HDT)

90-day feeding study : NOEL greater than 10,000
(rats) ppm or 800 mg/kg bwt/day

Chronic effects. Tests indicate no oncogenic or teratogenic potential and no reproductive toxicity at HDT, and negative activity in five mutagenicity studies:

1-year dietary toxicity : NOEL = 1,000 ppm; Lowest-
study (beagle dogs) observed-effect level (LOEL) = 5,000 PPM

2-year oral dietary (rat): NOEL = 10,000 ppm (HDT) or 500 mg/kg bwt/day

18-month oncogenicity : NOEL = 1,000 ppm (150
(mouse) mg/kg bwt); Lowest effect level (LEL) = 4,000 ppm

Three-generation repro- : NOEL = 10,000 ppm (1,000
duction (rat) mg/kg bwt) (HDT)

Teratology (rabbits) : Teratogenic NOEL = 500 mg/kg/day; embryotoxic NOEL = 500 mg/kg/day; maternal NOEL = 250 mg/kg/day; maternal LEL = 500 mg/kg/day

Teratology (rats) : Teratogenic: NOEL greater than 2,000 mg/kg bwt/day; fetotoxic: NOEL = 500 mg/kg bwt/day, and LOEL = 2,000 mg/kg/day maternal toxicity: NOEL = 500 mg/kg bwt/day, and LOEL = 2,000 mg/kg bwt/day

Single low-dose metab- : Almost entirely excreted
olism study (rats) in 48 hours; urine - 94%; feces - 4%

Mutagenicity - Ames test : Negative

Dominant lethal test (rats): Negative

In vitro cytogenetics (CHO): Negative

Unscheduled DNA synthesis : Negative
(rat hepatocytes)

CHO/HGPRT point mutation : Negative

Major route of exposure: Dermal, inhalation

Physiological and biochemical behavorial characteristics:

Foliar absorption: Absorption occurs through both the foliage and roots.

Mechanism of pesticidal action: When applied to soil, susceptible weeds emerge, growth stops and the weeds either die or are not competitive with the crop. When applied postemergence, adsorption occurs through both the foliage and roots, growth stops and the weeds either die or are not competitive with the crop. When applied preemergence, rainfall or irrigation is necessary to activate imazaquin.

Metabolism in plants and animals: The nature of the residue in plants and animals is adequately understood for the use of imazaquin on soybeans. The parent compound is the residue compound of concern. The residue analytical method is adequate for the determination of residues in soybeans and for enforcement purposes.

Environmental characteristics: Imazaquin is stable to hydrolysis at pH 3 and 5 and has an aqueous hydrolytic half-life of 5.5 months at pH 9. It is slightly soluble in distilled water, 60 ppm @ 25° C.

Absorption and leaching: Additional field dissipation studies must be submitted to further define the leaching potential of imazaquin.

Microbial breakdown: Imazaquin is decarboxylated slowly to CO_2, as well as degraded to the major metabolite CL 266,066 and at least 6 minor metabolites.

Loss from photodecomposition and/or volatilization: Volatilization does not occur.

Resultant average persistence: Imazaquin should not persist beyond 4 to 6 months.

Ecological characteristics: The following test results indicate that imazaquin is practically nontoxic to avian species, finfish, aquatic invertebrates, and honeybees:

Avian acute oral toxicity
(mallard duck and bobwhite quail) : Greater than 2,150 ppm

Avian dietary toxicity
(mallard duck and bobwhite quail) : Greater than 5,000 ppm

Fish acute toxicity--rainbow trout : Greater than 280 ppm
bluegill sunfish: Greater than 420 ppm
channel catfish : Greater than 320 ppm

Aquatic invertebrate toxicity
(Daphnia magna) : Greater than 280 ppm

Honeybee : Non-toxic at 100 ug/bee

Tolerance Assessment:

List of crops and tolerances (40 CFR 180.426):

Commodity	Part per million
Soybeans	0.05

Results of tolerance assessment: The accepted daily intake (ADI), based on the 1-year dog feeding study (NOEL of 1,000 ppm or 25 mg/kg bwt/day) and using a 100-fold safety factor, is calculated to be 0.25 mg/kg bwt/day. The maximum permissible intake (MPI) for a 60-kg person is calculated to be 15 mg/day. The theoretical maximum residue contribution (TMRC) for use on soybeans is calculated to be 0.0007 mg/day, which accounts for 0.00 percent of the ADI (0.25 mg/kg bwt/day).

4. SUMMARY OF REGULATORY POSITION AND RATIONALE

Position: The Agency has conditionally accepted the use of this chemical on soybeans.

Rationale: The Agency has reviewed the data submitted and, with the exception of field dissipation studies, found these data to be adequate for the proposed use of this chemical. The proposed use of this chemical, prior to completion of repeated field studies, is not expected to result in any adverse effects to humans or the environment and will fill a need for a herbicide to control weeds in soybeans.

Use restrictions: None

Unique label statements: None

5. SUMMARY OF MAJOR DATA GAPS

    164-1 Field Dissipation Studies (Soil) - due March 1988

6. CONTACT PERSON AT EPA

    Robert J. Taylor (PM-25)
    U.S. Environmental Protection Agency (TS-767C)
    401 M Street SW.
    Washington, DC 20460
    (703) 557-1800

DISCLAIMER: The information presented in this Pesticide Fact Sheet is for informational purposes only and may not be used to fulfill data requirements for pesticide registration and reregistration.

ISAZOPHOS

Date Issued: June 25, 1987
Fact Sheet Number: 138

1. Description of Chemical

    Generic name: 0-(5-chloro-1-[methylethyl]-1H-1,2,4-triazol-3-yl) 0,0-diethyl phosphorothioate

    Common name: Isazophos (BSI)

    Trade name: CGA-12223, Triumph

    EPA Shaughnessy code: 124101

    Chemical abstracts service (CAS) number: 42509-80-8

    Year of initial registration: 1987

    Pesticide type: Insecticide

    Chemical family: Organophosphate

    U.S. Producer: None

    Foreign Producer: Ciba-geigy, Switzerland

2. Use Patterns and Formulations

    Application sites: lawns

    Types of formulations: liquid

    Types/methods of application: ground spray

    Application rates: 0.75 to 1.5 fl. oz. of product in a minimum of 3 gals. of water per 1,000 sq. ft. of lawn. A maximum of 2.0 lbs a.i./A per year may be applied.

    Usual Carriers: xylene

3. Science Findings

Adequate studies are available to assess the toxicological and environmental hazards of both the technical product and the end-use emulsifiable concentrate product for use on lawns. No toxicological or environmental hazards of concern were identified from use of these products when used according to prescribed label directions.

Chemical characteristics

Physical state: slightly viscous liquid

Odor: characteristic organophosphorus odor

Color: pure material: colorless; technical material: yellow

Boiling point: 100°C at 0.001 mbar

Specific gravity: 1.22 g/cm^3 at 20°C

Solubility: water: 150ppm at 20°C

Vapor pressure: 4.3×10^{-5} mbar at 20°C

Miscibility: miscible with methanol, chloroform, dichloromethane, hexane, toluene, xylene.

Empirical formula: $C_9H_{17}ClN_3O_3PS$

Toxicology

Acute oral toxicity: from 33 mg/kg to 84 mg/kg in both male and female rats (Tox Cat. I)

Acute inhalation: from 0.245^C to 1.89^C for both male and female rats (Tox Cat. I)

Acute dermal (rabbit): intact: 870 mg/kg
abraded: 571 mg/kg (Tox Cat. II)

Primary eye irritation: caused slight eye irritation in rabbits which was found to be reversible within 7 days (Tox Cat. III)

Primary dermal irritation: caused slight dermal irritation in rabbits (Tox Cat. IV)

The no-observed-effect levels (NOEL) for the chemical with respect to cholinesterase inhibition are as follows:

Study	ppm	mg/kg
90-day rat feeding	20	0.1
6-month dog feeding	0.05	0.0125
21-day dermal in rabbits	----	0.1

Acute delayed neurotoxicity: The submitted study shows that the chemical does not induce delayed neurotoxicity in female chickens at a dose of approximately 25.4 mg/kg.

Teratogenicity: The NOEL for maternal toxicity in the rat and rabbit were 6 mg/kg/day and 5 mg/kg/day, respectively. The effects were clinical signs of cholinesterase inhibition. Fetotoxicity (runts) were observed in pregnant rats given the highest dose tested (9 mg/kg). No teratogenic effects were observed in the submitted studies.

Physiological and Biochemical Behavioral Characteristics:

Mechanism of pesticidal action: cholinesterase inhibitor

Metabolism and Persistence in Plants and Animals

The metabolism studies which were submitted suggest that CGA 12223 is almost completely excreted within 24 hours after dosing of rats (93 to 97%). The primary route of excretion is the urine (approximately 99% of the total excreted). The phosphorus-triazolyl ester bond is cleaved, and the major portion of excreted residues in the urine consists of glucuronic and sulfuric acid conjugates of the triazole moiety or free triazole.

Environmental Characteristics

Studies show CGA-12223 to be subject to relatively rapid degradation by both hydrolysis and photolysis in aqueous solution. The hydrolysis halflife at neutral pH is 48 days, while the half-life for photolysis is approximately six days. The major degradation product (CGA-17193) of both reactions is the same and results from cleavage of the phosphorus ester bond to produce the resulting hydroxy triazole. Metabolism studies indicate significant degradation of CGA-12223 in soil, with a half-life of approximately 14 days under field conditions. As in the hydrolysis and photolysis studies, the major degradation product in the soil metabolism studies was CGA-17193. CGA-12223 did not significantly affect the growth of soil microorganisms or, specifically, cellulose-decomposing microorganisms and did not alter soil nitrification. Soil column leaching studies indicate that the chemical has a high potential to leach in representative bare soils. Dissipation from turf was found to be relatively rapid, with a half-life of approximately 13 days. An in-house Pesticide Root Zone Model (PRZM) simulation of Triumph used on turf shows that the chemical has little, if any potential to leach, when used on turf.

Bioaccumulation ratios in catfish exposed to a soil/water system were very low, with maximum values of 11X in the edible portion after one day of exposure and 44X in the whole fish after 14 days.

Ecological characteristics:

Avian acute oral toxicity (mallard duck) = 61 mg/kg

Avian dietary toxicity (mallard duck) = 244 ppm
(bobwhite quail) = 81 ppm

Fish acute toxicity (rainbow trout) = .00636 ppm
(bluegill sunfish) = .00383 ppm
(flathead minnow) = 0.138 ppm

Aquatic invertebrate toxicity (48 hr. - Daphnia magna) = 1.40 ppb

Avian reproduction: No reproductive effects at highest level tested (30 ppm) in both mallard duck and bobwhite quail.

Toxicity to estuarine and marine animals:
96 hr.; sheepshead minnow = 6.02 ppb

Embryo-Larvae and life-cycle:
21-day life cycle (Daphnia magna) = MATC
> 0.198 and > 0.495 ug/L
early-life stage (fathead minnow) = MATC
> 2.5 and > 6.3 ppb

Tolerance assessments:

There are no tolerances for the chemical. The use on lawns is a non-food use.

4. Summary of Regulatory Position and Required Unique Labeling

Use classification - use on lawns restricted to use only by certified applications or persons under their direct supervision. Restriction is due to avian fish and aquatic organism toxicity.

Formulation: 46.8% emulsifiable concentrate.

An assessment of potential exposure to humans re-entering lawns after application of the pesticide was conducted. The direct comparison of dermal exposure estimates to a subchronic dermal toxicity NOEL provides margin of safety (MOS) ratios of 51.3, 25.6, and 17.3 for 1-, 2-, or 3-hour exposures.

Based on data submitted in support of a re-entry interval, the following restriction is required on the labeling for use of the chemical on lawns:

> Do not allow children or pets on the grass on the day of treatment until 1/2 inch of water has been applied and the grass is dry.

Based on results of the PRZM modeling for the use of the chemical on turf, the following label restriction is required for use of the product on lawns:

> Do not use on sandy soil.

5. Summary of major data gaps

There are no data gaps for the registration of the chemical on lawns.

6. Contact person at EPA

William H. Miller
Product Manager (16)
Insecticide-Rodenticide Branch
Registration Division (TS-767)
Environmental Protection Agency
Washington, DC 20460

Telephone Number (703) 557-2600

DISCLAIMER: The information presented in this Chemical Information Fact Sheet is for informational purposes only and may not be used to fulfill data requirements for pesticide registration and reregistration.

ISOMATE-M

Date Issued: April 30, 1987
Fact Sheet Number: 126

1. Description of Biochemical Pesticide (Pheromone)

    Chemical Names: 2-8-dodecen-1-yl-acetate
    E-8-dodecen-1-yl-acetate
    2-8-dodecen-1-ol

    Common Name: Oriental Fruit Moth Pheromone

    Trade Name: Isomate-M

    EPA/OPP Chemical Code (Shaughnessy) Number(s):

    2-8-Dodecen-1-yl actetate - 128906

    E-8-Dodecen-1-yl acetate - 128907

    2-8-Dodecen-1-ol - 128908

    Chemical Abstracts Service (CAS) Number: 410C

    Year of Initial Registration: 1986

    Pesticide Type: Biochemical Pesticide

    U.S. and Foreign Product(s): Biocontrol, Ltd. of Australia

2. Use Pattern and Formulation(s):

    Application Site: Six-inch polyethylene tie-ons for fruit trees in orchards (specifically peaches and nectarines).

    Pests Controlled: Oriental Fruit Moth (Grapholitha molesta).

3. Science Findings

    Summary Science Statement:

    The available information that has been reviewed suggests that Isomate-M is potentially safer than many conventional pesticides currently used for control of the same pest. Due to the fact that this product is a pheromone which will be used only in plastic dispensers, environmental exposure is expected to be minimal.

Biochemical Pesticide Characteristics:

Isomate-M is a pheromone of the oriental fruit moth. Control of the pest is obtained through mating disruption.

This pheromone is applied only in plastic dispensers which are twisted around the branches of peach/nectarine trees in the orchard. Application rate is four dispensers per tree in standard orchard spacing, or 400 dispensers per acre. This is equivalent to approximately 30 mg of active ingredient per hour, or 12.145 mg per acre per hour. Isomate-M is species-specific; therefore, it does not affect other nontargeted or beneficial species.

Data Requirements

1. Product Analysis:

    Product identity and disclosure requirements have been satisfied.

2. Residue Chemistry Data:

    Residue chemistry data are applicable only if Tier II or Tier III toxicology data are required as specified by 40 CFR 158.165(e). This was not the case with this product. Therefore, no residue data are required in order to register Isomate-M.

3. Ecological Effects Data:

    The applicable data requirements consist of the following:

    a. Avian single-dose oral toxicity test for one species, preferably bobwhite quail (Guideline No. 154-6);

    b. Fish acute bioassay for one species, preferably rainbow trout (Guideline No. 154-8); and

    c. Aquatic invertebrate acute bioassay, preferably Daphnia (Guideline No. 154-9).

These tests must be conducted with technical material, and the protocols, including formulas for calculating maximum dose levels, for conducting the studies can be found in Pesticide Assessment Guidelines, Subdivision M, Biorational Pesticides, EPA-540/9-82-028, October 1982.

The registrant has agreed to submit these data. The Agency has decided to issue a conditional registration contingent on the submission of these data.

Toxicology Data

The company submitted a battery of studies as required by 40 CFR 158.165(c).

These studies included the following data which were reviewed and classified as follows:

1. Acute oral LD_{50}, rat

    LD_{50} = > 20 mL/kg (17.12 g/kg)
    Toxicity Category: IV
    Classification: Core Minimum Data

2. Acute dermal LD_{50}, rat

    LD_{50} = > 2000 mg/kg
    Toxicity Category: III
    Classification: Core Minimum Data

3. Primary dermal irritation, rabbit

    P.I. score = 0.96, a slightly
      irritating agent
    Classification: Core Minimum Data

4. Primary eye irritation, rabbit

    No corneal opacity or iritis.
    Conjunctival redness in all eyes at
    1 and 24 hours, which persisted in
    three eyes to 48 hours.
    Toxicity Category: III
    Classification: Core Minimum Data

5. Acute inhalation toxicity, rat

    Acute inhalation LC_{50} = > 4.7 mg/L
    Toxicity Category: III
    Classification: Core Minimum Data

6. Ames Mutagenicity Assay

    No mutagenic potential shown by this
      assay.
    Classification: Acceptable study

Summary of Regulatory Position and Rationale:

Section 3(c)(7)(C) of FIFRA requires that a public interest finding be made before a conditional registration can be issued. The Agency has concluded that this requirement has been satisfied.

Isomate-M is a relatively low-toxicity pesticide which is encapsulated in a plastic tube. This product will be used as part of an Integrated Pest Management system. This will result in the reduction of the use of broad spectrum conventional pesticides of relatively higher toxicity.

Furthermore, the development and use of these innovative methods of pest control are deemed to be in the public interest, since they may serve to alleviate the use of larger quantities and frequency of use of the more conventional pesticides.

Contact Person at EPA:

Arturo E. Castillo
Product Manager (17)
Registration Division (TS-767C)
Office of Pesticide Programs
Environmental Protection Agency
401 M Street SW.
Washington, D.C. 20460

LACTOFEN

Date Issued: March 18, 1987
Fact Sheet Number: 128

1. Description of Chemical

   Generic Name: 1-(carboethoxy)ethyl-5-[2-chloro-4-(trifluoromethyl)phenoxy]-2-nitrobenzoate

   Common Name: Lactofen

   Trade Names: Cobra™, PPG-844

   EPA Shaughnessy Code: 128885

   Chemical Abstracts
   Service (CAS) Number: 77501-63-4

   Year of Initial
   Registration: 1987

   Pesticide Type: Herbicide

   Chemical Family: Diphenyl ethers

   Producer: PPG Industries, Inc.

2. Use Patterns and Formulations

   Application sites: Used for postemergent control of broadleaf weeds on the terrestrial food crop - soybeans.

   Types of formulations: Manufacturing-use product containing 60% active ingredient (ai). End-use product containing 23.2% ai formulated as an emulsifiable concentrate.

   Ususal carrier: Water. Crop oil concentrates and surfactants may also be used with the agricultural use formulation.

   Types and methods of application: Lactofen is applied by both ground and aerial application.

Application rates: Application rates for soybeans range from 10 to 12.5 fluid ounces per acre (0.16 to 0.2 lb/ai/A) depending on the target weed species.

3. Science Findings

Summary Science Statement: Lactofen has been found to be oncogenic in mice and rats and has been classified as a Group B2 oncogen (Probable Human Carcinogen). A quantitative risk estimate has been conducted for the use of lactofen on soybeans. Based on a $Q^* = 1.7 \times 10^{-1} (mg/kg/day)^{-1}$ and using a Theoretical Maximum Residue Contribution (TMRC) of 0.000017 mg/kg (1.5 kg diet) the "worst case" dietary risk was calculated to be 2.9 incidences in a million (2.9×10^{-6}). Using the TMRC provides a conservative estimate since it does not consider the effect of processing on residue levels in the raw agricultural commodity, that actual residue levels will be lower than the tolerance level (0.05 ppm), and that less than 100 percent of the crop is treated. The 0.05 ppm level is based on a conservative assumption that lactofen and its four metabolites could each be theortically present in an amount just below the level of dectection of the individual compounds (0.01 ppm).

Based on exposure estimates for use of lactofen on soybeans and the Q^*, the following ranges in risk numbers were calculated:

Private Applicators
  Ground boom application:
    Low 10^{-7}
    Mean 10^{-6}
    High 10^{-4} to 10^{-5}

  Mixing, loading, and spraying:
    Open loading system 10^{-5} to 10^{-6}
    Closed loading system 10^{-7}

Commerical Applicators
  Aerial application
    Pilots: 10^{-5} to 10^{-6}
    Flaggers: 10^{-5}

  Mixing/loading/aerial application:
    Open loading system 10^{-3} to 10^{-4}
    Closed loading system 10^{-5}

  Ground Boom application
  Mixing loading and spraying
    Open loading system 10^{-4}
    Closed loading system 10^{-4} to 10^{-5}

These estimates assume that workers are wearing long-sleeved shirts, long pants, and shoes; protective gloves are worn during mixing and loading, and 10 percent dermal absorption.

Lactofen is not considered to be teratogenic and the chemical did not significantly impair reproductive ability in a two-generation reproductive effects study in rats. Four mutagenicity studies with lactofen were negative. A second Ames test was positive.

Lactofen is not acutely toxic to humans or avian species. The pesticide is highly toxic to fish and moderately toxic to aquatic invertebrates. Environmental fate studies show that lactofen does not persist significantly in the environment, that it is relatively immobile, and therefore should not pose a risk of leaching to ground water.

An applicator carcinogenic warning and fish toxicity statements are required to appear on the product's labeling.

Chemical Characteristics:

Physical state: White crystalline solid (Pure Grade Active Ingredient [PGAI])

Molecular formula: $C_{19} H_{15} Cl F_3 NO_7$

Molecular weight: 461.8

Odor: Very faint aromatic

Solubility: 96 ppb (22 °C) in water (low solubility)

Melting point: 43.9 - 45.5 °C (PGAI)

Vapor pressure: 8×10^{-9} mmHg at 25 °C (extrapolated) (nonvolatile)

Toxicological characteristics:

Acute oral toxicity (rat): 5960 mg/kg (relatively nontoxic)

Acute dermal toxicity (rabbit): Greater than 2000 mg/kg (moderately toxic)

Acute Inhalation: Greater than 6.3 mg/L (moderately toxic)

Dermal sensitization: Nonsensitizing

Primary Dermal Irritation: Very slight erythema subsiding within 76 hours.

Primary Eye Irritation: Redness of iris and redness and chemosis of the conjunctiva disappearing by 72 hours post-administration.

Subchronic Effects: In a 90-day feeding study in the rat at doses of 40, 200 and 1000 ppm, the no-observed-effect level (NOEL) is 200 ppm (10 mg/kg/day) and lowest-observed-effect level (LOEL) is 1000 ppm (50 mg/kg/day).

Chronic effects:

Chronic Feeding/
Oncogenicity: An 18-month oncogenicity study in CD-1 mice at doses of 10, 50, and 250 ppm in the diet was positive for oncogenic response. A statistically significant increased combined incidence of liver adenomas and carcinomas was observed at 250 ppm in both sexes. The lowest dose, 10 ppm (1.5 mg/kg/day), was the LOEL with increased liver weight and hepatocylomegally.

A 2-year chronic feeding/oncogencity study in Sprague-Dawley rats at doses of 500, 1000, and 2000 ppm in the diet was positive for oncogenic response. A statistically significant increase of liver neoplastic nodules and foci of cellular alteration was observed in both sexes at 2000 ppm. The systemic NOEL is 500 ppm (25 mg/kg/day) and the LOEL is 1000 ppm (50 mg/kg/day) based on kidney and liver pigmentation.

In a 1-year feeding study with beagle dogs, the NOEL is 200 ppm (5 mg/kg/day) and the LOEL is 1000/3000 ppm (25/75 mg/kg/day) based on renal dysfunction, and decreased Hgb, Hct, RBC and cholesterol.

Mutagenicity: Unscheduled DNA synthesis, chromosomal aberration, DNA repair assay, and one Ames Study are negative. A second Ames test in strain TA 1538 of Salmonella typhimurium at 5000 and 7500 ug/plate (precipitates formed in plates) is positive.

Two-Generation Reproduction (rat): Charles River CD rats were dosed with 0, 50, 500, and 2000 ppm lactofen in the diets. The reproductive NOEL is 50 ppm (2.5 mg/kg/day) and the LOEL is 500 ppm (25 mg/kg/day) based on reduced mean pup weight, and increased pup heart and liver weight.

Developmental Toxicity: Sprague-Dawley rats were dosed with 0, 15, 50, and 50 mg/kg/day lactofen in the diet. The maternal and developmental toxicity NOEL is 50 mg/kg and the LOEL is 150 mg/kg/day based on maternal post implantation loss and reduced body weight and fetal bent ribs.

New Zealand White rabbits were dosed with 0, 1, 4, and 20 mg/kg/day lactofen in the diet. The NOEL is 4 mg/kg/day and the LOEL is 20 mg/kg/day based on reduction in maternal food consumption with no developmental toxicity occurring at any dose level tested.

Physiological and Biochemical Behavior Characteristics:

Foliar absorption: Contact activity results in relatively rapid knockdown of weeds.

Translocation: In presence of rainfall, plants absorb lactofen from soil by root uptake.

Environmental Characteristics:

Lactofen is relatively stable to hydrolysis at 25 °C in pH 5 and pH 7 but rapidly hydrolyzes in pH 9 solutions. Lactofen was found to degrade quickly under aerobic soil conditions and degrades under anaerobic conditions. Lactofen half-life under photolysis conditions was 23 days. Lactofen was found to be immobile in soil. Degradation products are highly mobile in sandy soils and have moderate to low mobilities in soils that have a high silt and clay content. However, in an aged column leaching study total radioactivity found in the leachate comprised only 0.27 percent of that applied and was determined to be extensively degraded products. Lactofen was found to have a field dissipation rate of less than or equal to 7 days in a variety of soils. Field rotational crop data indicate residues are not taken up by grains, leafy vegetable, or root crops. Lactofen was found to accumulate in bluegill sunfish with a bioaccumulation factor of 380X for whole fish after 30 days. Depuration was rapid.

Ecological Characterisics:

Avian acute oral toxicity:
Bobwhite quail   > 2510 mg/kg
                              (Practically nontoxic)

Avian dietary toxicity:
Mallard duck and bobwhite quail   > 5620 ppm
                                  (Practically nontoxic)

Freshwater fish acute toxicity
  Blugill sunfish: > 100 ppb[1]
                    0.46 ppm[2]
                    (Highly toxic)

  Rainbow trout: > 100 ppb[1]
                    0.81 ppm[2]
                    (Highly toxic)

Freshwater fish 7-day flow-through: > 100 ppb[1]

Aquatic invertebrate acute toxicity
  Daphnia magna: > 100 ppb[1]
                  2.0 ppm (TGAI above solubility)
                  4.8 ppm[2]
                  (Moderately toxic)

Honey Bee acute contact toxicity: > 160 ug/bee
                                   (Practically nontoxic)

Fish early life stage toxicity
  (Sheepshead minnow): Maximum acceptable toxicant
                       concentration > 0.78 < 1.6 ppm.

These data indicate that lactofen is essentially nontoxic to avian species and bees; and that it is highly toxic to fish and moderately toxic to aquatic invertebrates. Based on the acute and chronic data no significant problems to nontarget organisms are expected from lactofen's use on soybeans.

4. Tolerance Assessment

Tolerances have been established for the combined residues of lactofen and its metabolites containing the diphenyl ether linkage in or on the following raw agricultural commodity (40 CFR 180.):

Commodity	Tolerance (ppm)
Soybeans	0.05

There are no international tolerances/residue limits for lactofen.

---

[1]/Maximum solubility of technical grade active ingredient (TGAI)
[2]/Expressed as active ingredient derived from studies conducted with end-use product.

There are sufficient residue chemistry data available to support this tolerance, including plant and animal metabolism, storage stability (for both the parent compound and its metabolites), field residue studies, and analytical methods. Cattle and poultry feeding studies were not submitted. However, under the proposed conditions of use, measurable residues are not expected to be found in the raw agricultural commodities or fractions. These data are therefore not now necessary.

The Acceptable Daily Intake (ADI) and the Maximum Permissible Intake (MPI) are two ways of expressing the amount of a substance that the Agency believes, on the basis of the results of data from animal studies and the application of "safety" or "uncertainty" factors, may safely be ingested by humans without risk of adverse health effects. The ADI is expressed in terms of milligrams (mg) of the substance per kilogram (kg) of body weight per day (mg/kg/day). The MPI, a related figure, is obtained by assuming a human body weight of 60 kg, and is expressed in terms of mg of substance per day (mg/day).

The Agency has calculated an ADI for lactofen of 0.0015 mg/kg/day, based on a LOEL of 1.5 mg/kg/day in the mouse oncogenicity study and a thousandfold safety factor. The MPI for a 60 kg person is 0.09 mg/day. These tolerances have a theoretical maximum residue contribution (TMRC) of 0.000017 mg/day in a 1.5 kg diet and would utilize 1.13 percent of the ADI.

5. Contact Person At EPA

Richard F. Mountfort
U.S. Environmental Protection Agency
TS-767C
401 M Street, SW.
Washington, D.C. 20460

DISCLAIMER: The information presented in this Pesticide Fact Sheet is for informational purposes only and may not be used to fulfill data requirements for pesticide registration and reregistration.

LEAD ARSENATE

Date Issued: December 1986
Fact Sheet Number: 112

1. DESCRIPTION OF CHEMICAL

    Common Name: Lead Arsenate

    Chemical Name: Acid Orthoarsenate - $PbHAsO_4$
    Basic Orthoarsenate - $Pb_4(PbOH)(AsO_4)_3$

    Trade Name: Lead Arsenate, Gypsine, Security, Talbot

    EPA Shaughnessy Code: Standard (Acid)   013502
    Basic   013503

    Chemical Abstracts Service (CAS) Number: 7778-40-9

    Year of Initial Registration:

    Pesticide Type: Growth Regulator, Insecticide, Herbicide, and Fungicide

    Chemical Family: Inorganic Arsenicals

    U.S. and Foreign Producers: Mechema Chemicals Ltd. (Great Britian),

2. USE PATTERNS AND FORMULATIONS

    Lead arsenate is currently used as a growth regulator on 17% of the U.S. grapefruit crop. 10,000 pounds of lead arsenate are also used annually to control cockroaches, silverfish and crickets. The Agency is unaware of any current use as a foliar insecticide or as a herbicide.

    ° Types and Methods of Application: Airblast sprayer, foliar aerial dust, bait box.

    ° Application Rates: Growth Regulator - 1.3 lbs arsenic/A
    Foliar Insecticide - 1.7 lbs arsenic/A

    Types of Formulations: Dust, flowable liquid, wettable powder, granular, impregnated, wettable powder/dust

3. SCIENCE FINDINGS

   • Chemical Characteristics

   Lead arsenate is a pentavalent form of inorganic arsenic. It normally exists as white crystals with no discernable odor. Lead arsenate contains 22% arsenic and is very slightly soluble in cold water. The melting point of lead arsenate is 1042°C, the density is 7.80 and the molecular weight is 347.12. Technical lead arsenate consists of 95-98% lead arsenate. Under most conditions basic lead arsenate is more stable than acid lead arsenate.

   • Toxicological Characteristics

   Inorganic arsenical compounds have been classified as Class A oncogens, demonstrating positive oncogenic effects based on sufficient human epidemiological evidence.

   Inorganic arsenicals have been assayed for mutagenic activity in a variety of test systems ranging from bacterial cells to peripheral lymphocytes from humans exposed to arsenic. The weight of evidence indicates that inorganic arsenical compounds are mutagenic.

   Evidence exists indicating that there is teratogenic and fetotoxic potential based on intravenous and intraperitoneal routes of exposure; however, evidence by the oral route is insufficient to confirm lead arsenate's teratogenic and fetotoxic effects.

   Inorganic arsenicals are known to be acutely toxic. The symptoms which follow oral exposure include severe gastro-intestinal damage resulting in vomiting and diarrhea, and general vascular collapse leading to shock, coma and death. Muscular cramps, facial edema, and cardiovascular reactions are also known to occur following oral exposure to arsenic.

   • Environmental Characteristics: The environmental fate of lead arsenate is not well documented. Studies to demonstrate its fate must take into account the fact that inorganic arsenicals are natural constituents of the soil, and that forms of inorganic arsenic may change depending on environmental conditions. Based on very limited data lead arsenate is not predicted to leach significantly.

   • Ecological Characteristics: Lead arsenate is moderately toxic to birds, slightly toxic to fish and moderately toxic to aquatic invertebrate species.

- Metabolism: The metabolism of inorganic arsenic compounds in animals is well known. The pentavalent form, such as lead arsenate, is metabolized by reduction into the trivalent form, followed by transformation into organic forms which are excreted within several days via the urine. All animals exhibit this metabolism except rats, which retain arsenic in their bodies for up to 90 days.

- Tolerance Assessment: Tolerances were established in 40 CFR 180.194 for residues of lead arsenate.

- Reported Pesticide Incidents: The Agency's Pesticide Incident Monitoring System (PIMS) has many recorded incidents of accidental poisonings from the use of lead arsenate baits. Nine of these incidents involved hospitalizations and 16 involved child poisonings from "roach hive" products.

4. SUMMARY OF REGULATORY POSITION AND RATIONALE

The Agency is proposing to cancel all existing nonwood registrations of lead arsenate, with the exception of the growth regulator use on grapefruit. Measures to mitigate the inhalation risks including dust masks, respirators, which would be expected to reduce inhalation exposure by 80 and 90 percent, respectively, and restricting the use to certified applicators were considered by the Agency during the Special Review. The Agency has determined that these protective measures would not reduce risks to an acceptable level in light of the limited benefits. The Agency has further determined that the toxicological risks from all nonwood uses of lead arsenate, except the grapefruit use, outweigh the limited benefits. The growth regulator use on grapefruit is being deferred pending further evaluation by EPA's Risk Assesment Forum of the carcinogenic potency of inorganic arsenic from dermal and dietary exposure.

- Benefits Analysis: The economic impact from cancellation of the lead arsenate insecticide baits could range from $.84 to $6.7 million, the actual amount depending on whether the alternative chemical is applied by homeowners or professionals. No economic impact is expected as a result of cancellation of the herbicide and foliar insecticide uses of lead arsenate. Viable alternatives are available.

5. CONTACT PERSON

   Douglas McKinney
   Special Review Branch, Registration Division
   Office of Pesticide Programs (TS-767C)
   401 M Street, S.W.
   Washington, D.C. 20460
   (703) 557-5488

   DISCLAIMER: The information presented in this Pesticide Fact Sheet is for informational purpose only and may not be used to fulfill data requirements for pesticide registration or reregistration.

LINALOOL

Date Issued: April 15, 1986
Fact Sheet Number: 77.1

1. Description of Chemical

   Chemical Name: Linalool
   Trade Name: Linalool 925
   Common Names: linalool
   EPA Shaughnessy No.: 128838
   Chemical Abstracts Service (CAS) Number: 78-70-6
   Year of Initial Registration: 1985
   Pesticide Type: natural oil
   Chemical Family: acyclic terpene alcohols
   U.S. Producer: Pet Chemicals, Inc.

2. Use Patterns and Formulations

   Application Sites: indoor use only on dogs and cats and their bedding
   Types of formulations: liquid
   Types/methods of application: pump spray

3. Science Findings

   Chemical Characteristics: liquid with a fresh floral woody odor; boiling point, 379 °F; molecular weight, 154.24; specific gravity, 0.863; practically insoluble in water; miscible with alcohol and ether.

   Toxicity Characteristics: acute oral toxicity (4.9 and 4.13 g/kg for male and female rats, respectively); acute dermal toxicity (2 g/kg albino rabbits); acute inhalation toxicity (LC_{50} less than 2.95 mg/L); not a sensitizer. No subacute or chronic studies are required since linalool is generally recognized as safe under 21 CFR 182.60 (synthetic flavoring substances and adjuvants).

   Environmental Characteristics: none reported; no data required to support the manufacturing-use or indoor end-use product.

Ecological Characteristics: slightly toxic to rainbow trout and bluegill (28.8 and 36.8 ppm, respectively); slightly toxic to aquatic invertebrates (36.7 ppm); practically nontoxic to birds (LC_{50} greater than 5620 ppm in bobwhite quail).

4. Summary of Science Findings

   Use Classification: unrestricted.
   Formulation or Labeling Restrictions: indoor use only.

5. Summary of Data Gaps

   None

6. Contact Person at EPA

   William H. Miller (PM 16)
   Insecticide-Rodenticide Branch (TS-767C)
   401 M Street SW.
   Washington, DC 20460
   Telephone Number (703) 557-2600

   DISCLAIMER: The information presented in this Chemical Information Fact Sheet is for informational purposes only and may not be used to fulfill data requirements for pesticide registration and reregistration.

LINDANE

Date Issued: September 30, 1985
Fact Sheet Number: 73

1. Description of Chemical

    Generic Name: Gamma Isomer of 1,2,3,4,5,6-hexachloro-
                  cyclohexane
    Common Name: Lindane
    Trade Names: Exagamma, Forlin, Gallogamma, Gammaphex,
                 Gammex, Gexane, Grammapoz, Grammexane,
                 Inexit, Kwell, Lindafor, Lindagrain,
                 Lindagram, Lindagranox, Lindalo, Lindamul,
                 Lindapoudre, Lindaterra, Lindex, Lindust,
                 Lintox, Novigram, and Silvanol
    EPA Shaughnessy Code: 009001
    Chemical Abstracts Service (CA) Number: 58-89-9
    Year of Initial Registration: 1950
    Pesticide Type: insecticide/acaracide
    Chemical Family: chlorinated hydrocarbon
    U.S. Producer: None
    Foreign Producers: Celamerck GmbH KG
                         Ingelheim, Federal Republic of Germany
                       Rhone Poulenc Phytosanitaire
                         Lyon, France
                       Mitsui, Inc.
                         Fukuoka, Japan
                       Tianjin Interntl. Trust & Investment Corp.
                         Tianjin, China

2. USE PATTERNS AND FORMULATIONS

    Application sites: field and vegetable crops (including seed treatment) and non-food crops (ornamentals and tobacco), greenhouse food crops (vegetables) and non-food crops (ornamentals), forestry (including Christmas tree plantations), domestic outdoor and indoor (pets and household), commercial indoor (food/feed storage areas and containers), animal premises (including manure), wood or wooden structures, and human skin/clothing (military use only).

Percent of lindane used on various crops/sites:

Hardwood Lumber	19%
Seed Treatment	48%
Forestry	<1%
Livestock	20%
Pineapple	2%
Ornamentals	2%
Pecans	3%
Pets	3%
Structures	<1%
Household	1%
Cucurbits	1%

Types and methods of applications: dip tank solution (livestock, lumber, and pets), as a livestock spray, by ground equipment delivering a ground or foliar spray or dust, by soil incorporation, by soil injection in combination with a fumigant (for the pineapple use only), as a smoke (for greenhouse fumigation only), as a dust for human skin/clothing (military use only).

Application rates: ranged from 0.25 to 2.25 oz/100 lb of seed for seed treatment; 0.1 to 2.06 lb/A for foliar and soil treatment; 0.8 to 1.5 oz/50,000 ft^3 of greenhouse; 0.006 to 0.11 lb/gal for bark; 0.023 to 3% sprays, dips, and dusts for indoor and animal treatment; <0.01 lb/1,000 ft^2 for animal premises; <4 lb/1000 ft^2 (14.64% solutions for wood and wooden structures; and 1% dust for human skin/clothing treatment (military use only).

Types of formulations: 0.27%-11.2% impregnated formulations, 0.5-75% Dusts, 3%-73% wettable powders, 0.5-25% liquids, 0.25-3% pressurized liquids, 1-4% flowable concentrates, 0.45-40% emulsifiable concentrates,

3. SCIENCE FINDINGS

Summary Science Statement: Lindane is a chlorinated hydrocarbon of moderate mammalian acute toxicity. Lindane has been shown to be oncogenic in mice but it is not genotoxic. The Agency has concluded that lindane is a possible human carcinogen. The Agency is requiring that another rat chronic/oncogenicity bioassay be performed. Lindane has been associated with possible induction of blood dyscrasias (aplastic anemia). The Agency is requiring a laboratory animal study to permit assessment of lindane's potential to cause blood dyscrasias. Other toxicology studies demonstrate systemic toxicity, targeting the liver and kidney. Lindane's behavior in the environment is not well defined. The Agency is requiring a full complement of environmental fate studies. Lindane is slightly to moderately toxic to birds and highly toxic to some aquatic organisms. Lindane is highly toxic to honeybees and certain beneficial parasites and predacious insects. Additional studies on the ecological effects of lindane are required.

Chemical Characteristics

Technical lindane is a white crystalline solid.
Its melting point is 112° - 113°C.
It is soluble in most organic solvents and is relatively insoluble in water.
Lindane is stable to light, heat, air and strong acids, but decomposes to trichlorobenzenes and HCL in alkali.

Toxicology Characteristics

Acute Oral: 88 mg/kg, Toxicity Category II
Acute Dermal: 300 mg/kg Toxicity Category II
Acute Inhalation: Data gap
Primary Eye Irritation: Data gap
Primary Skin Irritation: Irritant Toxicity Category I
Skin Sensitization: Data gap
Major Routes of Exposure: Human exposure from lindane is greatest during mixing, loading, and application. Dermal, ocular, and inhalation exposures to workers may occur during application. Exposure can be reduced by the use of approved respirators, protective clothing, and goggles.
Oncogenicity: A two-year mouse oncogenicity study demonstrated increased incidences of liver tumors (male & female) when dosed at 400 ppm. An 80-week mouse feeding study demonstrated increased incidences of liver tumors at the 80 ppm level but not at the 160 ppm level. Two subchronic studies provide supportive evidence of oncogenicity. The mouse studies were referred to the Agency's Carcinogen Assessment Group (CAG) for evaluation. Based on the weight of the evidence, CAG classifed lindane in the range B2-C. OPP believes that the classification C is appropriate at this time and, therefore, will regulate lindane as a class C carcinogen, pending receipt of the required rat oncogenicity study.
Metabolism: Lindane does not appear to bioaccumulate in tissues.
Teratology: Teratology studies in the rat, rabbit, and mouse were negative for teratogenic effects.
Reproduction: A 3 generation rat reproduction study was negative at 100 ppm.
Mutagenicity: Available data show lindane to be negative for gene mutation in bacterial Ames assays, host mediated, and dominant lethal assays. Lindane has been reported as negative in other *in vitro* assays for DNA damage/repair in bacteria, rat and mouse hepatocytes, and mammalian cell transformation assays.

Physiological and Biochemical Characteristics

Mechanism of pesticidal action: Lindane acts in the nervous system through unknown mechanisms.
Metabolism and persistence in plants and animals: The metabolism of lindane in plants and livestock animals has not been adequately described. Additional data are being required.

Environmental Characteristics

Available data are insufficient to assess the environmental fate of lindane. Data gaps exist for all required studies. Preliminary adsorption data indicate that lindane has a low mobility in mineral soils and is relatively immobile in muck soils; however, the potential for lindane contamination of surface and ground water exists based on the results of a monitoring study conducted in certain southern states.

Ecological Characteristics

Avian acute oral toxicity: Data Gap
Avian dietary toxicity: 882 ppm for bobwhite quail, 561 ppm for ring-necked pheasant (moderately toxic), and >5000 ppm for mallard duck (practically nontoxic).
Freshwater fish acute (LC_{50}) toxicity: cold water species (rainbow trout) 27 ppb for technical lindane (very highly toxic), warm water species (bluegill) 68 ppb for technical lindane (very highly toxic).
Aquatic freshwater invertebrate toxicity: Daphnia 460 ppb (highly toxic). Additional data are required to fully characterize the ecological effects of lindane.
Available data are insufficient to fully assess the environmental fate of and the ecological effects from lindane.

Required Unique Labeling Summary

All manufacturing-use and end-use lindane products must bear appropriate labeling as specified in 40 CFR 162.10. In addition, the following information must appear on the labeling:

All manufacturing-use products must state that they are intended for formulation into other manufacturing-use products or end-use products only for registered uses.

All manufacturing-use products shall contain the following text in the Environmental Hazards section of the label:

"This pesticide is toxic to fish and aquatic invertebrates. Do not discharge effluent containing this product into lakes, streams, ponds, estuaries, oceans, or public waters unless this product is specifically identified and addressed in an NPDES permit. Do not discharge effluent containing this product into sewer systems without previously notifying the sewage treatment plant authority. For guidance, contact your State Water Board or Regional Office of the EPA."

All end-use products containing lindane that were classsified as restricted by the Final Notice of Determination concluding the RPAR shall continue to be classified for restricted use and the restricted use label must include the cancer hazard warning statement.

All end-use products shall continue to carry the applicator protection statements previously required by the Final Notice of Determination concluding the RPAR. Products with directions for foliar application to crops whose culture involves hand labor must bear the statements required under PR Notices 83-2 and 84-1 for farmworker safety, including a 24 hour re-entry interval.

End-use products with directions for spraying uninhabited buildings or empty storage bins must include protective clothing requirements, including the use of a respirator.

All end-use products for indoor use shall indicate that lindane is not to be applied to edible product areas of food processing plants or to serving areas while food is exposed.

All end-use products with uses on livestock or livestock premises must indicate not to contaminate food, feed, or water with the pesticide. Also, there must be a statement that indicates lindane is not to be applied to poultry houses, dairy barns, and milk rooms. All feed or water troughs must be covered and all livestock should be removed from animal shelters (barns, sheds, etc.) prior to treatment of the structure.

All end-use products for structural pest control must indicate that lindane is _not_ to be applied in currently occupied areas (i.e. regular living or working areas, including finished basements or finished attics) of homes or other buildings. The characterization of a use site depends upon its intended function and not upon whether there are occupants in the area at the time of treatment.

Tolerance Assessment

The Agency is unable to complete a full tolerance assessment because the metabolism of lindane in plants and livestock animals has not been adequately described. Also, seed treatment is now considered to be a food use requiring a tolerance unless results of a radiolabeled study indicate that there is no translocation to edible parts of the plant following seed treatment. No new tolerances, except those required to support the existing seed treatment uses of lindane, will be considered until the toxicology and residue chemistry data gaps identified in the Standard have been filled.

Established tolerances are published in 40 CFR 180.133.

A listing of U.S. tolerances includes the following:

Commodity	Tolerance expressed as parts per million
Apples	1.0
Apricots	1.0
Asparagus	1.0
Avocados	1.0
Broccoli	1.0
Brussels Sprouts	1.0
Cabbage	1.0
Cattle, fat	7.0
Cauliflower	1.0
Celery	1.0
Cherries	1.0
Collards	1.0
Cucumbers	3.0
Eggplant	1.0
Goats, fat	7.0
Grapes	1.0
Guavas	1.0
Hogs, fat	4.0
Horses, fat	7.0
Kale	1.0
Kohlrabi	1.0
Lettuce	3.0
Mangoes	1.0
Melons	3.0
Mushrooms	3.0
Mustard Greens	1.0
Nectarines	1.0
Okra	1.0
Onions, dry bulb only	1.0
Peaches	1.0
Pears	1.0
Pecans	0.01
Peppers	1.0
Pineapples	1.0
Plums, incl. Prunes	1.0
Pumpkins	3.0
Quinces	1.0
Sheep, fat	7.0
Spinach	1.0
Squash	3.0
Summer Squash	3.0
Strawberries	1.0
Swiss Chard	1.0
Tomatoes	3.0

The best available data for determining an interim acceptable daily intake level of lindane is a subchronic feeding study in rats (1983) which demonstrated a No Observed Effect Level (NOEL) of 4 ppm. Based on dietary analysis, food intake, and body weight data from this particular study, the NOEL of 4 ppm is equivalent to 0.3 mg/kg/day. Using this latter value and a safety factor of 1000, the Provisional Acceptable Daily Intake (PADI) is 0.0003 mg/kg/day and the Maximum Permissible Intake (MPI) for a 60 kg person is 0.018 mg/day. The Theoretical Maximum Residue Contribution (TMRC) for lindane, based on all established tolerances, is 1.4189 mg/day/1.5 kg of diet The percent of the MPI used by the TMRC is 7883%.

Although the theoretical concentration from existing tolerances greatly exceeds the MPI, FDA market basket surveys indicate that actual residues of lindane are much lower. The Agency believes that the actual risk to consumers from the daily consumption of lindane, based on FDA market basket data for 1978-1982, is only 0.000002 mg/kg/day. Under this scenario, only 0.7% of the Maximum Permissible Intake is actually used.

4. SUMMARY OF REGULATORY POSITION AND RATIONALE

The Agency has determined that it should continue the registration of all currently registered uses of lindane. The Agency concluded in the RPAR that most uses of lindane would be continued because a risk/benefit assessment demonstrated that the benefits from the uses outweighed the risks provided certain labeling restrictions, such as restricting some uses to certified pesticide applicators, requirements for protective clothing, and label statements describing necessary precautions were added to all lindane labels. The Agency has reevaluated that decision and concludes that, except as described below, the risks and benefits are substantially the same as those described in the RPAR process.

Based on FDA market basket residue levels, which the Agency believes in this case is more generally representative of actual residues than theoretical calculations, the estimate of the upper 95% confidence level for excess cancer risk is 2×10^{-6}. The estimate of the upper 95% confidence level for excess cancer risk to applicators for various uses is estimated to be from 10^{-4} to 10^{-7}, depending upon the site and method of application. The Agency has recalculated the exposures and margins of safety for applicators for 24 use patterns and has developed initial calculations for mixer/loaders or combination mixer/loader/applicator for 3 use patterns (forestry, cucurbits, and pecans). This reassessment is based on current Agency methods and models and consideration of a lower NOEL from the subchronic rat study for uses involving subchronic type exposure.

Based on these calculations, the Agency will initiate Special Review for the forestry and uninhabited buildings and empty storage bins spray uses on the basis of risks to applicators. Use of protective clothing, including an MSHA/NIOSH approved respirator, is now being required for spraying uninhabited buildings and empty storage bins while the Special Review is underway. Protective clothing requirements stipulated in the Final Notice of Determination of the lindane RPAR will continue for all other uses. No significant changes from the exposure values presented in PD-4 occurred for twelve of the uses. The calculated exposures for cucurbits, crawl spaces, dog dusts, dog dips, shelf paper, and commercial moth sprays increased by approximately an order of magnitude, but are still acceptable. Applicator exposure data is being required for seed treatment, structural treatment, livestock spraying, dog washes, dog shampoos, and dog dusts. Air monitoring data are required for the structural treatment and dog treatment uses. Exposure studies are required, in addition to toxicity studies, to support the registrations for application of lindane to human skin/clothing by the military.

Lindane seed treatments were registered many years ago as non-food uses not requiring tolerances. The Agency now considers seed treatment to be a food use and requires data to support a tolerance unless results of a radiolabeled study indicate that there is no translocation to edible parts of the plant following seed treatment. Data from one of these two alternatives must be submitted to support the seed treatment uses. No new tolerances, except those required to support the existing seed treatment uses, will be considered until the chronic feeding and residue chemistry data gaps identified in the Standard have been filled.

Available data are insufficient to fully assess the environmental fate of lindane and its ecological effects. A full complement of such studies is being required. Precautionary label statements will continue to be required.

5. Summary of Major Data Gaps   Due Date

   - An acute inhalation study   9 months
   - A 90-day inhalation study   15 months
   - A dermal sensitization study   9 months
   - A 21-day dermal toxicity study   12 months
   - A rat chronic/oncogenicity study   50 months
   - Laboratory animal blood dyscrasias study   pending
   - Full complement of environmental fate studies   39 months
   - Plant metabolism studies   24 months
   - Livestock animal metabolism studies   18 months
   - Residue chemistry studies on all crops,   48 months
       including seed treatment uses

Contact Person at EPA

George T. LaRocca
Product Manager (15)
Insecticide-Rodenticide Branch
Registration Division (TS-767C)
Office of Pesticide Programs
Environmental Protection Agency
401 M Street, S.W.
Washington, D.C. 20460

Office location and telephone number:
Room 204, Crystal Mall Building #2
1921 Jefferson Davis Highway
Arlington, VA 22202
703-557-2400

DISCLAIMER: THE INFORMATION PRESENTED IN THIS PESTICIDE FACT SHEET IS FOR INFORMATIONAL PURPOSES ONLY AND IS NOT TO BE USED TO FULFILL DATA REQUIREMENTS FOR PESTICIDE REGISTRATION AND REREGISTRATION.

LINURON

Date Issued: June 30, 1984
Fact Sheet Number: 28

1. Description of the chemical:

   Generic name: 3-(3,4-Dichlorophenyl)-1-methoxy-1-methylurea ($C_9H_{10}Cl_2N_2O_2$)
   Common name: Linuron
   Trade name: Alfanox®, Linurex®, Londax®, Lorox®
   EPA Shaughnessy Code: 035506
   Chemical Abstracts Service (CAS) Registry number: 330-55-2
   Year of initial registration: 1966
   Pesticide Type: Herbicide
   Chemical family: Substituted urea
   U.S. and foreign producers: E. I. duPont de Nemours and Company, Drexel
     Chemical Company, Griffin Corporation, Vertac Chemical Corp., Bayer AG,
     Makhteshim-Agan, Pennwalt Holland B. V., Rhone-Poulenc, Staveley
     Chemicals Ltd., and Universal Crop Protection Ltd.

2. Use patterns and formulations:

   Application sites: Linuron is a substituted urea compound registered
     for use as a herbicide to control a wide variety of annual and perennial
     broadleaf and grassy weeds on both crop and noncrop sites. Linuron is
     registered for use on numerous crop sites such as forage crops, field
     crops, fruits, vegetable, and ornamental crops. In noncrop applications,
     linuron is used on alleys, fencerows, fairways, highway right-of-way,
     sodfields, streets, and vacant lots.

   Types of formulations: Linuron is available as a wettable powder,
     granular flowable, and liquid suspensions.

   Types and methods of applications: Linuron is applied as follows:
     broadcast or band upon the soil surface using ground or aerial equipment.

   Application rates: 0.5 lbs. a.i./A to 3.0 lbs. a.i./A on crop sites;
     and 1.0 lbs. a.i./A to 3.0 lbs. a.i./A on noncrop sites.

   Usual carriers: Water, oil and clay.

3. **Science Findings:**

   Summary science statements:

   Linuron has low acute mammalian toxicity and it's uses are not expected to adversely affect avian and mammalian wildlilfe. The metabolism of linuron in plants and animals is adequately understood.

   Dietary exposures to linuron have induced dose related tumors in the rat testes and mouse liver. The available toxicology data are insufficient to fully assess the longterm reproductive and teratogenic potential of linuron.

   Chemical characteristics:

   Technical linuron is an odorless, white, crystalline solid. It is stable towards oxidation and moisture under conventional conditions and decomposes at 180-190°C. The chemical does not exhibit any unusual handling hazards.

   Toxicological characteristics:

   Acute toxicology studies on linuron are as follows:
   Oral LD_{50} in rats: 1,500 mg/kg body weight, Toxicity Category III
   Dermal LD_{50} in rats: > 2,000 mg/kg body weight, Toxicity Category III
   Inhalation LC_{50} in rats: 218 mg/l/hr, Toxicity Category IV
   Skin irritation in rabbits: slight irritant, Toxicity Category III
   Eye irritation in rabbits: slight irritant, Toxicity Category III.

   Chronic toxicology studies on linuron are as follows:
   A two year chronic feeding study on rats has shown that interstitial testicular (ISC) adenomas occured in all dosage groups(control, 50.0, 125.0, and 625.0 ppm) both during the two years and then at term.

   A chronic feeding study was conducted on male and female mice at diet levels of 0.0, 50.0, 150.0, and 1,500 ppm of linuron. The study showed a statistically significant increase of hepatocellular adenomas in the female mice from the highest dose group (1,500 ppm). A significant increase of hepatocellular adenomas was also observed among the males in the lowest dose group (50 ppm). The levels of methemoglobin were increased in treated mice of both sexes; this increase was related to the linuron administration.

   A two year dog study did not demonstrate carcinogenesis but showed hemosiderin deposition at 125 and 625 ppm.

   In several mutagenicity tests, Linuron did not affect DNA repair but may have inhibited mouse testicular DNA synthesis. Linuron has not been shown to be active in the Ames test. Linuron did not affect S. typhimurium in vivo in the mouse peritoneal cavity.

   Major routes of human exposure:

   The non-dietary exposure to linuron by a farmer as an applicator or mixer/loader is very high.

   The dietary exposure to linuron residues by the U.S. population is probable because of its consumption of treated crops.

Physiological and Biochemical Behavioral Characteristics:

　　Absorption characteristics: Linuron is most readily absorbed through the root system; less through foliage and stems.
　　Translocation: Linuron is translocated upward primarily in the xylem.
　　Mechanism of pesticidal action: It is a strong inhibitor of photosynthesis (Hill reaction).

Environmental characteristics:

　　Adsorption and leaching in basic soil types: Adsorption increases as clay content and/or organic matter content of soil increases; clays of high exchange capacity absorb more linuron than those of low exchange capacity.

　　Microbial breakdown: microbes are the primary factor in the breakdown of linuron in soils.

　　The available environmental fate data are insufficient to fully assess the degradation, metabolism, mobility, dissipation and accumulation activities of linuron. When additional studies are submitted, a complete environmental assessment can be made.

Ecological characteristics:

　　Avian LD_{50}: >3,000 mg/kg,
　　Fish LC_{50}: (96 hour), 16 ppm for bluegill and rainbow trout,
　　LC_{50}: (72 hour) >40 ppm for crawfish,
　　LC_{50}: (48 hour) >40 ppm for tadpole.

When additional ecological effects data are submitted, a complete hazard assessment can be made.

Tolerance assessments:

　　Since linuron and diuron have certain metabolites in common [1-(3,4-dichlorophenyl)-3-methylurea (DCPMU), and 3,4-dichlorophenylurea (DCPU)], the Agency will consider diuron's residue contribution in the tolerance reassessment of linuron for the following commodities: corn, sorghum, grains, wheat, asparagus, meat(red), and cottonseed.

　　If the complete tolerance reassessments for the above commodities are favorable, tolerances for residues of linuron and metabolites (which will hydrolyze to form 3,4-dichloroaniline) will have to be proposed for residues in milk and eggs at 0.05 ppm.

　　The tolerances listed below have not been revised:

Commodities	Parts per million
Asparagus	3.0
Carrots	1.0
Cattle, fat	1.0
Cattle, meat by-products	1.0
Cattle, meat	1.0
Celery	1.0
Corn, field, fodder	1.0
Corn, field, forage	1.0
Corn, fresh, (sweet)	0.25
Corn, grain (inc. pop)	0.25
Corn, pop, fodder	1.0
Corn, pop, forage	1.0
Corn, sweet, fodder	1.0
Corn, sweet, forage	0.25
Goats, fat	1.0
Goats, meat by-products	1.0
Goats, meat	1.0
Hogs, fat	1.0
Hogs, meat by-products	1.0
Hogs, meat	1.0
Horses, fat	1.0
Horses, meat by-products	1.0
Horses, meat	1.0
Parsnips (with or without tops)	0.5
Parsnips, tops	0.5
Potatoes	1.0
Sheep, fat	1.0
Sheep, meat by-products	1.0
Sheep, meat	1.0
Sorghum, fodder	1.0
Sorghum, forage	1.0
Sorghum, grain (milo)	0.25
Soybeans (dry or succulent)	1.0
Soybeans, forage	1.0
Soybeans, hay	1.0
Wheat, forage	0.5
Wheat, grain	0.25
Wheat, hay	0.5
Wheat, straw	0.5

Problems known to have occurred with use:

Exposure of humans to linuron through runoff contamination of surface water after heavy Spring precipitation has occurred in Northwestern Ohio.

4. **Summary of regulatory position and rationale:**

    Use classification

    Restricted use classification.

    Unique label warning statements:

    "The use of this product may be hazardous to your health. This product contains linuron, which has been determined to cause tumors in laboratory animals."

    "Do not reenter treated areas for 24 hours following application unless protective clothing is worn."

    Summary of risk/benefit review:

    The Agency has determined that linuron has exceeded the oncogenicity risk criteria and requires special review. Dietary exposure to linuron indicated clear evidence of oncogenicity for male rats using the NTP criteria. Using these data, the Agency calculated nondietary risk. The most realistic scenario is a farmer with no protection, who mixes, loads and applies this herbicide. This calculation resulted in a risk of 3.6×10^{-4} to 2.2×10^{-3}. It is possible that the actual risk may even be higher, because the commercial applicator exposure was not included. The Agency also calculated dietary risk. The most realistic scenario for dietary risk is the combination of maximum residue expected (MRE) and percent crop which resulted in a risk of 1.5×10^{-5}.

5. **Summary of major data gaps:**

    The following toxicology data are required:

    Two teratology studies, one in rat and one in another species (rabbit). A two-generation reproduction study in rats is required; this study must be designed to incorporate concerns regarding the significance of interstitial cell adenomas. Note that in the former studies (rat and dog), reticulocytes and erythroid precursors were not measured. This is a data gap, since at the high dose level (625 ppm), hemosiderin was observed in rats and also at 125 and 625 ppm in the dog. (This data may be filled by appropriate design inclusion into the required reproduction study above. The registrant must consult with the Agency on the appropriate protocol.)

Mutagenicity* and related data are required, which (1) satisfy
the 3 mutagenicity testing category requirements, (2) adequately
identify the risks, and where possible identify the mechanisms
associated with positive findings in rodent chronic studies.
The Agency is requiring data, relating levels of sulf- and
methemoglobin following dietary exposure for certain substituted
phenyl urea compounds such as linuron. This testing may be
combined with other testing involving dietary exposure, such as
the reproduction study. Dose levels must be such that a NOEL may
be established.

The following four mutagenicity studies* have been received and are
in Agency review:
1. "Mutagencity Evaluation In (Salmonella typhimurium)", HLR 1006-83, 5/5/83,
2. "Unscheduled DNA Synthesis/Rat Hepatocytes In Vitro", HLR 190-83, 6/3/83,
3. "CHO/HGPRT Assay for Gene Mutation", HLR 540-83, 12/16/83,
4. "In Vivo Bone Marrow Chromosome Study in Rats", HLO 376-83, 9/1/83.

The available toxicology data are insufficient to fully assess the
long-term reproductive, and teratogenic potential of linuron..
Long-term studies must be submitted from one to two years after
receipt of the guidance package. Please refer to the toxicology
data tables under § 158.135 for the specific dates for which long
term data must be submitted. Short-term studies must be submitted
within six months after receipt of this guidance package.

The following environmental fate data are required:

Hydrolysis test,
Photodegradation test in water,
Photodegradation test in soil,
Photodegradation test in air,
Metabolism test in aerobic soil
Metabolism test in anaerobic soil,
Leaching and adsosrption/desorption,
Mobility (volatility) test in the lab,
Mobility (volatility) test in the field,
Dissipation study in soil,
Dissipation study in soil (long tterm),
Accumulation study in fish,
Special Testing on applicator exposure,
Reentry data requirements.

Long-term studies must be submitted from one to four years after
receipt of the guidance package. Please refer to the environmental
fate data tables under § 158.130 for the specific dates for which
long term data must be submitted. Short-term studies must be
submitted within six months after receipt of the guidance package.

The following ecological effects data are required:

Acute avian toxicity,
Acute toxicity, freshwater fish,
Acute toxicity, freshwater invertebrates.

Acute studies must be submitted within six months after receipt of this guidance package.

The physical/chemical requirements listed in the §158.120, Product Chemistry data tables must be submitted, particularly:

Solubility,
Vapor pressure,
Octanol/water partition coefficient.

These studies must be submitted within six months after receipt of this guidance package.

The following residue data are required:

Residue data for asparagus, carrots, celery, corn, cottonseed, parsnips, potatoes, sorghum,, soybeans, and wheat are required to reflect uses of the 50%, dry flowable (DF) and 4 lb/gal, flowable concentrate (FIC) formulations. Data reflecting uses of the 50% DF are required for the following commodities: carrots (aerial applications), potatoes (aerial applications), soybeans (preemergence), sorghum (forage), wheat (forage and hay), asparagus (preemergence), and cottonseed (two applications per season). Data pertaining to residues in dehydrated potato products are required.

Long-term studies must be submitted within one year after receipt of this guidance package.

6. Contact Person at EPA:

Robert J. Taylor
Product Manager (25)
Environmental Protection Agency (TS-767C)
401 M Street, S.W.
Washington, D.C. 20460
(703) 557-1800

DISCLAIMER: This information presented in this Chemical Information Fact Sheet is for information purposes only, and may not be used to fulfill data requirements for pesticide registration and reregistration.

MANCOZEB

Date Issued: April, 1987
Fact Sheet Number: 125

DESCRIPTION OF CHEMICAL

Generic Name: Manganese ethylene bisdithiocarbamate

Common Name: Mancozeb

Trade Name: Dithane® M-45, Manzate® 200, Fore®

EPA Shaughnessy Code: 014504

Chemical Abstracts Service (CAS) Number: 8018-01-7

Year of Initial Registration: 1967

Pesticide Type: Fungicide (with minor insecticide use)

Chemical Family: Ethylene bisdithiocarbamate (EBDC)

U.S. and Foreign Producers: Rohm and Haas produces a formulating intermediate; E. I. Dupont deNemours Co. produces mancozeb for its own end-use products

USE PATTERNS AND FORMULATIONS

Application Sites: Terrestrial food and nonfood crops; aquatic (food); greenhouse (nonfood); forestry; outdoor domestic.

Major Crops Treated: Apples, potatoes, and tomatoes. Mancozeb is also used on approximately 80 percent of the onion acreage in the United States.

Formulation Types Registered: 80 percent active ingredient (ai) formulating intermediate; dust, wettable powder, flowable concentrate, and granular end-use products.

Methods of Application: Foliar applications by aerial or ground equipment. For ground equipment, mancozeb suspensions typically are made from a wettable powder or flowable concentrate that is applied by means of air blast sprayers or tractor-mounted boom sprayers. Dust formulations are typically applied by means of truck- or tractor-drawn duster or aerial equipment.

Treatment of the seeds may be accomplished by commercial seed treatment equipment by seed companies or by addition to the planter box at the farm site. Mancozeb may be used as a spray furrow treatment of soil at planting of onion sets. For treatment of surfaces of potato cut seed pieces or whole tuber seed pieces, mancozeb is applied by means of dip tanks or dusting equipment mounted over the seed pieces on a conveyor belt.

Application Rates: Rates range from 0.4-3.3 lb/100 gal. or 0.6-19.06 lb ai/A for foliar application (high rates are for turf use rather than food crops); 0.04-4.32 lb/100 lb. of seed; 0.78-27 lb/100 gal for preplant dip treatments and 2.4 lb/A for soil applications. Multiple applications may be made during the growing season.

Usual Carrier: Water

SCIENCE FINDINGS

Summary Science Statement: The major toxicological concern from exposure to mancozeb is the hazard to the human thyroid from presence of ethylenethiourea (ETU), a contaminant, degradation product, and metabolite present in mancozeb and other EBDC products. ETU is an acknowledged goitrogen, teratogen and oncogen. Additional chronic studies of mancozeb are required for further evaluation. Available data indicate that mancozeb itself is not a primary developmental toxicant or teratogen, however, an additional teratology study with mancozeb is required before its teratogenicity can be fully assessed.

Available data are not adequate to assess the environmental fate of mancozeb. However, studies do indicate that ETU has the potential to leach. A complete assessment of the environmental fate, including the potential for groundwater contamination, from the use of mancozeb products will be undertaken when data are available.

Available data are insufficient to completely evaluate the ecological effects of mancozeb. Based on available data, mancozeb is no more than slightly toxic to avian wildlife on an acute basis; is highly toxic to warmwater fish; and appears to be at least moderately toxic to coldwater fish.

Chemical Characteristics: Only limited product chemistry information is available, as follows:

  Physical state: Solid at room temperature, decomposes when heated.

  Empirical Formula: $(C_4H_6MnN_2S_4)\ x\ (Zn)\ y$

  Molecular Weight: $(265.3)\ x\ +(65.4)y$

Toxicology Characteristics:

  Acute Toxicity[1]:

Oral Toxicity:	4500 mg/kg, Category III
Dermal Toxicity:	> 5000 mg/kg, Category III
Inhalation Toxicity:	> 5.14 mg/L, Category IV
Eye Irritation:	Primary Irritation Score (PIS) – 2.3, Category III
Skin Irritation:	PIS – 0.5, Category IV
Dermal Sensitization:	Study required

  Subchronic Toxicity:

Oral (rats):	No observed effect level (NOEL) for kidney effects – 60 ppm (3.5 mg/kg/day, males; 4.4 mg/kg/day females)
	NOEL for thyroid effects – 125 ppm (7.9 mg/kg/day, males; 9.2 mg/kg/day, females)
Oral (dogs):	NOEL for systemic effects – 100 ppm (3.0 mg/kg/day, males; 3.4 mg/kg/day, females)
	NOEL for thyroid effects – 1000 ppm (29 mg/kg/day for both sexes)
Dermal:	Studies required
Inhalation:	NOEL for systemic effects – 20 mg/m^3
	NOEL for thyroid 80 mg/m^3

  Major Routes of Exposure: Dermal and inhalation

---

[1] Toxicity Categories are discussed in 40 CFR 162.10.

Chronic Toxicity:

    Chronic Toxicity: Studies required
    Oncogenicity: Studies required

    Teratology: Maternal toxicity: NOEL - 32 mg/kg/day; Lowest effects level (LEL) - 128 mg/kg/day (decreased body weight)

    Fetal toxicity: NOEL - 128 mg/kg/day; LEL - 512 mg/kg/day (increased resorptions)

    Teratogenic effects: NOEL - 128 mg/kg/day; LEL - 512 mg/kg/day (dilated ventricles, spinal cord hemorrhage, delayed/incomplete ossification of skull and ribs); A/D Ratio: 32/128 = 0.25[2/]

    Reproduction: Study required

    Mutagenicity: Negative - bacterial and in vitro mammalian cell systems, chromosome damage in vivo and in mammalian cell transformation

    Positive - sister-chromatid exchanges in CHO cells in vitro

Physiological and Behavioral Characteristics

Mechanism of Pesticide Action: Mancozeb inhibits enzyme activity by complexing with metal-containing enzymes including those involved with the production of adenosine triphosphate (ATP).

Metabolism: Mancozeb appears to be rapidly absorbed from the gastrointestinal tract, disbributed to target organs and excreted almost totally by 96 hours. The major metabolite is ETU, comprising almost 24% of the bioavailable dose in urine and bile. ETU residues in the thyroid and the liver were less than 1 ppm and were non-detectable after 24 hours.

Environmental Characteristics:

Hydrolysis: Mancozeb degrades with a half-life of 1 to 2 days at pH 5, 7, and 9. ETU is stable to hydrolysis at pH 5 and 7 and is very slowly hydroloyzed at pH 9.

Anaerobic Aquatic Metabolism: Mancozeb and ETU declined with half-lives of 92 days and 29 to 35 days, respectively.

[2/] An A/D Ratio of less than 1 indicates that developmental toxicity may be ascribed to secondary affects of maternal toxicity.

Environmental Fate/Groundwater Concerns: Available studies indicate that ETU has the potential to leach. A complete assessment of the environmental fate, including the potential for groundwater contamination, from the use of mancozeb products will be undertaken when data are available.

Exposure During Reentry Operations: Mancozeb is registered for use on crops which may involve substantial exposure to residues of the pesticide. Because ETU has demonstrated evidence of oncogenicity, mutagenicity, teratogenicity and thyroid effects and mancozeb has caused thyroid effects, reentry data are required. Until these data are received and evaluated, a 24-hour reentry interval is imposed.

Ecological Characteristics:

Avian Oral Toxicity:
- Japanese Quail — > 6400 mg/kg/day
- European Sparrow — 3000-6000 mg/kg/day
- English House Sparrow — 1500 mg/kg/day
- Mallard Duck — > 6400 mg/kg/day

Freshwater Fish Toxicity:
- Bluegill sunfish — 1.54 ppm
- Rainbow trout — 0.46 ppm
- Daphnia magna — 0.58 ppm

Tolerance Assessment:

Tolerances Established: Tolerances, expressed as zinc ethylene bisdithiocarbamate, have been established for residues of mancozeb in a variety of raw agricultural commodities and meat byproducts (40 CFR 180.176 and 40 CFR 180.319), and in processed food (21 CFR 193.460) and feed (21 CFR 561.410).

Results of Tolerance Assessment: The toxicology data for mancozeb are insufficient to determine an Acceptable Daily Intake (ADI) or whether the toxicity observed in the studies is due to mancozeb or ETU.

There are no acceptable chronic studies on which to calculate an ADI, therefore, a subchronic study has been used to calculate a Provisional ADI (PADI). Because a subchronic study was used, an uncertainty factor of 1000 (rather than 100 used with chronic studies) was employed. The PADI for mancozeb is 0.003 mg/kg/day based on the 90-day dog feeding study with a NOEL of 3 mg/kg/day.

The theoretical maximum residue contribution (TMRC), based on the assumption that 100 percent of each crop is treated and contains residues at the tolerance levels, is 0.028 mg/kg/day or approximately 900 percent of the PADI.

Based on a more realistic dietary assessment, using average field trial residues and theoretical percent of crop treated, the estimated average consumption for the U.S. population is 0.00097 mg/kg/day or 32.2 percent of the PADI.

SUMMARY OF REGULATORY POSITION AND RATIONALE

Summary of Agency Position: The Agency is currently evaluating the potential human health risks resulting from the use of mancozeb to determine whether additional regulatory action is warranted on mancozeb and the other EBDC pesticides containing the common contaminant, degradation product, and metabolite, ETU. At this time, the Agency will not establish any new food use tolerances or register any significant new uses. The Agency is also specifying precautionary labeling as set forth below.

Unique Warning Statements Required on Labels:

Manufacturing-Use Products

"This pesticide is toxic to fish. Do not discharge effluent containing this product into lakes, streams, ponds, estuaries, oceans, or public water unless this product is specifically identified and addressed in an NPDES permit. Do not discharge effluent containing this product to sewer systems without previously notifying the sewage treatment plant authority. For guidance, contact your State Water Board or Regional Office of the EPA."

End-Use Products

Outdoor Use Products (other than aquatic food and seed treatment)

"This pesticide is toxic to fish. Drift and runoff from treated areas may be hazardous to aquatic organisms in neighboring areas. Do not apply directly to water or wetlands (swamps, bogs, marshes, and potholes). Do not contaminate water by cleaning of equipment or disposal of wastes."

Seed Treatment Products

"This pesticide is toxic to fish. Cover or incorporate spilled treated seed. Do not contaminate water by cleaning of equipment or disposal of wastes."

Aquatic Food Use Products (cranberry, wild rice, taro)

> "This pesticide is toxic to fish. Drift and runoff from treated areas may be hazardous to aquatic organisms in neighboring areas. Do not contaminate water by cleaning of equipment or disposal of wastes."

All Home Use Products

> "PROTECTIVE MEASURES: Always spray with your back to the wind. Wear long-sleeve shirt, long pants, and rubber gloves. Wash gloves thoroughly with soap and water before removing. Change your clothes immediately after using this product and launder separately from other laundry items before reuse. Shower immediately after use."

Home Use Products with Food Uses

> "Preharvest intervals on this label are specified so that pesticide residues will be at an acceptable level when the crop is harvested."

All Agricultural Products

> "After (sprays have dried/dusts have settled/vapors have dispersed, as applicable) do not enter or allow entry into treated areas until the 24-hour reentry interval has expired unless wearing the personal protective equipment listed on the label."

> "WORKER SAFETY RULES

> "Keep all unprotected persons, children, livestock and pets away from treated area or where there is danger of drift.

> "Do not rub eyes or mouth with hands. See First Aid (Practical Treatment Section).

> "PERSONAL PROTECTIVE EQUIPMENT - For Mixers, Loaders, Applicators and Early Reentry Workers.

> "HANDLE THIS PRODUCT ONLY WHEN WEARING THE FOLLOWING PROTECTIVE CLOTHING AND EQUIPMENT: a long-sleeve shirt and long pants or a coverall that covers all parts of the body except the head, hands, and feet; chemical resistant gloves; shoes, socks, and goggles or a face shield. During mixing and loading, a chemical resistant apron must also be worn.

"During application from a tractor with a completely
enclosed cab with positive pressure filtration, or
aerially with an enclosed cockpit, a long-sleeve
shirt and long pants may be worn in place of the
above protective clothing. Chemical resistant gloves
must be available in the cab or cockpit and worn
while exiting.

"IMPORTANT! Before removing gloves, wash them with
soap and water. Always wash hands, face, and arms
with soap and water before eating, smoking or drinking.
Always wash hands and arms with soap and water before
using the toilet.

"After work take off all clothes and shoes. Shower
using soap and water. Wear only clean clothes. Do
not use contaminated clothing. Wash protective
clothing and protective equipment with soap and water
after each use. Personal clothing worn during use
must be laundered separately from household articles.
Clothing and protective equipment heavily contaminated
or drenched with mancozeb must be destroyed according
to state and local regulations.

"HEAVILY CONTAMINATED OR DRENCHED CLOTHING CANNOT BE
ADEQUATELY DECONTAMINATED.

"During aerial application, human flaggers are
prohibited unless in totally enclosed vehicles."

SUMMARY OF MAJOR DATA GAPS

Study	Due Date - From Issuance of Standard
Product Chemistry	6-15 months

Residue Chemistry:
  Plant and animal metabolism
  Residue studies
	18-24 months[3]

Toxicology:
  Dermal sensitization
  Subchronic dermal
  Chronic toxicity (rodent and
    nonrodent) (mancozeb and ETU)
  Oncogenicity (rat and mouse)
  Teratology (rabbit)
  Reproduction (rat) (mancozeb and ETU)
  Mutagenicity (mancozeb and ETU)
  Dermal (percutaneous) absorption
    (mancozeb and ETU)
	9-50 months

---

[3] Dates are determined based on beginning of planting season
after issuance of Standard.

Study	Due Date - From Issuance of Standard

Ecological Effects:
  Avian dietary
  Avian reproduction
  Estuarine and marine organism
  Fish early life stage
  Aquatic Invertebrate life cycle
	9-18 months

Environmental Fate:
  Hydrolysis (mancozeb and ETU)
  Photodegradation (mancozeb and ETU)
  Soil metabolism (mancozeb and ETU)
  Aquatic metabolism (mancozeb and ETU)
  Leaching and adsorption/desorption (mancozeb and ETU)
  Volatility
  Degradation (mancozeb and ETU)
  Rotational crops
  Irrigated crops
  Fish accumulation
	9-50 months

Reentry	27 months

CONTACT PERSON AT EPA

Ms. Lois Rossi
Product Manager (Team 21)
Fungicide-Herbicide Branch
Registration Division (TS-767C)
Office of Pesticide Programs, EPA
Washington, D. C. 20460

Telephone: 703-557-1900

DISCLAIMER: The information presneted in this Pesticide Fact Sheet is for information purposes only and may not be used to fulfill data requirements for pesticide registration and reregistration.

METHIOCARB

Date Issued: March, 1987
Fact Sheet Number: 120

1. Description of Chemical

    Chemical Name: 4-methylthio-3,5-xylylmethylcarbamate

    Common Name : methiocarb

    Trade Names : mercaptodimethur, metmercapturnon, mesurol, methiocarbe, Bay 37344, and H-321

    OPP (Shaughnessy) Number: 100501

    Chemical Abstracts Service (CAS) Number: 2032-65-7

    Year of Initial Registration: 1972

    Pesticide Type: Insecticide, acaricide; molluscicide; and bird and rodent repellent

    Chemical Family: Carbamate

    U.S. Producer: Mobay Chemical Corporation

2. Use Patterns and Formulations

    Application Sites: Corn and sunflower fields, fruit, orchards, blueberries, ginseng, avocadoes, peppers, ornamentals, greenhouses, lawns and turf

    Types of Formulation: Dust, granular, wettable powder and pelleted/tableted.

    Types/Methods of Application: Soil-incorporated, foliar, aerial and broadcast.

3. Science Findings

Summary Science Statement

Technical methiocarb is highly acutely toxic by the oral route and of relatively low acute toxicity by the dermal route. Methiocarb's primary mechanism of toxicity is cholinesterase inhibition. Methiocarb sulfoxide, a cholinesterase-inhibiting metabolite of methiocarb was shown to be more acutely toxic than methiocarb in an acute rat toxicity test (7 mg/kg in female rats and 9 mg/kg in male rats). Results of cholinesterase studies suggest that methiocarb sulfoxide may be more toxic than methiocarb. A 30-day dog feeding study conducted with methiocarb sulfoxide is required to assess the cholinesterase inhibition of this metabolite. Available studies are not sufficient to complete the assessment of methiocarb and its metabolites. Data gaps exist for acute inhalation; 21-days dermal; mouse oncogenicity; reproduction; mutagenicity; and general metabolism studies. Methiocarb is very highly acutely toxic to avian species, to both coldwater and warmwater fish, and to freshwater invertebrates on an acute basis. It is slightly toxic to practically non toxic to avian species on a subacute dietary basis. Field testing, an avian repellency study and an aquatic residue monitoring study are required for completion of the Agency's assessment of the potential risk to avian and aquatic species.

Chemical Characteristics:

Physical State: Crystalline solid

Color: White

Odor: Slight mercaptan-like

Boiling Point: Not distillable

Melting Point: 121°C

Bulk Density: 35-40 lb/cu ft.

ph: N/A (not soluble enough)

Toxicology Characteristics:

Acute oral toxicity (rat)　　　　14-30 mg/kg; Toxicity Category I

Acute dermal toxicity (rabbits)　　> 2000 mg/kg; Toxicity Category III

Acute delayed neurotoxicity (hen)	Negative at 380 mg/kg
Chronic feeding (rat)	NOEL for cholinesterase inhibition = 67 ppm (3.35 mg/kg/day)
Chronic feeding (dog)	NOEL for cholinesterase inhibition = 5 ppm (0.125 mg/kg/day)
Oncogenicity (rat)	not oncogenic up to 600 ppm
Teratology (rat)	negative up to 10 mg/kg/day
Teratology (rabbit)	negative up to 10 mg/kg/day

Physiological and Biochemical Characteristics:

Metabolism and persistence in plants and animals:

Radiolabeled studies on the uptake, translocation and metabolism of methiocarb in plants show that methiocarb undergoes two routes of metabolic breakdown in plants. Methiocarb may be oxidized to the sulfoxide (MSO) and thereafter hydrolyzed to methiocarb sulfoxide phenol (MSOP). These metabolites may be further oxidized and hydrolyzed to yield methiocarb sulfone (MSO_2) or methiocarb sulfone phenol (MSO_2P). A secondary metabolic route is the hydrolysis of methiocarb to the phenol (MP). The metabolism of methiocarb in animals is not well understood. Available information again suggests two pathways of metabolism. The major route in both chickens and ruminants appears to be hydrolysis to methiocarb phenol (MP) followed by oxidation to methiocarb sulfoxide phenol (MSOP). Additional metabolism studies in ruminants and poultry are required.

Mechanism of pesticidal action: cholinesterase-inhibition

Environmental Characteristics:

Adequate data are not available to assess the environmental fate of methiocarb; data are not adequate to assess methiocarb's potential for contaminating groundwater.

Ecological Characteristics:

Avian acute oral toxicity:	mallard duck	12.8 mg/kg
Subacute dietary toxicity:	mallard duck; 1071 ppm and ring-necked pheasant; greater than 5000 ppm.	

Avian reproductive effects: mallard duck; negative at 100 ppm and bobwhite quail; negative at 50 ppm.

96-hour fish toxicity: rainbow trout; 0.436 ppm and bluegill sunfish; 0.734 ppm

48-hour fresh water invertebrate toxicity: Daphnia magna; .019 mg/kg

Tolerance Reassessment

Tolerances have been established for residue of methiocarb and its cholinesterase-inhibiting metabolites in or on blueberries (5.0 ppm), cherries (5.0 ppm), citrus fruits (0.02 ppm), corn (0.03 ppm), and peaches (15.0 ppm). The tolerances for blueberries and cherries are interim tolerances, due to expire on March 31, 1989. Refer to 40 CFR 180.320.

The food additive tolerances listed for methiocarb under 21 CFR 193.145 and 21 CFR 561.175 were temporary tolerances established to cover residues in or on grape food and feed items resulting from application of the pesticide to grapes under an experimental use permit. These tolerances and experimental use permit expired December 31, 1980.

Tolerances are expressed in terms of methiocarb and its cholinesterase - inhibiting metabolites.

Available data are not sufficient to conduct a full tolerance assessment. Data gaps exist for the residue analytical method; field residue studies and animal metabolism studies.

4. Required Unique Labeling and Regulatory Position Summary

Use classification - All outdoor commercial and agricultural uses have been classified restricted use under this Standard; all on the basis of avian toxicity; and all, with the exception of the corn seed treatment, on the basis of fish and aquatic species toxicity. This is an interim precautionary measure pending submittal and evaluation of the avian repellency and field studies, and the aquatic residue monitoring studies.

Endangered species labeling - Labeling for seed treatment to corn; bait application to corn; and golf course and sod farm uses will be required pending concurrence on proposed labeling by the Fish and Wildlife Service, U.S. Department of the Interior.

Reentry interval - A 24-hour reentry interval is imposed pending submittal and evaluation of reentry data. This reentry labeling restriction is being imposed for use of methiocarb on commercial turf; commercially grown ornamentals; agricultural crops (except seed treatment); and in greenhouses.

Tolerances - no additional permanent tolerances will be established pending the approval of a residue analytical method. Tolerances above 5.00 ppm cannot be toxicologically supported and the existing tolerance of 15.0 ppm in/on peaches will be revoked.

5. Summary of major data gaps:

Toxicology - generic	Date Due
Acute inhalation - rats	October 1987
21-day dermal	October 1987
30-day feeding (dog) methiocarb sulfoxide	October 1987 (acceptable protocol)
Mouse oncogenicity study	October 1990
Reproductive effects	December 1988
Mutagenicity testing	February 1988
General metabolism study	April 1989

Residue Chemistry

Livestock metabolism - (ruminants and poultry)	October 1988
Analytical methodology for plants and animals	July 1988
Data on levels of residues in: citrus fruits, cherries, blueberries, and corn	**
Tolerance proposals and/or residue data to support the following uses: avocados; pepper and sunflower seed crops; preplant application for agricultural crops, and non-bearing deciduous fruit trees.	**

** Submittal due dates for these data are contingent on the submittal and evaluation of the analytical methodology data.

Ecological Effects

Field testing for mammals and birds	October 1987 (acceptable protocol)
Avian repellency test	April 1988
Acute toxicity for estuarine and marine organisms	August 1987
Fish early life stage	November 1987
Aquatic invertebrate life cycle	November 1987
Aquatic residue monitoring	April 1988

Environmental Fate:

Hydrolysis	January 1988
Photodegradation - water and soil	May 1987
Soil metabolism - aerobic and anaerobic	July 1989
Leaching and adsorption/desorption	August 1987
Soil dissipation	November 1988
Confined rotational crop	November 1989
Fish accumulation	April 1988
Reentry studies	December 1987

6. **Contact Person at EPA:**

   William H. Miller, (PM-16)
   Insecticide-Rodenticide Branch (TS-767)
   401 M Street, SW.
   Washington, DC 20460
   Tel. No. (703) 557-2600

DISCLAIMER: The information presented in this chemical Information Fact Sheet is for informational purposes only and may not be used to fulfill data requirements for pesticide registration and reregistration.

METHYL BROMIDE

Date Issued: August 22, 1986
Fact Sheet Number: 98

1. Description of Chemical

    Generic Name: Bromomethane

    Common Name: Methyl Bromide

    Trade Names   Brom-O-Gas®; Celfume®; Dowfume®; Embafume®; Kayafume®; Meth-O Gas®; Terr-O-Gas 100®

    EPA/OPP Pesticide Chemical Code: 053201

    Chemical Abstracts Service (CAS) Number: 74-83-9

    Year of Initial Registration 1961

    Pesticide Type. Acaricide; Fungicide; Herbicide; Insecticide; Nematicide; Rodenticide

    Chemical Family   Halogenated Hydrocarbons

    U.S. and Foreign Producers: Great Lakes Chemical Corporation; Ethyl Corporation; Ameribrom, Inc.

2. Use Patterns and Formulations

    Application Sites: Agricultural crops; ornamentals; soil, manure, mulch and compost fumigation; stored commodities (both raw agricultural commodities and processed foods/feeds); greenhouses; homes; grain elevators; mills; ships and transportation vehicles.

    Types of Formulations: Gaseous, liquid under pressure, or liquid.

    Types and Methods of Application: Chisel application to field soil; gravity distribution for smaller bins; forced (recirculation) distribution systems; tarpaulin.

Application Rates: For stored product pests infesting raw agricultural commodities. dosage rates are between 2-6 lbs/1000 cu. ft. with exposure times ranging from 12-24 hours for nuts and grains and from 2-4 hours for other commodities, for processed foods, dosage rates are between 1-3 lbs/1000 cu. ft. with exposure times ranging from 12-24 hours: for soil fumigation uses, dosage rates are between 180-870 lbs/A depending on the type of application with exposure times ranging from 24-48 hours; for structural pest control, dosage rates are between 1-3 lbs/1000 cu. ft. with exposure times ranging from 2-24 hours for insects and 12-18 hours for mice and rats.

Usual Carriers: None

3. Science Findings

Although methyl bromide has been widely used for many years, its chronic toxicity has not been adequately characterized. Uncertainty also exists in the areas of dietary exposure, applicator exposure, and groundwater contamination.

Chemical Characteristics:

    Physical State. Gaseous, liquid under pressure

    Odor: Odorless

    Boiling Point 4°C

    Melting Point: -94°C

    Flash Point Nonflammable

    Unusual Handling Characteristics: Non corrosive to metal containers; however, traces of water or acid may lead to corrosion of application equipment

Toxicology Characteristics:

Acute Oral Toxicity: Toxicity Category II; LD_{50} in the rat is 214 mg/kg

Acute Inhalation Toxicity: Toxicity Category I based on human experience and use history; LD_{50} is 2700 ppm for a 30 minute exposure in the rat; in humans, 1,583 ppm (6.2 mg/l) was lethal to adults exposed for 10-20 hours while 7,890 ppm (30.9 mg/l) was lethal after 1 1/2 hours.

Primary Eye Irritation: Toxicity Category I; Corrosive

Primary Skin Irritation: Toxicity Category I; Corrosive

Major Route of Exposure: Inhalation

Problems which are known to have occurred with use of methyl bromide: In California, methyl bromide ranks seventh in terms of systemic poisonings, second in terms of number of people hospitalized and first in terms of number of days hospitalized for 1982-1985.

The chronic and subchronic toxicity data base is limited. A teratogenicity study in the rat and several mutagenicity studies are negative. Under the Data Call-In Program for Grain Fumigants, reproduction and oncogenicity studies were required. Preliminary results of these studies indicate that they are both negative. Mutagenicity, rabbit teratology, subchronic inhalation in the rat and rabbit, and chronic feeding studies in the rat and dog are required to complete the toxicology data base for methyl bromide.

Physiological and Biochemical Behavioral Characteristics: N/A

Environmental Characteristics:

Hydrolysis data indicate that methyl bromide breaks down at a rate of 1.4 mg/liter water/day at 25°C. Methyl bromide is not expected to run off fields to surface water because of application methods.

Ecological Characteristics:

Based on the registered patterns of use, no exposure to endangered species is expected.

Efficacy Review Results: N/A

Tolerance Assessments:

Tolerances are established at §180.123 of 40 CFR for residues of inorganic bromide (iBr) (calculated as Br) in or on a wide variety of agricultural commodities which have been fumigated with methyl bromide after harvest. Tolerances range from 5ppm - 240ppm, with the majority at 50ppm or less.

Tolerances are established at §180.199 of 40 CFR for residues of inorganic bromides (calculated as Br) in or on various raw agricultural commodities grown in soil fumigated with combinations of chloropicrin and methyl bromide.

Tolerances are established at §193.225 of 21 CFR for residues of inorganic bromides (calculated as Br) in or on milled fractions derived from cereal grains which have been fumigated with methyl bromide from all fumigation sources, including fumigation of grain-mill machinery, not to exceed 125 parts per million.

Tolerances are established at §193.230 of 21 CFR for residues of inorganic bromides (calculated as Br) in or on corn grits and cracked rice used in the production of fermented malt beverages which have been fumigated with methyl bromide. not to exceed 125 parts per million. Residues of inorganic bromides (calculated as Br) in the fermented malt beverage cannot exceed 25 parts per million.

Tolerances are established at §193.250 of 21 CFR for residues of inorganic bromide (calculated as Br) in or on the following processed food which have been fumigated with methyl bromide:

   400 parts per million in or on dried eggs and processed herbs and spices.

   325 parts per million in or on parmesan cheese and roquefort cheese.

   250 parts per million in or on concentrated tomato products and dried figs.

   125 parts per million in or on processed foods other than those listed above.

Tolerances are established at §561.260 of 21 CFR for residues of inorganic bromide (calculated as Br) in or on the following processed feed which have been fumigated with methyl bromide:

   400 parts per million for residues in or on dog food.

   125 parts per million for residues in or on milled fractions for animal feed from barley, corn, grain sorghum (milo), oats, rice, rye and wheat.

Data gaps exist for storage stability and the metabolism of methyl bromide in plant commodities fumigated postharvest. The requirements for animal metabolism data and livestock feeding studies are reserved pending the results of plant metabolism, storage stability, and plant residue data.

Pending submission of the requested plant metabolism data (and the conditionally required animal metabolism data), inorganic bromides and methyl bromide per se (MeBr per se) will tentatively be considered the only residues of concern resulting from both preplant soil and stored commodity fumigations. If the requested metabolism data and residue data on feed items so indicate, livestock metabolism and/or feeding studies involving iBr and MeBr per se may be required. Upon receipt of these studies, the Agency will determine the necessity and magnitude of tolerances in animal products for iBr, MeBr per se, and perhaps other metabolites of concern if found in metabolism studies.

None of the iBr tolerances in or on raw agricultural commodities (RACs) or processed products are supported due to the inadequacy of available data (40 CFR 180.123 and 180.199; 21 CFR 193.225, 193.250, and 561.260). In addition, data are required for each registered RAC and processed product depicting the residues of MeBr per se resulting from stored commodity fumigation (and possibly preplant soil fumigation). If both preplant and stored commodity fumigations are registered uses on a given commodity, then data are required depicting both iBr and MeBr per se residues resulting from the combination of the two types of treatment.

No crop group tolerances, as specified under 40 CFR 180.34(f), may be established at this time.

The TMRC for inorganic bromides is 25 mg/day based on a 1.5 kg diet. The ADI is 1.1 mg/kg/day (set by FDA); currently the TMRC accounts for 37.35 percent of the ADI.

4. Summary of Regulatory Position and Rationale

Methyl bromide is not being placed in the Special Review process at this time. Although the Agency is concerned about the acute toxicity risks associated with the use of methyl bromide, it believes that the precautionary labeling measures required by this standard, monitoring to establish safe levels for reentry to enclosed spaces and the addition of chloropicrin to formulations as a warning agent will significantly reduce these risks.

The Agency is requiring that chloropicrin at a concentration between 0.25% and 2.0% be present as a warning agent in all formulations, except those used for commodity fumigation. Since methyl bromide is odorless, the use of chloropicrin with its ability to cause painful iritation to the eyes, producing tearing and its disagreeable pungent odor at low concentrations will warn workers of methyl bromide exposure and will promote its safe use.

Methyl bromide products that are restricted according to 40 CFR 162.31 will continue to be restricted. Additionally, the Agency will propose to amend 40 CFR 162.31 to require restricted-use classification of all other product containers of methyl bromide. The restricted-use classification is necessary to protect users from the acute toxic effects of methyl bromide.

Because of the limitations of space on labels and the complexity of the precautions necessary to safely use methyl bromide, the Agency will require the development of application manuals by the manufacturers.

The Agency will continue to require that enclosed spaces fumigated with methyl bromide be aerated until the level of methyl bromide is below 5 ppm. This level was established under the Label Improvement Program for Fumigants, PR Notice 85-6, August 30, 1985, because the Agency was concerned about the possible inhalation exposure of workers.

Based on all available data, the Agency has determined that during soil fumigation the concentration of MeBr in the working area will not generally exceed 5 ppm as a time weighted average and will not require approved respiratory equipment to be worn. Such equipment is required to be available on the premises in case of a spill or leak.

The Agency will not impose a special label advisory statement for endangered species at this time because there is no expected exposure based on the registered patterns of use.

The Agency will require the submission of supporting data to support the current tolerances for mangoes, papayas, pomegranates, and cumin seed. There are no registered uses for these commodities, but the tolerances may be required for importation purposes. If the data are not developed, these tolerances may be revoked.

Due to the inadequency of available data, the Agency is requiring the submission of residue chemistry data to support all current inorganic bromide tolerances and to establish tolerances for methyl bromide per se.

Since none of the current methyl bromide tolerances are supported by data, no crop group tolerances, as specified under 40 CFR 180.34(f) may be established at this time. When data are received, crop grouping will be considered.

Since tolerances for methyl bromide per se are now being required, the Agency will propose to delete 40 CFR 180.3(c), the section which currently exempts the organic molecule from tolerances.

As soon as the required inorganic bromide data have been submitted, the Agency will consider deleting all existing paragraphs in the 21 and 40 CFR concerning inorganic bromide tolerances and replacing them with a single paragraph for raw agricultural commodities (40 CFR 180) and processed products (21 CFR 193 and 561).

Because of the residue and toxicology data gaps for methyl bromide, the Agency will not consider significant new food uses until the data are submitted.

While data gaps are being filled, currently registered manufacturing-use and end-use product containing methyl bromide may be sold, distributed, formulated, and used, subject to the requirements of the methyl bromide registration standard. The Agency does not normally cancel or withhold registration for previously registered use patterns simply because data are missing or inadequate. Data required under that Standard will be reviewed and evaluated, after which the Agency will determine if additional regulatory changes are necessary.

Unique Warning Statements: Restricted-Use for retail sale to and use only by Certified Applicators or persons under their direct supervision, and only for those uses covered by the Certified Applicator's certification. Applicators must wear loose cotton long sleeve shirts and pants, shoes and socks that are cleaned after each wearing. Gloves and boots should not be worn, as methyl bromide is heavier than air and may be trapped inside and cause skin injury. Labeling for end-use products intended for structural, transportation, space, or commodity fumigation require the use of a self-contained breathing apparatus (SCBA) or combination air-supplied/SCBA respirator if the concentration in the working area exceeds 5 ppm (20 mg/m3). Labeling for end use products intended for outdoor (soil fumigation) uses require approved respiratory equipment to be available in case of emergency situations but do not require SCBA under normal conditions of use.

5. Summary of Major Data Gaps

Residue Chemistry:

Plant Metabolism: Due August 1987
Plant Residue Analytical Methods: Due February 1988
Storage Stability Data: Due February 1988
Magnitude of the Residue Studies: Due August 1989

Environmental Fate:

Photodegradation in Water: Due May 1987
Aerobic Soil: Currently under review
Anaerobic Soil: Currently under review
Leaching and Adsorption/Desorption: Currently under review
Field Dissipation Studies-Soil: Currently under review

Toxicology:

90-Day Inhalation-Rat: Due November 1987
90-Day Inhalation-Rabbit: Due November 1987
Chronic Toxicity-Rat (Gavage): Due October 1990
Chronic Toxicity-Dog (Gavage): Due October 1990
Oncogenicity-Rat (Inhalation): Due January 1987
Oncogenicity-Mouse (Inhalation): Due January 1987
Oncogenicity-Rat (Gavage): Due October 1990
Oncogenicity-Rat (Gavage): Due October 1990
Teratogenicity-Rabbit: Due November 1987
Reproduction-Rat (Inhalation): Currently under review
Reproduction-Rat (Gavage): Due November 1989
Mutagenicity: Due August 1987

Reentry:

Soil Dissipation: Due November 1988
Inhalation Exposure: Currently under review

6. Contact Person at EPA

Walter C. Francis, PM Team (32)
Registration Division (TS-767C)
Disinfectants Branch
401 M Street, S.W.
Washington, D.C. 20460
Telephone: (703) 557-3964

DISCLAIMER: The information present in this Chemical Information Fact Sheet is for informational purposes only and may not be used to fulfill data requirements for pesticide registration and reregistration.

METHYL PARATHION

Date Issued: December 1986
Fact Sheet Number: 117

1. DESCRIPTION OF CHEMICAL

    Generic Name: 0,0-dimethyl-0-4-nitrophenyl phosphorothioate

    Common Name: Methyl Parathion

    Trade Names and other names: 0,0-dimethyl 0-(4-nitrophenyl) phosphorothioate; 0,0-dimethyl 0-(p-nitrophenyl) phosphorothioate; parathion-methyl; metaphos; Cekumethion; Devithion; dimethyl parathion; E601; Folidol M; Fosferno M50; Parataf; Paratox; Partron M; Penncap M; Tekwaisa; Wofatox; Metacide; Bladan M; Metron; Dalf; Nitrox 80.

    EPA Shaughnessy Code: 053501

    Chemical Abstracts Service (CAS) Number: 298-00-0

    Year of Initial Registration: 1954

    Pesticide Type: Insecticide

    Chemical Family: Organophosphate

    U.S. and Foreign Producers: Monsanto in the U.S., and Bayer, AG in West Germany, Chemiekombinat Bitterfield VEB in East Germany, and A/S Cheminova in Denmark.

2. USE PATTERNS AND FORMULATIONS

    Application Sites: Field, vegetable, tree fruit and nut crops, tobacco and ornamentals, forestry, aquatic food crops, mosquito abatement districts, terrestrial and non-crop sites.

    Pests controlled: A wide variety of insects and mites as well as tadpole shrimp.

    Types and methods of application: Usual appication is foliar. May be applied by aircraft or ground equipment.

Types of Formulations: Dusts, wettable powders, micro-encapsulated, emulsifiable concentrates, and ready-to-use liquid.

Rates: 0.1 to 6.0 lbs. a.i. per acre

Usual Carriers: petroleum solvents, clay carriers

3. SCIENCE FINDINGS

Summary Science Statement:

Methyl parathion is a Toxicity Category I organophosphate compound which is highly toxic to laboratory mammals, humans, aquatic invertebrates, and birds. There is some evidence that methyl parathion may effect reproductive success in birds. Methyl parathion poses a hazard to many endangered species. In laboratory rats of the Wistar strain, oncogenicity could not be determined as the data were insufficient; the Agency is requiring additional information on this study as well as a repeat of the mouse study. Chronic toxicity data indicate that methyl parathion causes retinal and sciatic nerve damage in rats at high dose levels (50 ppm in diet). Because data are not available, the Agency is unable to determine a no observed effect level (NOEL) for sciatic nerve damage.

Chemical characteristics:

Little information is available. Technical methyl parathion has a vapor pressure of 0.14 mg/m^3 at 20 C, and an octanol/water parition coefficient of 3300. Methyl parathion is soluble in most organic solvents and is slightly soluble in aliphatic hydrocarbons. This compound is practically insoluble in water.

Toxicology Characteristics:

Acute Toxicity:

Methyl parathion causes cholinesterase inhibition. It is highly toxic to mammals by all routes of exposure and is classified in Toxicity Category I (LD_{50} 4.5 to 16 mg/kg).

Major Routes of Exposure:

The major route of exposure is acknowledged to be dermal with inhalation, ocular, and oral exposure being much smaller.

Information from the California Department of Food and Agriculture reported incidents of worker poisonings and illnesses during mixing, loading and application of methyl parathion. EPA is requiring additional "Worker Safety Rules," including protective clothing, to reduce exposure.

Delayed Neurotoxicity:

Methyl parathion is not believed to cause delayed neurotoxicity

Chronic feeding/Oncogenicity studies:

The Agency has two 2 year chronic feeding/oncogenicity studies in the rat, one in the mouse and a one-year dog study.

The Agency is unable to definitively evaluate oncogenicity at this time; additional information is required in the Wistar rat strain and another mouse study is required. Chronic effects noted; retinal and sciatic nerve damage at high dose levels(50 ppm) was observed in the rat.

Subchronic Studies:

Subchronic feeding studies show cholinesterase as the primary target for the toxic action of methyl parathion. A NOEL was established in the rat at 2.5 ppm or 0.25 mg/kg/day. The NOEL in the dog was 0.3 mg/kg/day (this NOEL was used to establish the current PADI). However, additional subchronic studies in both the rat and dog are required to determine the NOEL for retinal and sciatic nerve damage in the rat and retinal damage in the dog.

Metabolism:

Data gap; additional data are required.

Teratogenicity:

Some evidence of embryotoxicity and fetotoxicity at 1.0 mg/kg in rats. However, maternal toxicity was not established. Additional data are required. No signs of developmental toxicity were noted in the rabbit.

Reproduction:

No reproductive effects were observed in rats at dietary levels up to 25 ppm. No additional information is required.

Mutagenicity:

The Agency has evaluated the reports of a number of assays which address the three major categories of alterations, i.e., 1)gene mutation, 2), structural chromosomal aberrations, and 3) other mechanisms of genotoxicity. Although results of several of the individual tests are negative, other tests in each of these major categories provide limited evidence that methyl parathion is genotoxic. No additional information is required.

Physiological and Biochemical Characteristics:

Methyl parathion acts by causing irreversible inhibition

of cholinesterase enzyme, allowing accumulation of acetylcholine at cholinergic neuroeffector junctions and autonomic ganglia. Poisoning symptoms include headaches, nausea, vomiting, cramps, weakness, blurred vision, pinpoint pupils, tightness in the chest, drooling or frothing of mouth and nose, muscle spasms, coma, and death. The mechanism of pesticidal action is not known.

Environmental Fate and Exposure:

Insufficient information is available for the analysis of the environmental fate and the exposure of humans and nontarget organisms to methyl parathion. Additional data are required.

Methyl parathion, it is believed, does not bioaccumulate.

Dermal, ocular, and inhalation exposure can occur during mixing, loading, and application, cleaning and repair of equipment, and during early reentry. EPA is requiring additional "Worker Safety Rules," including protective clothing, to reduce exposure.

Methyl Parathion, it is believed, has little or no potential to contaminate ground water. This chemical was not included on the list of potential ground water contaminators.

Ecological Characteristics:

Avian Oral Toxicity: 6.6 mg/kg for mallard duck and 7.6 mg/kg for bobwhite quail.

Avian Dietary Toxicity: 336 ppm for mallard duck and 90 ppm for bobwhite quail.

Small Mammal Oral Toxicity: 57 to 379 mg/kg for microtine rodents

Avian Reproduction: Laboratory studies showed no direct reproductive impairment; however, significant depression of brain cholinesterase activity was observed (These studies were conducted with the Penncap M formulation.) Field studies indicate the possibility of reproductive impairment. Effects on the survival of nestlings were also noted.

Freshwater Fish Acute Toxicity: 3.7 ppm for rainbow trout and 4.4 ppm for bluegill.

Aquatic Invertebrate Acute Toxicity: *Daphnia magna* 0.14 ppb

Marine and Estuarine Toxicity: Mysid shrimp 0.98 ppb
Sheepshead minnow 12,000 ppb

Endangered Species:

Previous consultations with the Office of Endangered Species have resulted in jeopardy opinions and labeling for crops (alfalfa, apples, barley, corn, cotton, pears, peanuts, sorghum, soybeans, and wheat), rangeland and pastureland, silvacultural sites, aquatic sites, and noncropland use. Labeling is required in an effort to reduce the risk to endangered species.

Tolerance Assessment:

Present United States, Canadian, Mexican and Codex tolerances for methyl parathion in or on raw agricultural commodities are specified in Table 1. Established tolerances for residues of methyl parathion are also listed in 40 CFR Sections 180.121 (a) and (b). Because there are considerable gaps in both residue chemistry and toxicology, a tolerance assessment cannot be made at this time. The nature of the residue in plants and animals is not adequately understood because of inadequate metabolism data. When the required data are submitted to the Agency, the following will be evaluated: 1. the tolerance definition in plants; 2. the need for and nature of tolerances in or on animal commodities.

Because data gaps prevent the formulation of an acceptable daily intake, a provisional acceptable daily intake has been established. The PADI for methyl parathion is 0.0015 mg/kg/day with a safety factor of 200. This figure will be retained until additional data are received. The theoretical maximum residue contribution for methyl parathion is approximately 800% of the provisional acceptable daily intake.

Reported Pesticide Incidents:

Most of the pesticide incidents reported involve illnesses during mixer/loading, application, and drift from target areas.

4. Summary of Regulatory Position and Rationale

A review of the data available indicates that no risk criteria listed in 40 CFR 154.7 have been met or exceeded for methyl parathion.

The Agency is requiring avian reproduction and terrestrial full field testing and simulated or full field aquatic testing to better define the extent of exposure and hazard to wildlife.

B. No new tolerances or new food uses will be considered until the Agency has received data sufficient to assess existing tolerances for methyl parathion.

C. The Agency is concerned about the potential for human poisonings (cholinesterase inhibition) from the use of methyl parathion. The Agency will continue to classify for restricted use (due to very high acute toxicity). The Certified applicator must be physically present during mixing, loading, application, equipment repair, and equipment cleaning. Information from the California Department of Food and Agriculture reported incidents of worker poisonings and illnesses during mixing, loading, and application. EPA is requiring more stringent "Worker Safety Rules", including protective clothing, to reduce exposure.

D. A 48 hour re-entry interval, previously established under 40 CFR 170.3 (b)(2) will remain in effect.

E. The Agency has concluded that data are not adequate to determine the oncogenic potential of methyl parathion. and is requiring another mouse study and additional information on the Wistar rat.

F. The Agency is requiring glove permeability and drift studies because of the high acute toxicity of methyl parathion.

G. All manufacturing-use products and end-use products must bear appropriate labeling as specified in 40 CFR 162.10. Additionally, the following information must appear on the labeling:

  a. Labeling requirements have been imposed to protect fish and wildlife (including endangered species).
  b. Methyl parathion will continue to be classified Restricted Use and the labeling must state the reason, "Due to very high acute toxicity". Certified applicator must be physically present during mixing, loading, application, repair and cleaning of equipment.
  c. Effluent containing methyl parathion may not be discharged into lakes, streams, ponds, estuaries, oceans or public waters unless this product is specifically identified in an NPDES permit. Discharge of effluent containing this product is forbidden without prior notice to the sewage treatment plant authority.
  d. Protective clothing requirements are mandatory in order to protect applicators, fieldworkers, mixer/loaders, and persons who clean and repair application equipment.
  e. During aerial application, human flaggers are strictly prohibited.

6. SUMMARY OF MAJOR DATA GAPS

Animal and plant metabolism studies
Magnitude of residue in almost all crops
Full battery of Environmental Fate data
Additional subchronic toxicity testing to determine a NOEL for cholinesterase inhibition and other systemic effects (retinal degeneration, sciatic nerve effects, abnormal gait)

Additional oncogenicity and teratogenicity information
Glove permeability and drift studies
Aquatic accumulation studies
Avian reproduction and terrestrial full field testing
Simulated or full field aquatic testing
Early life stage and fish life cycle studies
Reentry studies
Applicator Exposure Monitoring studies

7. **CONTACT PERSON AT EPA**

   Dennis Edwards
   Acting Product Manager (12)
   Insecticide-Rodenticide Branch
   Registration Division (TS-767C)
   Office of Pesticide Programs
   Environmental Protection Agency
   401 M Street, SW
   Washington, DC 20460

   Office Location and telephone number:

   Room 202, Crystal Mall #2
   1921 Jefferson Davis Highway
   Arlington, VA 22202
   (703) 557-2386

DISCLAIMER: The information presented in this Chemical Information Fact Sheet is for informational purposes only and may not be used to fulfill data requirements for pesticide registration and reregistration.

METHYL PARATHION
TABLE I
Summary of Present Tolerances

Commodity	United States	Canada	Mexico[a]	International (Codex)[b]
			Tolerances (ppm)	(MRL)
Garden Beets	1.0	—	—	—
Carrots	1.0	—	—	—
*Parsnips	1.0	—	—	0.7
Potatoes	0.1	—	0.1	—
Radishes	1.0	—	1.0	—
*Rutabagas	1.0	—	—	—
Sugar Beets	0.1	—	—	0.5
Sweet Potatoes	0.1	—	0.1	—
Turnip	1.0	—	—	—
*Garlic	1.0	—	1.0	—
Onions	1.0	—	1.0	—
Celery	1.0	—	1.0	—
*Endive	1.0	—	—	—
Lettuce	1.0	—	1.0	—
*Parsley	1.0	—	—	—
Spinach	1.0	—	1.0	—
*Swiss Chard	1.0	—	—	—
Broccoli	1.0	—	1.0	0.2
Brussels Sprouts	1.0	—	—	0.2
Cabbage	1.0	—	1.0	0.2
Cauliflower	1.0	—	—	0.2
Collards	1.0	—	—	0.2
Kale	1.0	—	—	0.2
Kohlrabi	1.0	—	—	0.2
Mustard Greens	1.0	—	—	0.2
Beans	1.0	—	1.0	0.2
*Guar Beans	0.2	—	1.0	—
Peas	1.0	0.7	1.0	0.7
*Lentils	1.0	—	1.0	—
Soybeans	0.1	—	0.1	—
Eggplant	1.0	—	1.0	—
Peppers	1.0	—	1.0	—
Tomatoes	1.0	—	1.0	0.2
Cucumbers	1.0	—	1.0	0.2
Melons	1.0	—	1.0	0.2
Pumpkins	1.0	—	1.0	0.2
Squash	1.0	—	1.0	—
*Summer Squash	1.0	—	1.0	0.2
*Citrus Fruits	1.0	—	1.0	1.2
Apples	1.0	—	1.0	—
Pears	1.0	—	1.0	—
*Quince	1.0	—	—	—
Apricots	1.0	—	—	0.2
Cherries	1.0	—	—	0.2
Nectarines	1.0	—	—	0.2
Peaches	1.0	—	1.0	0.2
Plums	1.0	—	—	0.2
*Blackberries	1.0	—	—	—
Blueberries	1.0	—	—	0.2

METHYL PARATHION
TABLE I
Summary of Present Tolerances (con't)

Commodity	Tolerances (ppm) United States	Canada	Mexico[a]	(MRL) International (Codex)[b]
*Boysenberries	1.0	—	—	0.2
*Cranberries	1.0	—	—	0.2
*Currants	1.0	—	—	—
*Dewberries	1.0	—	—	0.2
Gooseberries	1.0	—	—	0.2
Grapes	1.0	—	—	0.2
*Loganberries	1.0	—	—	0.2
*Raspberries	1.0	—	—	0.2
Strawberries	1.0	—	1.0	0.2
*Youngberries	1.0	—	—	0.2
Almonds	0.1	—	—	—
*Filberts	0.1	—	—	—
Pecans	0.1	—	—	—
*Walnuts	0.1	—	—	—
Barley	1.0	—	—	—
Corn	1.0	—	1.0	—
Oats	1.0	—	—	—
Rice	1.0	—	1.0	—
Rye	0.5	—	—	—
Sorghum	0.1	—	0.1	—
Wheat	1.0	—	—	—
Forage Grass	1.0	—	—	—
Alfalfa Forage	1.25	—	1.25	—
Alfalfa Hay	5.0	—	5.00	—
Clover Forage & Hay	1.0	—	—	—
*Trefoil	1.25	—	—	—
*Trefoil Hay	5.0	—	—	—
Vetch Forage & Hay	1.0	—	—	—
Miscellaneous Crops				
Artichokes	1.0	—	1.0	—
Avocados	1.0	—	1.0	0.2
Cottonseed	0.75	—	0.75	—
*Dates	1.0	—	—	0.2
*Figs	1.0	—	1.0	0.2
*Guavas	1.0	—	1.0	0.2
Hops	1.0	—	—	0.05
*Mangos	1.0	—	1.0	0.2
Mustard Seed	0.2	—	—	—
*Okra	1.0	—	1.0	—
*Olives	1.0	—	—	0.2
Peanuts	1.0	—	1.0	—
*Pineapple	1.0	—	1.0	0.2
*Rape Seed	0.2	—	—	—
*Sugarcane	0.1	—	0.1	—
Sunflower Seed	0.2	—	—	—

a = The U.S., Canadian, and Mexican tolerances expressed in terms of residues of methyl parathion per se. b = The Codex Maximum Residue Levels expressed as residues of methyl parathion and its oxygen analog, methyl paraoxon. * These commodities have tolerances but no Federal Registrations.

METOLACHLOR

Date Issued: January, 1987
Fact Sheet Number: 106

1. Description of chemical

    Generic name: 2-chloro-N-(2-ethyl-6-methylphenyl)
    N-(2-methoxy-1-methylethyl)acetamide
    Common name: Metolachlor
    Trade names: Dual, CGA-24705, Ontrack, Pennant
    EPA Shaughnessy code: 108801
    Chemical Abstracts Service (CAS) number: 51218-45-2
    Year of initial registration: 1976
    Pesticide type: Herbicide
    Chemical family: Chloracetanilide
    U.S. producer: Ciba-Geigy Corporation, Agricultural Division (from active ingredient manufactured outside the U.S.)

2. Use patterns and formulations

    Application sites: For preemergence control of certain broadleaf and grassy weeds in terrestrial crop areas (corn, sorghum, cotton, potatoes, peanuts, soybeans, green beans, kidney and other beans, blackeye peas and other peas, stone fruits and tree nuts) and terrestrial noncrop areas (ornamental plants, railroad, and highway rights-of-way).
    Types of formulations: 95% active ingredient (ai) technical grade manufacturing-use product. 5%, 15%, and 25% granulars, 86.4% emulsifiable concentrate, 27.5%, 31.8%, and 36.1% flowable concentrate with atrazine; 36.3% flowable concentrate with propazine; and 73.6% emulsifiable concentrate with metribuzin.
    Types and methods of application: End-use product is applied by ground spray equipment, aircraft, or through center pivot irrigation systems.
    Application rates: 1.25 to 4 lb metolachlor ai per acre (A) for terrestrial crop and noncrop areas.
    Usual carrier: Water, fluid fertilizers.

3. Science findings

    Summary science statement:

    Metolachlor is not considered to be teratogenic or cause reproductive effects. It is not oncogenic in mice, but is considered an oncogen in rats, and is tentatively classified as showing limited evidence of carcinogenicity in animals. Metolachlor is not mutagenic in available studies, but mutagenicity and metabolism testing requirements not in effect at the time of issuance of the orginal Metolachlor Registration Standard in 1980 must be met. Metolachlor has been found in several surface water surveys, some tapwater samples, and in ground water in two States. Monitoring studies are required to determine the extent of contamination on a national scale. Metolachlor is slightly to moderately toxic to nontarget organisms. Available data are insufficient to assess the environmental fate of metolachlor.

Chemical characteristics:

    Physical state: Liquid
    Color: White to tan
    Molecular weight: 238.8
    Boiling point: 100 °C at 0.001 mmHg
    Solubility: 530 ppm in water at 20 °C, miscible with xylene toluene, dimethyl formamide, methyl cellusolve, butyl cellusolve, ethylene dichloride, and cyclohexanone. Insoluble in ethylene glycol and propylene glycol.
    Vapor pressure: About 10^{-5} mmHg at 20 °C

Toxicological characteristics:

  Acute effects:

    Acute oral toxicity (rat): 2780 mg/kg
    (Toxicity Category III - moderately toxic)
    Acute dermal toxicity (rabbit): > 10,000 mg/kg
    (Toxicity Category III - moderately toxic)
    Acute inhalation toxicity: > 1.752 mg/L with 4-hour exposure
    (Toxicity Category IV - non toxic)
    Primary eye irritation: Non-irritating
    Primary dermal irritation: Non-irritating
    Dermal sensitization: Sensitizer in guinea pig

    Subchronic oral toxicity (dog): NOEL = 100 ppm (2.5 mg/kg) Decreased gain in body weight in males and females. Failure of the serum alkaline phosphatase to decrease with increased age, and possible effects on blood clotting systems at 300 and 1000 ppm.

  Chronic effects:

    3-generation reproduction (rat): NOEL = 300 ppm (15 mg/kg) Reduced pup weights and reduced parental food consumption at 1000 ppm (50 mg/kg)

Teratogenicity: Rabbits - Not fetotoxic or teratogenic Maternal toxicity at high dose (360 mg/kg/day)

Rats - Not fetotoxic or teratogenic. Decrease in food consumption at high dose (360 mg/kg/day) in first third of study.

Chronic feeding/oncogenicity: Mice - Not oncogenic in two studies up to and including 3000 ppm (429 mg/kg)
Rat - Systemic NOEL of 30 ppm (1.5 mg/kg). Systemic LEL of 300 ppm (testicular atrophy). In one study a statistically significant increase in primary liver neoplasms in females of high dose group (3000 ppm). In repeat study a statistically significant increased incidence of neoplastic liver nodules and proliferative hepatic lesions in females of the high dose group (3000 ppm)

Mutagenicity: Negative in an Ames Test, and a mouse dominant lethal test.

Major routes of exposure: Dermal, ocular, and inhalation from mixing concentrates and applying spray mixtures.

Physiological and biochemical behavior characteristics:

Absorption: Generally applied prior to plant emergence. Absorbed through shoots just above seed and possibly roots.

Mechanism of pesticidal action: Member of group of chloracetamide herbicides which are general growth inhibitors, especially of root elongation. Metolachlor may disrupt the integrity of plant cell membranes and inhibit lipid synthesis.

Environmental characteristics: Available data are insufficient to assess the environmental fate of metolachlor. There are indications that metolachlor is essentially stable in loamy sand soil over 64 days. Absorption constants in sandy clay loam, loam, and two sand soils indicate mobility in these soils. Aged metolachlor ^{14}C residue was mobile in columns of loamy sand soil. Metolachlor ^{14}C residues were mobile in sandy loams, sand, and silt loam soil. Metolachlor ^{14}C residues were detected in plants grown in metolachlor-treated soil. Metolachlor has been found in several surfacewater surveys. Transient peaks of 1.2 to 4.4 ppb are reported in riverwater possibly as a result of runoff during spring and summer. Detectable levels of metolachlor were found in some tapwater samples. It has been found in ground water in two States.

Ecological characteristics:

Avian oral toxicity: Mallard duck > 2510 mg/kg
Avian dietary toxicity: Mallard duck > 10,000 ppm
   Bobwhite quail > 10,000 ppm
Avian reproduction: Mallard duck - showed no impairment at any test level - 1,000 or 10,000 ppm
   Bobwhite quail - NOEL = 10 ppm, impairment at 300 ppm but no effect at 1000 ppm
Freshwater fish toxicity: Bluegill sunfish - 10.0 ppm
   Rainbow trout - 3.9 ppm
Aquatic invertebrates: Daphnia magna - 25.1 ppm
Fish life cycle: Fathead minnow - Maximum acceptable toxicant concentration > 0.78 < 1.6 ppm

Available data suggest that the hazard to nontarget organisms on an acute basis is slight to moderate.

Tolerance assessment:

Tolerances have been established for residues of metolachlor and its metabolites in raw agricultural commodities, milk, eggs, meat, fat, and meat byproducts [40 CFR 180.368 (a)] as follows:

Commodities	Parts Per Million
Almond hulls	0.3
Cattle, fat	0.02
Cattle, kidney	0.2
Cattle, liver	0.05
Cattle, meat	0.02

Commodities	Parts Per Million
Cattle, meat byproducts (mbyp) (except kidney and liver)	0.02
Corn, fresh (inc. sweet, kernel plus cob with husk removed)	0.1
Corn, forage and fodder	8.0
Corn, grain	0.1
Cottonseed	0.1
Eggs	0.02
Goats, fat	0.02
Goats, kidney	0.2
Goats, liver	0.05
Goats, meat	0.02
Goats, mbyp (except kidney and liver)	0.02
Hogs, fat	0.02
Hogs, kidney	0.2
Hogs, liver	0.05
Hogs, meat	0.02
Hogs, mbyp (except kidney and liver)	0.02
Horses, fat	0.02
Horses, kidney	0.2
Horses, liver	0.05
Horses, meat	0.02
Horses, mbyp (except kidney and liver)	0.02
Legume vegetables group foliage (except soybean forage and soybean hay)	15.0
Milk	0.02
Peanuts	0.5
Peanut, forage and hay	30.0
Peanut, hulls	6.0
Peppers, chili	0.5
Potatoes	0.2
Poultry, fat	0.02
Poultry, liver	0.05
Poultry, meat	0.02
Poultry, mbyp (except liver)	0.02
Safflower seed	0.1
Seed and pod vegetables (except soybeans)	0.3
Sheep, fat	0.02
Sheep, kidney	0.2
Sheep, liver	0.05

Commodities	Parts Per Million
Sheep, meat	0.02
Sheep, mbyp (except kidney and liver)	0.02
Sorghum, forage and fodder	2.0
Sorghum, grain	0.3
Soybeans	0.2
Soybeans, forage and hay	8.0
Stony fruits group	0.1
Tree nuts group	0.1

Tolerances have been established [40 CFR 180.368 (b)] for indirect or inadvertent residues of metolachlor as a result of application of metolachlor to the growing crops listed in 180.368 (a) as follows:

Commodities	Parts Per Million
Barley, fodder	0.5
Barley, forage	0.5
Barley, grain	0.1
Buckwheat, fodder	0.5
Buckwheat, forage	0.5
Buckwheat, grain	0.1
Millet, fodder	0.5
Millet, forage	0.5
Millet, grain	0.1
Milo, fodder	0.5
Milo, forage	0.5
Milo, grain	0.1
Oats, fodder	0.5
Oats, forage	0.5
Oats, grain	0.1
Rice, fodder	0.5
Rice, forage	0.5
Rice, grain	0.1
Rye, fodder	0.5
Rye, forage	0.5
Rye, grain	0.1
Wheat, fodder	0.5
Wheat, forage	0.5
Wheat, grain	0.1

Canadian tolerances of 0.1 ppm have been established for residues of metolachlor in or on beans, corn, peas, potatoes, and soybeans. No Mexican tolerances or Codex Alimentarius Commission Maximum Residue limits have been established for residues of metolachlor.

Results of tolerance assessment:

Using the NOEL of 30 ppm (1.5 mg/kg/day) from the rat chronic feeding study and a safety factor of 100, the acceptable daily intake (ADI) is 0.015 mg/kg/day, and the maximum permissible intake (MPI) is 0.9 mg/day for a 60 kg person. The established tolerances result in a total theoretical maximum residue concentration (TMRC) of 0.0755 mg/day (1.5 kg diet) which corresponds to 8.38 percent of the MPI for a 60 kg person.

4. Summary of regulatory positions and rationales

Unique label precautionary statements:

Manufacturing-Use Products

Environmental Hazards

Do not discharge effluent containing this product into lakes, streams, ponds, estuaries, oceans, or public water unless this product is specifically identified and addressed in an NPDES* permit. Do not discharge effluent systems without previously notifying the sewage treatment plant authority. For guidance, contact your State Water Board or Regional Office of the EPA.

End-Use Products

a. Environmental hazards for emulsifiable concentrates and flowable concentrates

Do not apply directly to water or wetlands (swamps, bogs, marshes, and potholes). Do not contaminate water by cleaning of equipment or disposal of wastes.

b. Environmental hazards for granules

Cover or incorporate granules that are spilled during loading or are visible on soil surface in turn areas. Do not contaminate water by cleaning of equipment or disposal of wastes.

c. Ground water and surface water advisory

Metolachlor has been identified in limited sampling of ground water and there is the possibility that it may leach through soils to ground water, especially where soils are coarse and ground water is near the surface. Following application and during rainfall events that cause runoff, metolachlor may

---

* National Pollution Discharge Elimination System

reach surface water bodies including streams, rivers and reservoirs.

Care must be taken when using this product to prevent back siphoning into wells, spills or improper disposal of excess pesticide, spray mixtures or rinsates.

Check valves or antisiphoning devices must be used on all mixing and/or irrigation equipment.

d. Crop rotation statement

Crops other than beans (succulent and dry), fresh corn, grain corn, cotton, peanuts, peas (succulent and dry), chili peppers, potatoes, safflower, sorghum, soybeans, stone fruits, tree nuts, and barley, buckwheat, millet, milo, oats, rice, rye, and wheat may not be planted in metolachlor-treated soil until 12 months after application

e. Endangered species statements for products registered for crop uses

It is a violation of Federal laws to use any pesticide in a manner that results in the death of an endangered species or adverse modification of their habitat.

The use of this product may pose a hazard to certain Federally designated endangered species known to occur in specific areas within the CALIFORNIA counties of Merced, Sacramento and Solano. Before using this product in these counties you must obtain the EPA Endangered Species Crop Bulletin. The bulletin is available from either your County Extension Agent, the Endangered Species Specialist in your State Wildlife Agency Headquarters, or the Regional Office of either the U.S. Fish and Wildlife Service (Portland, Oregon) or the U.S. Environmental Protection Agency (San Francisco, California). THIS BULLETIN MUST BE REVIEWED PRIOR TO PESTICIDE USE. THE USE OF THIS PRODUCT IS PROHIBITED IN THESE COUNTIES UNLESS SPECIFIED OTHERWISE IN THE BULLETIN.

5. Summary of data gaps (data required and due date)

Toxicology

Mutagenicity studies	October 1987
General metabolism	October 1988
Effects on coagulation	April 1987 **

** Date protocols are due. After acceptance, the Agency will provide timeframe for submission of the reports.

Environmental fate

Hydrolysis	July 1987
Photodegradation studies	July 1987
Metabolism studies	January 1989
Mobility studies	January 1989
Accumulation studies	January 1990
Ground and surface water monitoring	January 1987 **

Product chemistry/residue chemistry

Product chemistry	January 1988
Plant metabolism	April 1988
Storage stability	October 1988
Selected residue studies	April 1989

6. Contact person at EPA

Richard F. Mountfort
U.S. Environmental Protection Agency
TS-767C
401 M Street SW.
Washington, DC 20460
(703) 557-1830

DISCLAIMER: The information presented in this Pesticide Fact Sheet is for information purposes only and may not be used to fulfill data requirements for pesticide registration and reregistration.

** Date protocols are due. After acceptance, the Agency will provide time frame for submission of the reports.

METRIBUZIN

Date Issued: June 30, 1985
Fact Sheet Number: 53

1. Description of the Chemical

   Generic name: 4-amino-6(1,1-dimethylethyl)-3-(methylthio)-1,2,4-triazin-5(4H)

   Empirical formula: $C_8H_{14}N_4OS$

   Common name: Metribuzin

   Trade name: Sencor, Lexone

   Chemical Abstracts Service (CAS) Registry Number: 21097-64-9

   Office of Pesticide Program's EPA Chemical Code Number: 101101

   Year of Initial Registration: 1973

   Pesticide Type: Herbicide

   Chemical Family: S-triazine

   U.S. Producer: Mobay Chemical Corporation

2. Use Patterns and Formulations

   Application sites:

   Metribuzin is registered for control of broadleaf weeds and grasses in soybeans, potatoes, barley, winter wheat, dormant established and sainfoin fields, asparagus, sugarcane, tomatoes, lentils, peas, and non-cropland.

   Type of formulation:

   Metribuzin is available as a 50 percent formulation intermediate, 94 percent technical for formulation into end-use products, wettable powder, flowable concentrate and dry flowable concentrate.

   Types and methods of application:

   Metribuzin may be soil incorporated, surface applied or applied foliarly, broadcast or band with ground equipment. It can be applied by aerial equipment or sprinkler irrigation (potatoes).

Application Rates:

0.25 to 4.0 ai/A on crop sites,
6.0 to 8.0 ai/A on railroad rights-of-way.

Usual Carrier:

Water

3. **Science Findings** (Rationale for Regulatory Position)

Summary Science Statement:

Metribuzin is not acutely toxic by oral, dermal, inhalation, or eye irritation routes of exposure. The available data do not indicate that any of the risk criteria listed in 162.11(a) of Title 40 of the U.S. Code of Federal Regulations have been met or exceeded for the uses of metribuzin at the present time. Data gaps include rat chronic study, teratology study, multigeneration reproduction study and two categories of mutagenicity testing. There are also extensive residue chemistry data gaps.

Metribuzin has been found in Ohio rivers and Iowa wells. Although there are extensive data gaps in the area of environmental fate, available data indicate that metribuzin has a potential to contaminate ground water in soils lower in organic matter and clay content.

Available data indicate that metribuzin is moderately toxic to upland bird species on an acute oral basis, no more than slightly toxic to birds in the diet, moderately toxic to freshwater fish and invertebrates. Metribuzin is slightly toxic to shrimp. A detailed ecological hazard assessment cannot be made until the acute dietary study on an upland gamebird, acute toxicity studies on a marine/estuarine fish species and an oyster species, and appropriate environmental fate data are fulfilled.

Chemical Characteristics:

Metribuzin is a solid at room temperature. Its molecular weight is 214.28. The melting point is 125.5 to 126.5 °C. Metribuzin is soluble in aromatic and chlorinated hydrocarbon solvents, and in water (at 20 °C) to 1220 ppm.

Toxicological Characteristics:

Acute toxicity effects of metribuzin are as follows:

Acute Oral Toxicity in rats: 2200 mg/kg body weight for males; 2345 mg/kg body weight for females, Toxicity Category III.

Acute Dermal Toxicity in rats: 20,000 mg/kg body weight, Toxicity Category IV.

Acute Inhalation LC_{50}-rat: > 20 mg/L/1 hour, Toxicity Category IV.

Skin Irritation in rabbits: PIS = 0.33/8.0, Toxicity Category IV.

Eye Irritation in rabbits: Not an irritant, Toxicity Category IV.

Dermal Sensitization in guinea pig: Not a sensitizer, Toxicity Category IV.

Subchronic and Chronic Effects:

The 2-year dog study indicated dogs dosed with 1500 ppm (37.5 mg/kg) had reduced weight gain, increased mortality, hematological changes and liver and kidney damage. The no-effect level is 100 ppm. The oncogenic potential of metribuzin is unclear at this time. The mouse oncogenicity study is negative for oncogenic effects. The chronic rat study indicates a statistically significant ($p < 0.05$) increase in the incidence of adenoma of liver bile duct and pituitary gland in females at the 300 ppm dose level. Additional histopathology and historical control data on the incidence of these tumors in this particular strain of rats are needed before it can be determined if the increase is compound related. A teratology study in rabbits indicated no evidence of teratogenic effects at 135 mg/kg/day, the highest dose tested (HDT) and a NOEL of 15 mg/kg/day for maternal and fetal toxicity. Data gaps include rat chronic study, rat teratology study and multigeneration reproduction study.

Mutagenic Effects:

Available data indicate that metribuzin is not mutagenic. Data gaps exist in two categories of mutagenicity testing, specifically gene mutation studies in mammalian cells and tests for primary DNA damage such as sister chromatid exchange or unscheduled DNA synthesis assay.

N-Nitroso Contaminants:

Available data do not provide grounds for concern at this time. The data are incomplete. The analysis for N-nitroso contaminants is requested.

Major Routes of Human Exposure:

Primary nondietary exposure to the farmer is expected to be dermal and to occur during mixing, loading, and application. Exposure through ocular, inhalation and ingestion routes are also expected.

PHYSIOLOGICAL AND BIOCHEMICAL BEHAVIOR CHARACTERISTICS:

Adsorption Characteristics:

Metribuzin is absorbed through the leaves from surface treatment, but the major and significant route for uptake is via the root system.

Translocation Characteristics:

Uptake through the roots is best described as osmotic diffusion. Metribuzin is translocated upward in the xylem and moves distally when applied at the base of the leaves. It concentrates in the roots, stems, and leaves.

Mechanism of Pesticidal Action:

Photosynthesis inhibitor.

Metabolism in Plants:

The major routes of detoxification are the action of oxidation and conversion to water soluble conjugated products.

ENVIRONMENTAL CHARACTERISTICS:

Adsorption and Leaching in Basic Soil Types:

Metribuzin is moderately adsorbed on soils with high clay and/or organic matter content. Metribuzin is readily leached in sandy soils low in organic matter content.

Microbial breakdown:

Microbial breakdown appears to be the major mechanism by which metribuzin is lost from soils. Breakdown occurs fastest under aerobic conditions and at comparatively high temperatures.

Loss from photodecomposition and/or volatilization:

Slight loss.

Average persistence at recommended rates:

Half-life varies with soil type and climatic conditions. Half-life of metribuzin at normal use rates is 1 to 2 months.

Potential Ground Water Problem:

Metribuzin has been found in Ohio rivers and Iowa wells. Available data show that metribuzin has a potential to contaminate ground water in soils low in organic and clay content. The Agency is requesting water monitoring studies on metribuzin and has determined that all uses of metribuzin should be classified for "restricted use" with appropriate labeling including a ground water advisory statement.

Ecological Characteristics:

Avian Acute Oral Toxicity: 169.22 mg/kg (moderately toxic).

Subacute Dietary Toxicity: > 4000 ppm for mallard duck and bobwhite quail (slightly toxic).

Acute Toxicity on Freshwater Invertebrate: 4.18 ppm (moderately toxic).

Acute Toxicity on Fish: 76.78 ppm for rainbow trout (slightly toxic), 75.96 ppm for bluegill sunfish (slightly toxic).

96-Hour LC_{50} on a Marine/Estuarine Shrimp: 48.3 mg/l (slightly toxic).

Potential Problem for Endangered Species:

The Agency evaluated metribuzin under the cluster/use pattern approach for use on corn, soybeans, and small grains. Available data indicate that metribuzin use on crops would probably not affect Federally listed animal species.

Consultation with Office of Endangered Species (OES) on use of sulfometuron methyl indicated several species of endangered plants which occur on or adjacent to rights-of-way would be jeopardized by exposure from its use. The Agency has concluded that these plants would be jeopardized by exposure to metribuzin. The Agency is imposing a statement concerning endangered plant species on all end-use products containing metribuzin and labeled for use on rights-of-way.

Tolerance Assessment:

The Acceptable Daily Intake (ADI) is based on a no-observable effect level of 100 ppm (2.5 mg/kg) from the 2-year dog study. Using a 100-fold safety factor the ADI is 0.025 mg/kg/day with a Maximum Permissible Intake (MPI) of 1.5 mg/kg for a 60 kg adult human. Theoretical maximum residue contribution (TMRC) for metribuzin based on established tolerances is 0.3508 mg/day for a 1.5 kg diet. Currently, the permanent tolerances utilize 23.39 percent of the ADI.

The Agency is unable to complete a full tolerance reassessment because the available metribuzin toxicology and residue data do not fully support the established tolerances listed below. The metabolism of metribuzin in animals is not fully understood. Therefore, the Agency is requiring data on metabolism of metribuzin and related metabolites in ruminants, poultry, and several crops. The additional data will be used to assess dietary exposure to metribuzin and may lead to revisions in the existing tolerances.

Commodities	Parts Per Million
Alfalfa, green	2.0
Alfalfa, hay	7.0
Asparagus	0.05
Barley, grain	0.75
Barley, straw	1.0
Cattle, fat	0.7
Cattle, mbyp	0.7
Cattle, meat	0.7
Corn, fodder	0.1
Corn, forage	0.1
Corn, fresh (inc sweet K + CWHR)	0.05
Corn, grain (inc popcorn)	0.05
Eggs	0.01
Goats, fat	0.7
Goats, mbyp	0.7
Goats, meat	0.7
Grass	2.0
Grass, hay	7.0
Hogs, fat	0.7
Hogs, mbyp	0.7
Hogs, meat	0.7
Horses, fat	0.7
Horses, mbyp	0.7
Horses, meat	0.7
Lentils (dried)	0.05
Lentils, forage	0.5
Lentils, vine hay	0.05
Milk	0.05

Commodities	Parts Per Million
Peas	0.1
Peas (dried)	0.05
Peas, forage	0.5
Peas, vine hay	0.05
Potatoes	0.6
Poultry, fat	0.7
Poultry, mbyp	0.7
Poultry, meat	0.7
Sainfoin	2.0
Sainfoin, hay	7.0
Sheep, fat	0.7
Sheep, mbyp	0.7
Sheep, meat	0.7
Soybeans	0.1
Soybeans, forage	4.0
Soybeans, hay	4.0
Sugarcane	0.1
Tomatoes	0.1
Wheat, forage	2.0
Wheat, grain	0.75
Wheat, straw	1.0

Food	Parts Per Million
Barley, milled fractions (except flour)	3.0
Potatoes, processed (inc potato chips)	3.0
Sugarcane molasses	2.0
Wheat, milled fractions (except flour)	3.0

Feed	Parts Per Million
Barley, milled fractions (except flour)	3.0
Potato waste, processed (dried)	3.0
Sugarcane bagasse	0.5
Sugarcane molasses	0.3
Tomato pomace, dried	2.0
Wheat, milled fractions (except flour)	3.0

International Tolerances - Canada

Commodities	Parts Per Million
Asparagus	0.1
Barley grain	0.1
Lentils	0.1
Peas	0.1
Potatoes	0.1
Soybeans	0.1
Tomatoes	0.1
Wheat grain	0.1

Although the above Canadian tolerances differ from those in the United States, it is inappropriate for the Agency to harmonize these tolerances at the present time because of extensive toxicology and residue chemistry data gaps.

There are no tolerances for residues of metribuzin in Mexico or Codex Alimentarius.

Problems Known to Have Occurred With Use:

The Pesticide Incident Monitoring System (PIMS) does not indicate any incident involving agricultural uses of metribuzin.

4. Summary of Regulatory Position

Based on the review and evaluation of all available data and other relevant information on metribuzin, the Agency has made the following determinations.

The available data do not indicate that any of the risk criteria listed in 162.11(a) of Title 40 of the U.S. Code of Federal Regulations have been met or exceeded for the uses of metribuzin at the present time.

The Agency will not allow any significant new uses to be established for metribuzin until the toxicology and residue chemistry deficiencies identified have been satisfied.

The Agency is requesting data on presence of nitroso-contaminants in metribuzin. Available data do not provide grounds for concern at this time.

Based on concern for ground water contamination, the Agency has determined that all uses of metribuzin should be classified as restricted use and carry appropriate labeling including a ground water advisory statement.

The agency is concerned about the exposure of endangered/ threatened plant species occurring on or asjacent to rights-of-way from the use of metribuzin. An Endangered Species statement is being required on the labeling.

The Agency is imposing a rotatonal crop restriction. The extent of this restriction will be reconsidered when additional data are recieved.

Specific Label Precautionary Statements:

Hazard Information:

The Human Precautionary Statements must appear on all MP labels as precribed in 40 CFR 162.10.

Environmental Hazards Statements:

All manufacturing-use products (MP's) intended for formulation into end-use products (EP's) must bear the following statements:

> Do not discharge effluent containing this product into lakes, streams, ponds, estuaries, oceans or public water unless this product is specifically indentified and addressed in an NPDES permit. Do not discharge effluent containing this product to sewer systems without previously notifying the sewage treatment plant authority. For guidance contact your State Water Board or Regional Office of EPA.

All end-use products with outdoor uses must bear the following statement.

> Do not apply directly to water or wetlands. Do not contaminate water by cleaning of equipmet or disposal of waste.

Ground Water Statement:

All end-use products (EP's) must be classified as "RESTRICTED USE" (Refer to 40 CFR 162.10(j)(2)(B)) and the labels must bear the following ground water advisory.

> Metribuzin is a chemical which can travel (seep or leach)though soil and can contaminate ground water which may be used as drinking water. metribuzin has been found in ground water as result of agricultural use. Users are advised not to apply metribuzin where the water table (ground water) is close to the surface and where

soils are very permeable; i.e., well drained soils such as loamy sands. Your local agricultural agencies can provide further information on the type of soil in your area and the location of ground water.

Endangered Species:

Notice: The use of this product on rights-of-way may pose a hazard to certain Federally designated endangered plant species. They are known to be found in specific areas within the locations noted below. Prior to making applications, the user of this product must determine that no such species are located in or immediately adjacent to the area to be treated. For information on protected species contact the Endangered Species Specialist of the appropriate Regional Office of the U.S. Fish and Wildlife Service listed below:

Region 1-Portland, Oregon
    California counties of Contra Costa, Solano, San Diego, Santa Barbara, Ventura, Los Angeles and Orange.
    Idaho, Idaho County.
    Oregon, Harney County.
Region 2-Albuquerque, New Mexico
    Arizona counties of Coconino and Navajo.
    New Mexico counties of San Juan, Otero, Chaves, Lincoln, Eddy and Dona Ana.
    Texas counties of El Paso, Pecos and Runnels.
Region 3-Twin Cities, Minnesota
    Iowa counties of Allamakee, Clayton, and Jackson.
Region 4-Atlanta, Georgia
    Florida counties of Clay, Gulf, Gadsden, Franklin and Liberty.
    Georgia counties of Wayne and Brantley.
    North Carolina, Henderson County.
    South Carolina, Greenville County.
Region 5-Newton Corner, Massachusetts
    New York, Ulster County.
Region 6-Denver, Colorado
    Utah counties of Emery, Piute, Garfield, Washington, Utah and Wayne.
    Colorado counties of Montezuma, Delta and Montrose.

Restrictions on Rotational Crops

Do not plant food and feed crops other than those which are registered for use on metribuzin treated soils.

METSULFURON METHYL

Date Issued: March 28, 1986
Fact Sheet Number: 71

1. Description of chemical

    Generic name: Methyl-2-]]]](4-methoxy-6-methyl-1,3,5-triazin-2-yl)
    amino]carbonyl]amino]sulfonyl]benzoate
    Common name: Metsulfuron methyl
    Trade name: DuPont Ally Herbicide, DuPont Escort Herbicide
    EPA Shaughnessy code: 122010
    Chemical Abstracts Service (CAS) number: 74223-64-6
    Year of initial registration: 1986
    Pesticide type: Herbicide
    Chemical family: Sulfonyl urea
    U.S. and foreign producers: E. I. du Pont de Nemours & Company

2. Use patterns and formulations

    Application sites: Wheat and barley, fallow or noncropland

    Formulations: 60% dry flowable (DF)

    Types and method of application: Foliar-postemergence application, soil application to fallow land. Application may be made by ground or aerial equipment.

    Application rates: 1/10 ounce active ingredient/acre (oz ai/A)
    for wheat and barley
    1/3 to 2 oz/ai/A for noncropland

3. Science findings

    Summary science statement: All data are acceptable to the Agency. It has low toxicity (Category III) for acute dermal and primary eye irritation and is even less toxic (Category IV) for all other forms of acute toxicity. It was not oncogenic to rats or mice, not teratogenic to rats or rabbits, and not mutagenic with Salmonella or unscheduled DNA synthesis in rats. The pesticide is foliarly and root absorbed, and translocated by the plant. It works by inhibition of cell division in the shoots or roots. The pesticide will leach in some soils. Metsulfuron methyl is practically nontoxic to birds, fish, invertebrates, and honeybees. The nature of the residue in plants and animals is adequately understood and adequate methodology is available for enforcements of tolerances in barley and wheat grain; green forage of barley and wheat; hay of barley and wheat; straw of barley and wheat; meat, fat, and meat byproducts of cattle, goats, hogs, horses, and sheep, and milk.

Chemical Characteristics

    Physical state: solid
    Color: Technical - white to pale yellow
           60% DF formulation - off white
    Odor: Technical - faint, sweet ester-like
           60% DF formulation - odorless
    Melting point: 158 °C
    Specific gravity: 1.47 grams per cubic centimeter
    Molecular weight: 381.40

Solubility: In distilled water at 25 °C = 109 milligrams/liter (mg/L).
In 0.05N sodium phosphate buffer at 25 °C as a function of pH:

pH 6.7	9500 mg/L
pH 5.4	1750 mg/L
pH 4.6	270 mg/L

In organic solvents at 20 °C

n-hexane	0.79 mg/L
methylene chloride	121000 mg/L
acetone	3600 mg/L
methanol	7300 mg/L
ethanol	2300 mg/L
xylene	580 mg/L

Unusual Handling Characteristics: No special handling needed.

Toxicology Characteristics:

  Acute Studies:

  Acute Oral Toxicity (Rat)
    Greater than (>) 5000 milligrams/kilogram (mg/kg)
    Toxicity Category IV[1]/

  Acute Dermal Toxicity (Rabbit)
    > 2000 mg/kg
    Toxicity Category III[1]/

  Primary Dermal Irritation
    Slightly Irritating
    Toxicity Category IV[1]/

Primary Eye Irritation
  Corneal opacity in one unwashed eye at 24 hours. Cleared in
    48 hours.
  Toxicity Category III[1]

Dermal Sensitization (Guinea pig)
  No sensitization shown
  Toxicity Category IV[1]

[1]Label statements for Toxicity Category III for acute dermal: "Harmful if absorbed through skin. Avoid contact with skin, eyes, or clothing. Wash thoroughly with soap and water after handling.
Toxicity Category III for primary eye irritation: "Causes (moderate) eye injury (irritation). Avoid contact with eyes or clothing. Wash thoroughly with soap and water after handling.
Toxicity Category IV: No precautionary labeling statements required.

Major Routes of Exposure:

The major routes of exposure are through dermal and eye contact.

Chronic Studies:

2-Year Feeding/Oncogenicity (Rat)
  Systemic no-observable-effect level (NOEL) = 500 parts per million (ppm)
  Systemic lowest effect level (LEL) = 5000 ppm
  No oncogenic effects noted at 5000 ppm [highest dose tested (HDT)].

18-Month Oncogenicity (Mice)
  Systemic NOEL = 5000 ppm
  No oncogenic effects noted at 5000 ppm (HDT)

1-Year Feeding (Dog)
  NOEL = 50 ppm
  LEL = 500 ppm

Teratology (Rat)
  Maternal toxic NOEL less than (<) 40 mg/kg/day
  Fetal toxic NOEL < 1000 mg/kg/day
  No teratogenic effects at 1000 mg/kg/day (HDT)

Teratology (Rabbit)
  Maternal toxic NOEL = 25 mg/kg/day
  Maternal toxic LEL = 100 mg/kg/day
  Fetotoxic NOEL   > 700 mg/kg/day
  No teratogenic effects at 700 mg/kg/day (HDT)

2-Generation Reproduction (Rat)
  Fetotoxic NOEL > 5000 ppm
  Maternal NOEL = 500 ppm
  Maternal LEL = 5000 ppm
  No reproductive effects at 5000 ppm (HDT)

Mutagenicity-Salmonella typhimurium - negative
Mutagenicity-chromosomal aberrations (CHO) in vitro - aberrations in
   Chinese hamster chromosomes were caused with and without activation.

Mutagenicity-unscheduled DNA synthesis (rat) - negative - not mutagenic

Metabolism (Rat)
   Rapid elimination of radioactivity: 91% or over 96 hours

Physiological and Biochemical Behavioral Characteristics

   Foliar absorption - rapid
   Translocation - systemic after absorption by foliage or roots.
   Mechanism of pesticidal action - inhibition of cell division
      in rapidly growing tips of roots and shoots by inhibition of amino
      acid synthesis.
   Potential to contaminate ground water: Metsulfuron methyl has the
      potential to contaminate ground water at very low concentrations.
   Metabolism in plants and animals: metabolized to several nonherbicidal
      compounds.
   Persistence in plants and animals: does not persist in either animals
      or plants.

Environmental Characteristics

   Absorption and leaching in basic soil types: moderately to very
      mobile depending on organic matter content and soil texture.
   Leaches through silt loam and sand soils.

   Microbial breakdown: degrades under anaerobic conditions to lower
      molecular weight compounds.
   Loss from photodecomposition and volatilization - none.
   Bioaccumulation - does not accumulate in fish.
   Resultant average persistence - half-life of 120 to 180 days in silt
      loam soil.

Ecological Characteristics

   Avian Acute Oral Toxicity (mallard duck) > 2150 mg/kg
   Avian 8-day Dietary Toxicity (mallard duck) > 5620 ppm
   Avian 8-day Dietary Toxicity (bobwhite quail) > 5620 ppm

   Fish Acute 96-hour Toxicity (rainbow trout) > 150 ppm
   Fish Acute 96-hour Toxicity (bluegill sunfish) > 150 ppm

   Invertebrate 48-hour Acute Toxicity (Daphnia magna) > 150 ppm

   Honeybee Acute Toxicity (Apis) > 25 µg/bee

   Available data indicate metsulfuron methyl is practically nontoxic
   to birds, fish, invertebrates, and honeybees.

Tolerance Assessment

The nature of the residue in plants and animals is adequately understood, and adequate analytical methods are available for enforcement purposes.

Tolerances are established for the combined residues of the metsulfuron methyl (methyl 2-[[[[(4-methoxy-6-methyl-1,3,5-triazin-2-yl)amino]carbonyl]amino]sulfonyl]benzoate) and its metabolite, methyl 2-[[[[(4-methoxy-6-methyl-1,3,5-triazin-2-yl)amino]carbonyl]amino]sulfonyl]-4 hydroxybenzoate, in or on the following raw agricultural commodities:

Commodities	Parts Per Million
Wheat, grain	0.05
Wheat, green forage	5.0
Wheat, hay	20.0
Wheat, straw	0.1
Barley, grain	0.05
Barley, green forage	5.0
Barley, hay	20.0
Barley, straw	0.1

Tolerances are established for residues of metsulfuron methyl (methyl-2-[[[[(4-methoxy-6-methyl-1,3,5-triazin-2-yl)amino]carbonyl]amino]sulfonyl]benzoate) in or on the following raw agricultural commodities:

Commodities	Parts Per Million
Cattle, fat	0.1
Cattle, meat	0.1
Cattle, meat byproducts	0.1
Goat, fat	0.1
Goat, meat	0.1
Goat, meat byproducts	0.1
Hogs, fat	0.1
Hogs, meat	0.1
Hogs, meat byproducts	0.1
Horses, fat	0.1
Horses, meat	0.1
Horses, meat byproducts	0.1
Milk	0.05
Sheep, fat	0.1
Sheep, meat	0.1
Sheep, meat byproducts	0.1

The acceptable daily intake (ADI) based on the 1-year dog feeding study (NOEL of 1.25 mg/kg/day) and using a hundredfold safety factor is calculated to be 0.0125 mg/kg/day. The maximum permissible intake (MPI) for a 60-kg human is calculated to be 0.75 mg/kg/day. The theoretical maximum residue contribution (TMRC) for these tolerances for a 1.5 kg diet is calculated to be 0.0455 mg/kg/day (1.5 kg). The current action will use 6.06 percent of the ADI. There are no published tolerances for this chemical.

4. SUMMARY OF REGULATORY POSITION AND RATIONALE

The Agency has accepted the use of metsulfuron methyl for control of weeds in wheat and barley. All major data requirements have been filled.

5. SUMMARY OF MAJOR DATA GAPS

None.

6. CONTACT PERSON AT EPA

Robert J. Taylor
Office of Pesticide Programs, EPA
Registration Division (TS-767C)
401 M Street, SW.
Washington, DC 20460
Phone (703) 557-1800

DISCLAIMER: The information presented in this Pesticide Fact Sheet is for informational purposes only and may not be used to fulfill data requirements for pesticide registration and reregistration.

MONOCROTOPHOS

Date Issued: September 30, 1985
Fact Sheet Number: 72

1. Description of Chemical

    Generic Name: Dimethyl phosphate of 3-hydroxy-N-methyl-cis-crotonamide

    Common Name: Monocrotophos

    Trade Names: Azodrin; Apodrin; Bilobron; Crisodrin; Glore Phos 36; Hozodrin; Monocil 40; Monocron; Nuracron; Pillardrin; Plantdrin; Susrin; and Ulvair.

    EPA Shaughnessy Code: 058901

    Chemical Abstracts Service (CAS) Number: 6923-22-4

    Year of Initial Registration: 1965

    Pesticide Type: Insecticide/acaracide

    Chemical Family: Organophosphate

    U.S. and Foreign Producers: Shell Chemical Co. (U.S.A.)

2. Use Patterns and Formulations

    Application Sites: Terrestrial, nondomestic food uses on cotton, peanuts and sugarcane. Terrestrial, nondomestic nonfood uses on tobacco, ornamental conifers (nursery stock), ornamental flowering plants (nursery stock), ornamental woody shrubs (nursery stock), and ornamental deciduous trees (nursery stock).

    Types of Formulations: Soluble concentrate liquid

    Types and Methods of Application: Aerial and ground application.

    Application Rates: 0.125 - 1 pound per acre (lb/a)

    Usual Carriers: CONFIDENTIAL BUSINESS INFORMATION

3. Science Findings

Summary Science Statement:

Toxicology, environmental fate, ecological effects, product chemistry and residue chemistry data gaps preclude the Agency from making a complete assessment for monocrotophos. However, based on available data, monocrotophos can be characterized as having very high acute oral toxicity to both humans and birds. Monocrotophos is a potent cholinesterase inhibitor (NOEL = 0.03 ppm in rats). Monocrotophos is also fetotoxic (NOEL = 1.0 ppm) but not teratogenic at the highest dose tested (2 mg/kg), decreases fertility at 9.0 ppm (NOEL = 2.7 ppm) and is weakly mutagenic in vitro.

Chemical Characteristics:

Technical monocrotophos is a reddish brown solid with a melting point of 25 to 30 °C. Its odor is characteristic of a mild ester. Monocrotophos is soluble in water, acetone, and alcohol. It is stable when stored in glass or polyethylene containers; stable in simple alcohols and glycols at room temperature; relatively stable in sunlight and nonvolatile at 100 °F; decomposes at 310 to 320 °F. At 20 °C, hydrolysis is quite slow. Its half-life in solution (2 parts per million (ppm)) at pH 7 and 38 °C is 23 days.

Toxicology Characteristics:

Current available toxicological studies on monocrotophos are as follows:

- Acute oral toxicity: rat, LD_{50} = 23 mg/kg (males) and 18 mg/kg (females), (Tox Category I).

- Acute dermal toxicity: rat, LD_{50} = 354 milligrams per kilogram (mg/kg) (Tox Category II).

- Primary eye irritation: rabbit, slight to moderate irritation and corneal opacity reversible by day 14 (Tox Category II).

- Primary Dermal Irritation: rabbit, PIS = 0.6 to 1.0, slightly irritating (Tox Category IV).

Major routes of exposure: Application by ground and aerial equipment increases the potential for exposure of humans, livestock and wildlife due to spray drift. Human exposure to monocrotophos from handling, application and reentry operations is minimized by the use of approved respirators and other protective clothing.

Chronic toxicity results:

- Rat chronic feeding and oncogenicity:

    Not carcinogenic at the highest dose tested (HDT) 9 ppm.
    No Observable Effect Level (NOEL)
    - ChE I = 0.03 ppm.
    Lowest Effect Level (LEL) - ChE I = 0.09 ppm.
    Systemic NOEL = 0.9 ppm.
    Systemic LEL = 9.0 ppm (body weight decrease in males; decreased survival in females). This study indicates that the rat is the most sensitive species for measuring cholinesterase inhibition (NOEL = 0.03 ppm) compared to the dog (NOEL = 1.6 ppm).

- Dog chronic feeding:

    NOEL - ChE I = 1.6 ppm.
    LEL - ChE I = 16.0 ppm.
    Systemic NOEL = 16.0 ppm.
    Systemic LEL = 100 ppm (salvation and tremors).

- Rat teratogenicity:

    Fetotoxic effects were found at 2 mg/kg. The effects consisted of runting, reduced fetal weight and length (NOEL = 1.0 mg/kg), and maternal toxicity in the form of reduced body weight gain at 1.0 mg/kg (NOEL = 0.3 mg/kg). No teratogenic effect was observed at the HDT (2.0 mg/kg/day).

- Rat Reproduction:

    Generated a reproductive (and offspring) NOEL of 2.7 ppm and an LEL of 9.0 ppm (as evidenced by decreased fertility, pup viability and weight, partly attributed to depressed maternal lactation).

- Mutagenicity:

    A total of 19 studies evaluating monocrotophos for mutagenicity are available, but only 10 are adequate (acceptable). Monocrotophos is weakly mutagenic *in vitro*, as determined mainly from studies assessing DNA damage/repair and sister chromatid exchange.

Physiological and Biochemical Behavioral Characteristics

- Mechanism of Pesticidal Action:

    Monocrotophos is a systemic and contact poison. As an organophosphate, monocrotophos exerts its toxic action by inhibiting certain important enzymes of the nervous system (cholinesterase).

- Metabolism and Persistence in Plants and Animals:

    The metabolism of monocrotophos in animals and plants has not been adequately described. Metabolism studies utilizing ruminants and poultry will be required to fill the animal metabolism data gaps. Currently, no tolerances for residues of monocrotophos in animal products exist; however, monocrotophos and some of its metabolites have been identified in the milk, muscle, and liver of cows and in the milk of goats following ingestion of this chemical.

    Additional plant metabolism data are required, including studies to reflect the potential for uptake of soil metabolites following foliar applications.

Environmental Characteristics

Monocrotophos hydrolyzes rapidly (half-life of 14-21 days at pH 9 and 25 ° C), with the rate decreasing at lower pH's and increasing at higher temperatures. Degradation on soil exposed to natural sunlight is rapid (half-life less than 7 days) and on dark control samples is slower (half-life approximately 30 days). Residues have a low potential for bioaccumulation in catfish and are depurated fairly rapidly.

Monocrotophos is mobil in soil and although it degrades rapidly, it may possess potential for groundwater contamination. Pertinent data (mobility, metabolism and dissipation) are necessary to fully assess monocrotophos's potential for ground water contamination.

Ecological Characteristics:

Avian Oral Acute Toxicity: Test results showed that acute oral toxicity for upland game birds ranges from 0.763 to 6.49 mg/kg; 1.58 to 4.76 mg/kg for waterfowl; 1.00 to 5.62 mg/kg for passerines and for the golden eagle, the value is 0.188 mg/kg (very highly toxic).

Avian Dietary Toxicity: Dietary studies on the ringed-neck pheasant and mallard duck resulted in dietary toxicity values of 3.1 and 9.6 ppm, respectively (very highly toxic).

Fish Acute Toxicity: Test results for warm water acute fish toxicity range from 12.1 ppm for bluegill sunfish to greater than 50 ppm for fathead minnows (moderately toxic).

Freshwater Invertebrate Acute Toxicity: Test results for acute toxicity to Daphnia magna were 0.034 ppm (very highly toxic).

Avian Reproduction: Test results are sufficient to characterize monocrotophos as not having an effect on the overall reproductive success of birds at levels of 0.1 to 3.0 ppm in the diet (non-toxic to reproduction). Typical reproductive effects in the field are unlikely from the use of monocrotophos. Rather more likely, breeding birds will be exposed to a toxic dose themselves or will lead/feed a toxic dose to their brood.

Honeybee Acute Toxicity: 0.350 micrograms per bee (highly toxic).

Monocrotophos is one of the most toxic pesticides to birds. Monitoring and incident reports contain numerous observations of avian mortality attributed to monocrotophos; thus, it has the potential for causing significant impacts on populations of avian wildlife. The field studies that have been submitted are inadequately designed and contain mostly cursory monitoring information; therefore, terrestrial field testing for effects on avian wildlife is needed. Monocrotophos has been reviewed under the cotton "cluster" for endangered species, and no jeopardy has been determined for endangered avian species. The Agency will initiate a formal consultation with the U.S. Fish and Wildlife Service Office of Endangered Species concerning potential adverse effects of monocrotophos on terrestrial species for the remaining uses.

Tolerance Assessment

Established tolerances for monocrotophos are published in 40 CFR 180.296 and 21 CFR 193.151 and are:

Commodity	Part per Million
potatoes	0.1
tomatoes	0.5
cottonseed	0.1
peanuts	.05
sugarcane	0.1
concentrated tomato products	2.0

The Agency is unable to complete a full tolerance assessment for the established tolerances because of residue chemistry data gaps including plant and animal metabolism studies and residue data to determine whether food/feed additive tolerances must be proposed for the processed products of all the registered crops.

The NOEL for cholinesterase inhibition (ChE) has been set at 0.03 ppm (and for systemic effects at 0.9 ppm), generating an ADI of 0.00015 mg/kg/day (systemically, 0.0045 mg/kg/day), which results in the TMRC for previously published tolerances occupying 397 percent of the ADI (132% based on systemic effects). Thus, on either basis (ChE or systemic), the margin of safety has been exceeded for those tolerances already published which precludes granting any new requests.

The Agency requested, and the registrant agreed, to voluntarily delete the use of tomatoes from currently approved labels. The registrant has since submitted an application with revised labels removing both tomatoes and potatoes from their section 3 product labels. The elimination of the use of monocrotophos on tomatoes and potatoes lowers the TMRC to 66 percent of the ADI. Nevertheless, the Agency will not allow any new uses to be established for monocrotophos until the required residue chemistry data (including animal metabolism studies) have been submitted and evaluated so that a tolerance reassessment can be made.

4. Summary of Regulatory Position and Rationale

The Agency has determined that it should continue to allow the registration of monocrotophos. None of the criteria for unreasonable adverse effects listed in the regulations [§162.11 (a)] has been met or exceeded. However, because of gaps in the data base a full risk assessment cannot be completed.

The Agency will not allow any significant new uses to be established for monocrotophos until the residue chemistry data deficiencies have been satisfied, a tolerance reassessment is made, and a well-designed field test in birds has been evaluated.

Because of the high acute toxicity of monocrotophos to humans:

° All end-use products containing monocrotophos shall continue to be classified for restricted use.

° The Agency is requiring applicators with a high exposure to monocrotophos, mixers, and loaders to wear protective clothing. The use of backpack or knapsack sprayers for application of monocrotophos is being prohibited.

° The Agency will continue to require the reentry interval of 48 hours established under 40 CFR 170 for all outdoor uses of monocrotophos in order to minimize exposure to

workers entering treated areas, pending the receipt and
evaluation of reentry data to assess the potential for
exposure to workers coming in contact with monocrotophos.

Because of the high acute toxicity of monocrotophos to birds:

º The Agency is requiring a well-designed field test on
birds. The field sites will include a number of sites in
each growing area and will investigate monocrotophos
exposure to birds from diet and drinking water and the
effect on young fledglings using nest boxes. The
design of the field testing protocols, selection of
sensitive indicator species (including avian predators),
and the selection of the test sites must be submitted
to the Agency. This protocol must be approved by the
Agency prior to the initiation of the study. Until the
study is conducted and reviewed, label precautionary
statements are required.

The Agency is requiring all end-use products registered
for outdoor use to bear a restriction on rotating food or
feed crops to monocrotophos treated soils unless monocrotophos
is registered for use on the rotated crop. This restriction
will remain in effect until such time as data are submitted
and reviewed which allow the Agency to determine a time
interval at which rotated crops planted in treated soil will
be free of pesticide residues.

The Agency is not requiring additional residue data to
determine if any detectable residues of the trimethyl phosphate
(TMP) contaminant persists in raw agricultural commodities
or processed foods unless the plant metabolism study being
required shows that TMP persists.

5. Summary of Major Data Gaps

Product Chemistry: Data on product identity, ingredients,
                     impurities, and physical and chemical
                     characteristics.

Residue Chemistry: Studies on plant and animal metabolism,
                     storage stability of samples, and
                     residue data to determine whether food/
                     feed additive tolerances are required
                     for processed products of all registered
                     crops.

Toxicology: Studies on acute and 21-day inhalation,
                     dermal sensitization, general metabolism,
                     and rabbit teratogenicity; and additional infor-
                     mation on the mouse oncogenicity study.

Wildlife and Aquatic Organisms: A field test on birds and acute toxicity to estuarine and marine organisms.

Environmental Fate: Soil metabolism, mobility, dissipation and accumulation studies and data on reentry protection and spray drift.

6. <u>Contact Person at EPA:</u> William H. Miller
Product Manager (16)
Insecticide-Rodenticide Branch
Registration Division (TS-767)
Environmental Protection Agency
Washington, DC 20460
Tel. (703) 557-2600

DISCLAIMER: The information presented in this Chemical Information Fact Sheet is for informational purposes only and may not be used to fulfill data requirements for pesticide registration and reregistration.

NABAM

Date Issued: April, 1987
Fact Sheet Number: 124

DESCRIPTION OF CHEMICAL

Generic Name: Disodium ethylene bisdithiocarbamate

Common Name: Nabam

Trade Names: Chem Bam®; Dithane® D-14; Dithane® A-40; DSE®; Nabasan®; Parzate®; Spring Bak®.

EPA Shaughnessy Code: 014503

Chemical Abstracts Service (CAS) Number: 142-59-6

Pesticide Type: Fungicide

Chemical Family: Ethylene bisdithiocarbamate (EBDC)

U. S. and Foreign Producers: Alco Chemical Corporation; Vinnings Chemical Company; Rohm and Haas

USE PATTERNS AND FORMULATIONS

Registered Uses: Industrial sites (cooling towers, evaporative condensers, air washer systems, secondary oil recovery water and drilling fluids, pulp and paper mills, cane sugar mills, beet sugar mills). All registered food (except for use in sugar mill flume water) and ornamental uses of nabam, a total of 35, are currently suspended.

Major Uses: Domestic usage of nabam as a pesticide (Typical current year basis (1983-1985)):

Sector	lbs. x 1,000	Percent
Cooling Water	475	40
Sugar	425	37
Paper	275	23
Totals	1175	100

Formulation Types Registered: Ready-to-use 3.75 to 22% nabam, Formulation Intermediates - 22.5 to 30% nabam, and Soluble Concentrates - 22% or 93% nabam.

Methods of Application: In most industrial uses, nabam is added directly to water either by a single dose or continuous feed.

Application Rates: Rates range from .09 to 3.0 parts per million (ppm) in sugar processing; 30 to 120 ppm in pulp and paper processing; 2.6 to 60 ppm in water cooling towers. General industrial preservative uses are in the range of 30 to 120 ppm.

Usual Carrier: Water

SCIENCE FINDINGS

Summary Science Statement: A major toxicological concern from exposure to nabam is the hazard to the human thyroid from presence of ethylenethiourea (ETU), a contaminant, degradation product, and metabolite present in nabam and other EBDC fungicides. Additional chronic studies on nabam are required for further evaluation. Moreover, ETU has caused developmentally toxic/teratogenic effects in rats and hamsters. There are no adequate teratology studies on nabam. Teratology studies with nabam are required before its teratogenicity can be fully assessed.

ETU has been classified as a Group B2 carcinogen in accordance with the Agency's Guidelines for Carcinogen Risk Assessment (September 26, 1986, 51 CFR 33992), based on studies which show that it induced an increased incidence of thyroid adenomas and adenocarcinomas in rats and hepatomas in mice. Because of the presence of ETU in nabam, the Agency is considering whether further regulatory considerations are warranted.

Available data are not adequate to assess the environmental fate of nabam. However, studies do indicate that ETU, the major degradate of nabam, has leaching potential in certain soil types. A complete assessment of the environmental fate, including the potential for groundwater contamination, from the use of nabam products will be undertaken when data are available.

Available data are insufficient to completely evaluate the ecological effects of nabam. Based on available data, nabam is moderately toxic to freshwater fish; moderately to highly toxic to certain estuarine organisms.

Physical/Chemical Characteristics:

    Physical state: Liquid at 25° C

    Empirical Formula: $C_4H_6N_2Na_2S_4$

    Molecular Weight: 256.3

Toxicology Charateristics:

    Major Route of Exposure: Dermal

    Acute Toxicity:

        Oral Toxicity: Toxicity category II

        Dermal Sensitization: Skin sensitizer

    Mutagenicity: Additional studies required

        Negative - Host-mediated assay in mice in Salmonella typhimurium.

        Positive - Sister-chromatid exchanges in CHO cells

        Negative - In Vivo bone marrow cytogenetic assay

Physiological and Behavioral Characteristics

    Mechanism of Pesticidal Action: Nabam inhibits enzyme activity by complexing with metal-containing enzymes including those involved with the production of adenosine triphosphate (ATP).

    Metabolism: Nabam appears to be rapidly absorbed from the gastrointestinal tract, distributed to target organs and excreted almost totally within 96 hours.

Environmental Characteristics:

    Photodecomposition and/or Volatility: Studies required

    Hydrolysis: Studies required

    Anaerobic Aquatic Metabolism: Studies required

    Leaching and adsorption/desorption studies: Nabam is slightly mobile in certain soil types. Additional studies required.

Ecological Characteristics:

    Freshwater Fish Toxicity: Moderately toxic. Studies required.

    Effects on Non-target Insects: Relatively nontoxic to honey bees.

Tolerance Assessment

There are no established tolerances for residues of nabam per se on any food or feed items. All food crop uses of nabam are currently suspended. When used on food crops nabam is tank mixed with zinc sulfate leading to the formation of zineb. Nabam is not mixed with zinc sulfate when used on onions to control smut. Therefore, for this use registrants are required to submit residue data on nabam. For the other food crop uses, application actually consists of a solution of zineb instead of nabam. Therefore, residues in agricultural commodities treated with nabam are expected to be zineb.

The use of nabam as an industrial biocide to treat flume water in sugar beet mills is considered to be a treatment of a raw agricultural commodity. Regulations have been established by the Food and Drug Administration under 21 CFR 173.320(b)(3) for nabam residues following application to sugar mill grinding, crusher and/or diffuser systems. The treatment of flume water, transporting and washing systems are under EPA's jurisdiction. Residue data are required in this Standard to determine if the residues concentrate in any of the processed sugar commodities.

SUMMARY OF REGULATORY POSITION AND RATIONALE

EPA is currently evaluating the potential human health risks resulting from the food, field and food crop, and terrestrial non-food uses of nabam to determine whether additional regulatory action is warranted on nabam and the other EBDC pesticides containing the common contaminant, degradation product, and metabolite, ETU. ETU is mutagenic, oncogenic and teratogenic, and the Agency has classified it as a Group B2 oncogen (Probable Human Carcinogen).

After considering exposure to ETU from the industrial, non-food uses of nabam, the Agency has concluded that the risks from exposure to ETU from use of nabam are not a concern for most industrial uses at this time because applicator exposure, based on available information, appears negligible. The Agency lacks data to assess exposure to nabam from the metalworking fluid and tanning uses. Additional information is being required for all industrial uses. The Agency does not believe that further regulatory action on nabam products registered for industrial non-food uses is warranted at this time.

The Agency will not consider any new food use tolerances for nabam at this time. The Agency will consider the need for establishment of tolerances for ETU and any intermediate metabolites when data are sufficient to permit such decisions.

While data gaps are being filled, currently registered manufacturing-use products (MP's) and end-use products (EP's) containing nabam as the sole active ingredient may be sold distributed, formulated, and used, subject to the terms and conditions of the nabam registration standard. However, significant new uses will not be registered. Registrants must provide or agree to develop and provide additional data in order to maintain existing registrations.

The Agency has determined that all data will be immediately reviewed as they are submitted.

Precautionary Statements Required on Labels:

Manufacturing-Use Products

"This pesticide is toxic to fish. Do not discharge effluent containing this product into lakes, streams, ponds, estuaries, oceans, or public water unless this product is specifically identified and addressed in an NPDES permit. Do not discharge effluent containing this product to sewer systems without previously notifying the sewage treatment plant authority. For guidance, contact your State Water Board or Regional Office of the EPA."

End-Use Products

Agricultural Use Products

"This pesticide is toxic to fish. Drift and runoff from treated areas may be hazardous to aquatic organisms in neighboring areas. Do not apply directly to water or wetlands (swamps, bogs, marshes, and potholes). Do not contaminate water by cleaning of equipment or disposal of wastes."

All Home Use Products

"PROTECTIVE MEASURES: Always spray with your back to the wind. Wear long-sleeve shirt, long pants, and rubber gloves. Wash gloves thoroughly with soap and water before removing. Change your clothes immediately after using this product and launder separately from other laundry items before reuse. Shower immediately after use."

Home Use Products with Food Uses

"Preharvest intervals on this label are specified so that pesticide residues will be at an acceptable level when the crop is harvested."

All Agricultural Products

"After (sprays have dried/dusts have settled/vapors have dispersed, as applicable) do not enter or allow

entry into treated areas until the 24-hour reentry
interval has expired unless wearing the personal
protective equipment listed on the label."

"WORKER SAFETY RULES

Keep all unprotected persons, children, livestock, and
pets away from treated area or where there is danger of
drift.

"Do not rub eyes or mouth with hands. See First Aid
(Practical Treatment Section)."

"PERSONAL PROTECTIVE EQUIPMENT

HANDLERS (MIXERS, LOADERS, AND APPLICATORS) AND
EARLY REENTRY WORKERS MUST WEAR THE FOLLOWING PROTECTIVE
CLOTHING AND EQUIPMENT: a long-sleeve shirt and long
pants or a coverall that covers all parts of the body
except the head, hands, and feet; chemical resistant
gloves; shoes, socks, and goggles or a face shield.
During mixing and loading, a chemical resistant apron
must also be worn.

During application from a tractor with a completely
enclosed cab with positive pressure filtration, or
aerially with an enclosed cockpit, a long-sleeve
shirt and long pants may be worn in place of the
above protective clothing. Chemical resistant gloves
must be available in the cab or cockpit and worn
while exiting.

IMPORTANT! Before removing gloves, wash them with
soap and water. Always wash hands, face, and arms
with soap and water before eating, smoking or drinking.
Always wash hands and arms with soap and water before
using the toilet.

After work take off all clothes and shoes. Shower
using soap and water. Wear only clean clothes. Do
not use contaminated clothing. Wash protective
clothing and protective equipment with soap and water
after each use. Personal clothing worn during use
must be laundered separately from household articles.
Clothing and protective equipment heavily contaminated
or drenched with nabam must be destroyed according
to state and local regulations.

HEAVILY CONTAMINATED OR DRENCHED CLOTHING CANNOT BE
ADEQUATELY DECONTAMINATED.

During aerial application, human flaggers are prohibited
unless in totally enclosed vehicles."

Industrial Use Products

"This pesticide is toxic to fish. Spills, sprays, and runoff from use site may be hazardous to aquatic organisms in neighboring areas. Do not apply directly to water or wetlands (swamps, bogs, marshes, and potholes). Do not contaminate water by cleaning of equipment or disposal of wastes."

"HANDLE (INCLUDING MIXING, LOADING, OR APPLYING) THIS PRODUCT ONLY WHEN WEARING THE FOLLOWING PROTECTIVE CLOTHING AND EQUIPMENT: A long-sleeve shirt and long pants or a coverall that covers all parts of the body except the head, hands, and feet; chemical resistant gloves; and goggles or a face shield. During mixing and loading, a chemical resistant apron must also be worn."

SUMMARY OF MAJOR DATA GAPS
─────────────────────────

Product Chemistry - All

Residue Chemistry - Plant and animal metabolism
                    Residue studies

Toxicology - Acute dermal and inhalation toxicity
             Primary eye and dermal irritation
             21-day subchronic dermal
             Chronic toxicity (rodent and nonrodent) (food uses)
             Oncogenicity (rat and mouse)(food uses)
             Teratology (rabbit and rat)
             Reproduction (rat)
             Mutagenicity (gene mutation, cell transformation,
                     unscheduled DNA synthesis)
             Percutaneous dermal absorption

Ecological Effects - Avian oral toxicity
                     Avian dietary toxicity
                     Freshwater fish toxicity
                     Freshwater invertebrate toxicity
                     Estuarine and marine organism toxicity
                     Aquatic organism accumulation
                     Simulated or actual field testing-aquatic
                         organisms
                     Nontarget area phytotoxicity-aquatic plant
                         growth

Environmental Fate - Hydrolysis
                     Photodegradation (water, soil)
                     Aerobic and anaerobic soil metabolism
                     Aerobic and anaerobic aquatic metabolism
                     Leaching and adsorption/desorption

Volatility (laboratory)
Field dissipation (terrestrial, aquatic)
Rotational crops (confined)
Fish accumulation

Reentry - Reentry protection studies

- Ethylenethiourea (ETU) -

Toxicology - Chronic (rodent and nonrodent)
Reproduction
Dermal (percutaneous) absorption
Mutagenicity (promotion)

Environmental Fate - Hydrolysis
Photodegradation (soil and water)
Aerobic and anaerobic soil metabolism
Leaching and adsorption/desorption
Degradation (soil)

CONTACT PERSON AT EPA

Ms. Lois Rossi
Acting Product Manager (Team 2')
Fungicide-Herbicide Branch
Registration Division (TS767C)
Office of Pesticide Programs, EPA
Washington, D.C. 20460

Telephone: 703-557-1900

DISCLAIMER: The information presented in this Pesticide Fact Sheet is for information purposes only and may not be used to fulfil data requirements for pesticide registration and reregistration.

NALED

Date Issued: June 30, 1983
Fact Sheet Number: 4

1. Description of Chemical

    Generic name: 1,2-dibromo-2,2-dichloroethyl dimethyl phosphate
    Common name: Naled
    Trade name: Dibrom®
    EPA Shaughnessy code: 034401
    Chemical abstracts service (CAS) number: 300-76-5
    Year of initial registration: 1959
    Pesticide type: insecticide - acaricide
    Chemical family: organophosphate
    U.S. and foreign producers: Chevron Chemical Co.

2. Use patterns and formulations

    Application sites: field, vegetable, and orchard crops; livestock and poultry, and their surroundings; greenhouses; forest and wasteland; agricultural, domestic, medical, and commercial establishments; and urban and rural outdoor areas (mosquito control).

    Types of formulations: dusts, impregnated materials, emulsifiable concentrates, soluble concentrates, liquid and ready-to-use liquids.

    Types and methods of application: aerial and ground as a spray or dust; fogging; ultra low volume (ULV)

    Application rates: varies from .01 to 10 lbs/acre

    Usual carriers: Confidential Business Information

3. Science Findings

    Summary science statement:

    Available acute oral and dermal LD_{50} data place naled in toxicity category II and toxicity category I based on primary eye irritation data.

    Naled is characterized as very highly toxic to bees and aquatic invertebrates. It is moderately to highly toxic to fish and slightly toxic to upland game birds and waterfowl.

Naled has numerous data gaps in areas of product chemistry, residue chemistry, toxicity, environmental fate and ecological effects. The Agency cannot conduct a full risk assessment until the data required in the Naled Standard are submitted and evaluated.

Chemical characteristics:

Manufacturing-use naled is a light, straw-colored, oily liquid with a slightly pungent odor. The pure compound is a white, low melting point solid. The boiling point for pure naled is 120°C at 0.5 mm Hg and the vapor pressure is 2×10^{-4} mm. Hg at 20°C. The empirical formula is $C_4H_7O_4PBr_2Cl_2$ and the molecular weight is 381. Naled has limited solubility in aliphatic solvents; is highly soluble in oxygenated solvents such ketones and alcohols; and a low solubility in water.

Toxicological characteristics:

Current available toxicology studies on naled are as follows:

- Acute oral LD_{50}: rat, 222-389 mg/kg; mouse, 160 mg/kg (Tox category II)
- Acute dermal LD_{50}: rabbit, 390-1100 mg/kg; rat, 800 mg/kg (Tox category II)
- Primary eye irritation: corneal opacities 7 days+ (Tox category I)
- Primary dermal irritation: rabbit, PIS 5.8 - 5.92; human volunteers, severe irritation
- Subchronic feeding: cholinesterase NOEL in dog, 0.25 mg/kg/day

Additional data are needed to fully assess the toxicity of naled.

Major routes of exposure: application by fog and mist sprayers, and aircraft increases the potential for exposure of humans and non-target organisms to naled. Human exposure to naled during mixing, handling, application, and reentry operations would be minimized by the use of approved respirators and other protective clothing. However, data are not available to fully assess such exposures. PR Notice 83-2 sets forth current Agency policy on required label changes for reentry and farmworker safety. A reentry level of 24 hours for the use of naled on crops is required.

Physiological and Biochemical Behavioral Characteristics:

Mechanism of pesticidal action: stomach and/or contact poison
Metabolism and persistence in plants and animals: no naled or DDVP was detected in tissues or milk of two goats

dosed with naled at 107 ppm in three equal daily portions and and sacrificed on day $^{-4}$. The limits of detection were 0.05 ppm for tissues and 0.005 ppm for milk. The dose is estimated to be about twenty times that which would normally occur in the goats diet.

Residues were non-detectable (<0.01 ppm) in milk from Holstein cows subject to body and premise sprays for 14 days with the 7.2 lb/gal EC formulation.

Environmental Characteristics:

Naled degrades fairly rapidly with half-lifes of ≤ 8 hours in soils and ≤ 25 hours in aqueous solutions. Dichlorvos (DDVP), a metabolite of naled is also rapidly degraded in soil with half-lifes of 2.3 - 8.0 hours. Naled exhibits low to intermediate mobility in soils, whereas dichlorvos is intermediately mobile to mobile.

Limited data indicate that the rapid dissipation and relatively low mobility of naled and intermedite mobility of dichlorvos in soil will mitigate contamination of ground water.

Naled did not accumulate in whole body tissues of killfish exposed to naled in static bioassy tests. Naled was not detected (<0.02 ppm) in any fish tissue samples taken over the 7-day test period. The degradate dichlorvos was found at a maximum concentration of 0.04 ppm, approximately twice the concentration in corresponding water samples, 1 hour after treatment, but was not found (<0.01 ppm) in tissue samples taken after 24 hours. Naled half-life in water samples was <24 hours. Dichlorvos was found in all water samples, at a maximum concentration of 0.02 ppm after 24 hours, but <0.01 ppm was found in samples taken at the end of the test period.

In summary naled and its degradate dichlorvos dissipate rapidly in aerobic soils. Naled exhibits low to intermediate mobility in soils, whereas, dichlorvos is intermediately mobile to mobile. Mobility appears to be related to soil organic matter content. Naled degrades rapidly in aqueous solution, with rates increasing at higher temperatures and pHs. Naled also rapidly degrades in sewage water to dichlorvos and dichloroacetaldehyde. Neither naled nor dichlorvos accumulate in fish tissues. In conclusion, naled does not appear to represent an environmental hazard based on the aforementioned data which indicate very rapid degradation and extremely low bioaccumulation potential.

However, available data are insuffcient to fully assess the environmental fate of naled.

Ecological Characteristics:

Currently available ecological effects studies on naled are as follows:

- Avian oral LD_{50}: 37-65 mg/kg
- Avian dietary LC_{50}: 2117-2724 ppm
- Freshwater fish LC_{50}: 160-900 ppb
- Acute LC_{50} freshwater invertebrates: 0.3 ppb

Based on studies available to assess hazards to wildlife and aquatic organisms, naled is characterized as very highly toxic to bees and aquatic invertebrates. It is moderately to highly toxic to fish and slightly toxic to upland game birds and waterfowl. Insufficient data are available to assess the toxicity of naled to estuarine and marine organisms.

Label precautionary statements required by the Standard should reduce the hazard to fish and other wildlife. After data gaps are filled, the potential hazards to terrestrial and aquatic species will be better defined and additional labeling requirements may be imposed.

Efficacy review results: none conducted.

Tolerance Assessment:

The following tolerances are established for combined residues of naled and 2,2-dichlorovinyl dimethyl phosphate (DDVP), expressed as naled, in or on raw agricultural commodities resulting from the application of naled formulations to growing crops livestock, and poultry (40 CFR 180.125).

Commodity	Tolerance (ppm)
Almonds (hulls, nuts)	0.5
Beans (dry, succulent)	0.5
Broccoli	1.0
Brussels sprouts	1.0
Cabbage	1.0
Cattle (fat, meat, meat by-products)	0.05
Cauliflower	1.0
Celery	3.0
Citrus fruits (grapefruit, lemons, oranges, tangerines)	3.0
Collards	3.0
Cottonseed	0.5
Cucumbers	0.5
Eggplant	0.5
Eggs	0.05

Goats (fat, meat meat by-products)	0.05
Grapes	0.5
Grasses, forage	10.0
Hogs (fat, meat, meat by-products)	0.05
Hops	0.5
Horses (fat, meat, meat by-products)	0.05
Kale	3.0
Legumes, forage	10.0
Lettuce	1.0
Melons	0.5
Milk	0.05
Mushrooms	0.5
Peaches	0.5
Peas (succulent only)	
Peppers	0.5
Poultry (fat, meat, meat by-products)	0.05
Pumpkins	0.5
Rice	0.5
Safflower seed	0.5
Sheep (fat, meat, meat by-products)	0.05
Spinach	3.0
Squash (summer, winter)	0.5
Strawberries	1.0
Sugar beets (roots, tops)	0.5
Swiss chard	3.0
Tomatoes	0.5
Turnips (tops)	3.0
Walnuts	0.5
All other raw agricultural commodities except those listed. (To account for area pest [fly and mosquito] control.)	0.5

The components of the residue metabolism in plants which are of concern are naled and DDVP, and to a lesser extent, organic bromide. Tolerances exist for combined residues of naled and DDVP (expressed as naled) and should continue to reflect the concern for these two components.

The components of the residue from the metabolism in animals which are of concern are the same as those in or on plants. However, data on the metabolism or naled in poultry are missing and this constitutes a data gap.

The Theoretical Maximum Residue Contribution (TMRC) is 1.1021 mg/day as naled, assuming a 1.5 kg diet, based on the tolerances and food factors for all of the commodities for which U.S. tolerances are established. No Acceptable Daily Intake (ADI) or Maximum Permissible Intake (MPI) figures have been established, due to the absence of

acceptable toxicological data for naled. Reassessment
of the established naled tolerances must await receipt
and evaluation of the required data.

Although the Agency is unable to complete a tolerance
reassessment for naled because of a number of residue
chemistry and toxicology data gaps, the Agency has concluded, based on available data, that no changes in
present tolerances are necessary at this time. The
Agency has also considered the residues of inorganic
bromide, resulting from the use of naled on crops and in
meat, milk, poultry and eggs, and does not anticipate
these residues to be of toxicological concern, and no
additional residue data on inorganic bromides are needed.

However, the Agency is concerned about organic brominated
metabolites of naled and its impurities. Accordingly,
additional data on this organic bromide in plants and
animals are being requested.

4. Summary of Regulatory Position and Rationale:

Use classification: not classified
Use restrictions: None
Unique warning statements required on labels:

The following environmental hazard statement must appear
on the manufacturing-use product labels:

"This product is toxic to fish, aquatic invertebrates,
and wildlife. Do not discharge into lakes, streams,
ponds or public water unless in accordance with NPDES
permit. For guidance contact your regional office of
the Environmental Protection Agency."

Labeling changes to end-use products are not required by
the Standard, however, based on data reviewed by the
Agency the following statements will be required for
end-use products under the Agency's Label Improvement
Program:

"This product is toxic to fish, aquatic invertebrates,
and wildlife. Do not apply directly to water or wetlands. Runoff from treated areas may be hazardous to
aquatic organisms in neighboring areas. Do not contaminate water by cleaning of equipment or disposal of
wastes."

"This product is highly toxic to bees exposed to direct
treatment on blooming crops or weeds. Do not apply
this product or allow it to drift to blooming crops
or weeds while bees are actively visiting the treatment area."

The following "General Warnings and Limitations" statements must appear on end-use product labels which bear directions for aquatic use on food or feed crops:

"Do not use with highly alkaline materials such as lime or bordeaux mixture. Shrimp and crabs may also be killed at application rates recommended. Do not appply to tidal or marsh waters which are important shrimp producing areas."

A reentry interval of 24 hours for the use of naled on crops is required on end-use product labels containing directions for use on crops.

The absence of reported fatalities from the Pesticide Incidence Monitoring System (PIMS) report, taken in conjunction with the apparent adequacy of medical and/or emergency room treatment suggests an acceptably low level or risk associated witn incidental or accidental exposure to naled products.

Summary of risk/benefit review:

Dichlorvos (DDVP), a metabolite of naled was originally referred to the Rebuttable Presumption Against Registration (RPAR) process because scientific studies indicated that dichlorvos was mutagenic, might cause cancer, nerve damage and birth defects in laboratory animals. The RPAR Decision Document on Dichlorvos, was issued September 30, 1982. It was concluded that the existing evidence does not suppport the issuance of an RPAR for dichlorvos and consequently, that an RPAR for naled as a precursor of dichlorvos is also not warranted.

However, the Decision Document concluded that additional data on carcinogenicity and mutagenicity are needed to complete the risk assessment for dichlorvos. Because the data base was incomplete, dichlorvos was removed from the RPAR process and returned to the registration process. A Data Call-in Notice under FIFRA Section 3(c)(2)(B) was issued March 23, 1983, requesting data on potential mutagenic effects of dichlorvos be submitted by March 23, 1985. Additionally, the Agency will wait until the ongoing National Cancer Institue dichlorvos bioassy on carcinogenicity is completed (currently scheduled for completion in 1984) and evaluated prior to determining if additional data on the carcinogenicity of dichlorvos will be required. Since dichlorvos is a metabolite of naled, evaluation of these studies will be necessary for the completion of the naled risk assessment.

No other human toxicological hazards of concern to the Agency have been identified in studies reviewed for the Standard.

5. **Summary of Major Data Gaps:**

   Product chemistry: data on the discussion of formation of ingredients; preliminary analysis; certification of limits; and analytical methods for enforcement of limits are the major product chemistry data gaps.

   Residue chemistry: additional data are required to support the tolerances for beans (dry and succulent), broccoli, Brussels sprouts, cabbage, cauliflower, celery, collards, cottonseed, cucumbers, eggplant, eggs, grapefruit, hops, kale, lemons, melons, mushrooms, oranges, pea forage, peaches, peas, peppers, poultry (fat, meat, meat by-products), pumpkins, soybean forage, spinach, strawberries, Swiss chard, tangerines, turnip tops, and winter squash.

   Data are required on residues in the processed products of citrus (any member fruit), cottonseed, grapes, hops, rice, and tomatoes. Data are also needed for turnip roots. A tolerances must be established for this commodity.

   Toxicology: inhalation LC_{50}, rat; 21-day dermal, rabbit; 90-day inhalation, rat; chronic toxicity, 2 species; oncogenicity, 2 species; teratogenicity, 2 species; reproduction, 2-generation rat; gene mutation; chromosomal aberration.

   Reentry protection: foliar dissipation.

   Wildlife and aquatic organisms: freshwater fish LC_{50} (on typical end-use product); acute LC_{50}, freshwater invertebrates (on technical grade of active ingredient, and on typical end-use product); acute LC_{50}, estuarine and marine organisms).

   All data must be submitted by June, 1986.

6. **Contact person at EPA:**

   William H. Miller
   Product Manager (16)
   Insecticide-Rodenticide Branch
   Registration Division (TS-767)
   Environmental Protection Agency
   Washington, DC 20460

   Tel. (703) 557-2600

DISCLAIMER: The information presented in this Chemical Information Fact Sheet is for informational purposes only and may not be used to fulfill data requirements for pesticide registration and reregistration.

NAPTALAM

Date Issued: March 31, 1985
Fact Sheet Number: 49

1. Description of chemical:

   Generic Name: naptalam, naptalam sodium
   Common name: naptalam (WSSA, BSI, ISO)
   Trade name: Alanap-L ®
   EPA Shaughnessy Code: 030702
   Chemical Abstracts Service (CAS) Number: 132-66-1
   Year of Initial Registration: 1949
   Pesticide Type: Herbicide
   Chemical family: Phthalic acid
   U.S. and Foreign Producers: Uniroyal

2. Use patterns and formulations:

   Application sites: Liquid and granular naptalam are used as selective preemergence herbicides for the control of a wide range of annual broadleaf weeds and grasses in soybeans, peanuts, cucumbers, melons, and established woody ornamental stock. Naptalam may also be applied after emergence on soybeans.
   Types of formulations: Liquid and granular forms.
   Types and Methods of Application: Planter-mounted preemergence herbicide sprayer, aerial spraying, or granular applicator. Applications may be preemergence, or postemergence in soybeans and cucurbits. One application per year, except on cucurbits, where there may be two.
   Application Rates: For the liquid formulations, rates range from 2.0-6.0 lbs ai/A, except on ornamental nursery stock, where 8.0 lbs ai/A may be used. For the granular formulation, the rate is 4.3 lbs ai/A, except on cranberries, where 8.1 lbs ai/A may be used.
   Usual carriers: Water is the most common carrier for liquid formulations.

3. Science Findings:

Summary science statement:

Chemical characteristics:

Physical state:	solid
Color:	purple
Odor:	no data
Vapor pressure:	not available and not required
Melting point:	185°C (acid) and 234°C (sodium salt)
Octanol/water partition coefficient:	not available
Stability:	not stable above pH 9.5 or above 180°C. Tends to form the imide at elevated temperatures.

Solubility:

Solvent	Naptalam	Naptalam, Sodium
Water	0.02	30.0
Acetone	0.59	1.68
Xylene	*	0.04
Benzene	*	0.05
Hexane	*	*
Chloroform	0.01	*
DMF	3.94	5.53
DMSO	4.31	140.0
Ether	0.11	0.01
2-propanol	0.21	2.09
MEK	0.37	0.59

* Insoluble in this solvent.

Unusual handling characteristics: Very hard water and water with low pH may cause precipitation of free acid from solution of naptalam sodium. May be incompatible with some pesticides. High electrolyte may be a problem in mixtures with soluble fertilizers.

Toxicological characteristics:

Acute Effects:

Acute Oral LD_{50} - (rats) 8.2 g/kg (naptalam), 1.7 g/kg (naptalam sodium)(Category IV)
Acute Dermal LD_{50} - not available
Acute Inhalation Toxicity - not available
Primary Eye Irritation - (rabbits) Corneal opacity in 5 of 6 animals: reversed in 4 of 6 animals at day 7. Conjunctival damage in 6 of 6 animals:, reversed in 3 of 6 animals at day 7. (Category I)

Major Routes of Exposure: Dermal is the major route, followed by ocular and inhalation.

Chronic Effects:

Oncogenicity - <u>Mice</u>: Negative at 5000 ppm, the highest dose tested.
<u>Rats</u>: data gap

Teratology - Sprague-Dawley rats: Increased maternal mortality and resorptions were noted at the mid dose level and above. NOEL for maternal and fetotoxic effects is therefore 15 mg/kg/day (lowest dose tested).

Reproduction - data gap
Mutagenicity - data gap
Chromosomal Abberation - data gap

Physiological and Biochemical Behavioral Characteristics:

Translocation: When applied to soil, naptalam is absorbed by the roots and translocated to the leaves.
Mechanism of Action: Blockage of indoleacetic acid (IAA) action.

Environmental Characteristics:

Absorption and leaching characteristics: Incomplete data indicate that naptalam is very mobile in a fine sand, a sandy loam, and a silt loam soil, and slightly mobile in muck soil. Retention of naptalam by soil is correlated with CEC and organic matter content. The submitted data are not sufficient to fill data requirements.
Microbial breakdown: Data gap
Loss from photodegradation and/or volatilization: Data gap
Resultant average persistance: Unknown
Half-life in Water: Unknown.

Ecological characteristics:

Hazards to Birds: Low toxicity, suggesting minimal hazards.
Hazards to Fish and Aquatic Invertebrates: Low toxicity, suggesting minimal hazards.
Potential Problems with Endangered Species: No hazards indicated.

Tolerance Reassessment:

   List of crops and tolerances: (40 CFR 180.297)

COMMODITY	(PPM)
Cantaloups	0.1N
Cranberries	0.1N
Cucumbers	0.1N
Muskmelons	0.1N
Peanuts	0.1N
Peanuts, hulls	0.1N
Peanuts, hay	0.1N
Soybeans	0.1N
Soybeans, hay	0.1N
Watermelons	0.1N

   List of food contact uses: Cantaloups, cranberries, cucumbers, muskmelons, peanuts, soybeans, watermelons.
   Results of tolerance assessment: No ADI can be set at this time. The Agency will recommend that the negligible (N) classification be removed from all tolerances.

4. Summary of Regulatory Position and Rationale:

   Use Classification: Not a restricted use pesticide. Low toxicity.
   Groundwater Monitoring: Not required at this time. Data may be required later if warranted by environmental fate data.
   New Uses: The Agency will not approve new tolerances for significant new crops until data gaps regarding acute toxicity, teratogenicity, oncogenicity, plant and animal metabolism, storage stability, reproductive effects, and environmental fate are filled.
   Use, Formulation or Geographic Restrictions: Manufacturing-use products (of which there are none registered at this time) may only be formulated into end-use products intended for use as an herbicide on cucurbits, soybeans, peanuts, cranberries, or woody ornamental stock.

   Unique Label warning statements:

1. Use Pattern Statements:

Labels of all MPs must bear the statement:

"For formulation into end-use herbicide products intended only for use on soybeans, peanuts, cucurbits, cranberries, or woody nursery stock."

2. Precautionary Statements

    Labels of MPs and EPs must bear the statements:

    a. Hazards to Humans Statements

    "DANGER: Harmful if swallowed. Corrosive. Causes irreversible eye damage. Do not get in eyes or on clothing. Wear a face shield or goggles. Wash thoroughly with soap and water after handling and before eating or smoking. Remove contaminated clothing and wash before reuse." and;

    b. Statements of Practical Treatment

    "<u>If in eyes</u>: Flush with water for fifteen minutes. Call a physician."

    "<u>If swallowed</u>: Call a physician or Poison Control Center. Drink promptly a large quantity of milk, gelatin solution, or, if these are not available, drink large quantities of water. Avoid alcohol. <u>NOTE TO PHYSICIAN</u>: Probable mucosal damage may contraindicate the use of gastric lavage."

    c. Environmental Hazard Statement

    The following specific statements must appear on the labels of all MPs:

    "Do not discharge effluent containing this product into lakes, streams, ponds, estuaries, oceans, or public waters unless this product is specifically identified and addressed in a NPDES permit. Do not discharge effluent containing this product to sewer systems without previously notifying the sewage treatment plant authority. For guidance contact your State Water Board or Regional Office of the EPA."

    The labels of EPs intended for outdoor use must bear one of the following statements, depending on the formulation of the product:

    Granular products must bear the statement:

    "Do not apply directly to water. In case of spills, collect for reuse or properly dispose of the granules. Do not contaminate water by cleaning of equipment or disposal of wastes."

    Non-granular products must bear the statement:

    "Do not apply directly to water. Do not contaminate water by cleaning of equipment or disposal of wastes."

576 Pesticide Fact Handbook

d. Grazing Restrictions Statement

The label of all EPs registered for use on peanuts must bear the statement:

"Do not graze or feed forage or hay from treated peanuts to livestock."

The label of all EPs registered for use on soybeans must bear the statement:

"Do not harvest soybeans or soybean hay earlier than 90 days posttreatment. Do not graze or feed soybean forage or hay from treated fields to livestock."

The labels of all products must bear the appropriate container disposal statement. See Appendix IV-4 of the guidance package.

The required statements listed in the Standard must appear on the labels of all MPs and EPs released for shipment after March 1, 1986. The labels of all MPs and EPs currently in the channels of trade must be modified to include all the listed statements by March 1, 1987. After review of data to be submitted under the Standard, the Agency may impose additional label requirements.

5. Summary of major data gaps

Dates when major data gaps are due to be filled.

Data Required	Due date (Time allowed after publication of the Standard)
Description of beginning materials and manufacturing process	twelve months
Discussion of formation of unintentional ingredients	twelve months
Preliminary analysis of samples	six months
Certification of limits	six months
Odor	six months
Density, Bulk Density, or Specific Gravity	six months
Dissociation Constant	six months
pH	six months
Livestock metabolism	twenty-four months
Plant metabolism	twenty-four months
Storage stability data	twenty-four months
Magnitude of the residue for each food use	twenty-four months
Accumulation Studies (confined rotational crops)	twenty-four months
(in fish)	six months
(in non-target aquatic organisms)	six months
Dermal Toxicity	six months
Inhalation Toxicity	six months
90-day feeding (rodent, non-rodent)	twelve months
Chronic toxicity	four years
Reproduction (2-generation)	twenty-two months
Chromosomal aberration	ten months
Other genotoxic effects (DNA repair)	ten months

Data Required Under a Data Call-in Letter Sent October 31, 1984	Approximate Due Date
Hydrolysis	August 1, 1985
Photodegradation (water, soil)	August 1, 1985
Metabolism studies in lab	November 1, 1987
Mobility studies - leaching and absorption/desorption	August 1, 1985
Dissipation studies in field (soil, water)	November 1, 1987
Octanol/water partition coefficient	August 1, 1985
Oncogenicity	November 1, 1988
Teratogenicity	January 1, 1987
Water Solubility	August 1, 1985
Vapor Pressure	August 1, 1985

6. <u>Contact person at EPA</u>: Robert Taylor, U.S. Environmental Protection Agency, TS-767-C, 401 M Street SW, Washington, DC 20460 (703) 557-1650

DISCLAIMER: The information presented in this Chemical Information Fact Sheet is for informational purposes only and may not be used to fulfill data requirements for pesticide registration and reregistration.

NITRAPYRIN

Date Issued: June 28, 1985
Fact Sheet Number: 54

1. Description of chemical

   Generic Name: 2-Chloro-6-(trichloromethyl)pyridine
   Common name: Nitrapyrin
   Trade name: N-Serve
   EPA Shaughnessy code: 069203
   Chemical Abstracts Service (CAS) number: 1929-82-4
   Year of initial registration: 1974
   Pesticide type: Bacteriostat
   Chemical family: Pyridines
   U.S. and foreign producers: Dow Chemical Company

2. Use patterns and formulations

   Application sites: Nitrapyrin is registered for use as a nitrogen stabilizer in corn, cotton, rice, sorghum, strawberries, and wheat
   Types of formulations: Emulsifiable concentrate and soluble concentrate
   Types and methods of application: Broadcast or band with soil injection or incorporation during or immediately after application
   Application rates: 0.25 lb ai/A to 1.0 lb ai/A
   Usual carrier: Water and/or fertilizer

3. Science findings

   Summary science statement: The existing data base is insufficient to fully assess the toxicology, residue chemistry, environmental fate or ecological effects of nitrapyrin. Previous toxicity testing had been conducted using the major metabolite 6-chloropicolinic acid and in most areas, previous testing was found, upon reevaluation, to be supplementary or invalid according to current testing guidelines.

   Chemical characteristics:

   Nitrapyrin is a white crystalline solid with a mildly sweet odor. The melting point is 62-63°C. The vapor pressure of nitrapyrin is 2.8×10^{-3} mm Hg at 23°C and its solubility is 0.004 grams (g)/100 g in water.

   Toxicological characteristics:

   Acute Toxicology:

   Nitrapyrin has moderate oral toxicity (Toxicity Category III) and is a moderate dermal irritant (Toxicity Category III).

Chronic Toxicology:

Chronic feeding/oncogenicity, teratogenicity and reproduction testing using nitrapyrin TGAI have been identified as data gaps. Previous chronic testing was conducted using 6-chloropicolinic acid (6-CPA), a major metabolite which was initially identified as the residue of concern. A chronic feeding/oncogenicity study in the rat using 6-CPA, although considered supplementary data, indicated (upon reevaluation) significant compound related microscopic changes (biliary hyperplasia) occurred in female rat livers at all doses tested.

Nitrapyrin did not demonstrate an increase in unscheduled rat hepatocyte DNA synthesis and was not genotoxic in mutagenicity tests. Gene mutation (Ames) and chromosomal aberration tests are data gaps.

Physiological and biochemical behavioral characteristics:

Mechanism of Action:

Nitrapyrin is an ammoniacal nitrogen stabilizer which inhibits the nitrification of ammoniacal and urea nitrogen fertilizer in the soil by selectively and temporarily, inhibiting Nitrosomonas spp. These bacteria convert ammonium nitrogen to nitrite nitrogen, which in turn is converted to nitrate nitrogen by other bacteria, e.g., Nitrobacter.

Environmental characterisitics:

Degradation:

Preliminary data indicate that hydrolysis of nitrapyrin is rapid, the rate increasing with increasing temperature. It is not affected by pH and degrades more rapidly in the light than in the dark.

Persistence:

In soil, nitrapyrin has a half life of < 3 to 35 days, depending on soil type.

Bioaccumulation:

Nitrapyrin accumulates in the edible and visceral tissues of bluegill sunfish with maximum bioconcentration factors of 33 x and 60 x, respectively. The residues depurate rapidly.

Ecological Characteristics:

Avian studies:

Acute oral (Mallard duck)>2708 mg/kg (slightly toxic).
Avian dietary (Japanese quail) >820 ppm (slightly toxic)

Aquatic species studies:
Daphnia magna 48-hour acute is 5.8 mg/l (moderately toxic).
Bluegill sunfish 96-hour acute is 7.876 mg/l (moderately toxic).
Rainbow trout 96-hour acute is 9.191 mg/l (moderately toxic).

Potential problems for endangered species:

Based on an estimated environmental concentration (EEC) of 0.1 ppm in water (due to runoff), the expected hazard to aquatic organisms is expected to be minimal. However, use in rice, although estimated to be negligible, may result in an EEC in floodwater of 0.4 ppm. This level may impact on endangered or threatened species, such as the Fat Pocketbook Pearly Mussel, which is bottom dweller, and as such would be exposed to nitrapyrin in the both the water and sediments. The aquatic invertebrate toxicity test data are not appropriate in assessing hazard to bivalves. Therefore, marine mollusc shell deposition testing is required. Formal consultation with the Office of Endangered Species, USFWS, will be considered after these data are received and evaluated.

Tolerance assessments:

U.S. tolerances for residues of nitrapyrin [2-chloro-6-(trichloromethyl) pyridine] and its metabolite, 6-chloropicolinic acid, in or on raw agricultural commodities are as follows [40 CFR §180.350]:

Commodities	Tolerance (ppm)
Cattle, fat	0.05
Cattle, mbyp	0.05
Cattle, meat	0.05
Corn, fodder	1.0
Corn, forage	1.0
Corn, grain	0.1
Corn, fresh	0.1
Cottonseed	1.0
Goats, fat	0.05
Goats, mbyp	0.05
Goats, meat	0.05
Hogs, fat	0.05
Hogs, mbyp	0.05
Hogs, meat	0.05
Horses, fat	0.05
Horses, mbyp	0.05
Horses, meat	0.05
Poultry, fat	0.05
Poultry, mbyp	0.05
Poultry, meat	0.05
Rice, grain	0.2
Rice, straw	2.0
Sheep, fat	0.05
Sheep, mbyp	0.05
Sheep, meat	0.05
Sorghum, fodder	0.5
Sorghum, forage	0.1
Sorghum, grain	0.1
Strawberries	0.2
Wheat, forage	0.5
Wheat, grain	0.1
Wheat, straw	0.5

No mexican tolerances or Codex MRLs have been established for residues of nitrapyrin. A Canadian tolerance of 0.1 ppm exists for, presumably, residues of nitrapyrin per se in or on corn.

Residue studies are not adequate to support existing tolerances. The plant and animal metabolism of nitrapyrin are not adequately understood and analytical methodology may need to be revised when required data are evaluated. Storage stability data and additional, more geographically representative residue data are required for most commodities.

Tolerances have been proposed for residues of nitrapyrin and its metabolite, 6-CPA, in or on the crop group cereal grains; lettuce; potatoes; soybeans and soybean forage and hay. The Agency has decided, however, that no new tolerances or exemptions will be established until all major toxicology and residue chemistry data requirements have been satisfied.

4. Summary of Regulatory Position and Rationale

Risk/benefit review: None of the risk criteria set forth in Title 40 Code of Federal Regulations §162.11 for commencing a special review have been met or exceeded by nitrapyrin.

Ground Water Potential: Nitrapyrin per se has not been previously identified as a potential ground water contaminant and in monitoring studies conducted to date nitrapyrin residues have not been found in ground water. However, one study using the 6-CPA metabolite indicated a moderate potential for 6-CPA to leach in loam soil. Therefore, environmental chemistry studies are being required in the minimum times necessary to conduct the tests and will be evaluated on an expedited basis when received.

5. Summary of major data gaps and when these are due to be filled

6 Months:
Product identity and composition
Physical and chemical characteristics
Avian subacute dietary and freshwater fish toxicity
Acute freshwater invertebrate toxicity

9 Months:
Hydrolysis, photodegradation
Acute dermal and inhalation toxicity
Primary eye irritation and dermal sensitization
Gene mutation (Ames)

12 Months:
Preliminary analysis of product samples
Leaching and adsorption/desorption
Structural chromosomal aberration
Acute toxicity to estuarine and marine organisms

15 Months:
    Teratology (2 species)
18 Months:
    Animal metabolism and residues
24 Months:
    Plant metabolism
    Storage stability
    Analytical methods and stability of residues under storage
    Crop residue studies for corn, sorghum, strawberries, wheat and cottonseed
    General metabolism
27 Months:
    Aerobic aquatic metabolism
    Field dissipation (soil) and dissipation aquatic (sediment)
39 Months:
    Rotational crops (confined)
    Irrigated crops
    Reproduction (2-generation)
50 Months:
    Chronic feeding (2 species)
    Oncogenicity (2 species)
    Rotational crops (field)

6. Contact person at EPA

    Richard F. Mountfort
    Product Manager (23)
    Environmental Protection Agency (TS-767C)
    401 M Street, SW
    Washington, DC 20460
    (703) 557-1830

DISCLAIMER: The information presented in this Chemical Information Fact Sheet is for informational purposes only and may not be used to fulfill data requirements for pesticide registration and reregistration.

NORFLURAZON

Date Issued: December 1, 1984
Fact Sheet Number: 60

1. Description of the chemical

    Generic name: 4-chloro-5-(methylamino)-2-(alpha, alpha, alpha-trifluoro-m-tolyl)-3(2H)-pyridazinone
    Common name: Norflurazon
    Trade name: Zorial®, Solicam® and 'Evital®
    'EPA Shaugnessy Number: 105801
    Chemical Abstract Service Registry Number (CAS): 27314-13-2
    Year of initial registration: 1974
    Pesticide Type: Herbicide
    Chemical Family: Fluorinated pyridazinone
    U.S. & foreign producer: Sandoz Inc.

2. Use patterns and formulations

    Application sites: Norflurazon is registered for use as a selective preemergent herbicide to control germinating annual grasses and broadleaf weeds in cranberries, cotton, soybeans, almonds, apples, apricots, cherries, citrus (all), filberts, hops, nectarines, peaches, pears, pecans, plums, prunes, walnuts, and noncrop areas such as storage areas, airports and rights-of-way.
    Types of formulations: Norflurazon is the sole active ingredient in the following: 97% active ingredient (a.i.) technical manufacturing-use-product, 80% a.i. wettable powder, 50% a.i flowable, and 5% a.i. granular.
    Types and methods of applications: Band or broadcast ground application to soil surface. Aerial application is registered for cotton, cranberry and soybean use.
    Application rates: 0.5 to 8 lb. a.i. per acre (A):
        cranberries 4-8 lb. a.i./A; cotton and soybeans 1-2 lb. a.i./A (split application 0.5-1.0 lb. a.i./A); tree fruit, nut tree, citrus and hops 2-4 lb. a.i./A; and noncrop sites 4-8 lb. a.i./A.
    Usual carrier: Water

3. Science Findings

Summary science statement:

Norflurazon has a low acute toxicity and is not an eye or skin irritant or a skin sensitizer. The subchronic, chronic feeding, and reproduction studies did not produce results of toxicological concern. Norflurazon is not considered to be an oncogen or a teratogen. The mutagenicity studies reviewed thus far are negative. Norflurazon appears to be mobile in mineral soils and immobile in soils with high organic material and is persistent in soil. Norflurazon is relatively non-toxic to avian test species and is moderately to slightly toxic to aquatic (fresh water and marine) organisms. Data are available to determine and establish tolerances for residues of norflurazon and its desmethyl metabolite in over half of the commodities with established tolerances. Based on the established tolerances and the 6-month dog feeding study the percent of the accepatable daily intake utilized is 39%.

Chemical characteristics:

Norflurazon is a buff-white odorless crystalline solid. The melting point is 177 ± 3° C. The solubility of norflurazon at 25° C is 5 grams (g)/ 100 milliliters (ml) in acetone, insoluble in carbon disulfide, 14.2 g/100 ml in ethyl alcohol, 0.25 g/100 ml in xylol, and 28 parts per million (ppm) (w:w) in water. The vapor pressure is $< 1 \times 10^{-5}$ Torr (25°C). Norflurazon is quite stable in dilute acidic or basic aqueous solution and storage stability is greater than 2 years. No unusual handling characteristics were noted.

Toxicological characteristics:

Acute studies indicate the following:
    Rat acute oral was 9,000 milligrams (mg)/kilogram (kg), Toxicity Category IV.
    Rabbit acute dermal was > 20,000 mg/kg, Toxicity Category IV.
    Male rat acute inhalation of 80% WP was > 200 mg/Liter (L)/ 1 hour, Toxicity Category IV.
    Not an eye or skin irritant, Toxicity Category IV.
    Not a skin sensitizer.

Subchronic studies indicated the following:
    In a 6-month dog feeding study, the primary effects seen were congestion of the liver, hepatocyte swelling, increased liver weight, and an increase in colloidal vacuole in thyroid at 450 ppm. The No Observed Effect Level (NOEL) was 150 ppm. Levels tested were 0, 50, 150 and 450 ppm.
    In a 90-day rat feeding study, the primary effects were hypertrophic change in the thyroid glands at 2,500 ppm. The NOEL was 500 ppm. Levels tested were 0, 250, 500 and 2,500 ppm.
    In a 28-day mouse feeding study, diffuse and smooth granular livers and increased liver/body weight ratios were observed at 2,520 ppm. The NOEL was 420 ppm. Levels tested: 0, 70, 210, 420 and 2,520 ppm.
    A 14-day inhalation study in the rat was submitted. Levels tested were 0.1, 1.0, and 10.0 mg/L. The NOEL was 10 mg/L.
    A 21-day dermal toxicity study in the rabbit was performed

on 80% WP (Wettable Powder) norflurazon. Levels tested were 750 mg/kg/day and 2000 mg/kg/day. The NOEL was > 2000 mg/kg/day. The chronic studies indicated the following:

Chronic-feeding studies:
  A 2-year rat feeding study was conducted using technical norflurazon. Rats were fed dietary levels of 2, 15, 125, 375 and 1025 ppm. In the high dose group histopathological alterations included an increase in the number of chromophobe adenomas of the pituitary, nodular or cortical hypertrophy in adrenals and nephritis and/or casts in kidneys of the male rats; and fatty changes in adrenals, edometritis and squamous metaplasia of the uterus, cystic ovaries and hyaline casts and/or nephritis in kidneys of the females. A NOEL was demonstrated at 375 ppm. Norflurazon domonstrated no tumorigenic effect in the test animals in any of the dose levels tested.

  In a 2-year feeding study, mice were fed 0, 0 (double control), 85, 340 and 1360 ppm of technical norflurazon. Histopathological alterations included hepatoma/hyperplasia - hypertrophy in the liver at 1360 ppm. The NOEL level observed was 340 ppm. There was no significant increase in these lesions in the lower levels over that of the control. The failure to induce such lesions in the other long term studies permits the conclusion that this is not a potential carcinogenic response, but a toxic response to rather high level of chemical insult.

Reproduction studies:
  A 3-generation reproduction study was conducted in the rat. Norflurazon was fed at dietary levels of 0, 125, 375 and 1025 ppm for three generations. At 1025 ppm norflurazon caused reduced fertility, gestation and viability indices. No teratogenic effects were seen at any dose tested. The NOEL was established at 375 ppm.

  In a 1-generation reproduction study in the mouse, norflurazon was fed at dietary levels of 0, 0, 85, 170 and 340 ppm. No adverse findings were observed in any of the doses tested. The NOEL was established at 340 ppm.

Teratogenicity studies:
  Pregnant rabbits were fed a diet containing 0, 10, 30, and 60 mg/kg norflurazon on gestation days 6 through 15. Norflurazon was not teratogenic at 60 mg/kg/day. Maternal body weight was decreased at 60 mg/kg. Fetotoxic effects seen at 30 and 60 mg/kg/day were decreased weight and incomplete ossified variations. Maternal toxic NOEL was observed at 30 mg/kg/day. Fetotoxic NOEL was observed at 10 mg/kg/day.

  In the second teratology study, pregnant rats were fed 0, 100, 200 and 400 mg/kg/day of norflurazon on gestation days 6 through 15. Norflurazon was not embryotoxic or teratogenic. The NOEL was 400 mg/kg/day.

Mutagenicity studies:
  In two Ames mutagenic assays, norflurazon was tested in _Samonella typhimurium_ strains, TA-1535, TA-1537, TA-1538, TA-98, TA-100 and D-4 _Saccharomyces cerevisiae_ strain. The doses employed ranged from 0.1 micrograms (ug) to 500 ug per plate. The compound was tested directly in the presence of liver microsomal enzyme preparation from Aroclor induced rats. Norflurazon did not demonstrate mutagenic activity.

A reverse mutagenicity assay using <u>Salmonella typhimurium</u> strains TA-1535, TA-1537, TAS-1538, TA-98 and T-100, also <u>E. coli</u>, WP2 hcr strain (tryptophanrequiring strain) was conducted. The doses employed were 5, 10, 50, 100, 1,000 and 5000 ug per plate. Norflurazon was negative in this test.

Metabolism studies:
(^{3}H, ^{14}C)- norflurazon was administed by gavage to 10 male Wistar strain rats at a dose of 10 mg per day for 15 days. Approximately 17 percent of the administed dose was excreted in the urine and about 57 percent in the feces. Small amounts of the parent compound were isolated from the urine (0.1%) and and larger amounts from the feces (5.4%). Only traces of radioactivity were present in the tissues examined. Three major pathways seem to be operative in detoxification of norflurazon in the rat: Desmethylation, yielding desmethyl metabolite of norflurazon; a hydroxylation process involving the replacement of chlorine on carbon-4 of the pyridazinon ring; and conjugation through sulfur introduced at carbon-4 of the pyridazinon ring.

Physiological and biochemical behavioral characteristics:

Norflurazon is absorbed by the roots of weeds as they germinate and is translocated to the growing parts where it inhibits carotenoid biosysnthesis resulting in chlorophyll photodegradation in susceptible species. On emergence from the soil, the weed seedlings turn white or pinkish, become necrotic and die.

Environmental characteristics:

Norflurazon residues appear to be relatively mobile in most mineral soils and immobile in soils with high organic matter. The half-life in soils ranges from 38 days to 731 days.

Ecological characteristics:

Avian studies:
    Acute oral (Mallard duck) > 2510 mg/kg.
    Acute dietary (Bobwhite quail) > 10,000 mg/kg.
    Acute dietary (Mallard duck) > 10,000 mg/kg.
    Reproduction (Mallard duck and bobwhite quail) was not affected up to 40 ppm dietary exposure (highest dose tested).
Aquatic species studies:
    <u>Daphnia magna</u> acute 48 hour no effect level was 15 ppm (the highest level tested due to solubility of norflurazon technical).
    <u>Daphnia magna</u> chronic life cycle minimum treshold concentration was > 1.0 < 2.6 ppm due to effect on offspring production.
    Bluegill sunfish 96-hour acute was 16.3 ppm.
    Rainbow trout 96-hour acute was 8.1 ppm.
    Fathead minnow partial chronic maximum toxicant concentration (MATC) was > 1.1 < 2.1 ppm based on growth.

Rainbow trout partial chronic MATC was > 0.77 < 1.5 ppm based upon survival and growth.
Atlantic oyster larvae acute NOEL was 10 ppm.

Tolerance assessments:

U.S. tolerances for residues of norflurazon and its desmethyl metabolite in or on raw agricultural commodities are as follows [40 CFR §180.356(a)]:

Commodities	Maximum Residue Limit (ppm)
Almond, hulls	1.0
Almonds, meat	0.1
Apricots	0.1
Apples	0.1
Cattle, fat	0.1
Cattle, meat	0.1
Cattle, meat-by-products (mbyp)	0.1
Cherries	0.1
Citrus fruit	0.2
Cottonseed	0.1
Cranberries	0.1
Filberts	0.1
Goats, fat	0.1
Goats, meat	0.1
Goats, mbyp	0.1
Hogs, fat	0.1
Hogs, meat	0.1
Hogs, mbyp	0.1
Hops, green	1.0
Horses, fat	0.1
Horses, meat	0.1
Horses, mbyp	0.1
Milk	0.1
Nectarines	0.1
Pecans	0.1
Peaches	0.1
Pears	0.1
Plums (fresh prunes)	0.1
Poultry, fat	0.1
Poultry, meat	0.1
Poultry, mbyp	0.1
Sheep, fat	0.1
Sheep, meat	0.1
Sheep, mbyp	0.1
Soybeans	0.1
Soybean forage	1.0
Soybean hay	1.0
Walnuts	0.1

U.S. tolerances for indirect residues of norflurazon and its desmethyl metabolite in raw agricultural commodities when present as a result of application to cotton when peanuts are a replacement or follow-up crop are as follows [40 CFR §180.356(b)]:

Commodities	Maximum Residue Limit (ppm)
Peanuts	0.2
Peanut, hay	0.5

588   Pesticide Fact Handbook

>   Peanut, hulls   0.5
>   Peanut, vines   0.5

A food additive tolerance has been established for residues of
    norflurazon and its desmethyl metabolite in dried hops at 3.0 ppm
    [21 CFR §193.324].
Feed additive tolerances have been established for residues of norflurazon
    and its desmethyl metabolite in citrus molasses at 1.0 ppm and dried citru
    pulp at 0.4 ppm [21 CFR §561.283].
No Codex Almentarius or Mexican or Canadian tolerances have been
    established for residues of norflurazon on the above commodities.
The acceptable daily intake (ADI) was established using the 6-month
    dog feeding study with a no observed effect level of 150 ppm
    (3.750 mg/kg/day). Using a 1,000 fold safety factor the ADI is
    calculated to be 0.0038 mg/kg/day. The maximum permitted intake
    (MPI) for a 60 kg human is calculated to be 0.2250 mg/day. The
    current theoretical maximum residue contribution (TMRC) for
    norflurazon, based on the established tolerances, is 0.0877
    mg/day for a 1.5 kg diet and the percent ADI utilized is 38.98%.
Residue studies are adequate to support tolerances established for
    almonds, apricots, cherries, cranberries, cottonseed, filberts,
    grapes, nectarines, peaches, peanut hulls, pears, pecans, walnuts,
    milk, and the fat, meat, and meat-by-products of cattle, goats, hogs,
    horses, poultry, and sheep.

4. Summary of Regulatory Position and Rationale

Use, formulation, manufacturing process or geographical restrictions: None
    are required.
Unique precautionary statements, protective clothing requirements or
    reentry intervals: None required.
Risk/benefit review: None of the risk criteria set forth in Title 40 Code
    of Federal Regulations §162.11 have been exceeded by norflurazon.
Ground Water Potential: Because of the mobility and long half-life,
    norflurazon presents a potential for ground water contamination.
    The Ground Water Studies will be requested in an accelerated time
    frame. Due to the inadequate data base and since norflurazon to date
    has not been found in ground water, no interim restrictions were imposed.
    Any future decisions depend on the results of the required studies.

5. Summary of major data gaps and when these are due to be filled

Ground Water Studies:
    Hydrolysis, photodegradation and mobility are required within
        6 months after receipt of the Guidance Package.
    Metabolism, soil and aquatic (sediment) dissipation are required
        within 2 years after receipt of the Guidance Package.
    Soil, long term dissipation is required within 4 years after
        receipt of the Guidance Package.
Short term studies required to be filled within 6 months after
receipt of the Guidance Package:
    Product Chemistry: Description of manufacturing process, discussion
        of formation of impurities, analysis of product, density, dissociation
        constant, octanol/water partition coefficient, oxdizing or reducing
        action, explodability, pH and stability.

Honeybee acute contact.
Female rat metabolism.

Mutagagenicity studies for chromosomal aberation and other mechanisms of mutagenicity are required to be filled within 1 year after the receipt of the Guidance Package.

Long term studies required to be filled within 2 years after the receipt of the Guidance Package:
Rotational crops.
Plant and animal metabolism.
Analytical methods and stability of residues under storage.
Crop residues studies for soybeans, citrus, apples, plums, hops and peanuts.

6. Contact person at EPA

Richard F. Mountfort
Product Manager (23)
Environmental Protection Agency (TS-767C)
401 M Street S.W.
Washington, D.C. 20460
(703) 557-1830

DISCLAIMER: The information presented in this Chemical Information Fact Sheet is for informational purposes only and may not be used to fulfill data requirements for pesticide registration and reregistration.

OXAMYL

Date Issued: June, 1987
Fact Sheet Number: 129

1. Description of Chemical

    Generic Name: Methyl N'N'-dimethyl-N-[(methylcarbamoyl)oxy)-1-oxamimidate.
    Common Name: Oxamyl
    Trade Names: Vydate, DPX 1410, thiooxamyl.
    EPA Shaughnessy Code: 103801
    Chemical Abstracts Service (CAS) Number: 2135-22-0
    Year of Initial Registration: 1974
    Pesticide Type: Insecticide/Acaricide/Nematocide
    Chemical Family: Carbamate
    U.S. and Foreign Producers: E.I. DuPont de Nemours

2. Use Patterns and Formulations

    - Application Sites: Soil and foliar treatments for fruit, vegetables, ornamentals and miscellaneous crops. Greenhouse applications for vegetables and ornamentals.
    - Types of Formulations: Soluble concentrate, granular
    - Types and Methods of Application: Foliar, row (at planting), broadcast incorporated, band incorporated, row band unincorporated, transplant water, drip irrigation.
    - Application Rates: 0.0625-2.00 lbs ai/A foliar
        0.5-8.0 lbs ai/A soil application
        0.31-0.5 lbs ai/A transplant/irrigation water
        1.0-8.0 lbs ai/A drip irrigation

3. Science Findings

   Chemical Characteristics

   - Physical state: crystalline solid
   - Color: white
   - Odor: slightly sulfurous
   - Molecular weight and formula: 219.3 $C_7H_{13}N_3O_3S$
   - Melting point: 100-102 degrees C
   - Boiling Point: 100 degrees C
   - Density: 1.12 at 25 degrees C
   - Vapor pressure: 25.2 mmHg at 30 degrees C
   - Solubility in various solvents: Very soluble in methanol, soluble in water, acetone and ethanol; practically insoluble in toluene.
   - Stability: Stable in solid form and in liquid formulations. Aqueous systems decompose slowly; aeration, sunlight, alkalinity, and higher temperatures increase the rate of decomposition in water.

   Toxicology Characteristics

   - Acute oral: Oral LD50 for the technical is 3.1 mg/kg and 2.5 mg/kg for male and female rats respectively. Toxicity Category I. Atropine has been shown to be an antidote in both rats and monkeys.
   - Acute dermal: data gap.
     Primary dermal irritation: data gap.
     Primary eye irritation: data gap.
   - Skin sensitization: data gap.
   - Acute inhalation: One hour LC50 for the technical is 0.17 mg/L and 0.12 mg/L for male and female rats, respectively. Four hour LC50 for male rats is 0.064 mg/L. Toxicity Category I.
   - Subchronic Oral: data gap, (superseded by chronic tox. data gap)
   - Subchronic Dermal: data gap.
   - Chronic toxicity: data gap, (rodent & non-rodent)
   - Oncogenicity: No increased incidence in tumors for mice. NOEL 25 ppm, LOEL 50 ppm. Rat study is data gap.
   - Teratogenicity: Rabbit study indicates developmental NOEL is 4 mg/kg. Rat study is data gap.
   - Reproduction: Data gap.
   - Mutagenicity: Studies indicate that oxamyl does not cause DNA damage. Gene mutation and chromosomal aberration assays are data gaps.

   Physiological and Biochemical Characteristics

   - The mode of action in biological systems is cholinesterase inhibition. Toxic signs result from excessive stimulation at cholinergic junctions in the autonomic and central nervous system and include tremors and salivation.
   - Metabolism: Data gap.

Environmental Characteristics

- Data gaps exist for most applicable studies. However, available supplementary data indicate general trends of oxamyl behavior in the environment. Oxamyl degrades to the oxime, and both have been shown to leach in field studies. Oxamyl residues have been found in well-water surveys in New York and Massachusetts. Additional data have been reviewed as part of a ground water data call-in and oxamyl will be monitored in the National Survey for Pesticides in Ground Water.

Ecological Characteristics

- Avian oral toxicity: Acute oral LD50 3.16 mg/kg (mallard)
- Avian subacute dietary toxicity: data gap
- Freshwater fish acute toxicity (48 hr or 96 hr. EC50): 5.6 ppm for bluegill; 4.2 ppm for rainbow trout.
- Freshwater invertebrate toxicity (48 hr. or 96 hr. EC50): 5.7 ppm for Daphnia magna.
- Estuarine and Marine Organisms, acute toxicity: data gap.
- Fish Early life stage: data gap.

Tolerance Assessments:

The NOEL, based upon the available supplementary 2 year rat feeding study, is 50 ppm. As the study was supplementary, the ADI is now considered provisional (PADI). Utilizing a safety factor of 100, the PADI is set at 0.025 mg/kg/day.
- Additional data are necessary to support tolerances for residues of oxamyl in or on carrots, ginger, potatoes, garlic, onions, celery, bananas, cottonseed, peanut forage and hay, pineapple forage, and citrus.
- Processing studies are required for the commodities: citrus, potatoes, soybeans, tomatoes, cottonseed, pineapple, and peanuts.
- Data are required to evaluate the necessity for tolerances to support the registered uses on nonbearing fruit trees and strawberries.
- Data are required to determine the exposure of man to residues of oxamyl on tobacco.

Summary Science Statement

Oxamyl is a water soluble systemic carbamate insecticide, nematocide, and acaricide with very high acute toxicity to both birds and mammals. Chronic feeding studies indicate decreases of body weights but no other compound related effects. Available data do not indicate any concerns about developmental effects. Oxamyl did not cause DNA damage in bacteria. Due to data gaps, a determination as to whether the ADI should be revised is postponed until such time as the requested data have been submitted and reviewed. Oxamyl has been shown to leach and has been found in well water. Data have been requested under the ground water data call-in and are pending review. Oxamyl will be monitored in the National Survey for Pesticides in Ground Water. A ground water advisory statement is now required for all products containing oxamyl. Incidents have been reported as a result of applicator and worker exposure to oxamyl. Data to adequately evaluate the hazards of oxamyl to applicators and field workers are being required. Oxamyl is toxic to non-target birds, mammals, fish and invertebrates, including endangered and threatened species. Special labeling relative to these hazards has been required through PR Notices 87-4 and 87-5.

4. Summary of Regulatory Positions and Rationales

The Agency is deferring actions relative to Special Review until such time as field monitoring studies for birds and mammals have been submitted.
- The Agency will continue to restrict the 24% SC formulation. The 10% granular formulation will be restricted pending the outcome of required studies for acute toxicity and hazards to wildlife.
- EPA is requiring more stringent precautionary labeling, including requirements to wear protective clothing and equipment to reduce exposure.
- Data are required to evaluate the potential for exposure both as a result of drift and as a result of dermal contact with residues on treated crops.
- A reentry interval of 24 hours is being imposed for all agricultural uses until such time as the appropriate data have been submitted and evaluated.
- Endangered species labeling is required for uses of oxamyl likely to result in exposure to these non-target organisms.
- The Agency is requiring a groundwater advisory to be added to all oxamyl end-use labels.
- A 6 month crop rotation interval is being imposed for those crops for which no tolerance has been established.
- The Agency is prohibiting the use of human flaggers during aerial application, unless they are in an enclosed vehicle.

5. Summary of Major Data Gaps

   - All product chemistry
   - Acute dermal toxicity
   - Primary eye irritation
   - Primary skin irritation
   - Dermal sensitization
   - 21 day subchronic dermal toxicity
   - Rodent and Non-rodent chronic toxicity
   - Oncogenicity (rat)
   - Teratogenicity (rat)
   - Reproduction
   - Gene mutation
   - Chromosomal aberration
   - General metabolism
   - Avian dietary LC50
   - Avian and mammalian field studies
   - Estuarine and marine fish LC50
   - Fish early life stage
   - Honeybee toxicity for residues on foliage
   - Hydrolysis
   - Photodegradation in water
   - Anaerobic soil metabolism
   - Soil dissipation
   - Rotational crops (confined)
   - Glove permeability
   - Droplet size/drift evaluation
     Leaching and adsorption/desorption
   - Storage stability
   - Plant residues
   - Processing studies
   - Reentry studies

6. Contact Person at EPA

   Dennis Edwards
   Acting Product Manager No. 12
   Insecticide-Rodenticide Branch
   Registration Division (TS-767C)
   Office of Pesticide Programs
   Environmental Protection Agency
   401 M Street, SW
   Washington, DC 20460
   - Office location and telephone number:
     - Room 202, Crystal Mall #2
     - 1921 Jefferson Davis Highway
     - Arlington, VA 22202
     - (703) 557-2386

DISCLAIMER:

THE INFORMATION PRESENTED IN THIS CHEMICAL INFORMATION FACT SHEET IS FOR INFORMATIONAL PURPOSES ONLY AND NOT TO BE USED TO FULFILL DATA REQUIREMENTS FOR PESTICIDE REGISTRATION AND REREGISTRATION.

PARAQUAT

Date Issued: June, 1987
Fact Sheet Number: 131

1. Description of Chemical

    Generic Name: 1,1'-dimethyl-4-4'-bipyridinium ion

    Common Name: Paraquat

    Trade Names: Actor Cekuquat, Crisquat, Dextrone, Dexuron, Esgram, Gramanol, Gramoxone, Gramuron, Hebaxon, Herboxone, Goldquat 276, Paracol Paraquat CL, Pillarquat, Pillarxone, Sweep, PP148 (dichloride) and Dextrone X.

    EPA Shaughnessy Code: 061601 (dichloride)

    Chemical Abstracts Service (CAS) Number: 1910-42-5 (dichloride)

    Year of Initial Registration: 1964

    Pesticide Type: Herbicide, defoliant, desiccant, and plant growth regulator

    Chemical Family: Bipyridinium or dipyridinium

    U.S. and Foreign Producers: ICI Americas, Inc.; Taiwan by Comlets Chemical Industrial Company, Ltd., and Shinung Corp., Italy by VisplantChrimiasero S.p.I.; Spain by Hightex S.A.; Great Britain by ICI Plant Protection Division.

2. Use Patterns and Formulations

    Application Sites: Terrestrial food crops; terrestrial nonfood crops; terrestrial noncrop, forestry and domestic outdoor sites.

    Major Crops Treated: Field crops (corn and soybeans), and fruit and nut crops.

    Types and Method of Application: Foliarly applied by broadcast, band, or directed spray by ground equipment or broadcast by aircraft. It may be applied preplant, preemergence, or preharvest in relation to the crop.

Application Rates: 0.25 lb active ingredient/A to 1.0 lb active ingredient/A (0.28 to 1.12 kg active ingredient/ha).

Types of Formulation: Aqueous solution containing 2 lbs (0.24 kg) paraquat cation per gal (L).

Usual Carrier: Water.

3. Science Findings

Summary Science Statement: Paraquat is extremely toxic (Category I) via oral, dermal, and inhalation exposure routes to mammals. It is not teratogenic to mice or rats. Paraquat is not oncogenic to mice but is oncogenic to rats. Paraquat is weakly genotoxic - it was negative for mutagenicity in eight studies, weakly positive in four studies and positive in four studies.

Paraquat is moderately toxic to birds, slightly toxic to freshwater fish, moderately toxic to aquatic invertebrates, and relatively nontoxic to honeybees. Paraquat is not believed to cause problems with stable wildlife populations but may be hazardous to unstable or endangered populations of plants and animals.

Paraquat dichloride was stable to hydrolysis and photolysis in soil, preliminary data indicate that paraquat has a half-life of greater than 2 weeks in water plus soil, is immobile in silt loam and silty clay loam, and slightly mobile in sandy loam and potentially mobile in sandy soils extremely low in organic matter. The half-life of paraquat in water is approximately 23 weeks. Paraquat is not readily desorbed from the soil and is not likely to contaminate ground water in agricultural soils. Preliminary data indicate that the paraquat degradate ^{14}C-carboxy-1-methyl pyridinium (QINA) chloride is loosely absorbed in the soil and is potentially mobile and has a leaching potential.

Chemical Characteristics: Analytical grade of paraquat dichloride is a colorless odorless hygroscopic powder, whereas the technical product is yellow. Paraquat is very soluble in water, slightly soluble in alcohols and insoluble in hydrocarbons. It is nonvolatile, corrosive to metals, stable at room temperature (either as a solid or an aqueous solution at neutral or acidic pH), but is hydrolyzed by alkali, decomposes photochemically, and melts (with decomposition) at 300 °C.

Toxicology Characteristics:

Acute Toxicology

    Acute Oral Toxicity = 100 and 126 mg paraquat cation/kg
      (rat)   for males; 112-150 mg paraquat
                                cation/kg for females
                                Toxicity Category II

    Acute Oral Toxicity = 22 mg paraquat cation/kg for
      (guinea pig)   males and females
                                Toxicity Category I

    Acute Dermal Toxicity = 59.9 mg paraquat cation/kg
      (rabbit)   Toxicity Category I

    Acute Inhalation = 0.6 to 1.4 ug paraquat ion/L
      (rat)   Toxicity Category I

    Eye Irritation = Severe Eye Irritant
      (rabbit)   Toxicity Category I

    Primary Dermal Irritation = Primary irritation score of
      (rabbit)   2.1 - Toxicity Category III

    Skin Sensitization = Not a sensitizer
      (guinea pig)

Major Routes of Exposure:

Acute Inhalation: Two inhalation studies were submitted. One study had an inhalation toxicity of about 1.0 ug/L (Toxicity Category I) with 90 percent of the particle diameters below 0.3 um. A second study was performed with particles of median diameter of 21.5 to 23 um. The inhalation toxicity for this study was 3.5 ug/L (Toxicity Category I). These studies show that inhalation toxicity is highly dependent on particle size.

Information received on particle size of paraquat droplets formed during aerial application and during knapsack spraying indicate that virtually no droplets smaller than 15 um were formed during either method of application.

Subchronic Dermal Exposure: A 21-day dermal toxicity study in rabbits was submitted with a NOEL of 1.15 mg paraquat cation/kg body weight (bwt) for local skin effects and a NOEL for systemic toxicity of 6 mg paraquat cation/kg bwt. Data are available indicating that a dermal absorption rate for humans is about 0.5 percent. These data are preferable to dermal data from other species, therefore, the dermal data in rabbits are not of concern.

Combination of Acute Inhalation and Dermal Exposure: Margins of safety were calculated for combined inhalation and dermal exposure of workers repeatedly exposed to paraquat. Combined inhalation and dermal exposures of several groups of workers were then compared to a NOEL of 0.5 mg paraquat cation/kg/day derived from a 90-day dog-feeding study. The lowest effect level (LEL) for this study was 1.5 mg paraquat cation/kg/day, at which dosage level toxic effects in the lung were observed. All but two of the margins of safety are greater than 100.

Chronic Toxicology:

Teratology and Reproduction

Paraquat was not teratogenic to mice. The fetotoxic NOEL is 5 mg paraquat cation/kg bwt and the maternal NOEL is 1.0 mg paraquat cation/kg bwt.

Paraquat was not teratogenic to rats. Both the fetotoxic and maternal NOEL are 1.0 mg paraquat cation/kg bwt.

Paraquat had no effect on reproduction in rats. The systemic NOEL for reproduction is 25 parts per million (ppm).

The teratology and reproduction studies are acceptable and show no reason to place paraquat in Special Review.

Chronic Feeding - Oncogenic Studies

In the rat chronic feeding study, the systemic NOEL is 25 ppm of paraquat cation per kilogram of body weight and the systemic LEL is 75 ppm.

In the 1-year dog study the systemic NOEL is 15 ppm paraquat cation per kilogram of body weight. The systemic LEL is 30 ppm.

Paraquat was not oncogenic to mice. The systemic NOEL is 12.5 ppm of paraquat cation per kilogram of body weight. This study is acceptable.

Paraquat was oncogenic to rats. Squamous cell carcinoma occurred in 51.6 percent of all rats having tumors of the skin and subcutis in the head region. In high dose males, the incidence was significantly increased over concurrent controls.

Mutagenicity: Paraquat was negative in eight studies (mostly in gene mutation and chromosomal aberration assays); weakly positive in two gene mutations, one chromosomal aberration and one DNA damage/repair assays; and positive in four DNA damage/repair assays.

Physiological and Biochemical Behavior Characteristics:

Foliar Absorption: Very rapidly absorbed by the foliage.

Translocation: Can occur via the xylem under certain conditions.

Mechanism of Pesticidal Action: Lipid peroxidation resulting in disruption of cell membrane.

Metabolism and Persistence in Plants: In sunlight, limited photochemical breakdown occurs for paraquat which remains on the outside of treated plants. Since plants are killed rapidly in bright sunlight, significant quantities of the breakdown products are formed only on surfaces of dead tissues and there is no movement of these substances from the dead tissues to other parts of the plant.

Environmental Characteristics:

Absorption and Leaching in Basic Soil Types: Langmuir adsorption maxima values (m) ranged from 17 to 46.8 mg/100 grams on seven clay soils and one silty clay loam soil. Adsorption is positively correlated with soil cation exchange capacity (CEC). Paraquat is not readily desorbed from the soil and is not likely to contaminate ground water in agricultural soils. Data are required to determine the potential for the QINA degradate to leach in soils.

Microbial Breakdown: Bound paraquat is degraded with difficulty or not at all.

Loss from Photodecomposition and/or Volatilization: Does not volatilize; limited photodecomposition may occur on sprayed leaf surfaces and dead vegetation.

Contamination of Ground and Surface Water: Paraquat binds tightly to the soil and does not leach in agricultural soils.

Exposure of Humans and Nontarget Organisms to Chemical or
    Degradates: Margins of safety for humans appear adequate when product is used according to label directions. Use of paraquat may have an effect on a few mammals or ground nesting birds under unusual circumstances. Wildlife populations should not be adversely affected when paraquat is used according to label directions.

Exposure During Reentry: Current precautionary labeling and worker safety rules adequately protect worker, mixer, loader, applicator.

Ecological Characteristics: Paraquat is moderately toxic to birds and aquatic invertebrates and slightly toxic to freshwater fish.

Avian Acute Oral Toxicity with Bobwhite quail: 176 mg/kg

Avian Dietary Toxicity with Bobwhite quail: 981 ppm
                         Ring-neck pheasant: 1468 ppm
                         Mallard duck: 4048 ppm

Avian Reproduction Studies: Bobwhite quail: 100 ppm
                            Mallard duck: 30 ppm

Acute Toxicology: Freshwater Fish:
                    Rainbow trout: 15-38.7 ppm
                    Bluegill sunfish: 13-156 ppm

Acute Toxicity to Invertebrates (Daphnia): 1.2 to 4.0 mg/L

Endangered Species: Although paraquat is not believed to cause problems with stable wildlife populations, its acute and subacute toxicity may be hazardous to unstable or endangered populations (primarily plant species).

Tolerance Reassessment: Tolerances are established for residues of the plant regulator, desiccant, defoliant, and herbicide paraquat (1,1-dimethyl-4,4'-bipyridinium ion) derived from application of either the bis(methylsulfate) or dichloride salt (both calculated as the cation) in or on the following raw agricultural commodities (40 CFR 180.205):

Commodities	Parts Per Million
Acerola	0.05
Alfalfa	5.0
Almond hulls	0.5
Apples	0.05
Apricots	0.05
Asparagus	0.5
Avocados	0.05
Bananas	0.05
Barley, grain	0.05
Beans, forage	0.1
Beans, hay	0.4
Beans, lima (succulent)	0.05
Beans, snap (succulent)	0.05
Beets, sugar	0.5
Beets, sugar (tops)	0.5
Birdsfoot trefoil	5.0
Broccoli	0.05
Cabbage	0.05
Carrots	0.05
Cattle, fat	0.01
Cattle, meat	0.01
Cattle, meat byproducts	0.01
Cauliflower	0.05
Cherries	0.05
Chinese Cabbage	0.05
Citrus fruit	0.05
Clover	5.0
Coffee beans	0.05
Collards	0.05
Corn, fresh, inc. sweet corn (K + CWHR)	0.05
Corn, fodder	0.05
Corn forage	0.05
Corn grain	0.05
Cottonseed	0.5
Cucurbits	0.05
Eggs	0.01
Figs	0.05
Goats, fat	0.01
Goats, meat	0.01
Goats, meat byproducts	0.01
Grass, pasture	5.0
Grass, range	5.0
Guar beans	0.5
Guava	0.05
Hogs, fat	0.01
Hogs, meat	0.01
Hogs, meat byproducts	0.01
Hops, fresh	0.1

Commodities	Parts Per Million
Hops vines	0.5
Horses, fat	0.01
Horses, meat	0.01
Horses, meat byproducts	0.01
Kiwifruit	0.05
Lettuce	0.05
Milk	0.01
Mint, hay	0.5
Nectarines	0.05
Nuts	0.05
Oat grain	0.05
Olives	0.05
Onions, dry bulb	0.05
Onions, green	0.05
Papayas	0.05
Passion fruit	0.2
Peaches	0.05
Pears	0.05
Peas, succulent	0.05
Peas, hay	0.8
Pineapples	0.05
Pistachio nuts	0.05
Plums (fresh prunes)	0.05
Potatoes	0.5
Poultry, fat	0.01
Poultry, meat	0.01
Poultry, meat byproducts	0.01
Rhubarb	0.05
Rye grain	0.05
Safflower seed	0.05
Sheep, fat	0.01
Sheep, meat	0.01
Sheep, meat byproducts	0.01
Small fruit	0.05
Sorghum forage	0.05
Sorghum grain	0.05
Soybeans	0.05
Soybean forage	0.05
Strawberries	0.25
Sugarcane	0.5
Sunflower seeds	2.0
Turnips (roots)	0.05
Turnips (tops)	0.05
Vegetables, fruiting	0.05
Wheat grain	0.05

A food additive tolerance of 0.2 ppm is established for residues of the defoliant, desiccant, and herbicide paraquat (1,1'-dimethyl-4,4'-bipyridinium ion) derived from the application of either the bis(methylsulfate) or dichloride salt (both calculated as the cation) in or on dried hops resulting from application of the pesticide to growing hops (21 CFR 193.331).

Feed additive tolerances are established for residues of the defoliant, desiccant, and herbicide paraquat (1,1'dimethyl4-4'-bipyridinium ion) derived from the application of either the bis(methylsulfate) or dichloride salt (both calculated as the cation) in the following processed feeds when present therein as a result of application of paraquat to growing crops (21 CFR 561.289):

Feeds	Parts Per Million
Mint, hay, spent	3.0
Sunflower, seed, hulls	6.0

The acceptable daily intake (ADI) based on the 1-year dog study (NOEL of 0.45 mg/kg/day) and using a safety factor of 100 is calculated to be 0.0045 mg/kg/day. The maximum permitted intake (MPI) for a 60-kg human is calculated to be 0.27 mg/day. The theoretical maximum residue contribution (TMRC) for paraquat, based on published tolerances is 0.1134 mg/day. The TMRC constitutes 42 percent of the MPI.

Reported Pesticide Incidents: The Agency's Pesticide Incident Monitoring System (PIMS) indicated that the majority of the poisoning incidents involving paraquat resulted from the purposeful ingestion of the chemical in apparent suicide attempts.

4. Summary of Regulatory Position and Rationale

The Agency has classified paraquat as a Group C oncogen (Possible Human Carcinogen). However, after considering applicator exposure to paraquat, the Agency has concluded that the risks posed by paraquat are not of concern in terms of the magnitude of risk to the individual applicators.

The Agency is continuing to require that an emetic cleared under 40 CFR 180.1001 be incorporated into all manufacturing-use products (MUP's) and end-use products containing paraquat. The emetic is needed in the formulations to induce rapid vomiting thereby reducing absorption of paraquat if swallowed.

The Agency is requiring those agricultural products containing paraquat already classified as "Restricted Use" maintain this classification. Based on submitted acute toxicity and subchronic toxicity data the "Restricted Use" classification and current precautionary statements are necessary to protect mixer-loaders and applicators from effects of dermal toxicity.

The Agency has determined that the homeowner product containing 0.276 percent paraquat presently unrestricted will remain unrestricted. The Agency believes that this relatively dilute formulation, when used according to label directions, is not likely to present a significant health hazard to humans.

The Agency is requiring additional residue data on several crops and processed commodities as well as some changes in the tolerance listings. The Agency is requiring certain label revisions pertaining to application and grazing restrictions, as determined by review of available residue data.

The Agency is requiring labeling to protect endangered species. This labeling is addressed in Pesticide Registration (PR) Notices 87-4 and 87-5.

The Agency will issue registrations for substantially similar products while data gaps are being filled. Significant new uses for paraquat will be considered on a case-by-case basis.

5. <u>Summary of Major Data Gaps</u>   <u>Date due (from issuance of Standard)</u>

Summary of Major Data Gaps	Date due (from issuance of Standard)
Environmental fate data	9-50 months
Subchronic inhalation toxicity	15 months
Residue chemistry data	24 months
Product chemistry	6 months

6. <u>Contact Person at EPA</u>: Robert J. Taylor
Office of Pesticide Programs, EPA
Registration Division (TS-767C)
401 M Street SW.
Washington, DC 20460
Phone: (703) 557-1800

<u>Disclaimer</u>: The information presented in this Pesticide Fact Sheet is for informational purposes only and may not be used to fulfill data requirements for pesticide registration and reregistration.

PARATHION

Date Issued: December 1986
Fact Sheet Number: 116

1. DESCRIPTION OF CHEMICAL

    Generic Name: O,O-diethyl-O-p-nitrophenyl phosphorothioate

    Common Name: Parathion

    Trade Names and other names: Alkron, Alleron, Aphamite, Bladan, Corothion, Ethyl Parathion, Folidol E-605, Fosferno 50, Niran, Orthophos, Panthion, Paramar, Paraphos, Parathene, Parawet, Phoskil, Rhodiatox, Soprathion, Stathion, Thiophos

    EPA Shaughnessy Code: 057501

    Chemical Abstracts Service (CAS) Number: 56-38-2

    Year of Initial Registration: 1948

    Pesticide Type: Insecticide

    Chemical Family: Organophosphate

    U.S. and Foreign Producers: Monsanto U.S.A.; Bayer A.G., West Germany; and Cheminova, Denmark

2. USE PATTERNS AND FORMULATIONS

    Application Sites: Vegetable crops, field crops orchard crops ornamentals, aquatic crop and non-crop sites

    Types of Formulations: Emulsifiable concentrates, granulars, dusts, baits, wettable powders and impregnated materials.

    Types of Methods of Application: Ground applications; broadcast, band, and aerial application

    Application Rates: 0.1 to 10.0 lbs/active ingredient per acre

Usual Carriers: petroleum solvents, clay carriers

3. SCIENCE FINDINGS

Summary Science Statement

Parathion is a Toxicity Category I organophosphate compound which is extremely toxic to laboratory mammals. The chemical has demonstrated adverse chronic effects such as tumors in the adrenal glands, retinal atrophy and degeneration, and degeneration of the sciatic nerve. The Agency has carried out a weight-of-the-evidence analysis and has concluded that parathion is a Category C oncogen (possible human carcinogen). Parathion was not shown to be teratogenic. Parathion is extremely toxic to fish and birds. Parathion is entering the Special Review process because of acute effects to both humans and birds. Human poisoning from parathion exposure have occurred during mixing/loading, application, early reentry into treated fields, equipment repair and hanling, and contact with spray drift.

Chemical Characteristics of the Technical Material

Physical State: Liquid
Color: Dark Brown
Odor: Garlic-like
Molecular weight and formula: 291.26 - $C_{10}H_{14}NO_5PS$
Boiling point: 157-162 at 0.6 mm Hg
Vapor Pressure: 0.42 u hg at 25' C (pure active ingredient)
Solubility in various solvents: Miscible in most all organic solvents and oils, only slightly soluble in water

Toxicology Characteristics

Acute toxicity: Parathion is extremely toxic to mammals by all routes of exposure and is classified in Toxicity Category I (1.75- 15.0 mg/kg). Additional acute toxicity tests have been waived because they would not alter this classification.

Major routes of exposure: Inhalation and dermal.

Delayed neurotoxicity: Data gap- To be filled via the published literature.

Oncogenicity: This chemical is classified as a Category C oncogen.

Chronic Effects: Retinal degeneration and sciatic nerve degeneration in life-time feeding studies in the rat.

Metabolism: Data gap.

Chronic Effects: Retinal and sciatic nerve damage at high
dose levels was observed in the rat.
Cholinesterase inhibition.

Teratogenicity: Parathion was not teratogenic at levels up
to 1.5 and 16 mg/kg in the rat and rabbit,
respectively

Reproduction: Data gap

Mutagenicity: Positive for induction of DNA repair in
Unscheduled DNA Synthesis (UDS) assay in
a human cells study. Additional genotoxicity
studies are required to determine the mutagenic
potential of parathion.

Physiological and Biochemical Characteristics

Mechanism of pesticidal action: This insecticide is active
by contact, ingestion, and vapor action. Parathion in humans
acts by causing irreversible inhibition of cholinesterase
enzyme, allowing accumulation of acetylcholine at cholinergic
neuroeffector junctions and autonomic ganglia. Poisoning
also impairs the central nervous system function. Symptoms
of poisoning include headaches, nausea, vomiting, cramps,
weakness, blurred vision, pin-point pupils, tightness in
chest, labored breathing, nervousness, sweating, watering of
eyes, drooling or frothing of mouth and nose, muscle spasms,
coma and death.

Environmental Characteristics

Available data provide insufficient information for the
analysis of the environmental fate of parathion.

Parathion has little or no potential to contaminate
ground water. This chemical was not included on the list
of potential ground water contaminaters.

Ecological Characteristics

Avian oral toxicity: 1.3 mg/kg for house sparrows and pigeons

Avian dietary toxicity: 76 to 336 ppm for mallard and ring-
necked pheasant, respectively

Avian reproduction: Data gap

Freshwater fish acute toxicity: 0.32 to 2.65 ppm for
mosquitofish and channel
catfish, respectively

Aquatic freshwater invertebrate toxicity: 0.04 to 5 ppb for
most species

Marine and estuarine toxicity: 1.0 (48-hr) to 17.8 (96-hr) ppb
for brown shrimp and striped
bass, respectively

Endangered Species: Previous consultations with the Office of Endangered Species have resulted in jeopardy opinions and labeling for crops (alfalfa, apples, barley, corn, cotton, pears, and wheat), rangeland and pastureland, silvacultural sites, aquatic sites, and noncropland use. Labeling is being imposed to reduce the risk to endangered species.

4. Summary of Regulatory Position and Rationale

   A. The Agency anticipates initiating a Special Review based upon its adverse acute effects to both humans and birds.

   B. The Agency is requiring avian reproduction and terrestrial full field testing and simulated or full field aquatic testing to better define the extent of exposure and hazard to wildlife.

   C. No new tolerances or new food uses will be considered until the Agency has received data sufficient to assess existing tolerances for parathion.

   D. The Agency is concerned about the potential for human poisonings (cholinesterase inhibition) from the use of parathion. The Agency will continue to classify for restricted use (due to very high acute toxicity to humans and birds). The Certified applicator must be physically present during mixing, loading, equipment repair, and equipment cleaning. Information from the California Department of Food and Agriculture reported incidents of worker poisonings and illnesses during mixing, loading, and application. EPA is requiring more stringent "Worker Safety Rules", including protective clothing, to reduce exposure.

   E. A reentry interval of 7 days is required for apple, citrus, peach, nectarine and grapes in the states of Arizona, California, Nevada, New Mexico, Oklahoma, Texas and Utah. A 48-hour reentry interval will remain on other crops and the crops listed above in all other states.

   F. The Agency has classified parathion as a class C oncogen.

5. Required Unique Labeling Summary

   All manufacturing-use and end-use parathion products must bear appropriate labeling as specified in 40 CFR 162.10. In addition, the following information must appear on the labeling:

   All parathion products will continue to be classified for restricted use and the restricted use label will specify the reason (High acute toxicity to humans and birds).

   A reentry interval of 7 days is required for apple, citrus, peach, nectarine and grapes in the states of Arizona, California, Nevada, New Mexico, Oklahoma, Texas and Utah.

A 48-hour reentry interval will remain on other crops and the crops listed above in all other states.

Additional labeling requirements are imposed to protect fish, birds and other wildlife, including endangered species.

Effluent containing parathion may not be discharged into lakes, streams, ponds, estuaries, oceans or public waters unless this product is specifically identified in an NPDES permit. Discharge of effluent containing this product is forbidden without prior notice to the sewage treatment plant authority.

Protective clothing requirements are being standardized to protect applicators, fieldworkers and mixer/loaders.

Human flaggers are prohibited during aerial application.

6. Tolerance Assessment

U.S., Canadian, Mexican, and Codex tolerances for negligible residues parathion in or on raw agricultural commodities can be reviewed on attachment I. U.S. parathion tolerances can also be found in 40 CFR 140.121.

Because of extensive data gaps in residue chemistry and toxicology areas, a tolerance assessment can not be made at this time. Plant and animal metabolites are required to be identified and their toxicological significance determined. Storage stability data are required, so some of the present tolerances may be affected. Sufficient data are available to ascertain the adequacy of the established tolerances for residues of parathion in or on the following raw agricultural commodities (RACs): beans (dried only), peas (dried and succulent), soybeans, mustard seed, safflower seed, sugarcane, and sugarcane forage.

There are no tolerances for residues of parathion in animal products. On receipt of the requested plant and animal metabolism data, the need for these tolerances will be determined.

Parathion's Acceptable Daily Intake (ADI) is 0.005 mg/kg/day, based on rat cholinesterase inhibition. The TMRC currently utilizes 192% of the ADI. Recent chronic feeding studies in rats strongly suggest that the present no effect level (NOEL) is not adequate and a new lower NOEL is needed. Additional data are required in order for the Agency to determine an adequate NOEL and ADI.

7. SUMMARY OF MAJOR DATA GAPS

' Animal and plant metabolism data
' Magnitude of the residues in various crops
' Full battery of Environmental Fate data

- Additional subchronic toxicity testing to determine a NOEL for cholinesterase inhibition and other systemic effects (retinal degeneration, sciatic nerve, abnormal gait)
- Reproduction studies
- Gene Mutation studies
- Reentry studies
- Avian reproduction and simulated field testing
- Acute toxicity to fresh water, estuarine and marine organisms
- Early life stage and fish life cycle studies
- Aquatic accumulation studies
- Glove permeability and drift studies
  Applicator Monitoring studies

7. CONTACT PERSON AT EPA

   Dennis Edwards
   Acting Product Manager (12)
   Insecticide-Rodenticide Branch
   Registration Division (TS-767C)
   Office of Pesticide Programs
   Environmental Protection Agency
   401 M Street, S. W.
   Washington, D. C. 20460

   Office location and telephone number:
   Room 202, Crystal Mall #2
   1921 Jefferson Davis Highway
   Arlington, VA   22202
   (703) 557-2386

DISCLAIMER: The information presented in this Chemical Information Fact Sheet is for informational purposes only and may not be used to fulfill data requirements for pesticide registration and reregistration.

ATTACHMENT I

Summary of Present Parathion Tolerances

Commodity	Tolerances (ppm) United States [a]	Canada [a]	Mexico [a]	(MRL) International (Codex) [b]
Garden Beets	1.0	0.7	---	0.7
Carrots	1.0	0.7	---	---
*Parsnips	1.0	0.7	---	0.7
Potatoes	0.1	---	0.1	0.7
Radishes	1.0	0.7	1.0	0.7
Rutabagas	1.0	---	---	0.7
Sugar Beets	0.1	---	---	---
Sweet Potatoes	0.1	---	0.1	0.7
Turnips	1.0	0.7	---	0.7
Garlic	1.0	---	1.0	0.7
Onions	1.0	0.7	1.0	0.7
Celery	1.0	0.7	1.0	0.7
Endive	1.0	0.7	---	0.7
Lettuce	1.0	0.7	1.0	0.7
Spinach	1.0	0.7	1.0	0.7
Swiss Chard	1.0	0.7	---	0.7
Broccoli	1.0	0.7	1.0	0.7
Brussels Sprouts	1.0	0.7	---	0.7
Cabbage	1.0	0.7	---	0.7
Cauliflower	1.0	0.7	---	0.7
Collards	1.0	---	---	0.7
Kale	1.0	0.7	---	0.7
Kohlrabi	1.0	0.7	---	0.7
Mustard Greens	1.0	---	---	0.7
Beans	1.0	0.7	1.0	0.7
Peas	1.0	0.7	1.0	0.7
Soybeans	0.1	---	0.1	---
Eggplant	1.0	0.7	1.0	0.7
Peppers	1.0	0.7	1.0	0.7
Tomatoes	1.0	0.7	1.0	0.7
Cucumbers	1.0	0.7	1.0	0.7
Melons	1.0	1.0	1.0	0.5
Pumpkins	1.0	0.7	---	0.5
Squash	1.0	0.7	---	0.7
Citrus Fruits	1.0	1.0	1.0	1.0
Apples	1.0	1.0	1.0	0.5
Pears	1.0	1.0	1.0	0.5
Quince	1.0	1.0	---	0.5
Apricots	1.0	1.0	---	1.0
Cherries	1.0	1.0	---	0.5
Nectarines	1.0	---	---	0.5
Peaches	1.0	1.0	1.0	1.0
Plums	1.0	1.0	---	0.5
Blackberries		1.0	---	0.5
Blueberries	1.0	---	---	0.5

Summary of Present Parathion Tolerances (con't)

Commodity	United States[a]	Tolerances (ppm) Canada[a]	Mexico[a]	(MRL) International (Codex)[b]
Boysenberries	1.0	---	---	0.5
Cranberries	1.0	1.0	---	0.5
Currants	1.0	---	---	0.5
Dewberries	1.0	---	---	0.5
Gooseberries	1.0	1.0	---	0.5
Grapes	1.0	1.0	---	0.5
Loganberries	1.0	1.0	---	0.5
Raspberries	1.0	1.0	---	0.5
Strawberries	1.0	1.0	1.0	0.5
*Youngberries	1.0	---	---	0.5
Almonds	0.1	---	0.1	---
Filberts	0.1	---	0.1	---
Pecans	0.1	---	0.1	---
Walnuts	0.1	---	0.1	---
Barley	1.0	---	---	---
Corn	1.0	0.7	1.0	---
Oats	1.0	---	---	---
Rice	1.0	---	---	---
Sorghum	0.1	---	0.1	---
Wheat	1.0	---	---	---
Forage Grass	1.0	---	---	---
Alfalfa Forage	1.25	---	5.0	---
Alfalfa Hay	5.0	---	1.25	---
Clover Forage & Hay	1.0	---	---	---
Vetch Forage & Hay	1.0	---	---	---
Miscellaneous Crops				
Artichokes	1.0	---	1.0	0.7
Avocados	1.0	---	1.0	0.5
Cottonseed	0.75	---	0.75	---
Dates	1.0	---	---	0.5
Figs	1.0	---	1.0	0.5
*Guavas	1.0	---	1.0	0.5
Hops	1.0	1.0	---	---
Mangos	1.0	---	1.0	0.5
Mustard Seed	0.2	---	---	---
Okra	1.0	---	1.0	0.7
Olives	1.0	---	---	0.5
Peanuts	1.0	---	1.0	---
Pineapples	1.0	---	1.0	0.5
Rape Seed	0.2	---	---	---
Sugarcane	0.1	---	0.1	---
Sunflower Seed	0.2	---	---	---

a/ The U.S., Canadian, and Mexican tolerances are expressed in terms of residues of parathion per se.
b/ The Codex Maximum Residue Levels are expressed as residues of parathion and its oxygen analog, paraoxon.
* Although these crops have tolerance there are no EPA registered uses.

PENDIMETHALIN

Date Issued: March 31, 1985
Fact Sheet Number: 50

1. Description of the chemical:

    Generic name: N-(1-ethylpropyl)-3,4-dimethyl-2,6-dinitro-benzenamine
    Empirical formula: $C_{13}H_{19}N_3O_4$
    Common name: Pendimethalin
    Trade name: Prowl®, Herbadox®, Stomp®, and AC 92553
    Chemical Abstracts Service (CAS) Registry number: 40487-42-1
    Office of Pesticides Program's EPA Chemical Code Number: 108501
    Year of initial registration: 1974
    Pesticide type: Herbicide
    Chemical family: Dinitroaniline
    U.S. producer: American Cyanamid Company

2. Use patterns and formulations:

    Application sites: Pendimethalin is registered for control of broad leaf weeds and grassy weed species on the following sites: soybeans, cotton, field corn, beans, peanuts, potatoes, rice, sorghum, sunflower, tobacco, ornamentals, non-bearing fruit and nut crops, and vineyards.

    One site, jojoba, is registered under Section 24(c) in Arizona.

    Types of formulations: Pendimethalin is available in granular, dispersable granular, and emulsifiable concentrate formulations.

    Types and methods of applications: Pendimethalin is mainly applied as a preplant incorporation (except in corn, rice and sorghum), preemergence spray, early postemergence (rice), and late postemergence "culti-spray" (field corn and sorghum) applications.

    Application rates: 0.5 to 2.0 lbs a.i./A on crop sites.

    Usual carriers: Attapulgite clay and water.

3. Science Findings:

   Summary science statements:

   Pendimethalin is not acutely toxic by the oral, dermal, inhalation and occular routes of exposure. The available data is insufficient to show that any of the risk criteria listed in § 162.11(a) of Title 40 of the U.S.Code of Federal Regulations have been met or exceeded for the uses of pendimethalin at the present time. There are no valid mutagenicity and chronic rat studies for pendimethalin and insufficient oncogenicity data. There are also extensive residue chemistry and environmental fate data gaps.

   Pendimethalin is highly toxic to coldwater fish, highly to moderately toxic to warmwater fish and highly to moderately toxic to freshwater invertebrates. A detailed ecological hazard assessment cannot be made until certain environmental chemistry data requirements, and a monitoring study of aquatic sites next to treated rice fields are fulfilled.

   Chemical characteristics:

   Pendimethalin is crystalline at room temperature and has a fruit like odor. Its molecular weight is 281.30. The boiling point is 330°C. Pendimethalin is soluble in water (at 20°C) to <0.50 ppm and soluble in aromatic and chlorinated hydrocarbon solvents.

   Toxicological characteristics:

   Acute toxicology effects on pendimethalin are as follows:

   Acute Oral Toxicity in rats: >1,250 mg/kg body weight, Toxicity Category III
   Acute Dermal Toxicity in rabbits: >5,000 mg/kg body weight, Toxicity Category III
   Acute Inhalation Toxicity in rats: >320 mg/l (4 hour exposure) Toxicity Category IV
   Skin irritation in rabbits: slight irritant, Toxicity Category III
   Eye irritation in rabbits: corneal irritation clearing in less than seven days, Toxicity Category III.

   Subchronic toxicology data except for the 90-day rat feeding study have met the current toxicity requirements.

   Chronic toxicology effects on pendimethalin have not been completely evaluated because there are no valid mutagenicity and chronic rat studies for pendimethalin, and insufficient oncogenicity data.

A 2-year dog feeding study indicated that pendimethalin tested at 12.5 mg/kg, 50.0 mg/kg and 200.0 mg/kg produced increases in alkaline phosphatase level and liver weight. The NOEL is 12.5 mg/kg/ day.

A teratology test in rats has shown that pendimethalin tested at 500.0 mg/kg, highest dose tested (HDT), failed to induce teratogenic, or fetotoxic effects. A teratology test in rabbits has shown that pendimethalin tested at 60.0 mg/kg (HDT), failed to induce teratogenic or fetotoxic effects.

A reproduction study (3-generation rat) indicated that pendimethalin tested at 500 ppm to 5,000 ppm induced slightly fewer offspring, with no corresponding increase in deaths and decreased weight gain from weaning to maturity. The NOEL is 500 ppm.

Major routes of human exposure:

Non-dietary exposure to pendimethalin by a farmer as an applicator during mixing, loading , spraying and flagging is probable.

Physiological and biochemical behavioral characteristics:

Absorption characteristics: Pendimethalin is absorbed in limited amounts by monocotyledonous plants and in moderate amounts by small dicotyledonous plants.
Translocation: Pendimethalin is taken up from the soil by plants in very limited amounts.
Mechanism of pesticidal action: Pendimethalin inhibits plant cell division and cell elongation.
Metabolism in plants: Pendimethalin is principally degradated in plants by oxidation of the 4-methyl group(benzene ring) and the N-1-ethylpropyl group in the amine moiety.

Environmental characteristics:

Adsorption and leaching in basic soil types: Pendimethalin is strongly adsorbed by soil organic matter and clay and does not readily leach through the soil.
Microbial breakdown: Soil microorganisms do not appear to play a significant role in degradation of pendimethalin.
Loss from photodecomposition and/or volatilization: Slight losses.
Average persistence at recommended rates: When used at recommended rates under normal environmental conditions, little if any pendimethalin occurs in the subsequent crop.

Ecological characteristics:

    Avian acute oral toxicity: 1,421 mg/kg.
    Avian eight-day dietary toxicity (Bobwhite Quail): > 3,149 ppm.
    Avian eight-day dietary toxicity (Mallard Duck): > 4,640 ppm.
    96-hour fish toxicity: 0.199 ppm for bluegill sunfish (highly toxic) and 0.138 ppm for rainbow trout (highly toxic).
    48-hour aquatic invertebrate toxicity: 0.28 ppm (highly toxic for Daphnia magna.

Potential problem for endangered species:

    The Slackwater darter and certain freshwater mussels are endangered species at risk from the use of pendimethalin on cotton. The Agency is addressing appropriate means of labeling pesticides that may threaten the continued existence of endangered species. The labeling should be completed by the 1986 growing season. If it is not, this standard may be amended to impose interim labeling to protect endangered species.

    The Agency believes that the conventional labeling approach may be inadequate to properly inform the users on how to protect the endangered species. The Agency anticipates that appropriate labeling will be developed in time for the 1986 growing season for cotton.

Tolerance assessments:

    The Agency is unable to complete a full tolerance reassessment of pendimethalin because of certain residue chemistry and toxicology data gaps. The additional data may cause specific tolerances to be revised in the future.

The Acceptable Daily Intake (ADI) for pendimethalin was originally based upon a 2-year feeding study on rats, which was subsequently declared invalid. Subsequently, the Provisional Acceptable Daily Intake (PADI) for pendimethalin was calculated, using the 90 day portion of the same study. The subchronic portion of this study was used instead of a 2-year dog study because the (P)ADI value calculated for the rat is the more conservative value on a mg/kg basis than for the dog:

	NOEL (mg/kg)	Safety Factor	(P)ADI mg/kg/day	(P)MPI mg/day (60 kg
Rat	25.0	2,000	0.0125	0.7500
Dog	12.0	100	0.1250	7.5000

The No-Observable-Effect-Level (NOEL) for the rat study is 25.0 mg/kg. A 2,000-fold safety factor was used and the PADI was calculated as 0.0125 mg/kg/day with a Maximum Permissible Intake (MPI) of 0.7500 mg/day for a 60 kg person. The Theoretical Maximum Residue Contribution (TMRC) for pendimethalin-based permanent tolerances is 0.0166 mg/day for a 1.5 kg diet. Currently, the permanent tolerances utilize 2.22 % of the PADI.

In the United States, tolerances are currently established in 40 CFR § 180.361 for the combined residues of the herbicide, pendimethalin: N-(1-ethyl propyl)-3,4-dimethyl-2,6-dinitrobenzenamine and its metabolite: 4-[1-ethyl propyl)aminol-2-methyl-3,5-dinitrobenzyl alcohol, in or on the raw agricultural commodities listed below:

Commodities	Parts per million
** Beans (lima, dry, snap)	0.1
** Beans, forage	0.1
** Beans, hay	0.1
Corn, fodder	0.1
Corn, forage	0.1
** Corn, fresh(including sweet, K+CWHR)	0.1
Corn, grain	0.1
Cottonseed	0.1
Peanuts	0.1
Peanut hay	0.1
Peanut forage	0.1
Rice grain	0.05
Soybeans	0.1
Soybeans, forage	0.1
Soybeans, hay	0.1
Sunflower seeds	0.1

** The tolerances for these commodities were established after the science reviews for the pendimethalin registration standard were completed.

International Tolerances

Presently, there are no tolerances for residues of pendimethalin in Canada, Mexico, or in the Codex Alimentarius.

Problems known to have occurred with use:

The Pesticide Incident Monitoring System (PIMS) indicates four incidents involving agricultural uses of pendimethalin alone from 1966 through 1980. Of the four incidents, three required medical attention, two involved dermal exposure during ground application, one involved mixer/ loader exposure and one involved a truck/ container spill. No fatalities were reported.

PIMS does not include any details or consequences of these exposures, nor does it attempt to validate these voluntarily submitted reports. The PIMS data do not provide information on chronic health effects from exposure to pendimethalin, but do support the need for precautions relating to careful handling of pendimethalin products.

4. Summary of regulatory position and rationale:

Based on the review and evaluation of all available data and other relevant information on pendimethalin the Agency has made the following determinations:

The available data do not indicate that any of the risk criteria listed in § 162.11(a) of Title 40 of the U.S.Code of Federal Regulations have been met or exceeded for the uses of pendimethalin at the present time.

Pendimethalin is not acutely toxic by the oral, dermal, inhalation and occular routes of exposure.

The chronic dog study, rat and rabbit teratology studies, and a 3-generation reproduction study did not indicate adverse chronic effects.

There are no valid mutagenicity and chronic rat studies for pendimethalin, and insufficient oncogenicity data. There are also extensive residue chemistry and environmental fate data gaps.

The Agency is requiring monitoring data for potential residues in aquatic sites next to treated rice fields. Once the Agency has evaluated these additional data, it will determine if EPA should impose more stringent measures to minimize exposure of aquatic organisms to pendimethalin. Ecological effect studies indicate that pendimethalin is highly toxic to certain coldwater and warmwater fish; moderately to highly toxic to marine and estuarine organisms.

The Agency is requiring that levels of N-nitroso-pendimethalin contaminant not exceed 60ppm in the technical product and that the technical be analysed for other impurities.

Specific label warning statement:

Hazard Information

The human hazard statements must appear on all EP labels as presribed in 40 CFR 162.10.

Environmental Hazard Statements

All manufacturing-use products (MPs) intended for formulation into end-use products (EPs) must bear the following statements:

"This pesticide is toxic to fish. Do not discharge effluent containing this product directly into lakes, streams, ponds, estuaries, oceans or public waters unless this product is specifically identified and addressed in a National Pollutant Discharge Elimination System (NPDES) permit. Do not discharge effluent containing this product into sewer systems without previously notifying the sewage treatment plant authority. For guidance, contact your State Water Board or Regional Office of the Environmental Protection Agency".

End-Use Product Statements

a. Non Aquatic Uses (Granular products)

The following environmental hazard statement must appear on all EPs:

"This pesticide is toxic to fish. Do not apply directly to water. Runoff from treated areas may be hazardous to aquatic organisms in neighboring areas. Do not contaminate water by cleaning of equipment or disposal of wastes. In case of spills, either collect for use or dispose of properly".

b. Non Aquatic Uses (Nongranular products)

The following environmental hazard statement must appear on all EPs:

"This pesticide is toxic to fish. Do not apply directly to water. Drift and runoff from treated areas may be hazardous to fish in neighboring areas. Do not contaminate water by cleaning of equipment or disposal of wastes".

c. Aquatic Uses (Rice)

The following environmental hazard statement must appear on all EPs:

"This pesticide is toxic to fish and aquatic organisms. Fish may be killed at application rates recommended on the label. Do not contaminate water by cleaning of equipment or disposal of wastes".

"Do not apply to rice fields if fields are used for catfish or crayfish farming".

Restrictions on Rotational Crops

"Pending the submission of rotational crop data, do not apply pendimethalin on rice fields in which crayfish or catfish farming are included in the cultural practices, and do not plant crops in pendimethalin-treated fields unless pendimethalin is registered for use on those crops".

Restrictions on Irrigated Crops

"Pending the submission of irrigated crop data, do not use water containing pendimethalin residues from rice cultivation to irrigate food or feed crops which are not registered for use with pendimethalin".

5. Summary of major data gaps:

The following toxicological studies are required:
A dermal sensitization study in guinea pig is required by October 30, 1985,
A 90-day feeding study in the rat is required by January 30, 1986,
A chronic toxicity study in rat is required by April 30, 1989,
An oncogenicity study in rat and in mouse is required by April 30, 1989.
The following mutagenicity data are required by October 30, 1985:
a. Gene mutation in bacteria,
b. Gene mutation in mammalian cells in culture,
c. Chromosomal aberration analysis in mammalian cells in culture, and
d. DNA damage in mammalian cells in culture.

The following environmental fate data are required:
A photodegradation study in water is required by October 30, 1985,
A photodegradation study on soil is required by October 30, 1985,
A metabolism test in aerobic soil is required by April 30, 1987,
A metabolism test under anaerobic aquatic conditions is required by April 30, 1987,
A metabolism test under aerobic aquatic conditions is required by April 30, 1987,

A mobility study involving leaching and adsorption/desorption is required by October 30, 1985,
A mobility study involving volatility in the lab is required by October 30, 1985,
A mobility study involving volatility in the field is required by October 30, 1985,
A soil dissipation study in the field is required by April 30, 1987,
A dissipation study in aquatic (sediment) is required by April 30, 1987,
An accumulation study in rotational crops (confined) is required by October 30, 1987,
An accumulation study in rotational crops (field) is required by April 30, 1987,
An accumulation study in irrigated crops (field) is required by October 30, 1987,
An accumulation study in fish is required by October 30, 1985.

The following ecological effects data are required:

An acute freshwater invertebrate toxicity study using a typical EP is required by October 1985.
An aquatic field study to monitor residues next to rice fields using a typical EP is required by April 1987.

The following product chemistry data are required by October 30, 1985:
The name and address of the manufacturer or producer of each starting material used in the 90% technical product,
A discussion of each impurity believed to be present at >0.1% based on the beginning materials, all chemical reactions and any contamination is required by April 30, 1986,
Five or more samples must be analyzed for the active ingredient (A.I.) and each impurity present for which a certified limit is required by April 30, 1986,
A current Confidential Statement of Formula,
Quantitative methods to determine the remaining impurities in the technical product by April 30, 1986.
Data are required for ppm solubility in various solvents at 20 C,
Dissociation Constant data,
Octanol/water partition coefficient data,
Data on the pH, and
The following data are required for chemical stability: discussion of sensitivity of the A.I. to metal and metal ions, stability of the A.I. at normal and elevated temperatures, and the sensitivity of the A.I. to sunlight.

The following residue chemistry data are required:

Additional plant metabolism data are required with radiolabeled pendimethalin by April 30, 1987.

Levels of metabolites remaining unextractable in plant tissues and in polar fractions must be determined for possible toxicological residue concerns by April 30, 1987.

Metabolism studies utilizing ruminants dosed with ^{14}C ring labeled pendimethalin required by April 30, 1986. Distribution and characterization of residues must be determined in milk, muscle, kidney, and liver. If the ruminant metabolism differs significantly from the rat data, then swine metabolism data will also be required. If the additional metabolism data show the presence of new metabolites, then additional methodology data may be required.

Additional data are required by April 30, 1986 to show the stability of pendimethalin and its 3,5-dinitrobenzl alcohol metabolite in or on representative plant and animal samples stored at freezing temperatures.

Residue data are required by July 30, 1986 for carrot, radish, sugar beet.

Residue data are required by July 30, 1986 beans and peas.

Additional data are required by July 30, 1986 to support the established tolerance for soybean hay.

Data are required by July 30, 1986 for pendimethalin and its metabolite in or on soybean hay and straw.

If new metabolites are found, then additional field residue data for field corn may be required. When necessary, data will be extrapolated from the soybean processing study to corn.

If new residue metabolites are found, additional metabolism and field residue data may be required for sorghum.

Additional processing data may be required for cottonseed. When necessary, data will be extrapolated from the requested soybean processing study.

Additional metabolism and processing data on peanuts may be required. When necessary, data will be extrapolated from the soybean processing study.

Additional processing data may be required for sunflower seeds. When necessary, data will be extrapolated from the requested soybean processing study.

Residues of pendimethalin and its metabolite in catfish and crayfish are required by July 30, 1986.

Lactating ruminants must be dosed with pendimethalin to determine residues levels in milk.

A study on metabolites of pendimethalin in poultry will be required if additional metabolites of concern are found in the plant metabolism studies. The need for a poultry feeding study will depend upon the results of a poultry metabolism study.

The following data are required by July 30, 1986 for tobacco: Residue data involving the metabolism of pendimethalin in tobacco. If residues exceed 0.1 ppm, additional data on pyrolysis products must be submitted.

6. Contact Person at EPA:

Robert J. Taylor   (703) 557-1800
Office of Pesticide Programs, EPA,
Registration Division (TS-767C)
Fungicide-Herbicide Branch
401 M Street., S.W.
Washington, DC 20460.

DISCLAIMER: The information presented in this Chemical Information Fact Sheet is for informational purposes only and may not be used to fulfill data requirements for pesticide registration and reregistration.

PERFLUIDONE

Date Issued: September 30, 1985
Fact Sheet Number: 74

1. <u>Description of the chemical</u>:

    Generic name: 1,1,1-trifluoro-N-[2-methyl-4-(phenylsulfonyl)-phenyl]-methanesulfonamide
    Empirical formula: $C_{14}H_{12}F_3NO_4S_2$
    Common name: Perfluidone
    Trade name: Destun®
    Chemical Abstracts Service (CAS) Registry number: 37924-13-2
    Office of Pesticides Program's EPA Chemical Code Number: 108001
    Year of initial registration: 1976
    Pesticide type: Herbicide
    Chemical family: Sulfonamide
    U.S. producer: 3 M Company

2. <u>Use patterns and formulations</u>:

    Application sites: Perfluidone is registered for control of nutsedge species, certain grasses, and broadleaf weeds in flue-cured tobacco.

    Type of formulation: Perfluidone is available in a wettable powder formulation.

    Types and methods of applications: Perfluidone is banded or broadcast applied to the soil surface with ground equipment as a preemergence spray.

    Application rates: 1.5 lbs a.i./A on crop sites.

    Usual carriers: Water.

3. <u>Science Findings</u>:

    Summary science statements:

       Perfluidone is not acutely toxic by the dermal and ocular routes of exposure. A 90-day dog feeding study showed liver disorders (hepatic lesions, hepatocyte vacuolation, hyalin degeneration and biliary stasis) at the two highest dose levels (400 and 800 ppm).

The available data are insufficient to show that any of
the risk criteria listed in § 162.11(a) of Title 40 of
the U.S. Code of Federal Regulations have been met or
exceeded for the uses of perfluidone at the present time.
There are no valid mutagenicity and teratogenicity studies
for perfluidone. There are also residue chemistry and
environmental fate data gaps.

Perfluidone is slightly toxic to freshwater fish species.
Studies regarding freshwater invertebrates are not acceptable and there are no marine/estuarine data. A detailed
ecological hazard assessment cannot be made until certain
environmental chemistry data requirements have been met.

Chemical characteristics:

Perfluidone is a solid at room temperature and is odorless.
Its molecular weight is 379.40. The melting point is 143-145°C. Perfluidone is soluble in water (at 20°C) to 60.0
ug/ml and soluble in aromatic and chlorinated hydrocarbon
solvents.

Toxicological characteristics:

Acute toxicology effects of perfluidone are as follows:

Acute Dermal Toxicity in rabbits: >4,000 mg/kg body
weight, Toxicity Category III
Skin irritation in rabbits: Not an irritant, Toxicity
Category IV
Eye irritation in rabbits: Moderate eye irritant, Toxicity
Category II.

Subchronic toxicology effects of perfluidone are as
follows:

A 90-day dog feeding study showed liver disorders (hepatic
lesions, hepatocyte vacuolation, hyalin degeneration and
biliary stasis) at the two highest dose levels (400 and
800 ppm).

Chronic toxicology effects on perfluidone have not been
evaluated because there are no valid teratogenicity
studies in either the rat or rabbit, and no mutagenicity
tests.

Major routes of human exposure:

   Non-dietary exposure of applicators to perfluidone during mixing, loading, spraying and flagging is probable.

Physiological and biochemical behavioral characteristics:
   Translocation: Perfluidone is mobile in the xylem but is of limited mobility in the phloem.
   Mechanism of pesticidal action: Perfluidone inhibits photosystem I of the photosynthetic process.

Environmental characteristics:

   Adsorption and leaching in basic soil types: Perfluidone will leach through wet, neutral or slightly alkaline soils, with a tendency toward greater leaching in soils having low clay and organic matter. Leaching occurs to a lesser extent in acidic soil.
   Microbial breakdown: Soil microorganisms play a significant role in the degradation of perfluidone in the soil.
   Loss from photodecomposition and/or volatilization: Photodecomposition and/or volatilization play an important role in the degradation of perfluidone on or in soil.
   Average persistence at recommended rates: When used at recommended rates under normal environmental conditions, the half life of perfluidone in the soil is approximately one month.

Ecological characteristics:

   96-hour fish toxicity: 147.5 ppm for bluegill sunfish (practically non-toxic) and 17.0 ppm for rainbow trout (slightly toxic).

Potential problem for endangered species:

   Perfluidone was reviewed by the Agency under the endangered species cotton cluster, but it did not exceed any trigger. On flue-cured tobacco, if the maximum rate of 1.5 pounds active ingredient/acre were applied to six inches of water, the resulting residues (0.754 ppm) are less than 0.900 ppm (1/20 of the LC_{50} of the most sensitive fish). Therefore, perfluidone's use on tobacco is not expected to adversely affect the endangered species and no endangered species label statements are required.

Tolerance assessments:

The Agency is not conducting a tolerance reassessment on perfluidone because the only use is on flue-cured tobacco. This use is a non-food and non-feed use which does not require a tolerance. Therefore, the Agency will not require residue chemistry data on the metabolism of perfluidone and related metabolite(s) in crops and animals.

Problems known to have occurred with use:

Since perfluidone has never been commercially manufactured or sold in the United States, it has not been identified in the Pesticide Incident Monitoring System (PIMS) nor implicated in any incident.

4. Summary of regulatory position and rationale:

Based on the review and evaluation of all available data and other relevant information on perfluidone, the Agency has made the following determinations:

The available data are insufficient to indicate that any of the risk criteria listed in § 162.11(a) of Title 40 of the U.S. Code of Federal Regulations have been met or exceeded for the uses of perfluidone at the present time.

Perfluidone is not acutely toxic by the dermal and ocular routes of exposure. A 90-day dog feeding study showed liver disorders (hepatic lesions, hepatocyte vacuolation, hyalin degeneration and biliary stasis) at the two highest dose levels (400 and 800 ppm).

The absence of other toxicological data prevents the Agency from determining the acute, subacute and chronic effects of perfluidone. Given the lack of data, the most appropriate action is to move quickly to fill the data gaps. When data are submitted and reviewed, the Agency will determine the registrability of the affected use pattern.

The data base supporting the perfluidone tolerance on cotton seed (0.01 ppm) has been reviewed and found to be inadequate. However, the data tables in the perfluidone registration standard will not include these data gaps attributable to the tolerance on cottonseed: § 158.135 Toxicology, Subchronic Testing (82-4), Chronic Testing (83-1, 83-2, 83-4), Special Testing (85-1); § 158.125 Residue Chemistry, Livestock (171-4), Animal Residues (171-4), Cottonseed (171-4) and Meat/Milk/Poultry/Eggs (171-4). Instead, the Agency will issue a Proposed Rule to revoke the perfluidone tolerance on cottonseed. If there is no response to support the tolerance, the Agency will issue a Notice of Tolerance Revocation.

End-use product (EP) labels will be required to bear a revised environmental hazard statement

EP labels will be required to bear a rotational crop restriction.

EP labels will be required to bear a protective clothing statement for mixers, loaders, and applicators.

Manufacturing-use product (MP) labels will be required to bear a statement regarding discharge to bodies of water and sewer systems.

The Agency will not require a ground water advisory statement at this time. If additional data indicate that perfluidone may cause ground water concerns, the Agency may reconsider this decision.

No endangered species label statements are required.

The Agency is not requiring a reentry interval for the registered use of perfluidone.

Specific label warning statement:

Hazard Information

The human hazard statements must appear on all EP labels as presribed in 40 CFR 162.10.

Environmental Hazard Statements

All MPs intended for formulation into EPs must bear the following statements:

"Do not discharge effluent containing this product directly into lakes, streams, ponds, estuaries, oceans or public waters unless this product is specifically identified and addressed in a National Pollutant Discharge Elimination System (NPDES) permit. Do not discharge effluent containing this product into sewer systems without previously notifying the sewage treatment plant authority. For guidance, contact your State Water Board or Regional Office of the Environmental Protection Agency".

End-Use Product Statements

The following environmental hazard statement must appear on all EP products:

"Do not apply directly to water or wetlands. Do not contaminate water by cleaning of equipment or disposal of wastes".

The following rotational crop restriction statement must appear on all EP products:

"Limitations: Replant only tobacco in DESTUN® herbicide-treated soil during the year of application and the following crop year".

The following protective clothing statement must appear on all EP products:

"During mixing/loading or application, wear gloves impermeable to perfluidone. When handling the concentrated product, wear a dust mask and chemical resistant apron in addition to the gloves. Wash hands thoroughly with soap and water after handling and before eating, urinating or smoking. Remove and wash clothing before reuse. Clothing should be laundered separately from household articles. Replace gloves frequently. Used gloves and clothing which has been drenched or heavily contaminated should be disposed of in accordance with state or local regulations".

5. <u>Summary of major data gaps</u>:

The toxicological studies are required on the following dates:

An acute oral toxicity study (June 30, 1986),
An acute inhalation toxicity study (June 30, 1986),
A dermal sensitization study (June 30, 1986),
A 21-day dermal toxicity study (September 30, 1986),
Two teratogenicity studies (December 30, 1986), and
Mutagenicity studies (June-September 1986).

The environmental fate data are required on the following dates:

An hydrolysis study (June 30, 1986),
A photodegradation study (June 30, 1986),
An aerobic soil study (December 30, 1987),
A mobility/leaching study (September 30, 1986), and
A dissipation/soil study (December 30, 1987).

The ecological effects data are required on the following dates:

An acute avian oral toxicity study (June 30, 1986),
Two avian subacute dietary toxicity studies (June 30, 1986),
Two freshwater fish toxicity studies (June 30, 1986), and
An acute freshwater invertebrate toxicity study (June 30, 1986).

Product chemistry data are required during 1986.

6. Contact Person at EPA:

Robert J.Taylor
Office of Pesticide Programs, EPA,
Registration Division (TS-767C)
Fungicide-Herbicide Branch
401 M Street., S.W.
Washington, DC 20460.
(703) 557-1800

DISCLAIMER: The information presented in this Pesticide Fact Sheet is for informational purposes only and may not be used to fulfill data requirements for pesticide registration and reregistration.

PHORATE

Date Issued: February 1, 1985
Fact Sheet Number: 34.1

1. Description of chemical

    Generic name: 0,0-diethyl S-[(ethylthio)methyl]phosphorodithioate
    Common name: Phorate
    Trade name: Thimet, Rampart
    EPA Shaughnessy Code: 057201
    Chemical Abstracts Service (CAS) Number: 298-02-2
    Year of Initial Registration: 1959
    Pesticide Type: Insecticide-nematicide
    Chemical Family: Organophosphate
    U.S. and foreign producers: American Cyanamid Co. (U.S.)

2. Use patterns and formulations

    Application Sites: Non-domestic terrestrial and aquatic food/feed crops; and greenhouse commercial nursery stock (both outdoor and greenhouse).
    Type of Formulations: Granular, emulsifiable concentrate
    Application Rates (lbs. active ingredient): Ornamentals - 8-20 lbs./A (10% granular product); Agricultural food/feed crops - 1-3 lbs./A (10-20% granular product) except sugarcane which is 4 lbs./A and potato which is 3.6 lbs./A. The emulsifiable concentrates are applied at 1 lb./A (6 lb/gal. emulsifiable concentrate) for Bermudagrass; and at 2-2.18 lb./150 lb. seed (8-8.7 lb./gal. emulsifiable concentrate) for treatment of cottonseed.

3. Science findings

    Summary science statement:

    Phorate has a very high acute toxicity to humans, fish and wildlife. Pertinent data are lacking, however, and the Agency cannot conduct a full risk assessment until the data required in this Standard are submitted and evaluated.

    Certain oxidation products of phorate are more toxic than phorate itself. The oxidation products were previously toxicologically discounted in the establishment of the tolerances for phorate.

    The Agency does not, however, have the data needed to determine the level at which the identified metabolites are present in the residues occurring in or on the raw agricultural commodities resulting from the current registered uses of phorate. Also, the metabolism of phorate in animals is not adequately understood. Adequate metabolism studies utilizing ruminants are needed to determine the distribution and characterization of residues in tissues and milk. In addition, the toxicological studies needed to establish the acceptable daily intake levels of the identified metabolites are also lacking. Depending on results of the residue studies required to be submitted under the Standard, the dietary risk may be greater than it earlier appeared to be, although it may also be the same as before.

Chemical Characteristics:

Physical State: Liquid
Color: Clear
Odor: Skunk-like
Boiling Point: 118-120°C at 0.8 mm Hg
Melting Point: N/A
Flammability: 160°C (tagliabue open cup)
Solubility in Water: 50 ppm
There are no unusual handling characteristics.

Toxicology Characteristics:

- Acute Oral rat LD_{50}; 3.7 mg/kg (male); and 1.4 mg/kg (female); Toxicity Category I.
- Acute Dermal rat LD_{50}; 9.3 mg/kg (male); and 3.9 mg/kg (female); Toxicity Category I.
- Acute Inhalation rat LC_{50}; 60 mg/m^3 (male); and 11 mg/m^3 (female); Toxicity Category I.

- The major routes of exposure in order of toxicological significance: inhalation, dermal.

- Chronic toxicology results: Adequate studies include subchronic feeding in the rat, oncogenicity studies in the rat and the mouse, three generation reproduction, and teratology study in the rat. The available mutagenic studies include tests in vitro microbial and mammalian cells, and in vivo dominant lethal. No adverse effects were found in any of these studies.

Physiological and Biochemical Behavioral Characteristics:

Foliar absorption: N/A
Translocation: Available metabolism studies indicate that phorate and its soil metabolites are absorbed from the soil by plant roots and translocated to above-ground portion of the plant.
Mechanism of pesticidal action: As as organophosphate, phorate exerts its toxic action by inhibiting certain important enzymes of the nervous system, cholinesterase (ChE).

Metabolism and persistence in plants and animals: The metabolism in plants is adequately understood. Phorate is metabolized in plants by rapid oxidation to the sulfoxide (some oxidation to the O-analog may also occur), followed more slowly by oxidation to the sulfone and/or the O-analog sulfoxide; phorate sulfone and phorate O-analog sulfoxide are then further oxidized to the O-analog sulfone. Available studies indicate that hydrolysis of the oxidized metabolites eventually occurs to yield non-toxic water-soluble products. A field study of corn treated at 1 lb. a.i./A with 10% granular formulation indicate that phorate residues were nondetectable (<0.002 ppm) after 14 days while residues of the sulfoxide and sulfone persisted to 28 days. After 83 days, there were no detectable residues occuring in the kernels, cobs or husks.

The metabolism of phorate in animals is not presently understood due to lack of sufficient data.

Environmental Characteristics:

Phorate has some potential to leach through soil and contaminate ground water, particularly where soils are sandy and aquifers are shallow. Simulation of the leaching potential of phorate using the Pesticide Root Zone Model (a computer model which predicts movement through the root zone and the unsaturated soil zone based on chemical and soil properties) predicts some mobility in sandy soils but none in loam soils under typical phorate use conditions. Because of lack of sufficient data regarding the environmental behavior of phorate, the Agency is unable to completely evaluate the leaching potential. Additional studies are being requested on an accelerated basis; these studies include hydrolysis, metabolism, and mobility studies and, in particular, a field dissipation study which is to be carried out in a potato-growing area of Long Island, New York.

Ecological Characteristics:

- Avian Oral LD_{50} = 0.62 mg/kg (mallard) and 7.12 mg/kg (pheasant).
- Avian Dietary LC_{50} = 24 to 77 ppm (upland gamebirds) and 712 ppm (waterfowl).
- Fish LC_{50} = 6 to 13 ppb (coldwater fish) and 2 to 280 ppb (warmwater fish).
- Aquatic Invertebrate LC_{50} = 4 ppb.

Based on these studies, phorate is very highly toxic to avian species, freshwater fish and aquatic invertebrates. Regarding endangered species, there is a potential risk to the Aleutian Canada goose, Attwater's greater prairie chicken and the Kern primrose sphinx moth.

Tolerance Assessments:

Refer to attached table for the list of current tolerances established for phorate. Available data are not sufficient to conduct a tolerance assessment.

4. Summary of Regulatory Position and Rationale

Use classification: All emulsifiable concentrate (EC) phorate formulations containing 65% and greater and all granular formulations of phorate for use on rice have been previously classified restricted use pesticides pursuant to 40 CFR 162.31. All granular formulations containing 5% phorate and greater have been classified as restricted use pesticides under this Registration Standard. In addition, all granular formulations containing less than 5% phorate are considered restricted use pesticides pending receipt and evaluation of data required to be submitted under this Standard. All products subject to the restricted use requirement which are released for shipment after September 1, 1985 must be labeled for restricted use. All products subject to the restricted use requirement which are in channels of trade after September 1, 1986 must be labeled for restricted use.

Though there are no EC formulations containing less than 65% phorate currently registered, such formulations would be considered restricted use pesticides.

Use, formulation or geographical restrictions: Products containing phorate are not to be used or stored in or around the home. Geographical restrictions of varying degrees currently exist on some or all of the uses on Bermuda grass, corn, lettuce, rice, sorghum, sugarcane, tomato, wheat, and lilies. End-use products may be granular or liquid formulations.

Unique warning statements required on labels: Phorate manufacturing-use products (MP) require the use of protective clothing and respirator. End-use products require the restricted use statement and use of protective clothing. The phorate products for greenhouse use require the use of a respirator. A restriction against reentering treated fields before 24 hours after application is also required.

No new uses or tolerances for phorate will be considered until the human and environmental concerns raised in this Document are satisfactorily resolved.

5. Summary of Major Data Gaps

Toxicology: Acute delayed neurotoxicity; 6-month feeding studies (on the identified oxidation metabolites); chronic feeding study (non-rodent); a teratology study in a second species (other than the rat)*; mutagenicity studies and product integrity studies and a general metabolism study.

* A rabbit teratology study has recently been submitted and is currently under review by the Agency.

Environmental Safety: Avian reproduction; fish life cycle (freshwater); full field studies including population monitoring (avian, mammalian and aquatic species); and secondary poisoning studies for one mammal and one avian species.

Environmental Fate: Hydrolysis; photodegradation (water, soil, air); metabolism (aerobic soil and aquatic, anaerobic soil or aquatic); leaching; volatility; dissipation; accumulation (rotational crop, irrigated crop, fish; and aquatic organisms); and re-entry studies (dermal and inhalation exposure; and soil and foliar dissipation).

Residue Chemistry: Plant metabolism (the quantification of the oxidation products of phorate contained in the residue in or on raw agricultural commodities, including meat, milk, poultry and eggs, for which tolerances have been established as a result of the maximum registered use of phorate); livestock metabolism; and additional residue data (for each established tolerance except Bermudagrass and milk).

All data are to be submitted by July, 1987.

6. Contact person at EPA:

William H. Miller, (PM-16)
Insecticide-Rodenticide Branch (TS-767)
401 M Street, SW.
Washington, DC 20460.

Tel. No. (703) 557-2600

DISCLAIMER: The information presented in this Chemical Information Fact Sheet is for informational purposes only and may not be used to fulfill data requirements for pesticide registration and reregistration.

PHOSMET

Date Issued: October 1, 1986
Fact Sheet Number: 101

1. DESCRIPTION OF CHEMICAL

   Generic name: N-(mercaptomethyl) phthalimide S-(O,O-dimethyl phosphorodithioate)

   Common name: Phosmet

   Trade names: Phthalofos, PMP, Appa, Imidan, Kemolate, Prolate, R-1504

   EPA Shaughnessy Code: 059201

   Chemical Abstracts Service (CAS) Number: 732-11-6

   Year of Initial Registration: 1966

   Pesticide Type: Insecticide-acaricide

   Chemical Family: Organophosphate

   U.S. Producer: Stauffer Chemical Company

2. USE PATTERNS AND FORMULATIONS

   Application Sites: Terrestrial food crops (field, vegetable and orchard crops such as alfalfa, apples, almonds, apricots, blueberries, citrus, corn, cotton, grapes, nectarines, pears, peaches, pecans, plums, and potatoes); terrestrial non-food crops (nursery and ornamental crops); domestic outdoor and indoor.

   Types of formulations: Dust [1% and 5% active ingredient (A.I.)], wettable powders (7.5%, 12%, 12.5%, 50%, and 70% A.I.), impregnated resins (15% A.I.) and emulsifiable concentrates [1 pound (lb.) per gallon (gal.), 3 lb/gal. and 12.5% A.I.]

   Methods of Application: Foliar applications, aerial applications, animal treatments, stored commodity treatments, and impregnated materials.

   Usual Carrier: Water

3. SCIENCE FINDINGS

Phosmet is a member of a chemical family known as the organophosphates (OPs). OP pesticides act on the nervous system by interfering with an enzyme acetylcholinesterase. This effect (known as cholinesterase inhibition) is reversible once exposure stops. There are antidotes for this type of poisoning (atropine and 2-PAM). Phosmet has a moderate to low acute oral, dermal, and eye/skin irritation toxicity. It is moderately toxic (Toxicity Category II) to humans by ingestion. Additional data (acute inhalation and dermal sensitization) is required to complete the acute toxicity profile for technical phosmet. Insufficient data exist to fully assess the subchronic dermal, mutagenicity, oncogenicity, and general metabolism of phosmet. Reentry data is necessary in order to establish permanent worker reentry intervals.

Phosmet has been classified as a "tentative" category C carcinogen. This conclusion was reached after review of a 2-year mouse oncogenicity study. Additional studies are being required to complete the oncogenic assessment of the chemical. Currently available data indicate that phosmet does not cause neutrotoxic, teratogenic, or reproductive effects.

The environmental fate of phosmet is not well documented. A review of preliminary data indicates phosmet is moderately mobile to immobile in soil and hydrolyzes rapidly in soil. The physical-chemical characteristics of the chemical indicate a potential for phosmet and possibly its degradates to contaminate groundwater. Hydrolysis, soil dissipation, anaerobic soil metabolism, leaching, photodegradation, and rotational crop and reentry data are required.

Phosmet is practically non-toxic to slightly toxic to birds, and mildly toxic to mammals. It is unlikely that phosmet would be lethal to birds or mammals after a single application. Available data indicates the possibility of reproductive effects in birds and mammals due to the buildup of phosmet on avian and mammalian food items (apples, corn, cotton, and alfalfa) from repeat applications. Residue monitoring data on these food items is required to determine the magnitude of exposure. Phosmet is highly toxic to honeybees, fish, aquatic and estuarine invertebrates. Field monitoring studies are being required to determine the magnitude of exposure from the major crop uses. Additional fish and aquatic invertebrate studies are being required to complete the evaluation of hazard.

<u>Chemical Characteristics</u>: Information listed below references the technical grade active ingredient unless specified as the pure active ingredient (PAI).

Physical state:	Crystalline solid
Color:	White to greyish-white
Odor:	Typical phosphorodithioate
Boiling point:	Not applicable - the technical is a solid at room temperature
Flash point:	Not available in Agency files
Melting point:	72.0 - 72.7°C (PAI)

Chemical Characteristics (continued)

Specific gravity:	1.03 at 20° C
Solubility:	At 20° C, in water, 25 ppm; acetone, $\geq$ 1,000 grams/liter (g/l); kerosene, 10 g/l. xylene, 200 g/l.
Stability:	Not available in Agency files

Toxicological Characteristics:

Acute toxicity. Phosmet has moderate to low acute oral, dermal, and eye/skin irritation. Phosmet, like other organophosphate chemicals, can be absorbed by inhalation and skin penetration.

Acute Oral (rat): 113-304 mg/kg.

Acute dermal (rabbit): >3,160 mg/kg

Primary eye irritation: Produced mild redness when instilled in the unwashed eyes of 3 rabbits at 24 hours after exposure. Phosmet also produced corneal opacity, redness, chemosis, and discharge in 1 of 3 rabbits. Eyes were normal within 7 days.

Inhalation: Undetermined

Primary dermal: Non-irritant

Dermal sensitization: Undetermined

Chronic toxicity:

Oncogenicity: Phosmet has been classified as a "tentative" Category C carcinogen. This conclusion was reached after review of two (a 2-year mouse and a 2-year rat) oncogenicity studies. Phosmet was associated with a significantly elevated incidence of liver tumors (adenomas, and adenomas plus carcinomas combined) in male $B6C3F_1$ mice at the highest dose tested. These incidences were associated with liver hyperplastic changes and a decrease in the time to tumor occurrence. In female $B6C3F_1$ mice, the chemical was associated only with positive dose-related trends for liver adenomas and carcinomas. A 2-year rat oncogenicity study was considered inadequate (the number of animals sacrificed at the end of the study were too small to fully evaluate tumor responses). The chemical was essentially non-mutagenic (only one positive result occurred in a limited and inadequate battery of tests) and no positive correlation with respect to oncogenicity and mutagenicity could be made with known structural analogs. After a 2-year rat oncogenicity study and additional mutagenicity studies are submitted and evaluated, the Agency will reassess the oncogenicity issue and determine if dietary and worker risk assessments are necessary.

Mutagenicity: Phosmet was evaluated in several mutagenicity assays. The chemical was found to be positive only when tested in S. typhimurium strain TA-100. No mutagenicity study of phosmet was performed in mammalian cells in culture. Additional mutagenicity studies are required.

Teratogenicity: No teratogenic effects were reported for phosmet in oral teratology studies in monkeys (NOEL = 8.0 mg/kg) and rabbits (NOEL = 60 mg/kg).

Reproductive effects: Phosmet had no adverse reproductive performance effects in a 3-generation oral reproduction study in rats. (NOEL = 80 ppm).

Neurotoxicity: Delayed neurotoxic effects were not observed at levels up to 2,050 mg/kg of phosmet. Body, weight, food consumption, and egg production were significantly decreased in the 2,050 mg/kg test group.

Metabolism: Data indicates that phosmet is rapidly eliminated, with 78% being eliminated in the urine and 19% in the feces within 72 hours after administration of a single oral dose in rats. However, the major water soluble urinary metabolites have only been "tentatively" identified. A general metabolism study will be required.

Physiological and Behavioral Characteristics: Mechanism of Pesticide Action - - organophosphate cholinesterase inhibition.

Environmental Characteristics and Groundwater Concerns: Few data are available on the environmental fate of phosmet. Phosmet appears to be moderately mobile to immobile in soils ranging in texture from sand to silty clay loam. Because of phosmet's physio-chemical properties the potential exists for phosmet, and possibly its degradates, to contaminate groundwater. To date the Agency is not aware of incidents where phosmet has contaminated groundwater. To fully assess and complete the environmental fate profile of phosmet, the Agency is requiring hydrolysis, soil dissipation, anaerobic soil metabolism, leaching, photodegradation, rotational crop and reentry data.

Ecological Characteristics:

Avian acute toxicity:	Mallard duck -	2009 mg/kg
Avian dietary toxicity:	Bobwhite quail -	501 ppm
	Japanese quail -	2000 ppm
	Mallard duck -	>5000 ppm
Avian reporduction:	Bobwhite quail -	60-150 ppm
	Mallard duck -	25-60 ppm
Fish:	Rainbow trout -	230 ppb
	Bluegill sunfish -	70 ppb
Aquatic invertebrates:	Daphnia magna -	5.6 ppb
	Gamma fasciatus - 2.0 - 4.2 ppb	

Sufficient data are available to characterize technical phosmet as very highly toxic to warmwater fish and highly toxic to coldwater fish. The chemical is also very highly toxic to aquatic and estuarine invertebrates. Monitoring data in runoff water following terrestrial applications of phosmet is being required to complete the hazard assessment.

Phosmet is practically non-toxic to slightly toxic to birds and mammals. Phosmet may cause reproductive impairment in birds and mammals due to a buildup of residues on avian food items. Residue monitoring of avian and mammalian food items (apples, corn, cotton, and alfalfa) will be required to complete an evaluation of the reproductive hazards.

Phosmet is very highly toxic to honeybees and displays extended residual toxicity.

Endangered Species: Use on apple and pear orchards, alfalfa, corn, and cotton crops, could place endangered species in the vicinity of treated areas at risk. Also two endangered insect species in the vicinity of food crop uses in certain counties of California could be threatened. Residue monitoring data will be required to aid in completion of the assessment of hazards to endangered species.

Tolerance Assessment: Tolerances have been established for residues of phosmet in raw agricultural commodities, meat, fat and meat byproducts (40 CFR 180.261) and in processed food (21 CFR 193.279) for phosmet and its oxygen analog at levels ranging from 0.1 to 40.0 ppm.

The metabolism of phosmet in plants and animals is not adequately understood. Additional residue data and metabolism data will be required to reassess the adequacy of existing tolerances and to issue new tolerances. Processing studies are also being required for potatoes, apples, plums, peaches, grapes, field corn grain, and cottonseed.

The acceptable daily intake (ADI) for humans was based on a 2-year chronic feeding study in rats. The ADI in humans was calculated to be 0.02 mg/kg/day and the maximum permitted intake (MPI) is equal to 1.2 mg/kg/day with a NOEL of 40 ppm and a safety factor of 100. Using these calculations the percent utilization of the ADI would be 98.29 percent. Since virtually all of the ADI has been used up by the TMRC and the Agency is aware of a potential oncogenic response to phosmet, new tolerances and/or new uses will not be issued if they contribute significantly to the TMRC and/or result in a significant increase in the dietary exposure.

Reported Pesticide Incidents: In the period from 1978 to 1979, 67 incidents involving a flea dip formulation (Paramite) of phosmet were reported to the Agency. Of these 67 incidents, 39 involved cats only, 16 involved dogs only, 2 involved cats and dogs, 8 involved human reactions, and 2 involved dogs and human reactions. Reported mortalities from these incidents were 20 cats (one leukemic) and 12 dogs. Additional incidents of adverse animal reactions (primarily cats) involving the same formulation have been reported up through 1985. The Agency is re-evaluating the use of phosmet on pets.

ESTABLISHED PHOSMET TOLERANCES

Commodity	Parts Per Million
Alfalfa	40.0
Almond, hulls	10.0
Apples	10.0
Apricots	5.0
Blueberries	10.0
Cattle, fat	0.2
Cattle, meat	0.2
Cattle, mby	0.2
Cherries	10.0
Citrus fruits	5.0
Corn, fresh (including sweet K + 6 WHR)	0.5
Corn, fodder	10.0
Corn, forage	10.0
Corn, grain	0.5
Cottonseed	0.1
Cranberries	10.0
Goats, fat	0.2
Goats, mbyp	0.2
Goats, meat	0.2
Grapes	10.0
Hogs, fat	0.2
Hogs, mbyp	0.2
Hogs, meat	0.2
Horses, fat	0.2
Horses, mbyp	0.2
Horses, meat	0.2
Kiwi fruit	25.0
Nectarines	5.0
Nuts	0.1
Peaches	10.0
Pears	10.0
Peas	0.5
Peas, forage	10.0
Peas, hay	10.0
Plums (fresh prunes)	5.0
Potatoes	0.1
Sheep, fat	0.2
Sheep, mbyp	0.2
Sheep, meat	0.2
Sweet potatoes	10.0
Tomatoes	2.0
Cottonseed oil	0.2

4. SUMMARY OF REGULATORY POSITIONS AND RATIONALES

- No referral to Special Review is being made at this time. A repeat rat oncogenicity study and additional mutagenicity studies must be submitted. The Agency will reassess the oncogenicity issue and determine if dietary and worker carcinogenicity risk assessments are necessary. The available data also indicate that phosmet is highly toxic to fish. Terrestrial residue analysis and aquatic runoff modeling indicate that certain use patterns could result in exposure of certain aquatic organisms to hazardous levels of the pesticide. Additional data are needed before the Agency can complete a full assessment of this hazard potential.

- The Agency will reassess the adequacy of the existing tolerances after required metabolism data and residue data are submitted.

- New tolerances and uses will be issued on a case-by-case basis.

- The tolerance expression for phosmet under 40 CFR 180.261 will be amended by deleting the reference to "cholinesterase-inhibiting" residues.

- The Agency has determined that endangered species label restrictions are necessary in order to prevent unreasonable adverse effects on the environment.

- In the absence of reports of fish kills following phosmet application and actual field monitoring data, the Agency will not restrict certain uses of phosmet to certified applicators, but has determined, based on the high toxicity of phosmet to aquatic organisms, that precautionary labeling will be required. The restricted use classification may be required if additional studies indicate that phosmet use poses risks to aquatic organisms that could be mitigated by increased controls in application.

- The Agency is imposing a 24-hour reentry interval. Foliar dissipation data are required on crops whose propagation requires human tasks that involve substantial, prolonged human contact.

- Protective clothing is required for mixers/loaders and applicators.

- The Agency will analyze the safety and efficacy data of a phosmet flea dip formulation (Paramite) to determine if further regulatory action is warranted. A warning statement indicating that improper dilution of the product could cause serious injury to pets is being required.

- The Agency has determined that the tolerance for cranberries should be revoked because there are no registered uses for phosmet on cranberries.

- The Agency is requiring processing data for the following agricultural commodities: potatoes, apples, plums, peaches, grapes, field corn grain, and cottonseed.

- The Agency is not requiring a rotational crop restriction. If required data demonstrate that follow-up crops take up phosmet residues from soil, rotational crop restrictions or tolerances in those crops may be necessary.

- The Agency is not imposing a ground water advisory statement on phosmet labeling at this time, but is requiring data to fully characterize the potential of this chemical to reach ground water.

- While data gaps are being filled, currently registered end-use products containing phosmet as the sole active ingredient may be sold, distributed, and used, subject to the terms and conditions specified in the Registration Standard.

5. SUMMARY OF MAJOR DATA GAPS

Product Chemistry

Product Chemistry	Feb.	1987

Residue Chemistry

Plant/Livestock Metabolism	Feb.	1988
Plant/Animal Residues	Feb.	1988
Storage Stability	Feb.	1988

Environmental Fate

Hydrolysis/Photodegradation	July	1987
Mobility Studies	Sept.	1987
Accumulation (Rotational) Crops	Dec.	1989
Glove Permeability	Mar.	1987 - protocol
	Nov.	1987 - final report
Anaerobic Soil Metabolism	Dec.	1988
Soil Dissipation	Dec.	1988

Toxicology

Acute Inhalation Toxicity (rat)	July	1987
Dermal Sensitization	July	1987
21-Day Dermal (rabbit)	Sept.	1987
Oncogenicity (rat)	Nov.	1990
Gene Mutation	July	1987
Structural Chromosome Aberration	Sept.	1987
Other Genotoxic Effects	Sept.	1987
General Metabolism	Sept.	1988

Ecological Effects

Acute Toxicity to Freshwater Invertebrates	July	1987
Acute Toxicity to Estuarine and Marine Organisms	Sept.	1987
Fish Early Life Stage and Aquatic Invertebrate Life Cycle	Dec.	1987
Field Monitoring (avian, aquatic, and mammalian)	Feb.	1988

6. CONTACT PERSON AT EPA

George T. LaRocca
Product Manager 15
U.S. Environmental Protection Agency
TS-767C
401 M Street S.W.
Washington, D. C. 20460
(703) 557-2400

DISCLAIMER: The information in this Pesticide Fact Sheet is for informational purposes only and may not be used to fulfull data requirements for pesticide registration and reregistration.

PICLORAM

Date Issued: March 31, 1985
Fact Sheet Number: 48

1. Description of Chemical

Common Name: Picloram
Chemical Name: 4-amino-3,5,6-trichloropicolinic acid
Trade Name: Tordon, Grazon, Amdon
EPA Shaughnessy Code: 005101
Chemical Abstracts Service (CAS) Number: 1918-02-1
Pesticide Type: Herbicide
Chemical Family: Picolinic Acid
U.S. and Foreign Producers: Dow Chemical Company

2. Use Patterns and Formulations

Picloram is used on permanent grass pastures in the eastern half of the United States, and is used primarily for the control of two woody plant species. These are hawthorn and multiflora rose. On rangeland in the western states, picloram is used to control bitterweed, knapweed, leafy spurge, locoweed, larkspur, mesquite, prickly pear, and snakeweed. In Nebraska, Montana, Wyoming, Minnesota and the Dakotas, it is used to control wild buckwheat and thistles in small grain such as wheat, oats, and barley. Formulations of picloram include potassium and amine salts with the potassium and triisopropanolamine salts being the most commonly used. There are nine EPA registered products currently on the market that either contain picloram as their sole active ingredient or contain mixtures of picloram and a phenoxy herbicide.

3. Science Findings

Summary science statement:

Picloram is highly phytotoxic, moderately toxic to cold water fish and certain combinations of picloram and 2,4-D may produce sensitizing reactions in humans. Water contamination is a major concern in the exposure of nontarget plants to picloram since this chemical has been detected in ground water apparently as a result of movement through soil or through contamination of wells and in surface waters from runoff from treated areas. Product chemistry information was generally satisfactory but the impurities need better quantification and analytical methodology. For chronic feeding studies, the acid form of picloram is considered equivalent to salts and ester forms. Although there is no evidence that picloram poses risks of unreasonable adverse

health effects, additional long-term studies have been
identified as being necessary to support this conclusion and
to support present and future tolerances. The Agency has
concluded that the dietary cancer risk to the general public
of HCB in the fat and milk of cattle fed picloram treated
grass is an acceptable risk; however, the HCB must not exceed
200 ppm in the technical product. In addition, nitrosamine, if
present, must not exceed 1 ppm.

Chemical Characteristics:

Picloram is a damp off-white to brown powder substance.
Data indicates that picloram has low acute toxicity, low
dermal irritation potential and is neither teratogenic
nor mutagenic. Picloram is highly phytotoxic, and easily
absorbed by roots and foilage. Technical picloram is a damp
powder substance, off-white to brown in color, with a chlorine-like
odor and a melting point of 215°C. Picloram is stable in
both acidic and basic media. It is subject to photodecomposition
by ultraviolet radiation in aqueous solution. At 25°C pure
picloram is soluble in water at 0.043 grams per 100 milliliters.
The vapor pressure of picloram is 6.2×10^{-6} mm at 45°C. The
empirical formula for picloram is $C_6H_3Cl_3N_2O_2$.

Toxicology Characteristics:

The data on short term effects, environmental effects and genetic
mutation, as well as one NCI cancer study, support the current
registration of picloram. The registrant is conducting a new
rat feeding study to clarify the ambiguous results of the
second NCI study. The Agency has no current evidence indicating
that use of picloram may result in unreasonable adverse effects
to human health or the environment, although more data are
needed on long term effects to support this conclusion.

Environmental Characteristics:

Photodegradation and aerobic soil degradation are the main
processes for dissipation of picloram in the environment.
Following normal agricultural, forestry, and industrial
applications, long-term accumulation of picloram in the soil
does not occur. The half-life of picloram under most field
conditions is a few months, but it may exceed one year or more,
especially in dry climates. Picloram has a moderate mobility
in soil and its relatively high water solubility and low soil
absorption indicate that it has the potential to leach in soil.

Ecological Characteristics

Picloram is highly phytotoxic and is easily absorbed by roots and foilage. In soils not subject to leaching, it is very persistent; phytotoxicity has been detected in some cases well over one year after application. Picloram appears to be practically non-toxic to birds, moderately toxic to cold water fish and slightly toxic to warm water fish. However, chronic studies on lake trout suggest that low concentrations of picloram will adversely affect the rate of yolk sac absorption and growth of fry.

Efficacy review results, where conducted:

No efficacy data was reviewed because no public health uses were involved.

Tolerance assessment:

The established tolerances for picloram are not supported by the data now available to the Agency. Until significant toxicological studies are submitted and reviewed and it is determined whether there are concerns, the Agency cannot consider any new petitions for tolerances. If the toxicological studies indicate that additional residue data are required, an assessment of existing tolerances and new tolerance petitions will be made.

4. <u>Summary of Regulatory Position and Rationale</u>:

°Use

Picloram is classified as a Restricted Use Pesticide due to possible groundwater contamination and hazard to fish and wildlife.

°Formulations

Formulations of picloram include potassium and amine salts with the potassium and triisopropanolamine salts being the most commonly used. Liquid product concentrations of picloram range from 0.25 to 2 pounds acid equivalent per gallon while pelleted formulations of picloram range from 2 to 10 percent acid equivalent by weight.

°Manufacturing process or geographical restrictions

The Agency will require precautionary label statements advising against the use of picloram in very permeable, i.e., well-drained soils such as karst limestone and loamy sands. The Agency will require the registrants to do ground-water monitoring studies. While the data gaps are being filled, currently registered manufacturing-use products containing picloram as the sole active ingredient may be sold, distributed, formulated and used in the United States, subject to the terms and conditions of this standard.

º Unique warning statements

Manufacturing-Use Products - "Do not discharge effluent containing this product into lakes, streams, ponds, estuaries, oceans or public waters unless this product is specifically identified and addressed in an NPDES permit. Do not discharge effluent containing this product into sewer systems without previously notifying the sewage treatment plant authority. For guidance, contact your State Water Board or Regional Office of the Environmental Protection Agency."

End-Use Products - "Restricted Use Pesticide. Potential ground water contaminant. Toxic to non-target plants. For retail sale to and use only by applied by certified applicators or persons under their supervision and only for those uses covered by the certified applicator's certification. Picloram is a chemical which can travel (seep or leach) through soil and can contaminate ground water which may be used as drinking water. Picloram has been found in ground water as a result of agricultural use. Users are advised not to apply picloram where the soils are very permeable, i.e., well-drained soils such as karst limestone and loamy sands. Your local agricultural agencies can provide further information on the type of soil in your area and the location of ground water."

For rotated crops: "Do not rotate food or feed crops on treated land if they are not registered for picloram."

For ditchbank uses: "Water contaminated with residues of picloram from ditch bank uses shall not be used to irrigate crops which are not registered for use with this chemical."

For picloram mixtures with 2,4-D: "Warning: Avoid contact with skin, eyes or clothing. Avoid repeated skin contact since sensitizing reactions may occur."

For non-aquatic uses: "Do not apply directly to water or wetlands. Do not contaminate water by cleaning of equipment or disposal of wastes."

For aquatic uses: "Consult your State Fish and Game Agency before applying to public waters. Permits may be required before treating such waters. Do not apply directly to water except as directed on the labeling. Do not contaminate water by cleaning of equipment or disposal of wastes."

5. Summary of major data gaps

Major Data gaps exist for all scientific disciplines:

Product Chemistry - Product Composition, Analysis of Ingredients.
Residue Chemistry - Plant, Animal Metabolism, Analytical Methods.
Toxicology - Acute, Chronic and Mutagenicity Testing.
Environmental Assement - Metabolism, Mobility, Degradation Studies
Ecological Effects - Avian and Mammalian, Aquatic Organism, Non-target phytotoxicity and Non-target Insect Testing.

6. Contact person at EPA

Robert Taylor
Registration Division (TS-767)
Environmental Protection Agency
401 M Street, SW
Washington, D.C.   20460
(703)-557-1800

DISCLAIMER: The information presented in this Chemical Information Fact Sheet is for informational purposes only and may not be used to fulfill data requirements for pesticide registration and reregistration.

POTASSIUM BROMIDE

Date Issued: September 30, 1984
Fact Sheet Number: 38

1. Description of Chemical

    Generic Name: Potassium Bromide
    Common Name: Potassium Bromide
    Trade Name: Potassium Bromide
    EPA Shaughnessy Code No.   041101
    Chemical Abstracts Service No.   775802-3
    Year of initial registration: 1960
    Pesticide Type: Disinfectant
    Chemical Family: Inorganic Halogen
    U.S. and Foreign Producers: Many U.S. chemical supply houses, e.g. J. T. Baker Chemical Co., Mallinckrodt, etc. probably many foreign producers.

2. Use Patterns and Formulations

    Application Sites: Food contact surfaces, food handling equipment, eating utensils in food processing establishments

    Types of formulations: granular powder

    Types and methods of application: Dipping utensils in a tank, or by spraying

    Application Rates: 1 oz in 10 gallons of water

    Usual Carriers: the granular product is mixed with other chemicals generally recognized as safe; the use solution is prepared with water

3. Science Findings:

    Chemical Characteristics:

    Physical State: solid
    Color: White granules, colorless crystals
    Odor: None
    Melting Point: 730°C
    Density: 2.75

4. Toxicological Characteristics

   Technical Grade of Potassium Bromide (99% KBr): large doses cause central nervous system depression; prolonged intake may cause mental deterioration, acneform skin eruptions (Merck Index, 4th Edition).

   Chronic Toxicity: Evaluation of available information from the published literature raises no conerns related to chronic adverse effects

   The registered product covered by the standard is in category II, based on the approved label.

5. Physiological and Biochemical Behavioral Characteristics

   N/A

6. Environmental Characteristics

   N/A

7. Ecological Characteristics

   N/A

8. Tolerance Assessments

   Potassium Bromide, together with other chemicals generally recognized as safe, is cleared for use as a final sanitizing rinse for food contact surtaces, food handling equipment and eating utensils, in food processing establishments, under Section 178.1010 of Title 21 of the Code of Federal Regulations, which is administered by the Food and Drug Administration. All end-use products proposed for registration under the standard must comply with the provisions of this section.

10. Problems with Chemical

    The technical grade of the active ingredient is not registered as a manufacturing-use product. To register it for for the use covered in the standard the registrant must satisfy the data requirements specified in Table A of the standard.

End-Use Products: If a registrant qualifies for the formulator's exemption he or she may register the end-use product for the use covered by the standard by satisfying the data requirements listed in Table B of the standard. If a registrant does not qualify for the formulator's exemption he or she must address and satisfy the data requirements specified in Table A, for the technical grade of the active ingredient, as well as those specified in Table B, for the end-use product.

10. Summary Science Statement

A review of available information from the published literature on Potassium Bromide has not raised any concerns related to long term or chronic adverse effects for the registered use.

11. Summary of Regulatory Position and Rationale

Use Classification: General
Use, formulation, or geographical restrictions: The registrered use is as a final sanitizing rinse on food contact surfaces; the formulated product must be substantially similar to the registered use covered in the standard; there are no geographical restrictions. There are no unique warning statements for the registered product.

The technical grade of Potassium Bromide is not registered as a manufacturing-use product. To register it for the use covered in the standard the registrant must address and satisfy the data requirements listed in Table A of the standard These include acute toxicity data and ecological effects data. End-use product registrants who qualify under the formulator's exemption need only address and satisfy the data requirements listed in Table B of the standard. End-use product registrants who do not qualify under the formulator's exemption must address both sets of data requirements. Registrants must respond to the standard within 90 days from the date the standard is issued; and commit themselves to address and satisfy the data requirements within 6 months from the 90 day response date.

12. Summary of data gaps and dates when these data gaps are due to be filled.

Acute toxicity data and fish and wildlife data for the technical grade of Potassium Bromide as specified in Table A of the standard.

Acute toxicity data to register end-use products as specified in Table B of the standard.

13. EPA Contact:

Arturo Castillo, PM 32, Registration Division (TS-767C)
EPA, 401 M St., S. W., Washington, D. C., 703-557-13964

DISCLAIMER: The information presented in this Chemical Information Fact Sheet is for informational purposes only and may not be used to fulfill data requirements for pesticide registration and re-registraltion.
used to fulfill data requirements for pesticide registration and re-registration.

POTASSIUM PERMANGANATE

Date Issued: September 30, 1985
Fact Sheet Number: 80

1. Description of Chemical

Generic Name: $KMnO_4$
Common Names: Potassium Permanganate
Trade Names: None
EPA Shaughnessy Code: 068501
Chemical Abstracts Service (CAS) Number: 7722-64-7
Year of Initial Registration: 1962
Chemical Family: Inorganic Compound
Pesticide type: Microbiocide
U.S. and Foreign Producers: There are no registered products containing potassium permanganate. Potassium permanganate used in pesticide products is derived from commercial producers.

2. Use Patterns and Formulations

Application sites: as a bactericidal, fungicidal, and algaecidal agent in cooling towers, evaporative condensers, air wash systems, ornamental ponds, cooling fountains, aquaria, human drinking water, and poultry drinking water as applicable. Potassium permanganage is also used as a surface disinfectant, sanitizer, and deodorizer.

Types of formulations: formulations include ready to use liquids (0.004% - 2.5%), pelleted/tableted (40%), crystals (80% - 99.99%), and powder (41%).

Types and methods of application: as a disinfectant, potassium permanganate is diluted according to label directions and applied to hard, nonporous, inanimate surfaces. For the other uses described above, potassium permanganate is added directly to the water to be treated.

Application rates: for cooling towers, air washers, and evaporative condensers, use 3.6 oz. of active ingredient (a.i.)/500-1000 gal. of water, or 27-62 ppm a.i. For aquaria and ornamental ponds use 4 ppm a.i. to control algae growth. For human drinking water treatment use 1.1 mg of product per gal. of water (i.e., 4.4 - 4.5 ppm). For poultry drinking water, use 0.648 to 1.296 grains of product per quart of water (i.e., 272 - 543 ppm a.i.).

3. **Science Findings**

   Summary of science statement: "bottom line" of science information listed below.

   Chemical Characteristics:
   Physical state: crystalline
   Color: dark purple
   Melting Point: decomposes at 240°C
   Solubility: 6.38 gm/100ml at 20°C 25 gm/100ml at 65°C in water. Is decomposed by organic solvents (e.g., alcohols) and reducing substances (e.g., iodate, oxalate, and ferrous salts, especially under acidic conditions.

4. Toxicology Characteristics:

   Acute Eye Irritation: Tox. CAT I
   Skin Irritation: Tox. CAT I
   Oral LD_{50} Tox. CAT III

   Major routes of exposure: dermal and oral

   Chronic Toxicology: there are no chronic toxicity data on file with the Agency. Based on considerations regarding the chemical reactivity of the permanganate ion and the ubiquity of manganese, there are no chronic or subchronic toxicity data requirements.

   Environmental Characteristics: no data are available to assess the environmental fate of potassium permanganate.

   Ecological Characteristics:

   Freshwater Fish LC_{50} (97.5 - 100% a.i. tested):
   Bluegill sunfish   2.7 - 3.6 mg/l

   Tolerance Assessments: under the Federal Food, Drug and Cosmetic Act (FFDC), a tolerance or an exemption from the requirement of a tolerance is required for the residues of manganese in poultry and eggs (which result from the poultry drinking water use pattern). The Agency is requesting that the registrant petition the Agency for an exemption from the requirement of a tolerance.

   No problems have been reported to the Agency (i.e., PIMS or data in reference to Section 6(a)(2) of the Act) with respect to the use of this chemical.

5. Summary of Regulatory Position and Rationale

   The Agency has determined that potassium permanganate products to be used in treating human drinking water must bear labeling directions which would limit the residues of manganese in the finished potable water to not more than 0.05 mg/l. Otherwise, there are no formulation, manufacturing process or geographical restrictions for Potassium permanganate registrations.

   Unique warning statement:

   Environmental Hazards
   Do not discharge effluent containing this product into lakes, ponds, streams, estuaries, oceans or public water unless this product is specifically identified and addressed in an NPDES permit. Do not discharge effluent containing this product to sewer systems without previously notifying the sewage treatment plant authority. For guidance contact your State Water board or Regional Office of the EPA.

   A risk assessment has not been conducted on this chemical.

6. Summary of Major Data Gaps: NONE

7. Contact Person at EPA
   Registration Division (TS-767C)
   U.S. Environmental Protection Agency
   Office of Pesticide Programs
   401 M Street, S.W.
   Washington, D.C. 20460
   Attn.: Marshall Swindell   Telephone: (703)-557-3675

DISCLAIMER: The information presented in this Chemical Information Fact Sheet is for informational purposes only and may not be used to fulfill data requirements for pesticide registration and reregistration.

PROMETRYN

Date Issued: March 20, 1987
Fact Sheet Number: 121

1. Description of Chemical

   Common Name: Prometryn

   Chemical Name: 2,4-Bis(isopropylamino)-6-(methylthio)-s-triazine

   Other Names: Prometryne, Caparol, G-34161, Gesagard, Primatol Q, Prometrex

   OPP (Shaughnessy) Number: 080805

   Chemical Abstracts
     Service (CAS) Number: 7287-19-6

   Empirical Formula: $C_{10}H_{19}N_5S$

   Molecular Weight: 241.4

   Year of Initial Registration: 1964

   Pesticide Type: Herbicide

   Chemical Family: Substituted triazine

   U.S. and Foreign Producers: Ciba-Geigy Corp. (United States); Aceto Chemical Company, Inc. (U.S.); I.Pi.Ci. (Italy); Makhteshim-Agan (Israel); Verolit Chemical Manufacturing Company, Ltd. (Israel)

2. Use Patterns and Formulations

   Application Sites: Cotton, celery, pigeon peas, corn, ornamental plants, and forest trees (nursery seed beds)

   Percent of crop treated with the Pesticide: (1) Cotton, 9 to 12 percent of total U.S. cotton crop; (2) celery, 95 percent of total U.S. celery crop

   Percent of Pesticide Applied to Crop: (1) Cotton, approximately 97 percent (1,100,000 - 1,500,000 pounds active ingredient, [ai]) of total domestic usage of prometryn; (2) celery, approximately 3 percent (31,000 - 37,000 pounds ai) of total domestic usage of prometryn

   Types and Methods of Application: Applied by broadcast or directed spray (preplant, preemergent, or postemergent) to crop and weeds using ground equipment or aircraft.

   Pests Controlled: Annual broadleaf and grass weeds

   Application rates: Rates range from 0.48 to 3.2 lb ai/A.

   Types of Formulations: 95%, 97% active ingredient (ai) manufacturing-use products; 80% ai wettable powder; 44.4% (4 lb ai/gal) emulsifiable concentrate; 45.4% (4 lb ai/gal) flowable concentrate.

   Usual Carriers: Water at 20 to 40 gal/A.

3. Science Findings

   Summary Science Statement:

   Based on available acute studies (oral, dermal), prometryn has low acute toxicity and falls within Toxicity Category III, signal word CAUTION. Data gaps exist for the technical in acute toxicity, subchronic and chronic toxicity, teratology, oncogenicity, reproduction, and mutagenicity. Prometryn is not acutely hazardous to birds. It is slightly toxic to freshwater invertebrates and moderately toxic to fish. The chemical may pose a risk to some endangered species. The environmental fate of prometryn is not adequately understood. Additional data are required on degradation, mobility, accumulation, field dissipation, and metabolism. Leaching studies indicate that prometryn has intermediate mobility in sandy loam soils and is very mobile in sandy soils. A ground water monitoring study may be required pending the results of additional leaching and soil field dissipation studies.

Chemical Characteristics:

    Physical state: Powder

    Color: White

    Odor: Odorless

    Melting Point: 118-120 °C

    Density: 1.15 ± 0.02 g/cm^3 at 20 °C

    Solubility: 33 ppm in water at 20 °C
           readily soluble in organic solvents

    Vapor Pressure: 1.3×10^{-6} mbar at 20 °C

    Dissociation Constant: $K_b = 4.11 \pm 0.5$

    Octanol-Water
      Partition Coefficient: $pK = 4.1$ at 20 °C

    Stability: Stable in neutral, slightly acidic or basic media. Half-life at 25°C of 22.2 days and 1200 days in 0.1 N HCl and 0.1 N NaOH, respectively. Stable for a minimum of 3 years at room temperature.

Toxicological Characteristics

Acute oral toxicity (rat):	1802 mg/kg (males) and 2076 mg/kg (females); Toxicity Category III (97% ai)
Acute dermal toxicity (rabbit):	Greater than 2000 mg/kg; Toxicity Category III (97% ai)
Primary dermal irritation (rabbit):	Nonirritating to intact and abraded skin Toxicity Category IV (97% ai)

Chronic effects: The only available studies were done using formulated products. Data gaps remain for the technical materials.

Chronic feeding (dog): No observable effect level
(NOEL) = 3.75 mg ai/kg/day
Lowest effect level
(LEL) = 37.5 mg ai/kg/day
(80% ai)

Teratology (rabbit): Maternal and fetotoxic
NOEL = 12 mg ai/kg/day;
Developmental toxicity
and embryotoxic NOEL
> 72 mg ai/kg/day
(Formulated prometryn)

Major Routes of Exposure:

Mixers, loaders and applicators would receive the most exposure via skin/eye contact and inhalation.

Physiological and Biochemical Behavioral Characteristics:

Foliar and Root Absorption: Absorbed through both roots and foliage

Translocation: Translocated from roots to leaves

Mechanism of Action: Prometryn interferes with electron transport in the photosynthetic process.

Metabolism and Persistence in plants and animals: The nature of the residue of prometryn in plants and animals is not adequately understood. However, available data indicate the following: (1) In cotton, prometryn appears to concentrate in lysigenous glands where it slowly degrades to hydroxypropazine and other conjugated metabolites. (2) In the rat, most of the administered prometryn is excreted within 48 hours in urine and feces.

Environmental Characteristics:

Insufficient data are available to fully characterize the environmental fate of prometryn. Leaching studies do indicate, however, that prometryn has intermediate mobility in sandy loam soils and has high mobility in sandy soils. In addition, data indicate that prometryn has hydrolytic stability and is persistent in the soil. Degradation, accumulation, metabolism, soil field dissipation and additional leaching studies are required. Pending the results of soil field dissipation and leaching studies, a ground water monitoring study may also be required.

Ecological Characteristics:

    Avian acute oral toxicity: Greater than 4640 mg/kg (mallard)

    Avian dietary toxicity: 34,512 ppm (adjusted to 100% ai, mallard)

    Aquatic invertebrate acute toxicity: 18.59 ppm (Daphnia magna)

    Fish acute toxicity: 10 ppm (bluegill sunfish)

    Fish acute toxicity: 2.9 ppm (rainbow trout)

Potential problems related to endangered species: The use of prometryn on corn and cotton may pose a hazard to endangered species. A single application per year of prometryn should not harm most animal species, due to its low toxicity, except by destruction or adverse modification of habitat. There is concern for the endangered species, Solano grass and Valley Elderberry longhorn beetle following a single application of prometryn on corn. Both species occur in California. Proposed labeling has been designed to protect these species.

If multiple applications per year are used on corn and/or cotton, then several endangered species may be exposed to potential harm (due to sufficiently high application rates and the half-life of prometryn). Proposed endangered species labeling has been designed to protect these species when there are multiple applications of prometryn.

The endangered species labeling, as mentioned above, will be required after the Agency receives concurrence from the U.S. Fish and Wildlife Service. The Agency will notify registrants, in a Pesticide Registration Notice, of the final label requirements for endangered species.

Tolerance Assessment:

Tolerances have been established for residues of prometryn in or on the following raw agricultural commodities:

Commodity	Tolerance (ppm)
Celery	0.5
Corn, fodder, field	0.25
Corn, fodder, pop	0.25
Corn, fodder, sweet 1/	0.25
Corn, forage, field	0.25
Corn, forage, pop	0.25
Corn, forage, sweet	0.25
Corn, fresh (inc. sweet K+CWHR)	0.25
Corn, grain	0.25
Cotton 2/	1.00
Cottonseed	0.25
Pigeon peas	0.25

1/ This entry will be deleted from the tolerance expression, since corn fodder, sweet, is not considered a raw agricultural commodity of sweet corn.

2/ A feeding restriction is currently in effect for cotton forage, thus the entry "cotton" (presumably intended to cover cotton forage) will be deleted.

Dietary Assessment:

The provisional acceptable daily intake (PADI) was based on a 2-year dog feeding study with a NOEL of 3.75 mg ai/kg/day. Applying an uncertainty factor of 1000,* the PADI was calculated to be 0.004 mg/kg/day. This is equivalent to a maximum permissible intake (MPI) of 0.24 mg/kg/day for a 60 kg individual. The theoretical maximum residue contribution (TMRC) of prometryn in the daily diet is 0.000205 mg/kg/day based on the existing tolerances and daily food intake of 1.5 kg, with 5.13 percent of the PADI being utilized.

Established tolerances are based on the parent compound. Additional metabolism studies are required in order to identify and quantify all metabolites of toxicological concern.

* It is not clear whether another species will prove to be more sensitive to prometryn than the dog, thus an uncertainty factor of 1000 was used.

4. Summary of Regulatory Position and Rationale:

   Use, Formulation or Geographic Restrictions: No significant new food or feed uses of prometryn will be permitted until residue chemistry and chronic toxicology data are available to assess existing uses.

   Unique Label Warning Statements: All end-use products shall bear the following statements:

   a. Feeding/Grazing Restriction

      Do not allow livestock to feed or graze on treated cotton crops.

   b. Environmental Precautions

      Do not apply directly to water or wetlands (swamps, bogs, marshes, and potholes). Do not contaminate water by cleaning of equipment or disposal of wastes.

5. Summary of Major Data Gaps:

   Product Chemistry

   All product chemistry data

   Residue Chemistry

   Metabolism studies (plants, livestock)
   Residue analytical method
     (plant & animal residues)
   Storage stability
   Residue studies

   Toxicology

   Acute inhalation (rat)
   Eye irritation (rabbit)
   Dermal sensitization (guinea pig)
   21-Day dermal (rabbit)
   Chronic toxicity (rodent)
   Oncogenicity (2 species)
   Teratology (rat)
   Reproduction (rat)
   Mutagenicity battery
   General metabolism

Fish and Wildlife

> Avian dietary (upland game bird)
> Avian reproduction
>   (upland game bird and waterfowl)
> Acute toxicity to estuarine
>   and marine organisms
> Fish early life stage and
>   aquatic invertebrate life-cycle
> Aquatic organism accumulation

Plant Protection

> Seed germination/seedling emergence
> Vegetative vigor
> Aquatic plant growth

Environmental Fate

> Hydrolysis
> Photodegradation (water, soil, air)
> Anaerobic soil metabolism
> Leaching, adsorption/desorption
> Volatility (lab)
> Volatility (field)
> Soil dissipation
> Soil dissipation, long-term
> Accumulation:
>   Rotational crops (confined)
>   Rotational crops (field)
>   Fish

6. Contact Person at EPA:

> Robert J. Taylor
> U.S. Environmental Protection Agency
> Registration Division (TS-767C)
> 401 M Street SW.
> Washington, DC 20460
> (703) 557-1830

Disclaimer: The information presented in this Pesticide Fact Sheet is for informational purposes only and may not be used to fulfill data requirements for pesticide registration and reregistration.

PRONAMIDE

Date Issued: April 1986
Fact Sheet Number: 70

1. DESCRIPTION OF CHEMICAL

    Generic Name: 3,5-dichloro-N(1,1-dimethyl-2-
                            propynyl) benzamide
                                - or -
                            [N-(1,1-dimethylpropynyl)-3,5-
                            dichlorobenzamide]

    Common Name: Pronamide (WSSA), RH-315

    Trade Name: Kerb®

    EPA Shaughnessy Code: 101701

    Chemical Abstracts
      Service (CAS) Number: 23950-58-5

    Year of Initial
      Registration: 1972

    Pesticide Type: Herbicide

    Chemical Family: Substituted benzamide

    U.S. and Foreign
      Producers: Rohm and Haas

2. USE PATTERNS AND FORMULATIONS

    Application sites: alfalfa, apples, globe artichokes, birds-
        foot trefoil, blackberries, blueberries, cherries, clover,
        crown vetch, endive, grapes, lettuce, nectarines, peaches,
        pears, plums, raspberries, sainfoin, azalea, Douglas
        fir, fir, forsythia, holly, juniper, pine, rhododendron,
        yew, Christmas tree plantations, bermudagrass, centipede-
        grass, St. Augustinegrass, and zoysiagrass.

    Types of formulations: 94% active ingredient technical grade;
        50% formulation intermediate; and 50% wettable powder,
        <1% granular and <1% granular formulations mixed with
        fertilizer end-use products.

Types and methods of application: End-use product is applied preplant, preemergence or early postemergence by ground spray, incorporation, or aerial application. Hand spray application is limited to ornamentals and nursery stock.

Application rates: 0.5 to 4.0 lbs. active ingredient per acre, depending on crop and weed problem.

Usual carrier: Water

3. SCIENCE FINDINGS

Summary science statement: Pronamide has been found to be oncogenic in male mice and has been tentatively classified as a Group C oncogen (Possible Human Carcinogen), pending consideration of additional data. Acute toxicity studies indicate Toxicity Category III.[1/] Pronamide is practically nontoxic to birds, mammals, and fish, and possibly as much as moderately toxic to aquatic invertebrates. Available data are insufficient to fully assess the environmental fate of pronamide.

Chemical characteristics:

Physical state: Crystalline solid

Color: White

Melting point: 155-156° C

Specific gravity: 0.48 gm/cc

Solubility:

Solvent	ppm at 25° C
Dimethyl sulfoxide	33
Dimethyl formamide	33
Mesityl oxide	20
Isophorone	20
Methyl ethyl ketone	20
Cyclohexanone	20
Methanol	15
Isopropanol	15
Chlorobenzene	12
Butyl cellosolve	10
Xylene	10
Acetonitrile	10
Kerosene	10
Nitrobenzene	5
Ethylene dichloride	5
Water	15

[1/] Chemicals classified as Toxicity Category III are those which are harmful if swallowed, inhaled or absorbed through the skin. Contact with skin, eyes or clothing requires immediate first-aid and may require medical attention.

Vapor pressure: 8.50 x 10^{-5} Torr at 25° C

Stability : considered relatively stable

Toxicological characteristics:

Acute effects[1]:

Acute oral toxicity (rat) — 16,000 (10,666-24,000) mg/kg
Toxicity Category IV

Acute dermal toxicity (rabbit) — greater than 10 g/kg
Toxicity Category III

Acute inhalation toxicity (rat) — greater than 3.2 mg/L
Toxicity Category III

Primary eye irritation (rabbit) — slight eye irritant
Toxicity Category III

Primary skin irritation (rabbit) — not a primary dermal irritant
Toxicity Category IV

Dermal Sensitization (guinea pig) — technical material is not a skin sensitizer
Toxicity Category IV

Subchronic effects:

Feeding study (dog) — Lowest effect level (LEL) is 4050 ppm (approximately 90 mg/kg/day); NOEL is 1350 ppm (approximately 30 mg/kg/day)

Chronic effects:

Chronic feeding (rat) — NOEL is 300 ppm (highest dose tested) (15 mg/kg)

Chronic feeding (dog) — NOEL is 7.5 mg/kg

Teratology (rabbits) — Maternal NOEL 5 mg/kg/day
LOEL 20 mg/kg/day
Developmental LOEL 80 mg/kg/day
NOEL 20 mg/kg/day

Oncogenicity (mice) — Positive

---

[1] Toxicity Category III = Harmful if swallowed, inhaled or absorbed through the skin. Contact with skin, eyes or clothing require immediate first aid and may required medical attention.
Toxicity Category IV = No precautions are required.

Major route of exposure: Mixers and applicators would be expected to receive the most exposure via skin contact.

Physiological and biochemical behavorial characteristics

Foliar absorption: To obtain activity, pronamide must move into the root zone of the weeds. Little activity is obtained from foliar contact alone.

Translocation: Pronamide is readily absorbed by plants through the root system, translocated upward, and distributed into the entire plant. The degree of translocation from leaf absorption is not appreciable.

Mechanism of pesticidal action: Pronamide appears to act as a cell division inhibitor. Its main activity is against root development in germinating seeds or young seedlings, or, in the case of perennial grasses, against both root and vegetative bud development. Pronamide is taken up by the roots and needs to be carried into the root zone by rainfall, irrigation or incorporation.

Environmental characteristics: Available data are insufficient to fully assess the environmental fate of pronamide. Pronamide is stable to hydrolysis at pH 4.7-8.8.

Ecological characteristics: Results of studies indicate that pronamide is practically nontoxic to birds, mammals, and fish, and possibly as much as moderately toxic to aquatic invertebrates. These results are:

Aquatic invertebrate toxicity (Daphnia magna)	- Greater than 5.6 ppm[1]
Avian acute toxicity (Japanese quail)	- 8,700 mg/kg[2]
Avian acute toxicity (mallard duck)	- 20,000 mg/kg[2]
Avian 8-day dietary toxicity (mallard duck)	- greater than 10,000 ppm[2]
Avian 8-day dietary toxicity (bobwhite quail)	- greater than 10,000 ppm[2]

[1] No more than moderately toxic to aquatic invertebrates.
[2] Practically nontoxic to birds.

Avian dietary toxicity
(bobwhite quail) — greater than 4,000 ppm[1]

Avian subacute toxicity
(bobwhite quail) — greater than 30 ppm[2]

96-hour acute fish toxicity[3]
- bluegill — greater than 100 ppm
- rainbow trout — 72 (47-110) ppm
- goldfish — 350 (312-392) ppm
- guppy — 150 (113-200) ppm
- catfish — less than 500 ppm but greater than 200 ppm

Tolerance assessment:

List of crops and tolerances (40 CFR 180.317):

Commodity:	Tolerance (ppm)
Alfalfa, fresh	10.0
Alfalfa, forage	10.0
Alfalfa, hay	10.0
Apples	0.1
Artichokes	0.1
Blackberries	0.05
Blueberries	0.05
Boysenberries	0.05
Cattle, fat	0.02
Cattle, kidney	0.2
Cattle, liver	0.2
Cattle, Meat Byproducts (MBYP)(exc. kidney and liver)	0.02
Cattle, meat	0.02
Cherries	0.1
Clover	5.0
Crown vetch	5.0
Eggs	0.02
Endive (escarole)	2.0
Goats, fat	0.02
Goats, kidney	0.2
Goats, liver	0.2
Goats, MBYP (exc. kidney and liver)	0.02

[1] Practically nontoxic to birds.
[2] 300 ppm was highest dose tested: no deaths occurred, therefore, no toxicity level could be determined.
[3] Practically nontoxic to warmwater fish; slightly toxic to coldwater fish.

Commodity	Tolerance (ppm)
Goats, meat	0.02
Grapes	0.1
Hogs, fat	0.02
Hogs, kidney	0.2
Hogs, liver	0.2
Hogs, MBYP (exc. kidney and liver)	0.02
Hogs, meat	0.02
Horses, fat	0.02
Horses, kidney	0.2
Horses, liver	0.2
Horses, MBYP (exc. kidney and liver)	0.02
Horses, meat	0.02
Lettuce	1.0
Milk	0.02
Nectarines	0.1
Peaches	0.1
Pears	0.1
Plums	0.1
Poultry, fat	0.02
Poultry, kidney	0.2
Poultry, liver	0.2
Poultry, MBYP (exc. kidney and liver)	0.02
Poultry, meat	0.02
Raspberries	0.05
Sainfoin	5.0
Sheep, fat	0.02
Sheep, kidney	0.2
Sheep, liver	0.2
Sheep, MBYP (exc. kidney and liver)	0.02
Sheep, meat	0.02
Trefoil	5.0

Results of tolerance assessment: The provisional acceptable daily intake (PADI) for pronamide was based on the NOEL of 7.50 mg/kg in the 2-year dog study. It should be noted that this PADI is based on systemic toxicity (nononcogenic). The current PADI is 0.0750 mg/kg/day and the current published tolerance for pronamide has a calculated Total Maximum Residue Concentration (TMRC) of 0.0409 mg/day (1.5 kg diet). The percentage of PADI utilized is 0.91%.

4. SUMMARY OF REGULATORY POSITION AND RATIONALE

Use classification: Wettable powder end-use products are being classified as restricted-use pesticides. Wettable powder formulations of pronamide pose risks, through dermal exposure, to mixers and applicators which can be reduced to an acceptable level by requiring that pronamide only be applied by certified applicators.

Unique label warning statements:

Manufacturing Use Products:

"CAUTION - Harmful if absorbed through the skin or inhaled. Causes moderate eye irritation. Avoid contact with the skin, eyes, or clothing. Avoid breathing dust. Wash thoroughly with soap and water after handling. Remove contaminated clothing and wash before reuse."

"Do not discharge effluent containing this product into lakes, streams, ponds, estuaries, oceans, or public water unless this product is specifically identified and addressed in an NPDES permit. Do not discharge effluent containing this product into sewer systems without previously notifying the sewage treatment plant authority. For guidance contact your State Water Board or Regional Office of the EPA."

All End-Use Products

"GENERAL PRECAUTIONS: Avoid contact with eyes, skin or clothing."

"PROTECTIVE CLOTHING: When mixing, loading or applying this product, wear midforearm waterproof gloves, long-sleeved shirts and long pants, preferably one piece (coveralls). Hand-spray or hand-spreaders also require the use of waterproof boots or shoe coverings. Wash non-disposable gloves, boots and shoe coverings thoroughly with soap and water before removing."

"If water-soluble packaging is used, mixers and loaders are exempted from protective clothing requirements."

"Protective clothing/equipment is not needed during application if an enclosed tractor cab with filtered air supply or enclosed cockpit is used."

"Any article of clothing worn while handling product must be cleaned before reusing. Clothing should be laundered separately from household articles. Clothing which has been drenched or heavily contaminated should be disposed of in accordance with state or local regulations."

"Do not apply directly to water. Do not contaminate water by cleaning of equipment or disposal of wastes."

"Crops other than those on which pronamide may be applied may not be planted in pronamide-treated soil."

"Hand-spray applications of pronamide can be made only to ornamentals and nursery stock."

All Wettable Powder End-Use Products. These products must be packaged in water-soluble packaging and bear the following statements:

"RESTRICTED USE PESTICIDE: Because pronamide has produced tumors in laboratory animals, this product is for retail sale to and use only by Certified Applicators or persons under their direct supervision, and only for those uses covered by the Certified Applicator's certification."

"Dilution Instructions"

"The enclosed pouches of this product are water soluble. Do not allow pouches to become wet before adding them to the spray tank. Do not handle the pouches with wet hands or gloves. Always reseal overwrap bag to protect remaining unused pouches. Do not remove water soluble pouches from overwrap except to add directly to the spray tank."

"Add the required number of unopened pouches as determined by the dosage recommendations into the spray tank with agitation. Depending on the water temperature and the degree of agitation, the pouches should dissolve completely within approximately five minutes from the time they are added to the water."

All Granular End-Use Products

"Site treated with this product must be thoroughly watered after application."

5. SUMMARY OF MAJOR DATA GAPS

Data required: Due date (or after issuance of Standard):

Toxicology:

Chronic toxicity (rodent)	50 months
Oncogenicity (rat)	50 months
Teratogenicity (rat)	15 months
Reproduction (rat)	39 months
Mutagenicity - gene mutation	9 months
Mutagenicity - structural chromosommal aberration	12 months
Mutagenicity - other genotoxic effects	12 months
Special testing - general metabolism	24 months
Special testing - dermal penetration	12 months
Environmental fate	9-39 months
Residue Chemistry	January 1988

6. CONTACT PERSON AT EPA

Robert J. Taylor
U.S. Environmental Protection Agency
TS-767C
401 M Street SW.
Washington, DC 20460
(703) 557-1800

DISCLAIMER: The information presented in this Pesticide Fact Sheet is for informational purposes only and may not be used to fulfill data requirements for pesticide registration and reregistration.

PROPACHLOR

Date Issued: February 11, 1985
Fact Sheet Number: 44

1. Description of the chemical:

    Generic name: 2-Chloro-N-isopropylacetanilide
      ($C_{11}H_{14}ClNO$)
    Common name: Propachlor
    Trade name: Ramrod®, Bexton®, and CP 31393
    Chemical Abstracts Service (CAS) Registry number: 1918-16-7
    Office of Pesticides Program's Internal Control Number: 019101
    Year of initial registration: 1965
    Pesticide type: Herbicide
    Chemical family: Alpha-chloroacetamide
    U.S. producer: Monsanto Company

2. Use patterns and formulations:

    Application sites: Propachlor is registered for use as a preemergence herbicide on corn (all types), soybeans (seed only), grain sorghum (milo), green peas, pumpkins, cotton and flax. In corn, propachlor can also be applied as an early postemergence control. Sorghum is the largest use site for propachlor, accounting for most uses of the wettable powder formulation. Corn is the second largest use site for propachlor, accounting for most uses of the granular formulation.

    Types of formulations: Propachlor is available in granular, wettable powder, and flowable liquid concentrate formulations.

    Types and methods of applications: Propachlor is applied as a preemergence broadcast spray or banded ground application, and only one application is allowed per year.

    Application rates: 3.0 to 6.0 lbs a.i./A on crop sites.

    Usual carriers: Attapulgite/montmorillonite clay and water.

3. Science Findings:

    Summary science statements:

    Propachlor is not acutely toxic by the oral and dermal routes of exposure. However, primary eye irritation data on propachlor show corrosivity and corneal opacity not reversible within seven days in rabbits. If propachlor came in contact with the eyes, it could cause irreversible eye injury to man. Propachlor may also induce photosensitivity or photosensitization due to the presence of aniline in its composition.

Propachlor is highly toxic to coldwater fish, highly to moderately toxic to warmwater fish and highly to moderately toxic to freshwater invertebrates.

Chemical characteristics:

Propachlor is a light tan solid at room temperature and its molecular weight is 211.70. The boiling point is 110°C at 0.3 mm Hg. Propachlor is soluble in most organic solvents and in water (at 23°C) to 693 ppm.

Toxicological characteristics:

Acute toxicology effects of propachlor are as follows:

Acute oral toxicity in rats: 1.80 g/kg body weight, Toxicity Category III,
Acute dermal toxicity in rabbits: >20.0 g/kg body weight, Toxicity Category IV,
Skin irritation in rabbits: slight irritant, Toxicity Category III,
Eye irritation in rabbits: Corrosivity and corneal opacity not reversible within seven days, Toxicity Category I.

Chronic toxicology effects of propachlor are as follows:

A teratology test in rats has shown that propachlor tested at the highest dose level (HDT), failed to induce teratogenic, fetotoxic or maternal effects. The HDT was 200 mg/kg/day.

Major routes of human exposure:

Non-dietary exposure to propachlor by a farmer as an applicator during mixing, loading, spraying and flagging is probable.

Exposure of humans to propachlor through contamination of ground water and runoff contamination of surface water after heavy Spring precipitation is probable.

The dietary exposure (mg/kg/day) to propachlor by the U.S. population from treated food crops is possible.

Physiological and biochemical behavioral characteristics:

Absorption characteristics: Propachlor is absorbed mainly by germinating seedling shoots, secondarily by roots.
Translocation: Propachlor is translocated throughout the plant, mainly in the vegetative tissues.
Mechanism of pesticidal action: Propachlor is a strong inhibitor of cell elongation and protein synthesis.
Metabolism in plants: Metabolized rapidly in plants.

Environmental characteristics:

> Adsorption and leaching in basic soil types: Propachlor is adsorbed by soil colloids.
> Microbial breakdown: Microbes are the primary factor in the breakdown of propachlor in soils.
> Loss from photodecomposition and/or volatilization: Low.
> Average persistence of recommended rates: Half-life of 4 to 6 weeks; longer in soils high in organic matter.

Ecological characteristics:

> Avian acute oral toxicity: 91 mg/kg.
> Avian eight day dietary toxicity (Bobwhite Quail): > 5,000 ppm.
> Avian eight day dietary toxicity (Mallard Duck): > 5,000 ppm.
> 96 hour fish toxicity: >1.40 ppm for bluegill sunfish (moderately toxic) and 0.17 ppm for rainbow trout (highly toxic).
> 48 hour aquatic invertebrate toxicity: 7.80 ppm (moderately toxic) for Daphnia magna.

Potential problem for endangered species:

> The Office of Endangered Species (USDI) has determined that propachlor use on corn, sorghum or soybeans may impact the following endangered species: Slackerwater darter; eleven freshwater mussels; Woundfin; and Salanograss.

Tolerance assessments:

> The Agency is unable to complete a full tolerance reassessment of propachlor because of certain residue chemistry data gaps. The additional residue data requirements may cause specific tolerances to be revised in the future.

The tolerances listed below have not been revised:

Commodities	Parts per million
Beets, sugar, roots	0.2
Beets, sugar, tops	1.0
Cattle, fat	0.02
Cattle, meat by-products	0.02
Cattle, meat	0.02
Corn, forage	1.5
Corn, fresh, (sweet)(K+CWHR)	0.1
Corn, grain	0.1
Cottonseed	0.1
Eggs	0.02
Flax, seed	3.0
Flax, straw	10.0
Goats, fat	0.02
Goats, meat by-products	0.02
Goats, meat	0.02
Hogs, fat	0.02
Hogs, meat by-products	0.02
Hogs, meat	0.02
Horses, fat	0.02
Horses, meat by-products	0.02
Horses, meat	0.02
Milk	0.02
Peas, pods removed	0.2
Peas, forage	1.5
Poultry, fat	0.02
Poultry, meat by-products	0.02
Poultry, meat	0.02
Pumpkins	0.1
Sheep, fat	0.02
Sheep, meat by-products	0.02
Sheep, meat	0.02
Sorghum, fodder	5.0
Sorghum, forage	5.0
Sorghum, grain(milo)	0.25

International Tolerances:

Presently, there are no tolerances for residues of propachlor in Canada, Mexico, or in the Codex Alimentarius.

Problems known to have occurred with use:

The Pesticide Incident Monitoring System (PIMS) did not show any incidents involving uses of propachlor.

4. Summary of regulatory position and rationale:

Based on the review and evaluation of all available data and other relevant information on propachlor, the Agency has made the following determinations:

The available data are insufficient to show that any of the risk criteria listed in § 162.11(a) of Title 40 of the U.S. Code of Federal Regulations have been met or exceeded for the uses of propachlor at the present time.

There are no valid long term chronic toxicity data for propachlor. Studies indicate that propachlor is structurally similar to alachlor which is oncogenic in laboratory rats and mice. Data on propachlor show that it is corrosive to the eye with corneal opacity irreversible after 7 days. Further, Propachlor has been shown to be highly toxic to coldwater fish, highly to moderately toxic to warmwater fish and highly to moderately toxic to freshwater invertebrates. In addition to the extensive subchronic and chronic toxicity data gaps, there are also residue chemistry and environmental fate data gaps. Because the existing data are insufficient to show that any of the risk criteria have been met or exceeded, the Agency is not initiating a Special Review at this time.

The most appropriate regulatory action is to move quickly to fill the data gaps. Then the Agency will make a determination as to the future registerability of the affected uses.

Specific label warning statement:

Hazard Information:

"DANGER. Corrosive, causes irreversible eye damage. Harmful if swallowed or absorbed. Do not get in eyes or on clothing. Wear goggles, face shield or safety glasses".

Protective Clothing Requirements:

"Required clothing and equipment for mixing/loading and applying propachlor":

"One-piece coveralls which have long sleeves and long pants constructed of laminated fabric as specified in the USDA/EPA Guide for Commercial Applicators".

"Goggles, face shield, or safety glasses".

"Liquid-proof hat such as a plastic hard hat with a plastic sweat sweat band".

"Heavy-duty liquid proof rubber (neoprene) work gloves".

"Any article worn while handling propachlor must be cleaned before reusing. Clothing which has been drenched or has otherwise absorbed concentrated pesticide from any significant spill must be disposed in a sanitary landfill, by incineration or, if allowed by state and local authorities, by burning. If burned, stay out of smoke".

"Instead of clothing and equipment specified above, the applicator can use an enclosed tractor cab which provides a filtered air supply (as described by Taschenberg and Bourke, 1975)".

Prohibition on Aerial Application

"Do not apply with aerial equipment."

Environmental Hazard Statements:

All manufacturing-use products (MUPs) intended for formulation into end-use products (EUPs) must bear the following statement:

"Do not discharge effluent containing this product directly into lakes, streams, ponds, estuaries, oceans or public waters unless this product is specifically identified and addressed in a National Pollutant Discharge Elimination System (NPDES) permit. Do not discharge effluent containing this product into sewer systems without previously notifying the sewage treatment plant authority. For guidance, contact your State Water Board or Regional Office of the Environmental Protection Agency".

The following environmental hazard statement must appear on all granular EUPs:

"This pesticide is toxic to fish. Do not apply directly to water. Runoff from treated areas may be hazardous to aquatic organisms in neighboring areas. Do not contaminate water by cleaning of equipment or disposal of wastes. In case of spills, collect for use or properly dispose of granules".

The following environmental hazard statement must appear on all non granular EUPs:

"This pesticide is toxic to fish. Do not apply directly to water. Drift and runoff from treated areas may be hazardous to aquatic organisms in neighboring areas. Do not contaminate water by cleaning of equipment or disposal of wastes. In case of spills, collect for proper disposal".

Registrants must revise the labeling of MUPs and EUPs as specified in the "Required Labeling" section on products released for shipment as of September 1, 1985.

5. Summary of major data gaps:

The following toxicological studies are required:

An acute inhalation toxicity study in the rat,
A dermal sensitization study in the rabbit, *
A 90-day dermal study in the rabbit,
A 90-day inhalation study in the rat,
A 90-day feeding study in rodent, and non-rodent,
A chronic toxicity study in rat and in dog is required by February 1987,
An oncogenicity study in rat and in mouse is required by February, 1987,
A teratology study in the rabbit is required by April 1985, **
A two generation reproduction study in the rat is required by February 1987.
The following mutagenicity data are required by January, 1987:
  In vitro mammalian cell point mutation [L5178Y (TK): or CHO(HGPRT) or V79(HGPRT)],
  In vitro cytogenetic damage: both chromosomal aberration and SCE (in CHO cells: or human lymphocytes: or other rodent/ human cell line/ strains),
  In vitro/ in vivo primary hepatocyte repair for UDS testing both in vivo and in vitro exposure of cells to alachlor,
  In vivo cytogenetics test for chromosomal aberrations using bone marrow preparations of rats, and
  Dominant lethal test in rats or mice.
A domestic animal safety study is required.

* A dermal sensitization study in guinea pigs (Acc. 255806, December 1984) has been received by the Agency and it is being reviewed.
** A teratology study in rabbits (Acc. 255758, November 1984) has been received by the Agency and it is being reviewed.

The following environmental fate data are required:

A hydrolysis/degradation study is required by July, 1985,
A photodegradation study in water is required by July, 1985,
A photodegradation study on soil is required by July, 1985,
An additional metabolism test in aerobic soil is required by July, 1986,
Additional data on mobility of degradates from aged propachlor are required by April, 1985,
Soil dissipation data are required by July, 1987,
An accumulation study in rotational crops (confined); and
A flow-through/accumulation test in fish is required.

The following ecological effects data are required:

An acute freshwater fish toxicity test on a cold water species may be required pending review of environmental fate data which will be submitted in 1985, 1986 and 1987.

The following product chemistry data are required:

A complete statement of composition,
A preliminary analysis,
A certification of limits,
Analytical Methods for Enforcement of Limits,
The Octanol/water partition coefficient, and
The pH of the TGAI is required.

The following residue chemistry data are required:

^{14}C ring labeled data with corn and sorghum up to the seed stage,
^{14}C ring labeled data on the nature of the residue in ruminants,
Metabolism data on propachlor in poultry,
Metabolism data on propachlor in non-ruminants (name/species),
A radiocarbon study in which samples are analyzed simultaneously by cold and hot analytical methodology for N-isopropylaniline moiety including a comparison of the results
[If other metabolites are found in or on plants and in animals, then an enforcement method to determine these residues would be required]:
Storage stability data on propachlor residues in plant and animal samples,
Residue data on the metabolites of propachlor in plants,
Residue data on the metabolites of propachlor in animals,
Residue data on the metabolites of propachlor in poultry and eggs,
A cow feeding study on propachlor,
Geographical representation of residue data are not adequate, therefore, additional field residue data are required:
Residue data on corn forage to support the application rate of 6 lb ai/A on soils with an organic matter content of more than 3%,
Additional residue data for corn fodder using the 20% granular formulation,
Residue data for corn grain using the 20% granular and 42% flowable liquid formulations,
Residue data for sweet corn using the flowable liquid concentrate (FlC),
Residue data for the oil and milling fractions from the processing of treated corn or sorghum grain,
Residue data in cotton forage are required, [but an acceptable alternative would be a restriction against the grazing of livestock on cotton forage]:
Residue data in flax seed hulls which are fed to livestock,
Residue data for peas using the flowable liquid concentrate,
Residue data for pumpkins using the 90-day pre-harvest or post-harvest interval (PHI); and
The feeding rate data in the sheep study must be clarified.

PROPARGITE

Date Issued: September 30, 1986
Fact Sheet Number: 99

1. DESCRIPTION OF CHEMICAL

    Generic Name: 2-(p-tert-butylphenoxy)cyclo-
                                 hexyl 2-propynyl sulfite

    Common Name: Propargite

    Trade Names: Omite®, Comite® and
                                 Uniroyal D014

    EPA Shaughnessy Code: 097601

    Chemical Abstracts
    Service (CAS) Number: 2312-35-8

    Pesticide Type: Acaricide

    Year of Initial
    Registration: 1966

    Chemical Family: Organosulfite

    U.S. Producer: Uniroyal Chemical Company

2. USE PATTERNS AND FORMULATION

    Application Sites: Terrestrial food crops (field, vegetable, and orchard crops); terrestrial nonfood crops (ornamentals), aquatic food crop (cranberry); and greenhouse nonfood crops (ornamentals).

    Types of Formulations: Single active ingredient (a.i.) formulations consist of 4% dust; 3% and 30% wettable powders; 5, 6 and 6.55 pounds per gallon emulsifiable concentrates; The technical formulation is 85% a.i., and the formulation intermediate is 25% a.i.

    Types and Methods of Application: End-use product is applied foliarly using air and/or ground equipment (including air blast).

Application Rates: Application rate ranges from 0.75 to 6.75 lb active ingredient per acre.

Usual Carrier: Water

3. SCIENCE FINDINGS

   Summary Science Statement: Propargite is not considered at this time to be oncogenic. Propargite has a low acute (Category III) oral, dermal, inhalation toxicity. It is in toxicity category I however, for primary eye and primary skin irritation, and cases of severe dermatitis afflicting workers reentering treated sites have been reported. Chronic testing reveals that propargite appears to have little effect on laboratory animals except at higher dosage levels; effects reported include depressed body weights and rates of weight gain. Oral subchronic test results appear to parallel those for chronic testing. Propargite is not teratogenic in rabbits and rats. Insufficient data exist to fully assess the dermal sensitization, subchronic dermal toxicity, subchronic inhalation toxicity, and mutagenicity of propargite. Additional metabolism testing is also necessary. Propargite is relatively nontoxic to honey bees and avian species. It is very highly toxic to freshwater fish. The actual threat to aquatic organisms at this time cannot be accurately assessed due to the insufficiency of environmental fate data. Field monitoring data to determine propargite residues in water from terrestrial applications is needed as is additional testing to determine the effects of end-use products on coldwater and warmwater species of fish, aquatic invertebrates, and estuarine and marine organisms. The metabolism of propargite in both plants and animals is not sufficiently understood; additional metabolism data are necessary. Storage stability data are also necessary, as well as additional crop residue data and processing studies for certain crops registered for propargite use.

   CHEMICAL CHARACTERISTICS

Physical State:	Viscous liquid.
Color:	Dark amber
Odor:	Faint solvent odor to very faint solvent odor.
Density:	Specific gravity = 1.085 - 1.115 at 25°C; Bulk density = 40.92 lb/ft.

Solubility: In water, about 0.5 ppm at 25°C, miscible with organic solvents such as acetone, benzene, and ethanol.

Stability: No evidence of breakdown in one year.

Flash Point: At least 38.5°C

TOXICOLOGICAL CHARACTERISTICS

Acute Effects

Adequate data are not available to fully assess the toxicity of Propargite. While, in general, propargite is not highly toxic (Category III oral, dermal and inhalation), it is in toxicity Category I for primary eye and skin irritation. Available data present only supplementary information (as set forth below), and additional data must be submitted.

Acute Oral Toxicity (Rat): 2.2 g/kg

Acute Dermal Toxicity (Rabbit): 3.16(1.63-6.15) ml/kg

Acute Inhalation: > 2.5 mg/l

Primary Eye Irritation: Corneal effects that were not reversible after 14 days were observed in four of six rabbits.

Dermal Sensitization: Inconclusive

Subchronic Dermal Toxicity: Inconclusive

Chronic Effects

Teratogenicity (Rabbit): Maternal NOEL = 2 mg/kg/day
Maternal LEL = 6 mg/kg/day (reduced body weight gain)
Developmental Toxicity NOEL= 2 mg/kg/day; Developmental (increased resorption, reduced body weight, and delayed ossification). A/D ratio = maternal LEL/Developmental = 2/2 = 1

3-Gen. Reproduction (Rat): NOEL > 300 ppm. Additional data is required. Only one dose used throughout the study.

Mutagenicity:	Inconclusive. Additional categories of mutagenicity testing are required.
Chronic Feeding/ Oncogenicity (Dog):	NOEL = 900 ppm (HDT). No adverse effects were observed by the hematology, blood chemistry determinations or urine examinations.
Feeding/Oncogenicity:	Inconclusive. The study is classified as Supplementary because too few animals were examined histologically at 900 and 2000 ppm. This study needs to be repeated.
Metabolism:	Inconclusive. Additional data is required.

OTHER TOXICOLOGICAL EFFECTS

Propargite is not an organophosphate chemical; therefore, it does not have a neurotoxic potential, and a neurotoxicity study is not required.

MAJOR ROUTES OF EXPOSURE

There is a potential for dermal, ocular and inhalation exposure from mixing concentrates and applying spray mixtures.

PHYSIOLOGICAL AND BIOCHEMICAL BEHAVIOR CHARACTERISTICS

Foliar Absorption:	Data are not available to evaluate the effects of propargite in plants.
Translocation:	Data are not available to evaluate the translocation in plants.
Mechanism of Pesticidal Actions:	Mode of activity involves residual killing action

ENVIRONMENTAL CHARACTERISTICS

Available data are insufficient to fully assess the environmental fate of propargite. From the data that exists, however there seems to be no reason for concern about the leaching of the parent chemical into groundwater.

ECOLOGICAL CHARACTERISTICS

Avian Oral Toxicity:	Mallard duck - > 4640 ppm
Avian Dietary Toxicity:	Bobwhite quail 3401 ppm Mallard duck - > 4640 ppm
Avian Reproduction:	Data are inconclusive to determine the effects on avian reproduction. Additional data are required.
Freshwater Fish Toxicity:	Bluegill Sunfish - 0.167 ppm Rainbow trout - 0.118 ppm
Aquatic Invertebrates (freshwater):	Daphnia magna - 0.092 ppm
Aquatic Invertebrates (lifecycle):	Daphia magna - 0.009 - 0.014 ppm

Available data indicate that propargite is practically nontoxic to avian species. Propargite is highly toxic to fish.

TOLERANCE ASSESSMENT

Tolerances have been established for residues of propargite in raw agricultural commodities, milk, eggs, meat, fat and meat by-products (40 CFR 180.259).

Commodity	Parts Per Million (ppm)
Almonds	0.1
Almonds, hulls	55.0
Apples	3.0
Apricots	7.0
Beans, dry	0.2
Beans, succulent	20.0
Cattle, fat	0.1
Cattle, MBYP	0.1
Cattle, meat	0.1
Corn, fodder	10.0
Corn, forage	10.0
Corn, fresh (incl. sweet) (K+CWHR)	0.1
Corn, grain	0.1
Cottonseed	0.1
Cranberries	10.1
Eggs	0.1
Figs (fresh)	3.0

Commodity	Parts Per Million (ppm)
Goats, fats	0.1
Goats, meat	0.1
Grapefruit	5.0
Grapes	10.0
Hogs, fat	0.1
Hogs, meat	0.1
Hops	15.0
Horses, fat	0.1
Horses, MBYP	0.1
Horses, meat	0.1
Lemons	5.0
Milkfat (0.08 in whole milk)	2.0
Mint	50.0
Nectarines	4.0
Oranges	5.0
Peaches	7.0
Peanuts	0.1
Peanuts, forage	10.0
Peanuts, hay	10.0
Peanuts, hulls	10.0
Pears	3.0
Plums (fresh prunes)	7.0
Potatoes	0.1
Poultry, fat	0.1
Poultry, MBYP	0.1
Poultry, meat	0.1
Sheep, fat	0.1
Sheep, MBYP	0.1
Sheep, meat	0.1
Sorghum, fodder	10.0
Sorghum, forage	10.0
Strawberries	7.0
Tea (dry)	10.0
Walnuts	0.1

Results of the Tolerance Assessment: Because chronic feeding/oncogenicity and reproduction studies are needed, the current PADI is set on a 2-year dog feeding study a systemic NOEL at the highest dose tested (900 ppm). At the highest dose tested, there were no effects observed. With a safety factor of 1000, the TMRC is currently 112% of the PADI. The TMRC, however, is based on the assumption that 100% of the crop for which a tolerance is established is treated with propargite, which is not the case. If the TMRC is adjusted to reflect the actual percentage of crop treated, the TMRC would be reduced to a percentage level of the PADI significantly lower than 112%. Therefore, the public should be in no danger from dietary exposure while the Agency awaits data.

SUMMARY OF REGULATORY POSITIONS AND RATIONALES

The following are warning statements that must be included on propargite pesticide labels.

MANUFACTURING USE PRODUCTS

Under the Environmental Hazard Statement, add the following precaution:

"This pesticide is toxic to fish. Do not discharge effluent containing this product into lakes, streams, ponds, estuaries, oceans, or public water unless this product is specifically identified and addressed in an NPDES permit. Do not discharge effluent containing this product into sewer systems without previously notifying the sewage treatment plant authority. For guidance, contact your State Water Board or Regional Office of the EPA."

Protective Clothing Statement

"Mixer/loaders must wear goggles or a face shield, chemical-resistant apron, long-sleeved shirt and long pants or coveralls, and mid-forearm to elbow length chemical-resistant gloves when mixing, loading, or otherwise handling the concentrate."

END-USE PRODUCTS

Products with Aquatic Use(s): Under the Environmental Hazard Statement, add the following precaution:

"This pesticide is toxic to fish. Do not apply directly to water except as specified on this label. Drift and runoff from treated areas may be hazardous to aquatic organisms in neighboring areas. Do not contaminate water by cleaning of equipment or disposal of wastes."

Products with Terrestrial Use(s): Under the Environmental Hazard Statement, add the following precaution:

"This pesticide is toxic to fish. Do not apply directly to water or wetlands (swamps, bogs, marshes, and potholes). Drift and runoff from treated areas may be hazardous to aquatic organisms in neighboring areas. Do not contaminate water by cleaning of equipment or disposal of wastes."

Reentry Statement (For All Products with Crop Uses)

"Do not allow worker reentry into treated fields within 3 days of application, for strawberries, and within 7 days, for all other agricultural uses of propargite, unless appropriate protective clothing is worn. Protective clothing means at least a hat or other suitable head covering, a long-sleeved shirt and long-legged trousers or a coverall-type garment (closely woven fabric covering the body, including the arms and legs), chemical-resistant gloves, socks, and shoes."

Crop Rotation Statement

"Do not plant any food or feed crop in rotation within 6 months after last application of propargite unless the crop is a registered use for propargite."

Irrigated Crops Statement

"Do not use water leaving propargite treated fields to irrigate crops used for food or feed that are not registered for use with propargite."

Protective Clothing Statement

"Mixer/loaders must wear goggles or a face shield, chemical-resistant apron, long-sleeve shirt, long pants, and mid-forearm to elbow length chemical-resistant gloves. Applicators must wear a long-sleeve shirt and long pants, and chemical-resistant gloves while applying this pesticide. Applicators must also wear a wide-brimmed hat during upward directed spraying.

Any article of clothing worn while applying product must be cleaned before re-use. Clothing should be laundered separately from household articles. Clothing that has been drenched or has otherwise absorbed concentrated pesticide must be disposed of in a sanitary landfill, incinerated, or burned if allowed by State and local authorities."

Endangered Species Statement (For Products with Use on Terrestrial and Aquatic Food Crops, by February 1988)

"The use of any pesticide in a manner that may kill or otherwise harm an endangered or threatened species or adversely modify their habitat is a violation of Federal laws. The use of this product is controlled to prevent death or harm to endangered or threatened species that occur in the following counties or elsewhere in their range.

Before using this pesticide in the following counties, you must obtain the EPA Cropland Endangered Species Bulletin. The use of this pesticide is prohibited in these counties unless specified otherwise in the Bulletin. The EPA Bulletin is available from either your local pesticide distributor, your County Agricultural Extension Agent, the Endangered

Species Specialist in your State Wildlife Agency Headquarters, or the appropriate Regional Office of the U.S. Fish and Wildlife Service (FWS). <u>THIS BULLETIN MUST BE REVIEWED PRIOR TO PESTICIDE USE</u>."

5. SUMMARY OF MAJOR DATA GAPS

The following list presents data required and the due date for submission of this data:

Product Chemistry	Due Dates
Product Chemistry	Feb 1987

Residue Chemistry

Plant/Livestock Metabolism	Feb 1988
Plant/Animal Residues	Feb 1988
Storage Stability	Feb 1988

Toxicology

Sensitization Study	July 1987
Subchronic Dermal Toxicity (21 days)	May 1987
Subchronic Inhalation Toxicity (90 days)	May 1987
Metabolism	Sept 1988
Mutagenicity	Sept 1987
Chronic Feeding/oncogencity	Dec 1990
Two-Gen. Reproduction	Dec 1989

Environmental Fate

Soil Dissipation (Field)	Dec 1988
Aquatic (Sediment)	Dec 1988
Rotational Crops (Confined)	Dec 1989
Rotational Crops (Field)	Nov 1990
Irrigated Crops	Dec 1989
Fish (Accumulation Studies)	Sept 1987
Hydrolysis/Photodegradation	July 1987
Aerobic Soil Metabolism	Dec 1988
Anaerobic Soil Metabolism	Dec 1988
Anaerobic Aquatic	Dec 1988
Aerobic Aquatic	Dec 1988
Leaching and Adsorption/Desorption	July 1987
Volatility (Lab)	July 1987
Foliar dissipation (Reentry)	Nov 1987
Soil Dissipation (Reentry)	Nov 1987
Glove Permeability	Nov 1987

Summary of major data gaps (continued)

Ecological Effects

Residue Level Monitoring (Aquatic)	Feb 1988
Avian Reproduction	Sept 1988
Freshwater Fish (Warmwater)	July 1987
Freshwater Fish (Coldwater)	July 1987
Acute LC_{50} Freshwater (Invertebrates)	July 1987
Acute LC_{50} Estuarine & Marine Organisms (Shrimps)	July 1987
(Fish)	July 1987
(Mollusk)	July 1987

6. CONTACT PERSON AT EPA

George T. LaRocca
U.S. Environmental Protection Agency
TS-767C
401 M Street, S.W.
Washington, D.C. 20460
(703) 557-2400

DISCLAIMER: The information presented in this Pesticide Fact Sheet is for informational purposes only and may not be used to fulfill data requirements for pesticide registration and reregistration.

PROPHAM

Date Issued: March 31, 1987
Fact Sheet Number: 123

1. DESCRIPTION OF CHEMICAL

    Generic Name: isopropyl carbanilate

    Common Name: Propham

    Trade Name: Chem-Hoe®, Birgin, Triherbide

    EPA Shaughnessy Code: 047601

    Chemical Abstracts Service (CAS) Number: 122-42-9

    Year of Initial Registration: 1967

    Pesticide Type: Herbicide

    Chemical Family: Carbamate

    U.S. and Foreign Producers: Propham is produced in the Federal Republic of Germany by Bayer AG; in the Netherlands by Pennwalt Holland B.V.; and in the United States by PPG Industries, Inc.

2. USE PATTERNS AND FORMULATIONS

Application sites: Propham is registered for use on terrestrial food crops such as sugar beets, lettuce, alfalfa, clover, peas, lentils, safflower, and spinach and nonfood crops including established grasses grown for seed, flax and established perennial grass and fallow land. Most propham usage is confined to the western United States.

Percent of pesticide applied to particular crops: The significant uses of propham are in forage legumes (alfalfa and clover), sugar beets, and lettuce, which account for nearly 100 percent of its use.

Types and methods of application: Propham may be applied preplant, preemergence, and postemergence by ground or aerial equipment. The flowable concentrate formulation may also be applied in irrigation water or through center pivot sprinkler irrigation systems for certain uses. In limited areas, propham may be tank mixed with other herbicides for application to lettuce, spinach grown for seed, and fallow land to be planted to wheat. Also, since herbicidal action is mainly through the roots, soil surface applications must be moved into the root zone of weeds by rainfall or irrigation soon after application.

Application rates: The flowable concentrate is applied at the rate of 1 to 6 pounds of a. i. per acre depending on the soil, site, and peat controlled. The granular formulation is applied at the rate of 4.1 to 5.25 pounds a.i. per acre, depending on soil type.

Types of fomulations: Flowable concentrations (43 and 31 percent active ingredient) and granular (15 percent active ingredient) formulations.

Usual carrier: Water

3. SCIENCE FINDINGS

Summary science statement: Propham has low acute oral and dermal toxicity and is classified in toxicity Category III.* Additional toxicological assessment is not possible at this time due to insufficient data. Based on the available information, application of propham on proposed use sites is unlikely to result in environmental hazard to nontarget organisms other than plants because of its low toxicity to wildlife. Additional data are required before the environmental fate, including potential to contaminate ground water, of propham can be assessed.

Chemical characteristics:

Physical state:	Solid
Color:	Tan to light grey
Odor:	Faint amine-like odor
Melting point:	87 to 88 °C
Octanol/water partition coefficient:	445 ± 17 over a propham concentration range of 0.5 to 100 ppm (pure active ingredient)

Toxicological characteristics: Only acute oral and dermal toxicity studies are available. Other studies are required.

Acute Oral Toxicity:	Category III 2360 ± 118 mg/kg (female rats) 3000 ± 232 mg/kg (male rats)
Acute Dermal Toxicity:	Category III Greater than 3000 mg/kg (rabbits)

Major route of exposure: Dermal.

Subchronic toxicological results: Only one supplemental study, a 90-day feeding study in rats, is available. In that study the no observed effect level (NOEL) was 250 ppm (approximately 12.5 mg/kg/day). Additional studies are required.

---

* Refer to 40 CFR 162.10 for a discussion of the toxicity categories.

Chronic and Development Toxicological results: The only acceptable study available is a rat teratology study. In that study, NOEL's for maternal toxicity and developmental effects were 375.8 and 37.6 mg/kg, respectively. Additional chronic and developmental studies are required.

Physiological characteristics:

    Foliar absorption - Intact leaf surfaces do not absorb an appreciable amount of propham.

    Translocation - Propham is absorbed by roots and a portion is translocated via the apoplastic system to the shoots and the balance metabolized in the roots. That which reaches the shoots may be metabolized or volatilized.

    Mechanism of pesticide action - Propham is a mitotic poison that kills roots by inhibiting cell division. Propham produced rapid inhibition of cell activity and contraction of chromosomes, probably at all stages of cell division.

Environmental characteristics: With the exception of an aerobic soil metabolism study, available data are insufficient to fully assess the environmental fate of, and the exposure of humans and nontarget organisms to, propham. Data are required.

Under aerobic conditions, propham degraded with a half-life of 7 to 14 days in sandy loam soil. Two degradates were isolated from the soil but not identified.

Ecological characteristics:

    Hazards to fish and wildlife - Propham has a very low acute oral toxicity to birds, it is slightly toxic to coldwater and warmwater fishes and moderately toxic to freshwater invertebrates. The following are toxicity levels from available data:

Avian oral toxicity	: > 2000 ppm; additional data requested
Avian dietary toxicity	: Data required

Freshwater fish toxicity
  (Bluegill)   :   29 to 34.8 ppm
  (Rainbow trout)   :   23.5 to 38 ppm

Freshwater invertebrate
  toxicity (Daphnia pulex):   8 ppm

Potential problems related to endangered species:

None.

Tolerance assessment:

List of crops and tolerances - Interim tolerances are established for residues of propham in or on raw agricultural commodities as follows (40 CFR 180.319):

Commodities	Parts Per Million
Hay of alfalfa, clover, and grass	5.0
Alfalfa, clover, and grass	2.0
Flaxseed, lentils, lettuce, peas safflower seed, spinach, and sugar beets (roots and tops)	0.1
Eggs; milk; and the meat, fat, and meat byproducts of cattle, goats, hogs, horses, poultry, and sheep	0.05

Results of tolerance assessment: The Provisional Limiting Dose (PLD) for propham is 0.0125 mg/kg. This PLD is based on a 3-month rat feeding study, with a NOEL of 12.5 mg/kg/day (approximate conversion from 250 ppm), and applying a safety factor of 1000. This is equivalent to a PLD of 0.75 mg/day for a 60 kg individual. The Theoretical Maximum Residue Concentration (TMRC) based on the total tolerances listed and a daily food intake of 1.5 kg, is 0.043 mg/day, utilizing 5.7 percent of the PLD.

4. SUMMARY OF REGULATORY POSITION AND RATIONALE

1. The Agency is not initiating a Special Review of propham at this time.

    Rationale: Since available data are limited, the Agency is not yet able to make a determination as to whether any of the criteria specified in 40 CFR 154.7 have been met or exceeded.

2. No new significant* tolerances will be considered until the Agency has received data sufficient to thoroughly evaluate propham.

    Rationale: The toxicology data base on propham is not sufficient to consider establishment of new significant tolerances. In addition, the metabolism of propham in plants and animals is not adequately defined.

3. The Agency is requiring the following residue chemistry data: plant and animal metabolism and storage stability studies; residue studies for sugar beet roots, sugar beet tops, lettuce, spinach, peas (succulent and dry), lentils, grass forage, grass hay, alfalfa forage, alfalfa hay, flaxseed, and safflower seed; and processing studies to determine residues in dried pulp, molasses, and refined sugar from sugar beets; meal and hulls from flaxseed; and meal and oil from safflower seed. Petitions for food/feed additive tolerances, will be required if residues concentrate.

    Rationale: Adequate data are not available to assess the adequacy of existing tolerances or to ascertain the need for food/feed additive tolerances in processed commodities.

4. The Agency is requiring the registrant to: (i) propose appropriate Pre Grazing Intervals/Pre Harvest Intervals (PGIs/PHIs) for clover, lettuce, spinach, and sugar beets; (ii) designate propham registration for grasses as either pasture or rangeland use, and propose an appropriate PGI and PHI if pasture use is designated; and (iii) propose tolerances and provide supporting residue data or propose feeding restrictions for pea vines, pea vine hay, lentil forage, lentil hay, and flax straw. These proposals must be submitted with the revised labeling and in accordance with the timeframe required for submittal of revised labeling or the Agency will impose appropriate feeding restrictions.

---

* Significant new use is defined in 44 FR 27934, May 11, 1979. In the case of a new food or feed use, the Agency will consider as significant an increase in the Theoretical Maximum Residue Contribution of greater than 1 percent.

Rationale: Data are either unavailable or do not support the existing PGI/PHI or do not demonstrate that no PGI/PHI is needed for the cited crops. Adequate information is not available regarding the use for grasses and this use must be clarified and fully supported. There is currently no protective mechanism (either tolerances or feeding restrictions) to prevent excessive residues of propham in pea vines, pea vine hay, lentil forage, lentil hay, and flax straw.

5. The Agency is requiring the registrant to propose that the interim tolerances under 40 CPR 180.319 be converted to "permanent" tolerances under a separate paragraph of the published tolerance expressions at the same or, if necessary, different concentrations and provide the requested residue data to support these tolerances.

    In addition, the registrant must propose the following changes to commodity definitions in the tolerance statement: (i) "alfalfa" to "alfalfa forage"; (ii) "clover" to "clover forage"; (iii) "grass" to "grass forage"; and (iv) "peas" to "peas (succulent and dry)".

    Rationale: Interim tolerances were established when petitions for tolerances for negligible residues were pending. Since available data do not support the interim tolerances, as established, permanent tolerances can not be set based on currently available data. Therefore, when data are submitted in accordance with this document, the registrant must request conversion to permanent tolerances, and revise the commodity definitions to conform to current terminology or the Agency will propose revocation of the tolerances.

6. The Agency is requiring additional toxicological data, as set forth in Table A of this document to assess the toxicity of propham. Certain Acute, Subchronic, and Chronic testing is required.

    Rationale: These data are normally required under 40 CFR 158 for products with propham's use patterns. Existing data are insufficient to permit the Agency to thoroughly assess the toxicity of propham.

7. The Agency is requiring additional ecological effects data (see Table A).

    Rationale: Available data are insufficient or lacking to fully assess the hazard from propham use to the avian population.

8. The Agency is requiring environmental fate data as set forth in Table A.

   Rationale: Because the requirements have not been fully satisfied, available data are insufficient to fully assess the environmental fate of propham. The leaching data that is available indicate a potential for ground water contamination. Hydrolysis, photodegradation, metabolism, leaching, dissipation, and accumulation studies are required.

9. The Agency is not establishing a reentry interval at this time.

   Rationale: Data adequate to assess the need for a reentry interval for field workers are not available. Once data are received and evaluated, the Agency will determine the need for such an interval. An interim interval is not required because of the low acute toxicity demonstrated by the available data.

10. The Agency is requiring environmental precautionary labeling.

    Rationale: The Agency's regulations (40 CFR 162.10) require environmental hazards labeling. Updated labeling consistent with 162.10 is required. Additional required labeling statements to protect wetlands are specified in the registration standard Section D.3.

11. The Agency has identified certain data that will receive priority review when submitted.

    Rationale: Certain data are essential to the Agency's assessment of this pesticide and its uses and/or may trigger the need for further studies which should be initiated as soon as possible. The following studies have been identified to receive priority review as soon as they are received by the Agency:

    §158.130   Environmental Fate

    161-1 Hydrolysis
    161-2,3 Photodegradation
    163-1 Leaching and Adsorption/Desorption
    164-1 Soil Dissipation (Field)
    165-1 Rotational Crops (confined)

12. While data gaps are being filled, registered manufacturing-use products (MPs) and end-use products (EPs) containing propham as the sole active ingredient may be sold, distributed, formulated and used, subject to the terms and conditions specified in this Standard. Registrants must provide or agree to develop additional data, as specified in the Data Appendices, in order to maintain existing registrations.

    Rationale: Under FIFRA, the Agency does not normally cancel or withhold registration simply because data are missing or are inadquate (see FIFRA section 3(c)(2)(B) and 3(c)(7)). The limited, available data do not indicate any immediate, serious concern.

    Issuance of this Standard provides a mechanism for identifying data needs. These data will be reviewed and evaluated, after which the Agency will determine if additional regulatory changes are necessary.

Unique warning statements required on labels:

a. Manufacturing-Use Products

    "Do not discharge effluent containing this product into lakes, streams, ponds, estuaries, oceans, or public waters unless this product is specifically identified and addressed in an NPDES permit. Do not discharge effluent containing this product to sewer systems without previously notifying the sewage treatment plant authority. For guidance, contact your State Water Board or regional office of EPA."

b. End-Use Products

    "Do not apply directly to water or wetlands (swamps, bogs, marshes, and photholes). Do not contaminate water by cleaning of equipment or disposal of wastes."

5. SUMMARY OF MAJOR DATA GAPS

Table A

Study	Due Date*
Toxicology:	
Acute Testing	9 months
90-Day Feeding (rodent)	15 months
90-Day Feeding (nonrodent)	18 months
21-Day Dermal	12 months
Chronic Toxicity (rodent and nonrodent)	50 months

Table A (cont.)

Oncogenicity (rat and mouse)	50 months
Teratogenicity (rabbit)	15 months
Reproduction	39 months
Gene Mutation	9 months
Chromosomal Aberration	12 months
Other Mechanisms of Mutagenicity	12 months
General Metabolism	24 months

Residue chemistry:

Metabolism	18 months
Storage stability	15 months
Residue studies	18 months
Processed Commodity Studies	24 months

Environmental fate:

Hydrolysis	9 months
Photodegradation (water, soil)	9 months
Anaerobic Soil	27 months
Mobility (leaching and adsorption/desorption)	12 months
Soil dissipation	27 months
Rotational crops	39 months
Fish accumulation	12 months
Reentry	27 months

Ecological effects:

Avian oral and dietary	9 months

6. CONTACT PERSON AT EPA

> Robert J. Taylor
> Product Manager 25
> Office of Pesticide Programs
> Registration Division (TS-767C)
> Environmental Protection Agency
> 401 M Street SW.
> Washington, DC 20460
> Telephone: (703) 557-1800

DISCLAIMER: The information presented in this Pesticide Fact Sheet is for informational purposes only and may not be used to fulfill data requirements for pesticide registration and reregistration.

---

* Indicates months due after issuance of Standard or, in some cases of residue chemistry data, after first planting season after issuance of Standard. Refer to Standard for more information.

SIMAZINE

Date Issued: March 30, 1984
Fact Sheet Number: 23

1. Description of the chemical

   - Generic name: 2-chloro-4,6-bis(ethylamino)-s-triazine
   - Common name: Simazine
   - Trade names: Algae-A-Way, Algaecide, Algidize, Algi-eater, Algi-gon, Amizine, Aquazine, Atomicide, Cekusan, Cimacide, Framed, Gesapun, Gesatop, Primatol S, Princep, Simadex, Simanex, and Sim-trol.
   - EPA Shaugnessy Number: 080807
   - Chemical Abstract Service Registry Number (CAS) 122-34-9
   - Year of initial registration: 1957
   - Pesticide type: Herbicide
   - Chemical family: Triazine
   - U.S. & foreign producers - Ciba-Geigy Corporation, Griffin Corporation, Aceto Chemical Company, Inc., Drexel Chemical Company, and Vertac Chemical Company

2. Use patterns and formulations

   - Application sites: Simazine is registered for use as a selective or nonselective herbicide and algaecide. It is registered for use on agricultural, noncrop, forest and aquatic sites.
   - Types of formulations: Wettable powder, granular, liquid, flowable concentrate, soluble concentrate, dry flowable and liquid-ready to use
   - Types and methods of applications: Broadcast, band, soil incorporated, and soil surface application using ground or aerial equipment.
   The specific method of application and type of equipment are determined by site, formulation, and equipment availability.
   - Application rates: 1.6 lbs. a. i./A to 9.6 lbs. a. i./A, generally 4 lbs. a.i./A.
   - Usual carrier: Water, oil, and clay

3. Science findings

   Chemical characteristics:

   - Simazine is a white, odorless, crystalline solid. It is stable to heat and the melting point is 225-227C. Simazine is nonflammable and does not present unusual handling characteristics. Storage stability is greater than three years at room temperature under dry conditions.

   Toxicological characteristics:

   - Simazine is a moderate eye and dermal irritant (Toxicity Category III)

and has low oral and dermal toxicities (Toxicity Category IV).
- There are no data available for inhalation toxicity (A Data Gap exists for this requirement).
- Toxicology studies on simazine are as follows:
    - - Oral LD50 in rats: > than 15.380 mg/kg body weight.
    - - Dermal LD50 in rats: 10.2 mg/kg body weight.
    - - Inhalation in rats: No data available for review.
    - - Skin Irritation in rabbits: Slight irritant
    - - Eye Irritation in rabbits: 5 test animals after 72 hours showed moderate irritation which was reversible in 7 days. No corneal capacity was observed.
    - - Teratology in rats: No data available for review. A data gap exists for this requirement in 2 species.
    - - Three-Generation Reproduction Study in rats: The reproduction NOEL is > 100 parts per million (ppm). No adverse effects on reproductive performance in rats at a dietary level of 100 ppm for three generations over a total study period of 93 weeks.
    - - Chronic feeding/oncogenicity in rats: Chronic toxicity and oncogenic potential could not be determined in this study. A data gap exists for oncogenic or chronic toxicity in the rodent (rat).
    - - Chronic feeding in dogs: Neither chronic toxicity nor oncogenic potential could be determined from this study. A data gap exists for the chronic toxicity in non-rodents (dogs).
    - - Oncogenicity in mice: An oncogenicity study in a second species is required.
    - - Mutagenicity: Mutagenicity studies were not available for review. Data gaps exist for the entire category of mutagenicity testing required for registration.
    - - General Metabolism: A data gap exists for a required general metabolism study which identifies and quantitates metabolites formed in the exposure of the mammalian species.

Physiological and biochemical behavioral characteristics:

- - Foliar absorption: Absorbed mostly through plant roots with little or no foliar penetration. It has low adhering ability and is readily washed from foliage by rain.
- - Translocation: Following root absorption it is translocated acropetally in the xylem, accumulating in the apical meristems and leaves of plants.
- - Mechanism of pesticidal action: A photosynthetic inhibitor; but may have additional effects.
- - Metabolism and persistence in plants: Simazine is readily metabolized by tolerant plants to hydroxysimazine and amino acid conjugates. The hydroxysimazine can be further degraded by dealkylation of the side chains and by hydrolysis of resulting amino groups on the ring and some CO_2 production. These alterations of simazine are major protective mechanisms in most tolerant crop and weed species. Unaltered simazine accumulates in sensitive plants, causing chlorosis and death.

Environmental characteristics:

- - Adsorption and leaching in basic soil types: Simazine is more readily adsorbed on muck or clay soils than in soils of low clay and organic matter content. The downward movement or leaching of simazine is limited by its low water solubility and adsorption to certain soil constituents. Tests have shown that for several months after surface application the greatest portion will be found in the surface 2 inches of soil. It has little if any lateral movement in soil but can be washed along with soil particles.
- - Microbial breakdown: Microbial breakdown is one of several processes involved in the degradation of simazine. In soils, microbial activity possibly accounts for decomposition of a significant amount of simazine.
- - Loss from photodecomposition and/or volatilization: Under normal climatic conditions, loss of simazine from soil by photodecomposition and/or volatilization is considered insignificant.
- - Bioaccumulation: Simazine has a low potential to bioaccumulate in fish.
- - Resultant average persistence: The average half-life of simazine under anaerobic soil conditions is greater than 12 weeks. The half-life of simazine under aerobic soil conditions is 8 to 12 weeks. The persistence of simazine in ponds is dependent upon many factors including the level of algae and weed infestation. The average half-life for simazine in ponds is 30 days.

Ecological characteristics:

- - Avian oral LD50: > 4640 mg/kg (practically non-toxic)
- - Avian dietary LC50: 2000 ppm to 11000 ppm (moderately to slightly toxic)
- - Fish LC50: 6.4 to 70.5 ppm (moderately to slightly toxic)
- - Aquatic invertebrate LC50: 3.2 to 100 ppm (moderately to slightly toxic)
- - The use of simazine could affect endangered aquatic species only if there was a direct application to the water where they dwell. Terrestrial endangered species may be affected, particularly for such uses as ditchbanks and rights-of-way. Formal consultation with the office of Endangered Species, USFWS may be initiated.

Tolerance assessments:

- - Tolerances have been established for simazine in a variety of food and forage crops, meat and poultry, milk and dairy products, and shellfish. The Agency has reevaluated the existing data base which revealed significant deficiencies. Evaluation of acute toxicity data did not reveal any adverse acute effects of simazine. Data are either lacking or insufficient to determine long-term chronic effects, oncogenicity potentials, teratogenicity and mutagenicity. These data are crucial and necessary for the continuation of existing tolerances and for the consideration of additional tolerances. The information specifying sameness or differences between metabolites formed in plants and animals is also necessary.

Problems which are known to have occurred with use of simazine:

- - The files of the Pesticide Incident Monitoring System (PIMS) indicate 71 incidents involving simazine during the period 1966 to June 1981. Two groups of reports were distinguished in these incidents in which alleged adverse effects were reported. One group containing 18 reports cited the involvement of simazine alone. The other contained 53 reports and cited simazine in combination with other ingredients. Humans were involved in 13 incidents in which simazine alone was cited as causing the alleged adverse effects. One person was hospitalized and 12 received medical attention in these incidents. Humans were involved in 43 incidents in which simazine was cited in combination with other ingredients. Nine people were hospitalized, more than 35 received medical attention and 412 were affected or involved and did not seek medical advice in these incidents.

Summary of science findings:

- - Simazine is a moderate eye and dermal irritant with low oral and dermal toxicities. The available toxicity data are insufficient to fully assess the long-term chronic effects or the oncogenic, teratogenic, and mutagenic potential of simazine. The key data gap for most treated agricultural commodities is the simazine metabolites. Most established residue tolerances for raw agricultural commodities are expressed in terms of the parent compound only. Available data are insufficient to fully assess the environmental fate of simazine and the exposure of humans and nontarget organisms to simazine. The available simazine product chemistry data are insufficient to totally assess the chemical's characteristics. Simazine is not very toxic to nontarget insects, birds, or estuarine and marine organisms.

4. Summary of regulatory position and rationale

- All terrestrial use are RESTRICTED: all other uses are classified GENERAL.
- No major use, formulation, or geographical restrictions are required except for groundwater as addressed below.
- No unique warning statements, protective clothing requirements, nor reentry interval statements are required on the labeling; however, the labeling must bear the following groundwater contamination precautionary statements:

"Simazine is known to leach through soil and has been found in groundwater. Users are advised to apply this product only where groundwater contamination is unlikely. Do not apply in recharge areas of designated Sole Source Aquifers, or in areas with well-drained soils as defined by Class A of the Soil Conservation Service classification system which overlay shallow aquifers or which are not protected by an overlying impervious layer. Consult the proper state regulatory officials in your area for information on the location of sole source recharge areas, and the local agent of the Soil Conservation Service for information on your specific soil characteristics."

- No risk assessments were conducted.

5. Summary of major data gaps and when these major data gaps are due to be filled

   - The available product and residue chemistry data are insufficient to fully assess the chemical's characteristics. After receipt of guidance package, data must be submitted within 6 months for short term studies and within 4 years for long term studies. Long term studies include simazine and its metabolites in meat, milk, poultry, eggs and other commodities.
   - The available toxicity data are insufficient to fully assess the long-term chronic effects and the oncogenic, teratogenic and mutagenic potential of simazine. Data gap also exists for a general metabolism study in mammalian species. These studies must be submitted within 4 years after receipt of guidance package.
   - The following data are required to fully assess the environmental fate and transport of, and the potential exposure to simazine: (a) photo-degradation studies on soil and in water; (b) aerobic soil metabolism; (c) anaerobic and aerobic aquatic metabolism; (d) leaching and adsorption; (e) field dissipation studies: aquatic, forestry and long-term; (f) accumulation studies on rotational and irrigated crops.

- Contact Person at EPA:

    Richard F. Mountfort
    Product Manager (23)
    Environmental Protection Agency (TS-767C)
    401 M Street, S. W.
    Washington, D. C. 20460
    (703) 557-1830

DISCLAIMER: The information presented in this chemical Information Fact Sheet is for informational purposes only and may not be used to fulfill data requirements for pesticide registration and reregistration.

SODIUM AND CALCIUM HYPOCHLORITES

Date Issued: February 1986
Fact Sheet Number: 79

1. Description of Chemical

    Generic Name: Sodium and Calcium Hypochlorites

    Common Name: Bleach

    Trade Name: Clorox, Purex

    EPA Shaughnessy Code: 014703 (Sodium Hypochlorite)
    014701 (Calcium Hypochlorite)

    Chemical Abstracts Service (CAS) Number: 778-54-3 (Sodium hypochlorite)
    7681-52-9 (Calcium Hypochlorite)

    Year of Intial Registration: 1957

    Pesticide Type: Disinfectant, Algaecide or Sanitizer

    Chemical Family: Inorganic Salts

    U.S. and Foreign Producers: Jones Chemical Inc., Allied Chlorine Corp., Kuehne Chemical Company, Olin Corporation, Surpass Chemical Company, Pennwalt Chemical Corporation, PPG Industries.

2. Use Patterns and Formulations

    Application sites: Porous and nonporous food contact and nonfood contact surfaces, swimming pool water, drinking water, waste water and sewage, pulp and paper mill process water.

    Percent of pesticide applied to nonporous food contact surfaces, 100 to 200 ppm available chlorine; porous food contact surfaces, 600 ppm; nonfood contact surfaces, 600 to 1000 ppm; swimming pools start-up, 5 to 10 ppm available chlorine and maintain at 1.0 ppm to 1.5 ppm; drinking water, 0.2 ppm available chlorine; waste water, 0.5 ppm; pulp and paper mill process water, 5 to 10 ppm initial dose and subsequent dose of 1 ppm available chlorine.

    Application rates: N/A

    Types of formulations: liquid (sodium hypochlorite) granular, powder and, tablet (calcium hypochlorite)

    Usual carriers: Water

3. Science Findings

Based on a review of the available product chemistry, toxicity, environmental fate, and ecological effects data on sodium and calcium hypochlorites, the Agency has determined that any hazards associated with the uses of sodium and calcium hypochlorite are relatively small.

Chemical Characteristics

Physical state: clear liquid, sodium hypochlorite) granular, powder, and tablet, (calcium hypochlorite)

Color: clear liquid, (sodium hypochlorite) white, (calcium hypochlorite)

Odor: chlorine

pH: Sodium Hypochlorite: 9.75-10.50
    Calcium Hypochlorite: 12.59-13.11, concentrated aqueous solutions

Boiling point: N/A

Melting point: N/A

Flash point: N/A

Toxicity Characteristics

Toxicity category and value for sodium hypochlorite
Acute oral toxicity: IV; LD50, 192 mg/kg
Acute dermal toxicity: III; LD50, >3,000 mg/kg
Primary eye irritation: I; Corrosive
Primary skin Irritation: I; Corrosive

Toxicity category and value for calcium hypochlorite
Acute oral toxicity: III ; LD50, 850 mg/kg
Acute dermal toxicity: II ; LD50, >2g/kg
Acute inhalation: III ; LC50, <20mg/l
Primary eye irritation: I ; Corrosive
Primary skin irritation: I ; Corrosive

Physiological and Biochemical Behavorial Charateristics: N/A

Environmental Characteristics: N/A

Ecological Characteristics:

Values for sodium hypochlorite
Acute oral-Bobwhite Quail ; LD50, >2510 mg/kg
Acute dietary-Mallard Duck ; LC50, >5220ppm
Acute dietary-Bobwhite Quail ; LC50, >5620ppm
Acute fish-Rainbow Trout ; LC50, 0.18-0.22 mg/l
Acute fish-Bluegill Sunfish ; LC50, 0.44-0.79 mg/l
Acute invertebrate-Daphnia ; LC50, 0.033-0.048 mg/l

These values are representative for calcium hypochlorite also since both chemicals generate hypochlorous acid when dissolved in water. The hypochlorous acid is the disinfecting agent which is also the toxic material to fish and other non-target organisms. This is traditionally measured in terms of available chlorine in parts per million.

Efficacy Results Review:

Product-by-product efficacy data requirements normally levied on disinfectants and sanitizers have been waived because the Agency has concluded that the published literature can reasonably be extrapolated to the full range of these products.

Tolerance Assessment:

An exemption from the requirement for a tolerance was established (40 CFR 180.1054) for residues of calcium hypochlorite which may occur in or on raw agricultural commodity potatoes resulting from the use of washing solutions containing calcium hypochlorite. After reexamining the exemption from the requirement for a tolerance, the Agency has determined that the exemption is still appropriate under current scientific standards.

Food processing plants, dairies, canneries, breweries, wineries, beverage bottling plants and eating establishments use hypochlorites for sanitizing premises and for disinfecting equipment and utensils. A incidental food additive regulation allowing the use of sodium or calcium hypochlorites as a terminal santizing rinse on food processing equipment has been established (21 CFR 178.1010). A food additive regulation permitting the use of sodium hypochlorites in washing or assisting in lye peeling of fruits and vegetables has been established (21 CFR 173.315) by the Food and Drug Administration (FDA).

The provisions of the Federal Food Drug and Cosmetic Act (FFDCA) (40 CFR Part 180 require the establishment of a tolerance for exemption from the need for a tolerance) for the use of calcium hypochlorite on mushroom pins (preharvest), sweet potatoes (postharvest), pimento seeds, tomato seeds, pecans (postharvest), fish fillets. EPA plans to propose to issue an exemption from the requirements of a tolerance for these uses.

The provisions of the Federal Food Drug and Cosmetic Act (21 CFR 173 Subpart D - Specific Usage Additives) require the establishment of a food additive regulation for calcium hypochlorite in sugar syrup and raw sugar. Applicants whose product labeling contains such uses must either obtain a food additive regulation for the uses from the Food and Drug Administration within 12 months from the date of issuance of this standard or delete the claims from the labeling.

Reported Pesticide Incidents: N/A

4. Summary of Regulatory Position and Rationale

   Uses: Final sanitizing rinses on food processing equiptment and utensils; disinfection of nonporous hard surfaces; algicides/slimicides for water treatment systems; disinfection of poultry drinking water; disinfection of human drinking water, swimming pool water, hubbard/immersion tank water, spas/hot tubs, hydrotherapy pools, and human drinking water systems; laundry sanitizers; toilet bowl sanitizers.

   Formulations: Liquid sodium hypochlorite solutions; solid (granular, powder, or tablet) calcium hypochlorite.

   Sodium and calcium hypochlorite single active ingredient products are eye irritants, but the potential for hazard from use of these products may be mitigated with appropriate precautionary labeling.

   The hypochlorites are the most widely used chemicals for disinfecting water supplies, are generally recognized as safe for use as post harvest fungicides on agricultural commodities, and are listed as sanitizers, for use as terminal sanitizing rinses of food handling equipment. In the absence of significant long term dietary exposure to the hypochlorites from these patterns of use, there is no need for chronic or subchronic studies. If it comes to the attention of the Agency that hazards due to long term dietary exposure may be significant, then long term studies will be required.

   Unique warning statements: None.

5. Summary of Major Data Gaps

   There are no data gaps.

6. Contact Person at EPA

   Jeff Kempter
   Acting Product Manager (32)
   Disinfectants Branch
   Registration Division (TS-767C)
   U. S. Environmental Protection Agency
   401 M. Street SW
   Washington, DC 20460
   (703) 557-3964

DISCLAIMER: The information presented in this Pesticide Fact Sheet is for informational purposes only and may not be used to fulfill data requirements for pesticide registration and reregistration.

SODIUM ARSENATE

Date Issued: December 1986
Fact Sheet Number: 114

1. DESCRIPTION OF CHEMICAL

    Common Name: Sodium Arsenate

    Chemical Name: Sodium Orthoarsenate - $Na_2HAsO_4 \cdot 7H_2O$

    Trade Names: None

    EPA Shaughnessy Code: 013505

    Chemical Abstracts Service (CAS) Number: 7778-43-0

    Year of Initial Registration:

    Pesticide Type: Insecticide

    Chemical Family: Inorganic Arsenicals

    U.S. and Foreign Producers: Osmose Wood Pres. Company of America, Inc.

2. USE PATTERNS AND FORMULATIONS

    Sodium arsenate is currently registered for use as an ant bait. These baits are used in approximately 1% of U.S. homes.

    ° Methods of Application: Liquid bait applied where ants are seen. Applied as a bait station using cardboard, waxpaper, cotton, or bottle caps; or apply directly across ant trails and at entry points as a thin line 3 to 4 inches long.

    ° Application Rates: Insecticide- The bait used is a 1.3% arsenic (metal) solution.

    Types of Formulations: Ready to use solution, granular.

3. SCIENCE FINDINGS

- Chemical Characteristics

    Sodium arsenate is a pentavalent form of inorganic arsenic. It is a heptahydrate which normally exists as colorless crystals with no discernible odor. Sodium arsenate contains 24% arsenic and is soluble in 5.6 gm at 0°C and 100 gm at 100°C in 100 cc of water, soluble in glycerol and slightly soluble in alcohol. The melting point of calcium arsenate is 130°C , the density is 1.88 and the molecular weight is 312.01. The technical chemical contains 98% and the formulations contain from 0.92% to 3.08% sodium arsenate.

- Toxicological Characteristics

    Inorganic arsenical compounds have been classified as Class A oncogens, demonstrating positive oncogenic effects based on sufficient human epidemiological evidence.

    Inorganic arsenicals have been assayed for mutagenic activity in a variety of test systems ranging from bacterial cells to peripheral lymphocytes from humans exposed to arsenic. The weight of evidence indicates that inorganic arsenical compounds are mutagenic.

    Evidence exists indicating that there is teratogenic and fetotoxic potential based on intravenous and intraperitoneal routes of exposure; however, evidence by the oral route is insufficient to confirm sodium arsenate's teratogenic and fetotoxic effects.

    Inorganic arsenicals are known to be acutely toxic. The symptoms which follow oral exposure include severe gastrointestinal damage resulting in vomiting and diarrhea, and general vascular collapse leading to shock, coma and death. Muscular cramps, facial edema, and cardiovascular reactions are also known to occur following oral exposure to arsenic.

- Environmental Characteristics: The environmental fate of sodium arsenate is not well documented. Studies to demonstrate its fate must take into account the fact that inorganic arsenicals are natural constituents of the soil, and that forms of inorganic arsenic may change depending on environmental conditions. Based on very limited data sodium arsenate is not predicted to leach significantly.

- Ecological Characteristics: Sodium arsenate is moderately toxic to birds, slightly toxic to fish and moderately toxic to aquatic invertebrate species.

- **Metabolism:** The metabolism of inorganic arsenic compounds in animals is well known. The pentavalent form, such as sodium arsenate, is metabolized by reduction into the trivalent form, followed by transformation into organic forms which are excreted within several days via the urine. All animals exhibit this metabolism except rats, which retain arsenic in their bodies for up to 90 days.

- **Tolerance Assessment:** A tolerance for residues of the insecticide sodium arsenate on grapes was established at 3.5 ppm in 40 CFR 180.196.

- **Reported Pesticide Incidents:** The Agency's Pesticide Incident Monitoring System (PIMS) contains many recorded incidents of accidental poisonings from the use of sodium arsenate baits. 190 children were involved in 186 reported incidents; five of these children died and 43 were hospitalized.

4. SUMMARY OF REGULATORY POSITION AND RATIONALE

The Agency is proposing to cancel all existing nonwood registrations of sodium arsenate. Based upon the risk of acute toxicity poisonings and the other toxicological characteristics described above the Agency has determined that in light of the limited benefits for nonwood uses of sodium arsenate the risks of continued use outweigh the benefits.

- **Benefits Analysis:** No economic impact is expected as a result of cancellation of this use. Comparatively priced alternatives are available.

5. CONTACT PERSON

Douglas McKinney
Special Review Branch, Registration Division
Office of Pesticide Programs (TS-767C)
401 M Street, S.W.
Washington, D.C. 20460
(703) 557-5488

DISCLAIMER: The information presented in this Pesticide Fact Sheet is for informational purposes only and may not be used to fulfill data requirements for pesticide registration or reregistration.

SODIUM ARSENITE

Date Issued: December 1986
Fact Sheet Number: 113

1. DESCRIPTION OF CHEMICAL

    Common Name: Sodium Arsenite

    Chemical Name: Sodium Metaarsenite - $NaAsO_2$

    Trade Names: Chem Pels C, Chem-Sen 56, Kill-All, Penite Prodalumnol Double

    EPA Shaughnessy Code: 013603

    Chemical Abstracts Service (CAS) Number: 77784-46-5

    Year of Initial Registration:

    Pesticide Type: Acaricide, Fungicide, Herbicide, Insecticide, Termiticide

    Chemical Family: Inorganic Arsenicals

    U.S. and Foreign Producers: Agtrol Chemicals, Fasey & Besthoff, Inc.

2. USE PATTERNS AND FORMULATIONS

    Sodium arsenite is used as a broad spectrum herbicide for weed control in industrial areas, lots and tank farms. However, this use has been declining in recent years. Sodium arsenite is also used as a fungicide in California to control black measles, phomopsis shoot, and leaf necrosis on 5% of the U.S. grape crop. There is currently no known usage of sodium arsenite as an acaricide, insecticide, or termiticide.

    ° Types and Methods of Application: Wind spray machine, hand application equipment, injection, dip, and liquid bait.

    ° Application Rates: Fungicide - 1.5 lbs arsenic/A
                         Herbicide - 3.0 lbs ai/A

    Types of Formulations: Flowable liquid, soluble concentrate, wettable powder/dust

3. SCIENCE FINDINGS

   ° Chemical Characteristics

   Sodium arsenite is a trivalent form of inorganic arsenic. It normally exists as a gray-white powder with no discernible odor. Sodium arsenite contains 58% arsenic and is very soluble in water and alcohol. The boiling point of sodium arsenite is 100°C, the density is 1.87 and the molecular weight is 129.91. Sodium arsenite is a highly toxic substance with an acute oral LD_{50} (mammalian) of 10-50 mg/kg.

   ° Toxicological Characteristics

   Inorganic arsenical compounds have been classified as Class A oncogens, demonstrating positive oncogenic effects based on sufficient human epidemiological evidence.

   Inorganic arsenicals have been assayed for mutagenic activity in a variety of test systems ranging from bacterial cells to peripheral lymphocytes from humans exposed to arsenic. The weight of evidence indicates that inorganic arsenical compounds are mutagenic.

   Evidence exists indicating that there is teratogenic and fetotoxic potential based on intravenous and intraperitoneal routes of exposure; however, evidence by the oral route is insufficient to confirm sodium arsenite's teratogenic and fetotoxic effects.

   Inorganic arsenicals are known to be acutely toxic. The symptoms which follow oral exposure include severe gastrointestinal damage resulting in vomiting and diarrhea, and general vascular collapse leading to shock, coma and death. Muscular cramps, facial edema, and cardiovascular reactions are also known to occur following oral exposure to arsenic.

   ° Environmental Characteristics: The environmental fate of sodium arsenite is not well documented. However, because of its extreme toxicity and its solubility in soil moisture and hence the hazard to water supplies, sodium arsenite is not used now for many purposes formerly common practice. Studies to demonstrate its fate must take into account the fact that inorganic arsenicals are natural constituents of the soil, and that forms of inorganic arsenic may change depending on environmental conditions.

   ° Ecological Characteristics: Sodium arsenite is toxic to birds, fish and aquatic invertebrate species.

° Metabolism: The metabolism of inorganic arsenic compounds in animals is well known. The pentavalent form is metabolized by reduction into the trivalent form, followed by transformation into organic forms which are excreted within several days via the urine. All animals exhibit this metabolism except rats, which retain arsenic in their bodies for up to 90 days.

° Tolerance Assessment: Tolerances were established in 40 CFR 180.335 for residues of the insecticide sodium arsenite (expressed as As_2O_3) resulting from dermal application to animals under the supervision of the U.S. Department of Agriculture. An interim tolerance of 0.05 ppm (as As_2O_3) for grapes was established in 40 CFR 180.319.

° Reported Pesticide Incidents: The Agency's Pesticide Incident Monitoring System (PIMS) has many recorded incidents of accidental poisonings from the use of sodium arsenite. From 1966 to 1979, sixty-one reports involving humans were reported. Eleven of these incidents involved hospitalizations and 24 involved child or adult fatalities.

4. SUMMARY OF REGULATORY POSITION AND RATIONALE

The Agency is proposing to cancel all existing nonwood registrations of sodium arsenite, with the exception of the fungicidal use on grapes. Measures to mitigate the inhalation risks including dust masks, respirators, which would be expected to reduce inhalation exposure by 80 and 90 percent, respectively, and restricting the use to certified applicators were considered by the Agency during the Special Review. The Agency has determined that these protective measures would not reduce risks to an acceptable level in light of the limited benefits. The Agency has further determined that the toxicological risks from all nonwood uses of sodium arsenite, except for the use on grapes, outweigh the limited benefits. The fungicide use on grapes is being deferred pending further evaluation by EPA's Risk Assesment Forum of the carcinogenic potency of inorganic arsenic from dermal and dietary exposure.

° Benefits Analysis: No economic impact is expected as a result of cancellation of the herbicide and insecticide registrations of sodium arsenite.

5. CONTACT PERSON

   Douglas McKinney
   Special Review Branch, Registration Division
   Office of Pesticide Programs (TS-767C)
   401 M Street, S.W.
   Washington, D.C. 20460
   (703) 557-5488

   DISCLAIMER: The information presented in this Pesticide Fact Sheet is for informational purposes only and may not be used to fulfill data requirements for pesticide registration or reregistration.

SODIUM OMADINE

Date Issued: July 16, 1985
Fact Sheet Number: 61

1. Description of Chemical

    Generic name: 1-hydroxy-2(1H)-pyridinethione, sodium salt (CA)
    Common names: Sodium omadine
    Trade names: Sodium 2-pyridinethiol 1-oxide, Sodium 2-mercaptopyridine,
                Sodium 1-hydroxypyridine-2-thione, and Omadine® sodium.
    EPA Shaughnessy Code: 088004
    Chemical Abstracts Service (CAS) number: 15922-78-8
    Year of initial registration: 04-10-68
    Pesticide Family: Pyridine
    Pesticide type: Microbiocide
    U.S. and Foreign producers: Olin Corp., Research Center, 270 South
                                Winchester Ave., P.O. Box 30275,
                                New Haven, CT 06511

2. Use Patterns and Formulations

    Application sites: in metalworking fluids, paints, inks, adhesives,
        plastics, laundry rinse additive, polymers, floor finishes.

    Types of Formulations: liquid formulations include the following per-
        centages of sodium omadine: 4%, 5%, 5.9%, 6.4%, and 40%. Powder
        formulations include percentages 2% and 90% sodium omadine.

    Types and methods of application: for manufactured materials (i.e.,
        plastics, paints, adhesives, etc.,), add at any point during the
        manufacturing process. As a preservative (i.e., in metalworking
        fluids, and inks) add directly to the solution to preserved.

    Application rates: for metalworking fluids use 4 - 1280 oz./100 lbs.
        of fluid; for inks use 0.05 - 0.75% of product/lb. of ink; for
        paints use 0.1 - 0.25% of product/lb. of paint; for plastics and
        polymers use 0.1 -0.2% of product/lb. of plastic or polymer.

3. Science Findings

Summary of science statement: "bottom line" of science information listed below.

Chemical Characteristics:
Physical state: powder
Color: pale yellow
Melting Point: between 252 - 257°C

Solubility: 45.4% in water (25°C)
5.4% in ethanol
19.1% in ethylene glycol
28.6% in methanol

3. Toxicology Characteristics:

Acute Eye Irritation: Tox. CAT III
Skin Irritation: Tox. CAT III

Major routes of exposure: dermal and inhalation

Chronic Toxicology:
Teratology study (rat; dermal application): no teratogenic effects at 7 mg/kg/day.
Teratology study (rabbit; dermal application): sodium omadine displayed dose-related maternal toxic effects and potental teratogenic effects at all dosage levels tested (0.5, 2.0, and 8.0 mg/kg/day). A NOEL could not be determined and malformations cannot be defined as teratogenic until effects are seen in the absence of maternal toxicity.
Gene Mutation Study: no mutagenic effects noted at the highest dosage tested (25ug/plate [activated] and 3ug/plate [nonactivated]).

Environmental Characteristics: as there are no food/feed uses, or other outdoor uses, environmental fate data do not have to be submitted to the Agency.

Ecological Characteristics:

Avian Dietary LC_{50} (40% a.i. tested):
Bobwhite quail   3246 ppm
Mallard duck   9119 ppm

Freshwater Fish LC_{50} (4% a.i. tested):
Bluegill sunfish   66 ppm
Rainbow trout   0.28 ppm

Aquatic invertebrates LC_{50} (40% a.i. tested):
Daphnia magna   23 ppb

Tolerance Assessments: the registered uses of sodium omadine do not include direct application to a food or feed crop. Therefore, no tolerances have been established for this chemical.

No problems have been reported to the Agency (i.e., PIMS or data in reference to Section 6(a)(2) of the Act) with respect to the use of this chemical.

4. Summary of Regulatory Position and Rationale

There are no use, formulation, manufacturing process or geographical restrictions for sodium omadine registrations.

Unique warning statement:

Environmental Hazards

This pesticide is toxic to fish and freshwater invertebrates. Do not discharge effluent containing this product into lakes, ponds, streams, estuaries, oceans or public water unless this product is specifically identified and addressed in an NPDES permit. Do not discharge effluent containing this product to sewer systems without previously notifying the sewage treatment plant authority. For guidance contact your State Water board or Regional Office of the EPA.

A risk assessment has not been conducted on this chemical.

5. Summary of Major Data Gaps

Data Gap	Date Study Due
Acute Oral LD_{50}	12-31-85
Dermal Sensitization	12-31-85
Teratology (rabbit)	12-31-85
Mutagenicity Studies:	
Chromosomal	12-31-85
Other Mechanisms	12-31-85
Reproduction - 2 generations	08-31-87

6. Contact Person at EPA
Registration Division (TS-767C)
U.S. Environmental Protection Agency
Office of Pesticide Programs
401 M Street, S.W.
Washington, D.C. 20460
Attn.: Marshall Swindell   Telephone: (703)-557-3675

DISCLAIMER: The information presented in this Chemical Information Fact Sheet is for informational purposes only and may not be used to fulfill data requirements for pesticide registration and reregistration.

SODIUM SALT OF FOMESAFEN

Date Issued: April 1987
Fact Sheet Number: 132

1. Description of Chemical

    Generic Name: Sodium Salt of 5-[2-Chloro-4-(trifluoromethyl)-phenoxy]-N-(methylsulfonyl)-2-nitrobenzamide

    Common Name: Sodium Salt of Fomesafen

    Trade Name: Reflex, Flex

    EPA Shaughnessy Code: 123802

    Chemical Abstracts Service (CAS) Number: 72178-2-0

    Year of Initial Registration: 1987

    Pesticide Type: Herbicide

    Chemical Family: Diphenyl ethers

    U.S. and Foreign Producers: ICI Americas Inc.
    New Murphy Road & Concord Pike
    Wilmington, DE 19897

2. Use Patterns and Formulations:

    Application Sites: Fomesafen is proposed for use on soybeans.

    Types of formulations: 21.7% Liquid Concentrate End-Use Product

    Types and methods of application: End-use product is applied postemergence by ground application. Contact activity results in relative rapid knockdown of weeds in 3 to 5 days. Residual activity may result through root uptake from the soil if rainfall occurs soon after application.

Application rates: The proposed maximum application rate is 0.375 lb/ai/acre with one application per growing season.

Usual carrier: Water

3. Science Findings

Summary Science Statement:

Fomesafen has been found to be oncogenic in mice and has been classified as a Group C oncogen (possible human carcinogen). A quantitative risk estimate has been conducted for the use of fomesafen on soybeans. Based on a $Q^* = 1.9 \times 10^{-1}$ (mg/kg/day) and using a Theoretical Maximum Residue Contribution (TMRC) of 0.0000115 mg/kg (1.5 kg diet) the "worst case" dietary risk was calculated to be 2.2 incidences in a million (2.2×10^{-6}). Using the TMRC provides a conservative estimate since it does not consider the effect of processing on residue levels in the raw agricultural commodity, that actual residue levels will be lower than the level of detection (0.05 ppm), and that less than 100 percent of the crop is treated.

Based on exposure estimates for use of fomesafen on soybeans and the Q^*, the following ranges in risk numbers were calculated:

Private applicators:
- Farmers in South $\quad 10^{-5}$
- Farmers in North Central $\quad 10^{-4}$ to 10^{-5}

Commercial applicators:
- In South $\quad 10^{-4}$
- In North Central $\quad 10^{-4}$

These estimates assume that workers are wearing long-sleeved shirts, long pants, and shoes; protective gloves are worn during mixing, loading, and application; and 10 percent dermal absorption.

Fomesafen is not considered to be teratogenic and the chemical did not significantly impair reproductive ability in a two-generation reproductive effects study in rats. Four mutagenicity studies with fomesafen were negative. Two rat bone marrow cytogenic tests were positive.

Fomesafen is not acutely toxic to humans, avian species, freshwater fish and invertebrates, honey bees or marine species. The pesticide is slightly toxic to marine invertebrates. A minimum adverse effect is expected on nontarget organisms. Fomesafen is relatively stable to hydrolysis, moderately to very mobile in some soils and may persist at significant levels beyond 1 year; it may contaminate ground water.

An applicator carcinogenic warning, ground water advisory, and crop rotation restrictions are required to appear on the product's labeling.

Chemical Characteristics:

> Color: White (PGAI)
> Physical State:
>> Aqueous paste (technical)
>> Solid (PGAI)
>
> Odor: Faint, sweet
> Melting point: 220-221°C PGAI
> Specific gravity: 1.28 PGAI
> Solubility:
>> H_2O - 50 mg/L
>> Acetone - 300 g/L
>> Methanol - 25 g/L
>> Xylene - 1.9 g/L
>> Hexane - 0.5 g/L
>> Cyclohexane 150 g/L
>
> Stability: > 6 months at 50°C

Toxicology Characteristics:

> Acute effects [1]:
>
> Acute oral toxicity (rat):
>> 1,250-2000 mg/kg (males);
>> 1,595-5,203 mg/kg (females);
>> Toxicity Category III
>
> Acute dermal toxicity: > 780 mg/kg
> Primary eye irritation (rabbit): Corneal opacity, iritis and
>> conjunctivitis with remission before 7th day;
>> Toxicity Category II
>
> Primary skin irritation (rabbit):
>> Slight erythema graded PIS = 0.58
>> Toxicity Category III
>
> Dermal Sensitization: Acid form produced dermal sensitization.
>> Sodium salt produced no sensitization.
>
> Subchronic effects:
>
> Subchronic oral toxicity studies in the rat and the dog
> show that the liver is the primary target of toxicity in
> both sexes. Rats were dosed at 1, 5, 100, and 1000 ppm in
> the diet. The lowest-observed-effect level (LOEL) in this
> study was 100 ppm (5 mg/kg/day) and the no-observed effect
> level (NOEL) was 5 ppm (0.25 mg/kg/day). The dogs were dosed
> at 0.1, 1, and 25 mg/kg/day. The LOEL in this study was
> 25 mg/kg/day and the NOEL was 1 mg/kg/day.
>
> A 21-day subchronic dermal toxicity study in the rabbit, at
> doses of 10, 100, and 1000 mg/kg/day, showed moderate to
> severe skin irritation at the application site but no

---

[1] See 40 CFR 162.10 for discussion of toxicity categories and companion labeling requirements.

systemic effects at doses up to 1000 mg/kg/day. The LOEL for skin irritation was 100 mg/kg/day and the NOEL was 10 mg/kg/day.

Chronic-effects:

Chronic Feeding/Oncogenicity-- A 2-year feeding study in Wistar albino SPF rats at doses of 1, 5, 100, and 1000 ppm in the diet also identified the liver as the target organ. The LOEL in this study was 100 ppm based on increased hyalinization of liver cells and pigmentation of Kupffer cells at 100 ppm in males, and the NOEL is 5 ppm. No evidence of oncogenicity was reported in this study.

The 2-year feeding study in Charles River DC-1 mice, at doses of 1, 5, 10, 100, and 1000 ppm in the diet, was positive for oncogenic response. A statistically significant increased incidence of liver adenomas were observed in males at 1, 100, and 1,000 ppm and females at 100 and 1000 ppm. A statistically significant increased incidence of liver carcinomas in both sexes was observed at 1000 ppm.

Developmental toxicity:

Rats were dosed with 0, 50, 100, and 200 mg/kg/day fomesafen in the diet and in a second study the same strain of rats at the same laboratory were dosed with 0, 1.0, 7.5, and 50 mg/kg/day fomesafen. In the first study alterations of the 14th rib and increased ossification of the heel bone were observed only at 50 mg/kg/day. These effects were not repeated in the second study. The NOEL is 7.5 mg/kg/day. The maternal LOEL is 200 mg/kg/day with increased post-implantation loss and decreased body weight. No terata were observed under conditions of the studies.

Rabbits were dosed with 0, 2.5, 10, and 40 mg/kg/day fomesafen in the diet. The NOEL is 10 mg/kg/day and there were no adverse effects on offspring under conditions of the study.

Two-Generation reproduction - rat:

Wistar-derived rats (30/sex/dose) were dosed with 0, 50, 250, and 1000 ppm fomesafen in the diet. Significant effects in survival index and litter weight gain at the 1000 ppm were seen. The NOEL for this study is the mid-dose 25 ppm. No meaningful effects were seen at 50 ppm.

Mutagenicity;

Gene mutation tests in Salmonella and hamster kidney fibroblasts were negative. In vivo chromosomal aberration studies in the rat bone marrow were considered positive.

Physiological and biochemical characteristics:

Foliar absorption: Contact activity results in relatively rapid knockdown of weeds in 3 to 5 days.

Translocation: In the presence of rainfall, plants absorb fomesafen from the soil by root uptake.

Environmental characteristics:

Persistence:

Fomesafen is relatively stable to hydrolysis. The half-life was estimated to be about 3 years at 25 °C and does not appear to be pH dependent. In Northern aerobic soils the half-life is approximately 1 year. In Southern anaerobic soils fomesafen's half-life is less than 5 weeks. Laboratory data indicate rapid degradation of parent compound under anaerobic conditions, but degradates were not adequately identified, and where identified, were not monitored in field studies. Therefore, half-lives in Southern anaerobic soils do not provide an accurate estimate of total soil residues.

Absorption and leaching:

Fomesafen aged residues are moderately mobile in loams and clay loams, mobile in sands, and very mobile in coarse sands. The only residue identified was parent. Most of the radiolabeled residues at 20 to 30 cm depth of the coarse sand were unidentified.

Bioaccululation:

[^{14}C] Fomesafen residues have a low potential for bioaccumulation in bluegill sunfish exposed in a flowthrough system. Maximum bioconcentration factors of < 6X occurred in viscera after 7 and 14 days of exposure to [^{14}C] fomesafen at 1 ppm. Accumulated residues were depurated rapidly: < 50% of the residues in viscera were eliminated after 1 day of depuration.

Crop rotation:

Confined and field rotation studies with wheat are conflicting. Significant uncharacterized residues are present in wheat chaff and straw at about a 1-year rotation interval, but in the field study where only parent was analyzed, no residues were present. Lack of confirmation that parent represents 25 percent of residues in the confined study preclude acceptance of the field data.

Sodium Salt of Fomesafen

Environmental fate and surface and ground water contamination concerns:

Fomesafen may possibly contaminate ground water. Additional data are required before the Agency can fully assess the potential for ground water contamination.

<u>Ecological Characteristics:</u>

Avian acute oral toxicity (Mallard duck):	> 5000 mg/kg
Avian subacute dietary toxicity (Mallard duck and bobwhite quail):	> 20,000 ppm
Freshwater fish acute toxicity -	
Rainbow trout:	680 ppm
Bluegill sunfish:	6030 ppm
Freshwater invertebrate acute toxicity (Daphnia):	330 ppm
Marine fish acute toxicity (Sheepshead minnow):	> 163 ppm
Marine invertebrate acute toxicity	
Fiddler crab:	> 163 ppm
Pink shrimp:	> 212 ppm
Marine embryolarvae acute toxicity (Pacific oyster):	> 96.6 ppm
Marine invertebrate acute toxicity (Mysid shrimp)	22.1 ppm
Beneficial insects acute oral and contact toxicity (honey bees):	> 50 ug/bee
Avian reproduction:	
Mallard duck:	NOEL = 46 ppm
Bobwhite quail:	NOEL = 50 ppm
Fish early life stage toxicity (Sheepshead minnow):	54-89 ppm
Freshwater invertebrate life-cycle (Daphnia):	Maximum Acceptable Toxic Concentration (MATC) = 50-100 ppm
Marine invertebrate life-cycle (Mysid shrimp):	MATC = 0.69-1.71 ppm

Field chronic effects for soil
    annelida (earthworm):   NOEL (-1 yr) = 5 kg/ha

Field chronic effects for micro-
    arthropods (mites, collembola):   NOEL (-2 yr) = 5 kg/ha

These data indicate that fomesafen is essentially nontoxic to avian, freshwater fish, and invertebrate species and bees; and that it is slightly toxic to aquatic invertebrates. Based on the acute and chronic data no significant problems to nontarget organisms are expected from fomesafen's use on soybeans.

4. Benefits

Fomesafen is one of a number of postemergent herbicides recently registered, or for which registration is pending, for soybeans. Projecting use of these herbicides is difficult due to lack of data on distribution of various weeds controlled, comparative efficacy and lack of information on the failure of preemergent herbicides to provide adequate weed control. Based on the reasonable assumption that 30 percent of soybeans will be treated with postemergent herbicides, lower chemical cost, and on estimated market penetration, benefits range from $10 million to $15 million per year.

Registration of fomesafen would also provide another postemergent herbicide which would increase competition resulting in lower prices. These lower prices could result in benefits up to $30 million per year.

If additional weed control were to result in small yield increases or prevent some weeds from becoming problems in the future, further benefits would occur. However, no data exist to justify such a prediction, or quantify it at this time.

5. Tolerance Assessment

Tolerances have been established for residues of the sodium salt of fomesafen in or on the following raw agricultural commodity (40 CFR 180.):

Commodity	Tolerance (ppm)
Soybeans	0.05

There are no international tolerances/residue limits for fomesafen.

There are sufficient residue chemistry data available to support this tolerance, including plant and animal metabolism, storage stability (for both the parent compound and its metabolites), field residue studies, and analytical methods. Cattle and poultry feeding studies were not submitted. However, under the proposed conditions of use, measurable residues are not expected to be found in the raw agricultural commodities or fractions. These data are therefore not now necessary.

The Acceptable Daily Intake (ADI) and the Maximum Permissible Intake (MPI) are two ways of expressing the amount of a substance that the Agency believes, on the basis of the results of data from animal studies and the application of "safety" or "uncertainty" factors, may safely be ingested by humans without risk of adverse health effects. The ADI is expressed in terms of milligrams (mg) of the substance per kilogram (kg) of body weight per day (mg/kg/day). The MPI, a related figure, is obtained by assuming a human body weight of 60 kg, and is expressed in terms of mg of substance per day (mg/day).

The Agency has calculated an ADI for fomesafen of 0.0025 mg/kg/day, based on a NOEL of 0.25 mg/kg/day in the rat oncogenicity study and a hundredfold safety factor. The MPI for a 60 kg person is 0.15 mg/day. These tolerances have a theoretical maximum residue contribution (TMRC) of 0.0000115 mg/day in a 1.5 kg diet and would utilize 0.46 percent of the ADI.

6. Summary of Major Data Gaps

   Data required:

Guideline Number	Study	Time Generally Allowed for Response to Data Request
Environmental Fate		
161-2	Photodegradation - water	9 months
161-3	Photodegradation - soil	9 months
162-2	Anaerobic soil metabolism	27 months
163-1	Leaching (degradates)	12 months
164-1	Soil field dissipation	27 months
165-1	Rotational crops (confined)	39 months
	Ground water monitoring	27 months Interim reports due at 6 months and 18 months

Contact Person at EPA:

Richard F. Mountfort
Product Manager (23)
Registration Division (TS-767C)
Environmental Protection Agency
401 M Street SW.
Washington, D.C. 20460
(703) 557-1830

DISCLAIMER: The information presented in this Pesticide Fact Sheet is for informational purposes only and may not be used to fulfill data requirements for pesticide registration and reregistration.

SULFURYL FLUORIDE

Date Issued: June 30, 1985
Fact Sheet Number: 51

1. Description of Chemical

    Generic Name: Sulfuryl Fluoride
    Common Name: Vikane
    Trade Name: N/A
    EPA Shaughnessay Code: 078003
    Chemical Abstracts Service (CAS) Number: 2699-79-8
    Year of Initial Registration: December 15, 1959
    (EPA Reg. No. 464-236)
    Pesticide Type: Structural Fumigant
    Chemical Family: Inorganic Acid Halides
    Producers: US __ Dow Chemical Company
              Foreign ___ None

2. Use Patterns and Formulations

    Application Sites: Domestic dwellings and contents; Wood or
        wood structure treatment;
        Commercial, Institutional and Industrial areas;
        Surface Ships in ports; Vehicles
    Types of Formulations: Liquified Gas under pressure
    Application Rates: Metered release at concentrations
        measured in ounce-hours, i.e. ounces of fumigant
        per cubic feet of space, multiplied by the time
        in hours. The ounce-hour requirements depend on
        temperature, degree of leakage, etc.

    Usual Carriers: N/A

3. Science Findings

   Chemical Characteristics:

   Physical State: liquified gas under pressure
   Color: Colorless
   Odor: None
   Melting Point: -136.67°C at 760 mmHg
   Density: 3.72 g/l
   Solubility:

In Water (25°C)	0.075 g/100 g
In Wesson Oil (20°C)	0.780 g/100 g
In Acetone (22°C)	1.740 g/100 g
In Chloroform (22°C)	2.120 g/100 g

   Toxicity Characteristics

   Sulfuryl fluoride is an extremely hazardous gas or liquid under pressure. Inhalation of vapors may be fatal. Contact with the liquid may cause freezing, injury, or burn to eyes, mucous membranes, or skin, with delayed onset of symptoms. Sulfuryl fluoride has no warning properties such as odor or lacrimation or eye irritation. Early symptoms of over exposure are respiratory irritation (distress), pulmonary edema, nausea, and abdominal pain. Repeated exposure to high concentrations can result in lung and kidney damage. Single exposures at high concentrations have resulted in death. Respiratory equipment and protective clothing are required to use sulfuryl fluoride safely.

   Available Data:

   Inhalation LC_{50} 17.5 mg/liter (20 mg/cubic meter)
   Toxicity Category by this route is III.

   No other acute toxicity available or required by the guidelines because the pesticide is a gas.

Threshold Limit Value (TLV)[1]/ is 5 ppm (20 mg/m^3);
the level to which persons may be exposed daily
for an 8 hour work day without adverse effects
Short Term Exposure Limit (STEL) is 10 ppm (40 mg/m^3);
the level to which persons may be exposed continuously for 15 minutes without adverse effects

Teratology (rabbit, inhalation)
Teratogenic NOEL more than 225 ppm (HDT)
Maternal NOEL 75 ppm
Maternal LEL 225 ppm (decreased body weight gain)
Fetotoxic NOEL 75 ppm
Fetotoxic LEL 225 ppm (decreased body weight and decreased rump-crown length)
Levels tested: 0, 25, 75, 225 ppm during days 6-18
of gestation for 6 hrs/day by inhalation

Teratology (rat, inhalation)

Teratogenic NOEL more than 225 ppm (HDT)
Maternal NOEL more than 225 ppm (HDT)
Fetotoxic NOEL more than 225 (HDT)
Levels tested: 0, 25, 75, 225 ppm during days 6-15
of gestation for 6 hours per day, via inhalation

Physiological-Biochemical Behavioral Characteristics N/A

Environmental Characteristics

Indoor uses only; No expected exposure.

Ecological Characteristics

Indoor uses only; No expected exposure.

Efficacy Data: N/A. Non-Health related uses only.

---

[1]/ Threshold Limit Values for Chemical Substances in
the Work Environment adopted by the American Conference
of Government Industrial Hygienists, 1983-84.

4. Summary of Regulatory Position and Rationale:

   Because of the inhalation hazard associated with this chemical it is being classified as a restricted use pesticide.

   Unique Warning Statements:

   RESTRICTED USE PESTICIDE DUE TO ACUTE INHALATION HAZARDS

   For sale to and use only by Certified Applicators or persons directly under their direct supervision, and only for those uses covered by the Certified Applicator's certification. Special respiratory equipment, protective clothing, and training are required for safe use of this pesticide. Placarding of treated areas required. Re-entry only permitted after work area is determined to contain no residuals greater than 5 ppm of sulfuryl fluoride as determined by direct reading monitoring device.

   Tolerance Assessments: N/A The registered uses of Sulfuryl fluoride do not include direct application to a food or feed crop.

   Problems which are known to have occurred with use of the chemical (e.g., PIMS): none.

5. Summary of major data gaps:

   Residue data from household articles required to establish residues in treated articles and surfaces in habitable structures to which residents may be exposed upon re-entry of fumigated structures (dwellings).

6. Contact person at EPA :   A. E. Castillo
                             Product Manager (32)
                             Disinfectants Branch
                             Registration Division (TS-767c)
                             Tel. (703) 557-3964

   DISCLAIMER: The information presented in this Chemical Information Fact Sheet is for informational purposes only and may not be used to fulfill data requirements for pesticide registration and re-registration.

TEBUTHIURON

Date Issued: July 31, 1987
Fact Sheet Number: 137

1. Description of Chemical

    Chemical Name: N-[5-(1,1-dimethylethyl)-1,3,4-thiadiazol-2-yl]-N,N'dimethylurea

    Common Name : Tebuthiuron

    EPA Shaughnessy Code: 105501

    Chemical Abstracts Service (CAS) Number: 34014-18-1

    Year of Initial Registration: 1974

    Pesticide Type: Herbicide

    Producers: Elanco Products Company, Division of Eli Lilly and Company

2. Use Patterns and Formulations

    Type of pesticide: A relatively nonselective,"soil-actived" herbicide for the control of broadleaf weeds, grasses and brush in noncrop areas, and for spot treatment of woody brush on rangelands. It is readily absorbed through roots of broadleaf weeds, grasses and brush.

    Pests controlled: Broadleaf weeds, grasses and brush. Terrestrial uses: food crop (rangelands and pastures), noncrop areas (airport runways, fencerows, firebreaks, industrial sites, paved surfaces, and highway, railroad and utility rights-of-way).

    Aquatic Uses: ditchbanks.

    Predominant uses: Terrestrial noncrop areas

    Mode of activity: Photosynthesis inhibitor

Formulation: (95% a.i.) technical; 1%, 2%, 3%, & 5% granular; 10%, 13.8%, 15.2%, 20%, 30.5%, 40% pelleted/ tableted; 80% wettable powder; 85% flowable concentrate; and 0.36% soluble concentrate/liquid.

Method of application: Applied as broadcast or band by ground or aerial equipment, spot treatment, drop zone or drip zone treatment or grid pattern treatment.

Application rates: Terrestrial food and nonfood crops: 0.5 to 4.0lb. a.i. per acre, 1-4 grams a.i. per 100 sq. ft., 1 gram a.i. per inch of trunk diameter or 0.25 to 0.50 gram a.i. per ft. of plant height. Terrestrial noncrop areas and aquatic nonfood crop areas: 1 to 16 lb. a.i. per acre; 0.13 to 0.14 lb. a.i. per 1,000 sq. ft.; 0.1 lb. or 0.1 oz. a.i. per 2-4 inch of trunk diameter; 1 to 10 oz. finished spray per 2 to 4 inch of stem diameter; 0.5 oz. a.i. per 1 to 2 inch of stem diameter.

3. Science Findings

Tebuthiuron has low acute toxicity by inhalation in rats, and is Category III by this route. Toxicity Category III.* Subchronic feeding in rats and chronic feeding in dogs indicate only mild effects on the liver, kidneys, gonads, spleen, prostate and thyroid gland. No compound related histological effects were seen.

A rat multigeneration study showed no adverse effects on reproductive performance per se; although parent females in the high dose group failed to gain weight as those in the lower doses and control group in the pre-mating phase of the study.

The Agency is concerned about the potential for ground water contamination by tebuthiuron, based on tebuthiuron's ability to resist environmental degradation and its relatively high mobility (leachability) in a variety of soils. The Agency has determined that additional data are needed to characterize the potential for tebuthiuron to enter ground water.

---

* Toxicity categories are based on the acute toxicity of the chemical (LD_{50} or LC_{50} values) and are used to determine the appropriate signal word and precautionary language for product labeling. Category III requires the signal word CAUTION and precautions against swallowing, inhaling, or contact with the skin and eyes, along with appropriate first aid instruction.

Terrestrial, avian and aquatic vertebrates and invertebrates are not susceptible to Tebuthiuron, and it does not pose a hazard to them. However, numerous endangered or theatened plant species are at risk from the range and pastureland uses of tebuthiuron and products released for shipment after Feburary 1, 1988, and which recommend such uses must bear Endangered Species Labeling.

Chemical Characteristics:

Physical State: Crystalline solid

Color: Colorless, white

Melting Point: 159 to 161° C

Solubility: (at 25° C in mg/mL) 60 in acetonitrile; 70 in acetone; 170 in methanol; 20 in ethanol; 250 in chloroform; 60 in methyl cellosolve; 3.7 in benzene; 6.1 in hexane; and 2.5 in water.

Vapor Pressure: 2×10^{-6} mmHg at 25° C

Octanol/Water Partition Coefficient: 61 (Log K = 1.79)

Stability: Stable

Acute Toxicology

Acute oral	No valid studies. A study is required.
Acute dermal	No valid studies. A study is required.
Acute Inhalation (rat)	3.696 mg/L; Toxicity Category III.
Primary Eye Irritation	No valid studies. A study is required.
Primary Dermal Irritation	No valid studies. A study is required.
Dermal Sensitization	No valid studies. A study is required.

Chronic Toxicology

Oncogenicity	The oncogenic potential of tebuthiuron cannot be determined from the available study. Studies are required.
Reproduction (Rat)	Reproductive effects NOEL= 400 ppm (20 mg/kg bwt/day) systemic NOEL=100 ppm (5.0 mg/kg bwt/day)
Chronic Feeding (Beagle Dog)	1- year study. (NOEL) = 25.0 mg/kg bwt/day.
Mutagenicity	Not mutagenic in Ames assay with or without metabolic activation. Only slight signs of mutagenicity in a mouse lymphoma assay.
Chromosomal Aberration	Data limited but negative; additional data required.
Metabolism	Technical tebuthiuron and/or its metabolites appear in the milk of lactating rats. General metabolism data are not available, but the data are required.

Physiological and Biochemical Behavioral Characteristics

Foliar and Root Absorption - Tebuthiuron is bsorbed through roots, less so through the foliage.

Mechanism of Pesticidal Action - Phytotoxicity symptoms suggest that tebuthiuron inhibits photosynthesis.

Environmental Characteristics

<u>Degradation</u>

Tebuthiuron did not undergo significant degradation at pH 5, 7 and 9 at 25° C in 64 days and is considered stable in sterile water.

Preliminary data indicate that tebuthiuron is also quite stable under aerobic and anaerobic soil conditions. Tebuthiuron only degraded from 8 ppm to 5.7 ppm after 273 days (half life > 1 year) when incubated in loam soil at 25° C and was reported to degrade in an identical soil under anaerobic conditions with a half-life of >48 weeks. Similarly, irradiation with an artificial light that did not quite simulate sunlight resulted in only 42% decomposition after 15 days.

Tebuthiuron appears stable to biological and chemical degradation under environmental conditions and can be considered persistent.

Leaching

Preliminary data indicate that tebuthiuron is mobile to very mobile in loam, loamy sand, and lakeland sand soils and slightly mobile in silty loam soil. Kd values of lower than 2 were reported for clay, sandy loam and sand soils. About 40% of residues of a 30 days sandy loam aged tebuthiuron were found in the leachate.

Based on the above information, tebuthiuron has the potential to leach through a variety of soils and contaminate groundwater. Tebuthiuron was flagged as a groundwater contaminant through the GWDCI (Groundwater Data Call In) screen and has been found in shallow groundwater in Texas. Tebuthiuron will be further analyzed in the Agency's National Pesticides in Well Water Survey.

Existing data which were submitted in response to a Ground-Water Data Call-In have been found to be inadequate to fullfill Agency guidelines requirements. Therefore, additional data are necessary to fully characterize tebuthiuron's ability to contaminate ground water.

Ecological Characteristics

As a dietary administration, tebuthiuron is no more than slightly toxic to birds, however, avian dietary data are not complete. Avian reproductive studies show that tebuthiuron has no effect on reproduction at dietary levels up to 100 ppm.

Tebuthiuron is practically non-toxic (acutely) to fish and aquatic invertebrates. A fish early life-stage study gives a MATC (Maximum Allowable Toxic Concentration) between 9.3 and 18 mg/l based on impaired growth. Aquatic invertebrates show significant reductions of growth and fecundity at 44 mg/l. The MATC for aquatic invertebrates is between 21.8 and 44.2ppm.

In 1972 tebuthiuron was conditionally registered for control of brush in rangeland in Texas and Oklahoma. As a requirement of the registration the registrant was requested to perform a field monitoring study which would better define this chemical's actions in aquatic and terrestrial environments. Instead of a single monitoring study, several studies (EPA Accession No. 246373) were submitted by the registrant to fulfill the conditional requirement. While each study had several deficiencies that would preclude their applicability for singley satisfying the requirement, the series of studies were considered sufficient to satisfy the monitoring condition of the 1979 registration. In 1982, in response to a request for the addition of 17 states to the tebuthiuron registration, the Agency requested that the registrant continue (into a second year) the monitoring of water and hydrosoil at four study sites. The Agency is now requiring that these data be submitted to support the registration of tebuthiuron.

Because of tebuthiuron's extreme persistence, these monitoring data are still necessary to determine the long-term availability of the chemical for runoff into aquatic systems and the likelihood of long-term buildup of tebuthiuron in the soil. Monitoring data are being required through this Registration Standard. If information from past monitoring is not available or is determined to be unsatisfactory, a new monitoring study will be required.

Although tebuthiuron is not expected to pose a hazard to endangered or theatened terrestrial or aquatic animal species, its use on range and pastureland will pose a hazard to endangered or threatened plant species.

Products containing tebuthiuron with range and pasture uses which are released for shipment after February 1, 1988 must bear Endangered Species Labeling.

Endangered Species Labeling for non-crop, wide area, and general indoor/outdoor treatments is deferred until completion of the analysis by OES and the Agency of the non-crop uses.

Results Of Tolerance Assessment

Tolerances are established for residues of the herbicide tebuthiuron and its metabolites containing the dimethylethyl thiadiazole moiety in or on the following raw agricultural commodities:

Commodities	Tolerance(ppm)
Cattle, fat	2
Cattle,*mbyp	2
Cattle, meat	2
Goat, fat	2
Goat, mbyp	2
Goat, meat	2
Grass, hay	20
Grass,rangeland,forage	20
Horse,fat	2
Horse,mbyp	2
Horse,meat	2
Milk	0.3
Sheep, fat	2
Sheep, mbyp	2
Sheep, meat	2

The most recent Provisional Acceptable Daily Intake (PADI) is 0.017 mg/kg bwt/day based on the 5.0 mg/kg bwt/day systemic NOEL derived from the most sensitive study, (the reproductive NOEL was greater than 20 mg/kg bwt/day, the highest dose tested), and using an uncertainty factor of 300 (the 300-fold uncertainty factor is used because a chronic rodent study is missing).

---

*Mbyp "meat by product"

4. Summary Of Regulatory Position And Rationale

Summary of Agency Position: The Agency is requiring registrants of Tebuthiuron to submit additional data as identified in the Registration Standard and summarized in the following section. The Agency will not establish any new food use or register any significant new uses until adequate data are available to fully assess tebuthiuron.

Unique Warning Statements Required on Labels: Unique labeling is not imposed in the Registration Standard. Endangered species labeling, however, is required for rangeland and pastureland. This labeling is addressed in Pesticide Registration Notice (PR) 87-4 dated May 1, 1987.

5. Summary of Data Gaps

Data	Due
Product Chemistry	6 to 15 Months
Residue Chemistry	
Nature of Residues (Metabolism)	18 Months
Residue Analytical Method	15 Months
Storage Stability	6 Months
Environmental Fate	
Photodegradation (water/soil)	9 Months
Metabolism (Anaerobic/aerobic soil)	27 Months
Leaching and Adsorption/Desorption	12 Months
Dissipation (Soil/Aquatic)	27 Months
Accumulation in Fish	39 to 50 Months
Soil, Long Term	50 Months
Irrigated Crops	39 Months
Toxicology	
Acute Dermal	9 Months
Acute Oral	9 Monts
Primary Eye Irritation	9 Months
Primary Dermal Irritation	9 Months
Dermal Sensitization	9 Months
Chronic Toxicity (Rodent)	50 Months
Oncogenicity (Rat and Mouse)	48 to 50 Months
Teratogenicity (Nonrodent)	12 Months
General Metabolism	24 Months
Mutagenicity (Chromosomal Aberrations)	12 Months
Other Genotoxic Effects	12 Months

Wildlife and Aquatic Organisms

Nontarget Insect Testing	9 Months
Avian Dietary LC50	12 Months
Field Monitoring (Special Studies)	36 Months
Phytotoxicity	9 Months

Contact Person at EPA

Robert J. Taylor, PM 25
Registration Division (TS-767C)
U.S. Environmental Protection Agency
401 M Street SW.
Washington, DC 20460
(703) 557-1800

DISCLAIMER: The information presented in this Pesticide Fact Sheet is for information purposes only and may not be used to fulfill data requirements for pesticide registration and reregistration.

TERBUFOS

Date Issued: February 1, 1985
Fact Sheet Number: 5.1

1. Description of Chemical

    Generic name: S-[[(1,1-dimethyl-ethyl)thio]methyl] 0,0-diethyl phosphorodithioate
    Common name: Terbufos
    Trade name: Counter
    EPA Shaughnessy code: 105001
    Chemical Abstracts Service (CAS) number: 13071-79-9
    Year of initial registration: 1974
    Pesticide type: insecticide-nematicide
    Chemical family: organophosphate
    U.S. and foreign producer: American Cyanamid Company

2. Use Patterns and Formulations

    Application sites: Corn, sugar beets, and grain sorghum
    Types of formulations: granular
    Types and methods of application: soil incorporation
    Application rates: vary according to formulation and crop
    Usual carriers: Confidential Business Information

3. Science Findings

    Summary science statement:

    Terbufos is highly toxic to humans, fish and wildlife. Due to a number of major data gaps, the Agency cannot complete a full risk assessment until the data required under this Standard have been submitted and reviewed.

    Reassessment of established tolerances must await receipt and evaluation of required toxicological studies. Available chronic toxicity studies are only supplementary data, and a "No Observable Effect Level" (NOEL) cannot be established at this time, and a maximum permissible intake (MPI) cannot be calculated.

    Chemical characteristics:

    Terbufos, an organophosphate, is a clear slightly brown liquid. It is relatively stable in water under neutral or slightly acidic conditions but is subject to hydrolysis under alkaline conditions. It decomposed on prolonged heating at temperatures greater than 120°C. The chemical does not present any unusual handling hazards.

Toxicology characteristics:

Current available toxicology studies on terbufos are as follows:

- Oral LD_{50} in rats: from 1.3 to 1.57 mg/kg in females and 1.6 to 1.74 mg/kg in males (Tox category I)
- Dermal LD_{50} in male rats: 1.0 mg/kg (Tox category I)
- Eye irritation: due to the high acute toxicity of the chemical and the rapid death of animals, eye irritation scores were not reported
- Dermal irritation: due to the high acute toxicity of the chemical and the rapid death of animals, skin irritation scores were not reported

Symptoms of acute cholinesterase inhibition were reported in all acute studies.

- Multigeneration reproduction study in rats: NOEL of 0.25 ppm and LEL of 1.0 ppm based on noted increase in the percentage of litters with offspring death in each of the three generations as compared to the controls.

- Chronic feeding/oncogenicity in rats: both the ChE NOEL and the systemic NOEL appear to be lower than 0.25 ppm (LDT). The oncogenic potential of terbufos could not be assessed because too few animals of each group were histologically examined. Exophthalmus was noted in treated females in a dose-related fashion during the first year. This effect appeared to subside during the second year of the study.

- 6-month chronic feeding in dogs: a ChE of 0.0025 mg/kg/day for both plasma and red blood cells cholinesterase inhibition. However, it is not clear how the dosage was mixed with the feed and how homogeneous the distribution of the test substance was in the diet. Also, the raw data were not available. A one-year study is now required.

- 18-month oncogenicity in mice: the oncogenic potential of terbufos could not be determined because too few animals were histologically examined. Dose-related exophthalmia was noted in treated males during the first year. This effect subsided in the second year.

- Two 90-day subchronic feeding studies in rats: both studies reflected a ChE NOEL of 0.25 ppm; one study reflected a systemic NOEL of 0.25 ppm. A systemic NOEL could not be determined for the other study because histological data were not reported.

- 28-day feeding study in dogs: only one dose (0.05 mg/kg) was tested at 6 and 7 days/week exposure. Both groups of animals showed similar levels of ChE activities with plasma ChE being significantly inhibited (79% inhibition) while no inhibition was observed in RBCChE activity.

- 30-day dermal study in rabbits: systemic NOEL was determined to be 0.02 mg/k for the technical material.

- Acute delayed neurotoxicity in hens: terbufos did not produce signs of neurotoxicity after the second 21-day dosing period using the technical material at 40 mg/kg (the LD_{50} dosage).

- Mutagenicity: the Ames test on bacterial system was performed with and without microsomal activation - with negative results.

Additional data are needed to fully assess the toxicity of terbufos.

Environmental characteristics:

Based on available data, terbufos is not expected to leach into ground water. However, additional data are needed to fully assess the environmental fate of terbufos.

Ecological characteristics:

The following data are available:

- Avian oral LD_{50}: 28.6 mg/kg (highly toxic)
- Avian dietary LC_{50}: 143 ppm (highly toxic)
- Fish LC_{50}: 0.77-3.8 ppb (bluegill sunfish); 9.4-20.0 ppb (trout)
- Aquatic invertebrate LC_{50}: 0.31 ppb for Daphnia magna (very highly toxic)

The Agency is requiring further monitoring of water, sediment, and fish; in ponds adjacent to treated fields, to fully assess the potential hazard to nontarget aquatic species. Avian and mammalian field testing are required to assess the potential hazard to terrestrial organisms.

Efficacy review results:

None required.

Tolerance assessments:

Tolerances have been established, under 40 CFR 180.352, for combined residues of terbufos and its cholinesterase-innibiting metabolites in or on the following raw agricultural commodities:

Commodity	PPM
Beets, sugar; roots	0.05 negligible residues (N)
Beets, sugar; tops	0.1
Corn, field; fodder, forage	0.5
Corn, pop; fodder, forage	0.5
Corn, grain	0.05(N)
Corn, sweet (kernel + cob with husk removed)	0.05(N)
Corn, sweet; fodder, forage	0.5
Sorghum; fodder, forage	0.5
Sorghum; grain	0.05

There are currently no food or feed addivitve tolerances and it has been determined that none are required for food and/or feed byproducts of these commodities.

There are no tolerances for meat, milk, poultry and eggs, nor are any required since there is no reasonable expectation of finite residues occuring in these foods from feed use of the raw agricultural commodity including their processing byproducts. Request for new livestock feed crops, however, may require tolerances for meat, milk, poultry and eggs.

The established tolerances for terbufos are presently expressed in terms of terbufos and its cholinesterase-inhibiting metabolites without specifying the latter as phosphorylated metabolites. The Agency will proceed towards revising 40 CFR 180.352 by changing the wording to read ".....terbufos ..... and its phosphorylated (cholinesterase inhibiting) metabolites".

Reassessment of the established terbufos tolerances must await receipt and evaluation of pertinent toxicological studies. Available chronic toxicity studies are only supplementary data and thus may not be used as a basis for tolerance assessment. Consequently, a "No Observable Effect Level" (NOEL) cannot be established at this time, hence a maximun permissible intake (MPI) cannot be calculated.

The Pesticide Incident Monitoring System (PIMS) reports through June, 1981, include 31 reports involving terbufos, of which 19 involved terbufos alone. Of these 19 incidents, 9 involved humans, 8 involved domestic livestock and 2 involved wildlife. No human fatalities resulted. In those human exposure incidents which were reported with some detail, it appears that carelessness or negligence were important factors. In two of these incidents, the granular pesticide was reported to have been handled with bare hands during loading and application procedures.

In those instances involving livestock, one resulted in the death of about 600 cattle, another in the death of 127 cattle. The accidental contamination of livestock feed was reported as the cause in these incidents.

The two wildlife instances involved fish kills which were reportedly due to runoff from treated fields. Only one included analysis of the water samples. Though the sampling agnecy was unable to test for terbufos, no evidence of organophosphorus compounds was found in the sampled water.

Careless and/or negligence appear to have been important factors in most incidents. Strict adherence to proper storage and application techniques as prescribed in the label directions and precautions will minimize the risk of potential adverse effects to humans and domestic animals.

4. Summary of Regulatory Position and Rationale:

   Use classification: Based on the acute oral and dermal toxicity, granular end-use products containing 15% or more terbufos are classified for "Restricted Use". All such products released for shipment on September 1, 1985, or thereafter, must be labeled for restricted use. Similarly, all such products which are in channels of trade on or after September 1, 1986 must bear restricted use labeling.

   Restrictions: end-use products formulated from manufacturing-use products under the standard must be granular formulations for ground incorporated application only.

   Unique warning statements:

   - Labeling of manufacturing-use products must contain the following statements:

     "Wear protective clothing, rubber gloves and goggles."

"Wear a pesticide respirator jointly approved by the Mining Enforcement and Safety Administration (formerly the U.S. Bureau of Mines) and by the National Institute for Occupational Safety and Health under the provision of 30 CFR Part 11 for organic phosphate protection."

- Labeling of end-use products must contain the statements:

"Restricted Use"

"Wear protective clothing, gloves and goggles."

"Not for use or storage in or around the home."

5. <u>Summary of major data gaps</u>:

Toxicology: inhalation LC_{50}, 2 chronic feeding studies, 2 oncogenicity studies, 2 teratogenicity studies, and additional mutagenicity studies. These studies must be submitted no later than June, 1986.

Wildlife and Aquatic Organisms: avian reproduction, simulated and actual field testing - mammals and birds, fish early life stage and aquatic invertebrate life-cycle; and acute LC_{50} for estuarine and marine organisms. These studies must be submitted no later than June, 1986.

Environmental Fate: photodegradation in water, lab volatility study, rotational crop field study, and monitoring studies (soil and water, sediment and fish). These studies must be submitted no later than June, 1986.

6. <u>Contact Person at EPA</u>

William H. Miller
Product Manager (16)
Insecticide-Rodenticide Branch
Registration Division (TS-767)
Environmental Protection Agency
Washington, DC 20460

Tel. No. (703) 557-2600

DISCLAIMER: The information presented in this Chemical Information Fact Sheet is for informational purposed only and may not be used to fulfill data requirements for pesticide registration and reregistration.

TERBUTRYN

Date Issued: September 1986
Fact Sheet Number: 104

1. Description of Chemical

    Generic Name: 2-(tert-butylamino)-4-(ethylamino)-
                              6-(methylthio)-s-triazine

    Common Name: Terbutryn

    Trade Names: Igran, Prebane (Great Britain only),
                              Terbutrex, Terbutryne, Clarosan,
                              GS 14260, and Short-stop (discontinued

    EPA Shaughnessy Code: 080813

    Chemical Abstracts
      Society (CAS) Number: 886-50-0

    Year of Initial
      Registration: 1969

    Pesticide Type: Herbicide

    Chemical Family: s. Triazine

    Producers: Ciba-Geigy Corp. (U.S.);
                              Verolit Chemical Manufacturing Co.
                              Ltd.; and Makhteshim Agam (Israel)

2. Use Patterns and Formulations

    Application Sites: Winter wheat, winter barley, grain
        sorghum, fallow areas, and non-crop areas, including
        railroad rights-of-way, to control broadleaf weeds and
        grasses

    Types and Methods of Application: Foliar application or
        soil incorporation; broadcast application by ground
        or aerial equipment or band application

Application Rates: Ground application with water as carrier at a minimum of 20 gal/A; 1.2 to 1.8 lb a.i./A for post emergence in barley; 0.8 lb a.i./A for broadleaf control to 2.9 lb a.i./A for broadleaf and grass control in sorghum; and 4.0 lb a.i./A for short-term industrial weed control. May be tank mixed.

Formulations: 95-96 % active ingredient (a.i.) manufacturing-use product; 80% a.i. wettable powder and 80% a.i. dry flowable end-use products

Usual Carrier: Water (minimum 187 L/ha or 20 gal/A)

3. Science Findings

   Summary of Science Statement

   Terbutryn has been found to be oncogenic in rats and has been classified as a Group C oncogen (possible human carcinogen). Acute toxicity studies indicate that Terbutryn is relatively nontoxic (toxicity categories* III and IV). Dietary risk is insignificant since residues are 0.01 ppm or less and less than 1% of the U.S. wheat crop and less than 2% of the U.S. sorghum crop are treated. Worker exposure hazard from dermal exposure is the effect of greatest concern, but risks can be reduced substantially through the use of protective clothing and equipment and by packaging in water soluble bags.

   Terbutryn's potential for contamination of groundwater appears to be low; however, its major metabolite appears to be more persistent and mobile. Additional data are being required to evaluate the environmental fate of metabolites.

   The Agency has determined that the registered uses of this chemical will not generally cause unreasonable adverse efects to humans or the environment if used in accordance with the approved use directions and revised precautionary statements prescribed by the Registration Standard.

---

* Toxicity Categories are based on the acute toxicity of the chemical (LD_{50} or LC_{50} values) and are used to determine the appropriate signal word and precautionary language for product labeling. Toxicity Category III requires the signal word CAUTION and precautions against swallowing, inhaling, or contact with the skin and eyes, along with appropriate first aid instructions. Toxicity Category IV also requires the signal word CAUTION, but no precautionary statements are required. See 40 CFR 162.10.

Chemical Characteristics

Physical State:	crystalline solid or powder
Color:	white
Odor:	odorless to slightly aromatic
Melting point:	101-105° C
Density:	1.12-1.302 g/cc
Solubility	(at 20-25° C) 58 ppm in water 25 g/100 ml in isopropanol 10 g/100 ml in xylene 30 g/100 ml in ethylglycolmonoethylethe 30% in diethylalcohol
Vapor Pressure	9.6×10^{-7} mm Hg at 20° C
Dissociation constant	pka = 4.3 + or - 0.1 at 21° C
Stability	Stable to dilute aqueous alkaline and acidic solutions. Decomposed by ultraviolet irradiation. Hydrolyzed in strongly acidic or basic medium. Half life (at 25° C), >5 years in 0.1 N NaOH; 22+ or -3 days in 0.1 N HCl; and >6 years in water (ph=7). Stable at room temperature for at least 3 years when dry.

Toxicological Characteristics

With the exception of one study, the acute toxicology data base is complete. Terbutryn acute studies place it in the toxicity categories III and IV, or relatively non-toxic by oral, dermal, and inhalation routes, and it is not irritating to the eyes or skin.

Acute oral toxicity (rat)	1.9 g/kg (males) 2.1 g/kg (females) Toxicity Category III
Acute inhalation toxicity	Not available for technical, but tests on 80% a.i. formulation resulted in Toxicity Category III
Acute dermal toxicity (rabbits)	>20.0 g/kg Toxicity Category IV

Primary dermal No irritation at 72°
  irritation (rabbits) Toxicity Category IV
Dermal sensitization Not a sensitizer
  (guinea pigs)

Primary eye irritation Toxicity Category III

Chronic effects: The results of teratology and reproduction tests in animals indicate that the use of terbutryn is not expected to produce significant effects to humans in these areas. No mutagenic effects were observed in the available studies; however additional mutagenic testing is required. Terbutryn has been classified a group C oncogen, possible human carcinogen, based on available oncogenicity studies.

Chronic feeding/ Mouse: no evidence of oncogenicity
  oncogenicity Rat: two year feeding--at 3000 ppm, positive for oncogenicity (Group C, or possible human carcinogen).
  Beagle dog: (6 month study)
  NOEL: 10mg/kg/day
  Q_1^*: 10^{-2} *

Teratogenicity Not teratogenic to rabbits or rats.

Reproduction Three generation reproduction study
  (rats) showed decreased body weights and food consumption.

Mutagenicity Data limited but negative; additional data required.

Major Routes of Exposure

The primary potential for exposure is through the skin during mixing and loading or for flaggers who could be subject to direct dermal exposure. Therefore, protective clothing is required for mixers and loaders, and flaggers must be in enclosed vehicles.

4. Physiological and Biochemical Behavioral Characteristics

Foliar and Root Absorption: Absorbed through both foliage and roots with rapid foliar penetration.

---

* Q_1^*: The mathematical factor for the potency of a hazard, such as oncogenicity--the Q_1^* is a parameter of the linearized multistage extrapolation model. It is used as a multiplier of the estimated exposure (in units of mg/kg/day) to obtain the estimated 95% upper bound on risk. A change in the Q_1^* will result in a proportional change in risk.

Translocation: Translocated through xylem from roots and foliage, accumulating in the apical meristems.

Mechanism of Action: Terbutryn inhibits the photolysis of water in the photosynthetic process.

Persistence: Terbutryn degrades by oxidation into hydroxy-metabolites in plants and soils. Degradation is slow in plants and ranges from 3 to 10 weeks in soil.

5. Environmental Characteriatics

The available data indicate that terbutryn will not leach in agricultural soils. However, its major metabolite, hydroxy-terbutryn, appears to be more mobile and persistent and has the potential to leach to groundwater. The data deficiencies have been identified and additional data required to fill the data gaps.

Absorption and Leaching in Basic Soil Types: Terbutryn is readily adsorbed in soils with a high and organic matter or clay content. Adsorption is not irreversible and depends on factors such as pH, temperature, and moisture.

Microbial Breakdown: Microorganisms may play an important role in the degradtion of terbutryn.

Loss from Photodecomposition and Volitilization: Photodecomposition and volitilization are not significant factors in dissipation of terbutryn the soil.

6. Ecological Characteristics

Available acute toxicity data indicate that terbutryn is moderately toxic to warmwater fish and highly toxic to coldwater fish. However, except for the direct application to 6 inches of water, residues have been calculated to be insignificant, even to the most sensitive aquatic animal species. Since the current uses of terbutryn do not include direct application to water, its use is unlikely to result in significant acute adverse effects to aquatic animal species.

In terms of acute toxicity, data indicate that terbutryn is practically non-toxic to waterfowl, upland gamebirds, and honeybees.

The data base regarding toxicity data and non-target organism effects is incomplete for both aquatic and terrestrial organisms.

The following toxicity figures apply to technical terbutryn.

Hazards to Birds:

   Avian acute oral toxicity: Greater than 4,640 mg/kg in the mallard duck; greater than 2,000 mg/kg in the mallard duck and pheasant.

   Avian dietary toxicity: Greater than 4,640 ppm in the mallard duck; greater than 20,000 ppm, and greater than 2,000 ppm (two different studies) in bobwhite quail.

Hazards to Aquatic Organisms:

   Fish acute toxicity: 2.4 ppm in rainbow trout; 4.7 ppm in crucian carp; and 4.8 and 2.7 ppm in bluegill sunfish.

   Aquatic invertebrate toxicity: 2.66 ppm for Daphnia magna.

Hazard to Honeybees:

   Relatively nontoxic to honeybees; 2.9% mortality at 236 micrograms per bee.

7. Endangered Species Hazard Assessment

   There are sufficient toxicity and exposure data to indicate that the currently registered uses of terbutryn are unlikely to pose a hazard to endangered aquatic or avian species. Since it is a relative non-selective herbicide, EPA has an opinion from the Office of Endangered Species (OES) which indicates that certain endangered plant species are likely to be present in registered crop areas. Therefore, interim protective labeling for endangered species in treated crop areas is required. Potential impacts on endangered plant species in non-crop areas have not yet been evaluated. Protective labeling for non-crop areas may be required after the Agency has reviewed additional required data and consultations with OES are complete.

8. Worker Exposure and Risk Analysis

   A worker exposure evaluation and a dermal exposure risk assessment have been conducted by EPA. The risk assessment estimates were arrived at by using an estimate of the number of man hours associated with various application techniques, estimates of the hourly exposure for workers, and the calculated oncogenic potency (Q_1^*) of terbutryn. Additional studies have been required to further evaluate worker exposure.

   In order to reduce risks to workers EPA is requiring protective clothing and equipment, soluble bag packaging, and restricted use classification.

10. Tolerance Reassessment

Tolerances for terbutryn have been established for the raw agricultural commodities listed below:

Crop	Tolerance	Food Factor	mg/day (1.5 kg)
barley	0.1 ppm	0.03	0.000045
sorghum	0.1 ppm	0.03	0.000045
wheat	0.1 ppm	10.36	0.015540

Dietary Assessment

The Provisional Acceptable Daily Intake (PADI) was set using a six month dog feeding study with a no observed effect level (NOEL) of 10 mg/kg/day, based on mucosal thickening of various segments of the small intestine and submucosal lymphoid hyperplasia in the pyloric region of the stomach seen at 25 and 50 mg/kg/day.

Applying a safety factor of 1,000, since a NOEL was not determined in the chronic rat study, a Provisional Acceptable Daily Intake (PADI) of 0.0100 mg/kg/day can be calculated. This is equivalent to a Maximum Permissible Intake (MPI) of 0.6 mg/day for a 60 kg individual. The Theoretical Maximum Residue Contribution (TMRC) for terbutryn in the daily diet, based on the total tolerances and daily food intake of 1.5 kg, is 0.000260 mg/kg/day. Under these conditions 2.6% of the PADI has been utilized. Daily dietary exposure to terbutryn is thus substantially less than the calculated acceptable daily intake for humans.

Currently tolerances for residues are expressed as terbutryn per se, and the nature (identity) of metabolites is not adequately understood. Therefore, additional metabolism studies have been required of the registrant.

11. Summary of EPA Positions and Rationales

The Agency has determined that initiation of a Special Review is not warranted because the oncogenic risk can be reduced substantially by the incorporation of various protective measures (described below) on all product labels. Although terbutryn is a Group C oncogen (possible human carcinogen), dietary risk is insignificant, and risks to workers can be sufficiently reduced through protective measures.

In order to protect workers, terbutryn will be classified as a restricted use pesticide, and workers will be required to wear protective clothing; flaggers will be required to be in enclosed vehicles; and products will be packaged in water soluble bags.

Because additional data are needed to support existing tolerances as well as to establish tolerances for meat, milk, poultry, and eggs, feeding and grazing restrictions will be placed on all raw agricultural commodities.

Metabolism in plants and animals is not adequately understood but metabolites of terbutryn which contain an intact triazine ring are of toxicological concern. Therefore, data are required on plant and animal metabolism, and registrants are required to provide the appropriate validated mehtodology as well as storage stability and residue data for all metabolites with an intact triazine ring.

The major environmental metabolite of terbutryn, hydroxy-terbutryn, appears to have a high potential to reach groundwater. Additional field studies have been required.

EPA will issue registrations for substantially similar terbutryn products. However, new uses will be issued only on a case-by-case basis after considering the effects on the theoretical maximum residue contribution (TMRC), the maximum permissable intake (MPI), and the oncogenic risks.

12. Summary of Data Gaps

PRODUCT CHEMISTRY

Manufacturing Process	8 months
Discussion of impurities	8 months

ENVIRONMENTAL FATE

Photodegradation (water, soil)	9 months
Anaerobic Soil Metabolism	27 months
Field Dassipation Study (soil)	27 months
Rotational Crop Study (confined)	39 months
Fish Accumulation Study	6 months

FISH AND WILDLIFE

Avian Dietary (upland gamebird)	6 months
Freshwater Fish LC_{50}	6 months
Acute Estuarine and Marine Organism LC_{50} Studies	9 months
Fish Early Life-Stage and Aquatic Invertebrate Life Cycle Studies	12 months
Non-target Phytotoxicity	9 months

TOXICOLOGY

Acute inhalation (Rat)	8 months
21-Day Dermal	8 months
Chronic Toxicity (Rodent)	36 months
Mutagenicity Battery	12 months
General Metabolism	12 months

RESIDUE CHEMISTRY

Metabolism Studies (ruminants, poultry)	18 months
Uptake, Distribution, and Metabolism	18 months
Storage Stability Data	15 months
Processed Commodity Data	24 months
Residues of Concern on Grain and Milled Products	24 months

13. Contact Person at EPA

Robert J. Taylor
U.S. Environmental Protection Agency
Registration Division (TS-767 C)
401 M Street, S.W.
Washington, D.C. 20460
(703) 557-1830

DISCLAIMER: The information presented in this Pesticide Fact sheet is for informational purposes only and may not be used

THIODICARB

Date Issued: February 27, 1984
Fact Sheet Number: 18

1. Description of Chemical

    Generic Name: dimethyl N, N'-thiobis(methylimino) carbonyloxy bis ethanimidothioate

    Common Name: thiodicarb

    Trade Name: Larvin®

    EPA Shaughnessy Code: 114501

    Chemical Abstracts Service (CAS) Number: 900

    Year of Initial Registration: 1984

    Pesticide Type: insecticide

    Chemical Family: carbamate

    U.S. and Foreign Producers: Union Carbide Corporation

2. USE PATTERNS AND FORMULATIONS

    Application Sites and Rates:

    Thiodicarb is currently registered for use on sweet corn (fresh market only in the state of Florida). The application rates range from 0.5 lbs. active ingredient per acre to 0.75 lbs. active ingredient per acre per acre, not to exceed 7.5 lbs. active ingredient per acre per use season.

Types of Formulations:

Thiodicarb is commercially formulated into five (5) flowable products and one (1) wettable powder. There is also a 95% technical product federally registered.

Types and Methods of Application:

Application to sweet corn is made with both ground and air equipment.

3. Science Findings

Summary Science Statement

Thiodicarb is a cholinesterase inhibiting pesticide. Studies on the formulated products demonstrate a moderate toxicity to man (Toxicity Category II). The metabolism of thiodicarb is adequately understood. One of the metabolic by-products of thiodicarb in animals is acetamide, a potential carcinogen. Thiodicarb is not expected to leach and reach ground water or to bioaccumulate in the environment.

Chemical Characteristics:

Thiodicarb is a white crystalline powder with a slight sulfurous odor. It has a melting point of 173-174°C. Thiodicarb is stable in light and ambient conditions and unstable in alkaline conditions. Its main degradation product is methomyl.

Toxicology Characteristics:

Thiodicarb technical is moderately toxic (Toxicity Category II) via the oral and inhalation routes of exposure with LD_{50} values of 325 milligrams (mg)/kilogram (kg) and >0.32 mg/liter (L), respectively. The acute dermal LD_{50} for thiodicarb in rabbits is >2000 mg/kg (Toxicity Category III). Corneal opacity and conjunctival redness, chemosis and discharge were observed in the eyes of rabbits administered 44 mg of thiodicarb; however, all lesions cleared by day 7.

The toxicological data submitted in support of the established tolerance for residues in or on sweet corn includes a 2-year rat feeding/oncogenicity study which was negative for oncogenic effects at the levels tested (1.0, 3.0 and 10.0 mg/kg/day) and had a cholinesterase (ChE) and chronic toxicity no-observed-effect level (NOEL) of 10.0 and 3.0 mg/kg/day, respectively; a mouse oncogenicity study which was negative at the levels tested (1.0, 3.0 and 10.0 mg/kg/day); a 6-month dog feeding study with a ChE and subchronic NOEL of 15.0 mg/kg/day; a rat teratology study which was negative at 30.0 mg/kg/day and had a fetotoxic NOEL of 3.0 mg/kg/day; a mouse teratology study which was negative at 200 mg/kg/day and also had a NOEL of 200 mg/kg/day for fetotoxicity; a 3-generation rat reproduction study with a NOEL of 10.0 mg/kg/day (HDT); and an acute delayed neurotoxicity study which was negative at 660 mg/kg. Studies on mutagenicity showed negative potential. Based on the 2-year rat feeding study with a chronic toxicity NOEL of 3.0 mg/kg/day and using a safety factor of 100, the acceptable daily intake (ADI) for humans is 0.03 mg/kg of body weight (bw)/day. The theoretical maximum residue contribution (TMRC) from the established tolerance on sweet corn utilizes 2.38 percent of the ADI.

The oncogenic potential of acetamide has been demonstrated

in four different studies, the first being a study conducted by F. I. Dessau and B. Jackson in 1955, where two groups of Rockland albino rats were treated with a 40% solution of acetamide at a rate of 4000 mg/kg (equivalent to 40,000 ppm for younger rats or 80,000 ppm for older rats) by intubation 5 days/week for a period of 117 days for Group I and 205 days for Group II. Histopathological examination showed cytologic irregularities consisting of a greater variability of cellular and nuclear size, giant nuclei, and the presence of numerous mitoses, some of unusual appearance. Benign hepatocellular adenomas were also found in 2 treated animals in Group II.

Doctors Dessau and Jackson conducted a second study in 1961, with 3 groups of Wister albino rats. The test duration was 12 months. Group I animals were administered via diet a 5% (50,00 ppm) solution of acetamide continuously. Group II animals were divided into three subgroups receiving 5%, 2.5% (25,000 ppm) and 1.25% (15,000 ppm) acetamide. Test material was administered in a diet of ground Wayne Laboratory Blox. Each week, two rats from treatment Group III were taken off the acetamide diet and placed on a control diet for the remainder of the testing period. Hepatomas were noted in four of forty eight animals in Group I, one of eighteen animals, six of twenty-two animals and four of twenty four animals tested in Group II, subgroups 1-3 respectively, and twenty-two of eighty one animals tested in Group III. In a study conducted by J. H. Weisburger, R. S. Yamamoto, R. M. Glass and H. H. Frankel in 1969, 2 groups of male Wister rats were administered 2.5% (25,000 ppm) acetamide in a diet of Wayne Laboratory Blox. Group I animals were sacrificed after twelve months. Test animals in Group II were removed from the acetamide diet after twelve months and continued on a controlled diet for an additional three months. Animals in both test groups were administered 75 milligrams/liter of oxytetracycline (Terramycin) for one week every sixth week of the study. Hepatomas were noted in 2 of eight animals tested in Group I and 7 of sixteen animals tested in Group II. No effect was noted in the fifteen control animals.

The fourth study, a carcinogenesis bioassay of acetamide in rats and mice was conducted by R. W. Fleischman, et. al. in 1980. This study included 8 compound-dosage groups per sex for rats and 10 such groups for mice. Rats received 2.36% (23,600 ppm) of acetamide via diet. The mice were divided into two groups with Group I receiving 1.18% (11,800 ppm) of acetamide and Group II receiving 2.36% of acetamide. Test material was administered to animals in a diet of ground Wayne Lab Blox for a 12 month period and was then replaced with a controlled diet of Wayne Blox pellets for an additional 4 months. There were no apparent compound related effects noted in male and female mice. However, 41 liver carcinomas and 1 neoplastic nodule were noted in male rat test animals and 33 liver carcinomas and 3 neoplastic nodules were noted in female rat test animals.

The Agency has evaluated the four acetamide studies and have found the studies inappropriate for addressing the tumorigenicity potential of acetamide in accordance with today's standards for oncogenicity testing. Only a small number of male rats were used in 3 of the 4 studies in either the test groups or the controls or both. A single dietary level was administered to rats in 3 of the 4 studies which does not allow the determination of a dose related effect. In all studies, the exposure rates were extremely high which may have been responsible for the excessive weight loss and mortality noted in several of the studies. The administration of oxytetracycline (Terramycin to test animals in the study conducted by Weisburger et. al. (1969) raises questions on the quality of the animals used and the possibility of adversely influencing the results of the experiment. A time related dose response which may or may not be real from a biological point of view was noted in the Dessau and Jackson study of 1961. Also the results of this study which indicated that there were no tumor effects in similar rats receiving a diet of acetamide in Purina Laboratory Chow versus effects in test animals receiving a diet of acetamide in ground Wayne Laboratory Blox casts doubts on the certainty of acetamide's oncogenic potential, as well as its potential hazard to humans. In the study conducted by R. W. Fleischman, et. al. in 1980, test animals (mice) used came from different lots and suppliers. Data describing weight gain, survival and intercurrent disease were not provided. Also, in this study the number of tissues examined varied between study groups.

Based on the conduct of the available studies on acetamide and in consideration of the available oncogenicity testing in the rat and mouse for thiodicarb which demonstrated a negative oncogenic potential, the Agency has not determined that thiodicarb is oncogenic under normal agricultural practices. However, the Agency has conducted a risk assessment of the proposed tolerance request based on the four acetamide studies. The estimated maximum daily human exposure to acetamide from conversion of consumed thiodicarb residues is 1.4×10^{-4} mg/day for a 60 kg person and with an exposure risk of 3.07×10^{-3} the resulting life-time carcinogenic risk estimate is 7×10^{-9}. This life-time carcinogenic risk assessment is based on the following assumptions:

- Acetamide is presumed definitely to be carcinogenic.
- Carcinogenic effects noted in experimental animals at acetamide dietary levels of 10,000-80,000 ppm are applicable to humans exposed at a maximum level of 9.3 ppb.
- The mathematical relationship between dose and response that holds in the low dose region is based on the application of the one-hit model of carcinogenesis which yields the highest risk of any of the plausible models of dose response relations.
- The metabolic pathway of thiodicarb in humans is presumed to be the same as that found in test

animals and the highest value of risk obtainable from the animal data is applicable to humans.
- The conversion ratio of thiodicarb to acetamide in test animals is 306, based on metabolism studies, and this is the same in humans.
- Total production of sweet corn components in the United States will contain thiodicarb residues at the tolerance level.

Physiological and Biochemical Behavior Characteristics:
The metabolic pathway of thiodicarb in livestock has been demonstrated to be thiolysis to methomyl, followed by hydrolysis to the methomyl oxime and subsequent metabolization to acetonitrile. Acetonitrile is then metabolized to acetamide, a potential carcinogen, and further hydrolyzed to acetic acid which enters the intermediary metabolism cycle of the animal and is ultimately expired as carbon dioxide.

Plant metabolism studies show that thiodicarb is likewise metabolized to the methomyl oxime followed by acetonitrile and carbon dioxide, both of which are then volatilized.

Environmental Characteristics:
Thiodicarb is very stable at pH 6 and unstable in alkaline conditions. It is subject to decomposition by eight. The major by-product of photolysis is methomyl. Light textured soils causes more rapid degradation than heavy textured soils.

Thiodicarb exhibits low mobility in all soils. Degradation is also influenced by increasing temperatures, degree of aeration and microbial activity. The half-life on soil and plant surfaces is less than one week. Thiodicarb is non-persistent in the environment.

Ecological Characteristics:
Thiodicarb is moderately toxic to fish with a LC_{50} value of 2.55 ppm for the rainbow trout and 1.21 ppm for the bluegill sunfish. The avian acute LD_{50} for the bobwhite quail is 2023 ppm. The subacute dietary LC_{50} for the bobwhite quail and the mallard duck is 5620 ppm. The 48-hour acute LC_{50} for aquatic organisms is 0.0053 ppm.

Tolerance Assessment:
A tolerance of 2.0 parts per million (ppm) has been established to cover residues of thiodicarb and its metabolite methomyl in or on sweet corn grain (kernels plus cob with husk removed (K+WHR)) under the provisions of the Federal Food, Drug and Cosmetic Act (FFDCA). The 2.0 pm tolerance level is adequate to cover anticipated residues in or on sweet corn as a result of application under the currently registered use pattern.

4. Summary of Regulatory Position and Rationale:
Geographical Restrictions:
Thiodicarb is currently registered for use on sweet corn only. Products containing thiodicarb are also limited to application to sweet corn only in the State of Florida. Grazing and feeding

of treated corn fodder and forage is prohibited.

Summary of Risk Assessment

On the basis of the available studies on acetamide and the chronic oncogenicity studies for thiodicarb, the Agency has concluded that the human risks posed by the use of thiodicarb on sweet corn does not raise prudent concerns of unreasonable adverse effects.

5. Summary of Major Data Gaps

All data requirements have been addressed for thiodicarb. Therefore, all products containing thiodicarb have been uncondi tionally registered.

6. Contact Person at EPA

Jay S. Ellenberger
Product Manager (12)
Insecticide-Rodenticide Branch
Registration Division (TS-767C)
Office of Pesticide Programs,
Environmental Protection Agency,
401 M St., SW.,
Washington, D.C. 20460.
Office location and telephone number:
RM. 202, CM #2,
1921 Jefferson Davis Highway,
Arlington, VA 22202,
(703-557-2386).

THIOPHANATE ETHYL

Date Issued: February 1986
Fact Sheet Number: 84

1. Description of chemical

    Generic name: Diethyl 4,4'-o-phenylenebis[3-thioallophanate]
    Common name: Thiophanate, thiophanate ethyl
    Trade name: Topsin, Cleary's 3336, Cercobin
    EPA Shaughnessy code: 103401
    Chemical Abstracts Service (CAS) number: 23564-06-9
    Year of initial registration: 1973
    Pesticide type: Fungicide
    Chemical family: Thiophanate
    U.S. and foreign producers: Nippon Soda Co., Ltd., Japan
    Pennwalt Corporation

2. Use patterns and formulations

    Application sites: turf (golf courses), roses, flowers, ornamentals, and shade trees.

    Formulations: Wettable powder, flowable concentrate.

    Types and methods of application: Applied as a spray by means of hand-held equipment such as compressed air sprayers, hose end sprayers, spray guns or possibly sprinkling cans or by boom sprayers mounted from a tractor or trailer pulled by a tractor.

    Application rates: 1.36 to 10.9 lb/ai/A for turf uses and 0.25 to 0.75 lb ai for per 100 gallons for ornamental uses.

    Usual carriers: water

3. Science findings

    Summary science statement: Although no human toxicological hazards of concern have been identified in studies reviewed by the Agency for the standard, extensive environmental data gaps exist for thiophanate ethyl. The Agency has no information that indicates continued use will result in any unreasonable adverse effects to man or the environment during the time required to develop the data.

    Chemical characteristics: Technical thiophanate ethyl is a white to pale brown crystalline powder with a faint sulfur odor. It has a melting point of 191.7° C and a specific gravity of 1.44. It is stable in acid and slightly decomposes in base.

Toxicology Characteristics: Acute toxicity studies indicate that thiophanate ethyl has moderate to low acute toxicity to humans.

Study	Results	Toxicity Category
Acute inhalation	$LC_{50} = 6.7$ mg/L	III
Acute dermal	$LD_{50} > 15,000$ mg/kg	III
Primary skin irritation	Non-irritating	III
Acute oral	$LD_{50} > 15,000$ mg/kg	IV
Primary eye irritation	Non-irritating	IV

Studies conducted in rats and mice indicate that thiophanate ethyl is not oncogenic in laboratory animals. No decision can be made concerning the teratogenic and mutagenic potential of thiophanate ethyl until such studies are submitted to the Agency.

Biochemical Behavioral Characteristics: Thiophanate ethyl slowly converts to ethyl-2-benzimidazole carbamate (EBC) which is considered to be the fungicidally active agent. The mode of action of EBC involves its interference in the biosynthesis of DNA in the fungal cell division process.

Environmental Characteristics: The environmental fate and transport of and the potential exposure to thiophanate ethyl cannot be characterized until the required environmental fate data are submitted.

Ecological Characteristics: Thiophanate ethyl has extremely low toxicity to birds with an avian dietary LC_{50} greater than 5620 ppm and avian oral LD_{50} greater than 2510 mg/kg. Thiophanate ethyl has moderate toxicity to fish and aquatic invertebrates with a fish LC_{50} of 2.26 to 2.6 ppm and an aquatic invertebrate LC_{50} of 2.6 ppm.

Tolerance Reassessment: To date there have been no U.S. tolerances or registrations on food/feed items for thiophanate ethyl.

4. Summary of Regulatory Position and Rationale

Based on the review and evaluation of all available data and other relevant information on thiophanate ethyl, the Agency has made the following determinations:

The available data are insufficient to indicate that any of the risk criteria set forth in 40 CFR §162.11(a) have been met or exceeded for the uses of thiophanate ethyl at the present time. For example, (1) Thiophanate ethyl is not oncogenic in rats or mice, (2) Potential exposure to the active metabolite EBC from uses currently covered by the standard does not raise concern at this time, and (3) thiophanate ethyl has moderate to low acute toxicity to humans. The available data, however, are incomplete and no decision concerning the teratogenicity, mutagenicity, or environmental fate of thiophanate ethyl can be made until the required data are submitted. Thiophanate ethyl has extremely low toxicity to birds and moderate toxicity to fish and aquatic invertebrates. Additional acute toxicity testing with channel catfish is being required because the chemically related compounds thiophanate methyl and benomyl are highly toxic to the ictalurid (catfish) family. A detailed ecological hazard assessment cannot be made until certain environmental fate data requirements have been met.

End-use product (EP) labels will be required to bear a revised environmental hazard statement.

EP labels will be required to maintain a grazing/feeding restriction.

Manufacturing-use product (MP) labels will be required to bear a statement regarding discharge to bodies of water and sewer systems.

Specific label warning statements:

Hazard Information

> The human hazard statements must appear on all labels as prescribed in 40 CFR 162.10(h).

Environmental Hazard Statement

> All MPs intended for formulation into EPs must bear the following statements:

> "Do not discharge effluent containing this product directly into lakes, streams, ponds, estuaries, oceans or public waters unless this product is specifically identified and addressed in a National Pollutant Discharge Elimination System (NPDES) permit. Do not discharge effluent containing this product into sewer systems without previously notifying the sewage treatment plant authority. For guidance, contact your State Water Board or Regional Office of the Environmental Protection Agency."

End-Use Product Statements

> The following environmental hazard statement must appear on all EP products:

> "Do not apply directly to water or wetlands (swamps, bogs, marshes, and potholes). Do not contaminate water by cleaning of equipment or disposal of wastes."

The following feeding/grazing restriction statement must appear on all EP products:

"Do not graze treated areas or feed clippings to livestock."

5. Summary of Major Data Gaps and Dates for Submission

Product Chemistry
Product Identity (December 1986)
Analysis and Certification of Product Ingredients (June 1986)
Physical and Chemical Characteristics (June 1986)

Environmental Fate
Hydrolysis (June 1986)
Photodegradation (June 1986)
Metabolism (December 1987)
Mobility Studies (June 1986)
Soil Dissipation Studies (December 1987)
Fish Accumulation Studies (June 1986)

Toxicology
21-Day Dermal (June 1986)
Teratogenicity (September 1986)
Mutagenicity Testing (June 1986)

Wildlife and Aquatic Organisms
Freshwater Fish and Freshwater Invertebrates Acute Toxicity (September 1986)

6. Contact Person at EPA

Henry M. Jacoby
EPA (TS-767C)
401 M. St., S.W.
Washington, D.C. 20460

DISCLAIMER: The information presented in this Pesticide Fact Sheet is for informational purposes only and may not be used to fulfill data requirements for pesticide registration and reregistration.

THIOPHANATE-METHYL

Date Issued: October 1986
Fact Sheet Number: 92

1. DESCRIPTION OF CHEMICAL

   Common Name: Thiophanate-methyl

   Trade Names: NF44, AC 87,844, Topsin M, Cercobin M, Mildothane, Cycosin, Fungo, Fungo 70, Spot-kleen, TD-1771, Thiophanate-methyl

   Empirical Formula: $C_{12}H_{14}N_4O_4S_2$

   Chemical Abstracts
      Service (CAS) Number: 23564-05-8

   Pesticide Chemical Code (Shaughnessy): 102001

   Pesticide Type: Systemic fungicide

   Chemical Family: Carbamates

   U.S. and Foreign
      Producers: Agchem Division of the Pennwalt Corporation

2. USE PATTERNS AND FORMULATIONS

   Application sites: Thiophanate-methyl is registered for control of plant diseases on almonds, apples, apricots, bananas, beans, celery, cherries, cucumbers, melons, nectarines, onions, ornamentals, pecans, peaches, peanuts, plums, potatoes, pumpkins, soybeans, squash, strawberries, sugar beets, sugarcane, turf, and wheat.

   Type of formulations: 94.3% technical grade, dusts (2.5, 5%), granular (1.75, 2.3, 5.5%), wettable powder (1.75, 15, 24.4, 25, 50, 70%), a 19.65% liquid, and a 46.2% flowable.

   Types and methods of applications: End use products are applied aerially, banded, broadcasted, and applied over-the-top with ground equipment. Thiophanate-methyl is also used as a post harvest dip for fruits, as a seed piece treatment, as a soil drench for ornamentals, and as a soil furrow spray.

Application rates: 0.26 to 2.8 lb ai/A on crop sites, 0.033 to 0.36 lb ai/1000 sq ft on turf, 0.35 to 0.70 lb ai/100 gal for post-harvest treatments, and 0.175 to 1.16 lb ai/100 gal for seed piece treatments.

Usual carriers: Water, spray oil, or wax.

3. SCIENCE FINDINGS:

Summary science statements: Thiophanate-methyl is readily absorbed by animals and partly metabolized to methyl-2-benzimidazole carbamate (MBC). Eighty to ninety percent of thiophanate-methyl orally administered to laboratory animals is recovered in the excreta within 24 hours as parent compound and metabolites (including MBC).

Thiophanate-methyl did not show any positive effects in oncogenic studies in the rat and mouse. However, the fungicidal activity of thiophanate-methyl depends on its conversion to MBC in the plants. MBC has been shown to cause tumors solely in the mouse liver and has been tentatively classified as a Group C oncogen (possible human oncogen).

No compound-related maternal or fetal effects were observed in a rat teratology study. A mouse teratology study of thiophanate-methyl is incomplete. Therefore, another teratology study must be conducted. In a reproduction study in rats the only signs of toxicity observed were decreased pup and litter weights at birth and during lactation for pups and dams administered the high dose (640 ppm or 32 mg/kg/day). In a study with male mice given 5 consecutive oral daily doses of 192 mg/kg body weight, there was no indication that thiophanate-methyl affected the function of the testes or accesssory sex organs.

Acute toxicity studies of thiophanate-methyl indicate Toxicity Category III based on dermal toxicity. Thiophanate-methyl is virtually non-toxic to avian species and of low to moderate toxicity to freshwater fish with the exception of catfish. Studies conducted with thiophanate-methyl show that it is highly toxic to catfish.

Available data are insufficient to fully assess the residues of thiophanate-methyl and MBC in raw agricultural commodities. Available data are insufficient to fully assess the environmental fate of thiophanate-methyl.

Chemical characteristics: Thiophanate-methyl is a light-tan crystalline powder that has a sulfidic odor at room temperature. Its molecular weight is 342.4. It is stable at neutral pH at 25° C in aqueous solution up to 25 days.

Toxicological characteristics:

Acute toxicology effects of thiophanate-methyl are as follows:

Acute Oral Toxicity: (rat)	male: 7,500 mg/kg body weight female: 6640 mg/kg body weight Toxicity category IV
Acute Dermal Toxicity: (rat)	>10,000 mg/kg body weight Toxicity category III
Dermal Sensitization: (guinea pig)	non-sensitizing

Subchronic toxicology effects of thiophanate-methyl are as follows:

Subchronic Oral (rats)   NOEL = 64 ppm (3 mg/kg/day)
                         LEL = 320 ppm (16 mg/kg/day)
                             (decreased body weight gain)

At the highest dose tested (8000 ppm) thyroid histopathology included decreased coloidal content and epithelial hypertropphy in the follicles.

Chronic toxicology effects of thiophanate-methyl are as follows:

Thyroid histopathology similar to that observed in the subchronic rat feeding study was found in the long-term studies in male rats (32 mg/kg/day) and female dogs (250 mg/kg/day). Decreased body weight gain in female rats and testicular histopathology in male rats was noted at the 640 ppm (32 mg/kg/day) diet. NOEL's with respect to these effects were 8 mg/kg/day in rats. A NOEL of 50 mg/kg/day in dogs was based on decreased weight gain in males and changes in thyroid histology in females.

Oncogenic effects of thiophanate-methyl are as follows:

There were no compound related effects or tumor incidence in mice after two years of feeding diets containing 0, 40, 160, or 640 ppm (0, 1.5, 6, 24, or 96 mg/kg/day) thiophanate-methyl (a MTD was not reached in this mouse study). No oncogenic potential was observed in the rat study at the dietary levels up to 32 mg/kg/day, which did approach a MTD.

Teratogenic effects of thiophanate-methyl are as follows:

No compound-related maternal or fetal effects were observed in a limited teratology study in rats. In a second study, pregnant rats received thiophanate-methyl in their diets during gestation, and no fetal effects were observed. This study was limited because the palatability of diets containing the

highest dose was decreased by the fungicide. The highest
dose tested (1000 mg/kg/day) in a mouse study did not cause
toxicity in the dams or their fetuses. However, the report
was incomplete and, therefore, a teratology study in a second
species is needed.

Reproductive effects of thiophanate-methyl are as follows:
In a three-generation reproduction study in rats, there was
no effect on fertility, viability, gestation, or lactation
indices. There was a decrease in pup and litter weights
at the 640 ppm diets. The NOEL for this effect was 8 mg/kg/day.

Mutagenic effects of thiophanate-methyl are as follows:

In mutagenicity tests, thiophanate-methyl did not induce
reverse mutations in bacteria or affect fertility.

Major routes of human exposure:

Dermal, ocular, and inhalation exposures to workers
may occur during application. Exposure from mixing,
handling, and application of the dusts, granulars and
wettable powders is expected to be mainly dermal.

Physiological and biochemical behavioral characteristics:

Translocation: Thiophanate-methyl is mobile in the xylem
and the phloem.
Mechanism of pesticidal action: Thiophanate-methyl inhibits
the growth of plant pathogens.

Environmental characteristics: Available data are insufficient
to fully assess the environmental fate of thiophanate-methyl
and the potential exposure of humans and nontarget organisms.
An aerobic soil metabolism study is the only environmental fate
study available for thiophanate-methyl. In order to assess the
environmental fate and transport of, and the potential exposure
to thiophanate-methyl, the Agency is requiring a full range of
environmental fate studies.

Ecological characteristics: Available data indicate that
thiophanate-methyl has very low toxicity to birds, is slightly
toxic to sunfish and trout, and very highly toxic to catfish.
MBC has been shown to be highly toxic to fish. Thiophanate-methyl
has been shown to be moderately toxic to aquatic invertebrates.

Aquatic invertebrate toxicity:
Daphnia magna   27 ppm

96 hour fish toxicity:
Bluegill sunfish   15.8 - 58 ppm
Rainbow trout   8.3 - 25.2 ppm
channel catfish   0.030 ppm

Avian dietary toxicity:
  Mallard duck   >10,000 ppm
  Bobwhite quail   >10,000 ppm

Potential problem for endangered species:
  A review by the Agency indicates that two endangered catfish species inhabit areas where soybeans are grown and which may be treated with thiophanate-methyl.

Tolerance Reassessment:

Listed below are tolerances which have been established for residues of thiophanate-methyl on a wide range of raw agricultural products listed in 40 CFR 180.371.

The most appropriate NOEL's for this tolerance reassessment are those from the chronic and reproduction studies in rats (160 ppm or 8 mg/kg/day). There were no effects in any other study that evaluated doses at or below that NOEL. Using a safety factor of 100 and the 8 mg/kg/day NOEL, the ADI for humans was calculated as 0.08 mg/kg/day. The maximum permissible intake (MPI) for a 60 kg adult is 4.8 mg/1.5 kg diet.

To date the tolerances granted have accounted for approximately 22.85% of the acceptable daily intake with a theoretical maximum residue contribution (TMRC) to the daily diet of 1.0969 mg/day (for an average 1.5 kg daily diet).

Commodity	Tolerance (ppm)
Almonds pre-H	0.2
Almonds (hulls) pre-H	1.0
Apples (pre- and post-H)	7.0
Apricots (pre- and post-H)	15.0
Bananas (pre-H)	2.0
Bananas, pulp (pre-H)	0.2
Beans (snap and dry) pre-H	2.0
Bean (forage and hay) pre-H	50.0
Cattle, fat	0.1
Cattle, kidney	0.2
Cattle, liver	2.5
Cattle, meat byproducts (exc. kidney and liver)	0.1
Cattle, meat	0.1
Celery (pre-H)	3.0
Cherries (pre- and post-H)	15.0
Cucumbers	1.0
Eggs	0.1
Goats, fat	0.1
Goats, kidney	0.2
Goats, liver	2.5
Goat, meat byproducts (exc. kidney and liver)	0.1

Commodity	Tolerance (ppm)
Goat, meat	0.1
Hogs, fat	0.1
Hogs, liver	1.0
Hogs, meat byproducts (exc. liver)	0.1
Hogs, meat	0.1
Horses, fat	0.1
Horses, liver	1.0
Horses, meat byproducts (exc. liver)	0.1
Horses, meat	0.1
Melons	1.0
Milk	1.0
Nectarines (pre- and post-H)	15.0
Onion, dry	3.0
Onion, green	3.0
Pecans (pre-H)	0.2
Peaches (pre- and post-H)	15.0
Peanuts pre-H	0.2
Peanuts (hulls) pre-H	2.0
Peanuts (forage and hay) pre-H	15.0
Plums (pre- and post-H)	15.0
Potatoes (seed treatment)	0.05
Poultry, fat	0.1
Poultry, liver	0.2
Poultry meat byproducts (exc. liver)	0.1
Poultry, meat	0.1
Prunes (pre- and post-H)	15.0
Pumpkins	1.0
Sheep, fat	0.1
Sheep, kidney	0.2
Sheep, liver	2.5
Sheep, meat byproducts (exc. kidney and liver)	0.1
Sheep, meat	0.1
Soybeans (pre-H)	0.2
Squash	1.0
Strawberries (pre-H)	5.0
Sugar beets (roots pre-H)	0.2
Sugar beets (tops pre-H)	15.0
Sugarcane (seed piece treatment pre-H)	0.1
Wheat, grain	0.05
Wheat, hay	0.1
Wheat, straw	0.1

Problems known to have occurred with use:

The Pesticide Incident Monitoring System (PIMS) did not identify any incidents involving agricultural/domestic uses of thiophanate-methyl from 1966 to present.

4. SUMMARY OF REGULATORY POSITION AND RATIONALE:

Based on the review and evaluation of all available data and other relevant information on thiophanate-methyl, the Agency has made the following determinations:

a. The Agency will not place thiophanate-methyl and its metabolite MBC into Special Review [Section 162.11(a) of CFR 40].

Thiophanate-methyl was previously placed in Special Review by the Agency in December, 1977, because of its potential adverse effects on nontarget organisms and because its metabolite, MBC, had the potential to cause mutagenic effects. It was removed from the special review process in October, 1982 because it was determined that the risks were not unreasonable and were exceeded by the benefits associated with the use of thiophanate-methyl products.

MBC has been classified as a Group C oncogen (possible human oncogen). The current risk analysis performed for thiophanate-methyl, based on its metabolite MBC, is of the same order of magnitude as those calculated in the PD 4 for benomyl (i.e. 10^{-4} to 10^{-5}).

Although MBC is associated with liver tumors in Swiss and Swiss derived mice (CD-1), MBC is not associated with liver tumors in NMRKf strain of mouse , which has a low background incidence of liver tumors. MBC was not oncogenic in the rat and only weakly mutagenic as a possible result of inhibition of the cellular spindle apparatus rather than direct gene mutation or alteration of DNA repair.

The Agency has concluded that the risks to humans posed by thiophanate-methyl are minimal and of the same magnitude as estimated in the 1982 decision. The Agency has also reviewed the benefits of thiophanate-methyl and has concluded that the benefits have not changed significantly since the 1982 decision. The benefits of thiophanate-methyl outweigh its risks with the protective measures on the label. Hence inititation of an additional special review is not necessary at this time.

b. The Agency does not intend to establish new food additive regulations pursuant to Section 409 of the Federal, Food, Drug, and Cosmetic Act (FFDCA). The Agency is deferring action on the presently established food additive regulations until receipt and evaluation of comments in response to a Federal Register notice discussing this issue. This notice

c. The Agency is not requiring a reentry interval for currently registered uses of thiophanate-methyl. The acute toxicity for thiophanate-methyl is low (Category IV) except for dermal irritation (Category III). Additionally, exposure and the resultant risks to field workers are not expected to be significant because thiophanate-methyl is expected to be poorly absorbed dermally. The SAP is being asked if a dermal absorption study should be required.

d. The Agency is imposing labeling restrictions on rotational crops. The extent of the restrictions will be reconsidered when additional data are submitted and reviewed. The restrictions will serve to protect the public from impermissable residues in food and feed. In addition this restriction will protect subsequent planted crops from possible effects due to persistent residues of thiophanate-methyl in the soil.

e. The Agency has determined that the available data are insufficient to fully assess the environmental fate of thiophanate-methyl. Preliminary laboratory data show that thiophanate-methyl and its degradate MBC are moderately mobile in Lakeland sand and Sultan silt loam columns.

f. The Agency determined that two endangered catfish species inhabit areas where soybeans may be grown. To protect these species, endangered species labeling is required.

Use Site: The Agency has not proposed changes in thiophanate-methyl's uses.

Specific Label Warning Statements:

The following required label statements must appear on the labels of all products in channels of trade within two years [April 30, 1988] of issuance of this Standard. After review of data to be submitted under this Standard, the Agency may impose additional label requirements or take further regulatory actions as necessary.

1. Manufacturing-Use Product Statements

All products intended for formulation into EPs must bear the following environmental hazard statement:

"Do not discharge effluent containing this product directly into lakes, streams, ponds, estuaries, oceans or public waters unless this product is specifically identified and addressed in a National Pollutant Discharge Elimination System (NPDES) permit. Do not discharge effluent containing this product into sewer systems without previously notifying the sewage treatment plant authority. For guidance, contact your State Water Board or Regional Office of the Environmental Protection Agency."

2. End-Use Product Statements

The following environmental hazard statement must appear on all EP's:

"This pesticide is toxic to catfish. Do not apply directly to water or wetlands (swamps, bogs, marshes, and pot holes). Drift and runoff from treated areas may be hazardous to catfish in adjacent areas. Do not contaminate water by cleaning of equipment or disposal of wastes."

--for granular products add "cover or incorporate spills."

Restrictions on Rotational Crops

"Do not plant food or feed crops in thiophanate-methyl treated fields for 18 months after the last application unless thiophanate-methyl is registered for use on those crops."

Endangered Species Information for use on Soybeans
The following must appear on all EP's registered for use on Soybeans:

"ENDANGERED SPECIES RESTRICTIONS

The use of any pesticide in a manner that may kill or otherwise harm an endangered or threatened species or adversely modify their habitat is a violation of federal laws. The use of this product is controlled to prevent death or harm to endangered or threatened species that occur in the following counties or elsewhere in their range.

STATE Species	COUNTY
OHIO Scioto madtom	CHAMPAGNE FRANKLIN LOGAN MADISON PICKAWAY UNION
TENNESSEE Yellowfin madtom	CLAIBORNE HANCOCK
VIRGINIA Yellowfin madtom	LEE RUSSELL SCOTT

Before using this pesticide in these counties you must obtain the EPA Cropland Endangered Species Bulletin (EPA/ES-CROP). The use of this pesticide is prohibited in these counties unless specified otherwise in the Bulletin. The EPA Bulletin is available from either your County Agricultural Extension Agent, the Endangered Species Specialist in your State

Wildlife Agency Headquarters, or the appropriate Regional Office of either the U.S. Fish and Wildlife Service (FWS) or the U.S. Environmental Protection Agency (EPA). THIS BULLETIN MUST BE REVIEWED PRIOR TO PESTICIDE USE."

5. SUMMARY OF MAJOR DATA GAPS:

The toxicological studies are required on the following dates:

Acute Inhalation Toxicity - rat	January 31, 1987
Eye Irritation - rabbit	January 31, 1987
Dermal Irritation - rabbit	January 31, 1987
Teratogenicity	April 30, 1987
Oncogenicity - mouse	June 30, 1990
Structural Chromosomal Aberration	April 30, 1987
Mechanisms of Mutagenicity	April 30, 1987
Dermal Absorption Study	April 30, 1987

The environmental fate data are required on the following dates:

Hydrolysis	January 31, 1987
Photodegradation in water and on soil	January 31, 1987
Anaerobic Soil	July 31, 1988
Leaching and Adsorption/Desorption	April 30, 1987
Volatility (lab)	April 30, 1987
(field)	July 31, 1987
Soil dissipation	July 31, 1989
Accumulation - rotational crops (confined)	July 31, 1989
- rotational crops (field)	June 30, 1990
- fish	April 30, 1987

The ecological effects data are required on the following dates:

Acute Avian Oral Toxicity	January 31, 1987
Avian Subacute Dietary Toxicity - waterfowl	January 31, 1987
Acute Toxicity to Aquatic Invertebrates	January 31, 1987

The residue chemistry data are required on the following dates:

Nature of residue (animal metabolism)	October 31, 1987
Residue Analytical Methods	October 31, 1987
Storage Stability Data	July 31, 1987
Residue studies on crops, processed food/feed commodities	April 30, 1988

Product chemistry data required within a year of issuance of standard.

6. CONTACT PERSON AT EPA

Henry M. Jacoby, PM 21
Office of Pesticide Programs, EPA
Registration Division (TS-767C)
Fungicide-Herbicide Branch
401 M Street., 20460
(703) 557-1900

DISCLAIMER: The information presented in this Pesticide Fact Sheet is for informational purposes only and may not be used to fulfill data requirements for pesticide registration and reregistration.

THIRAM

Date Issued: June 1, 1984
Fact Sheet Number: 29

1. Description of Chemical

    Generic Name: Bis-(dimethylthiocarbamyl) disulfide

    Common Name: Thiram

    Trade Names: Trade names and other names for thiram are AAtak, Arasan, Delsan, Mercuram, Nomersan, Polyram-Alltra, Pomarsol, Spatrete, Tersan, Thimer, Thiramad, Thirasan, Thiuramin, Trametan, Triampa, Tripomol, Tuads, Vancide, Tetramethylthiuram disulfide, TM-95, TMTD, and TMTDS.

    EPA Shaughnessy Code: 079801-7
    Chemical Abstract Service (CAS) Number: 137-26-8
    Year of Intitial Registration: August 5, 1948, DuPont's Tetramethyl Thiuramdisulfide, EPA Reg. No. 352-114
    Pesticide Type: Fungicide and Rodenticide
    Chemical Family: Organo-sulfur
    U.S. and Foreign Producers:

    1. E.I. du Pont de Nemours and Company

    2. UCB Societe Anonyme, Belgium, for Virginia Chemical Inc., Prochimie International, Inc., and UCB Chemical Corp.

    3. R.T. Vanderbilt Company, Inc.

    4. Aagrunol Chemicals, B.V., for Aceto Chemical Co., Inc.

2. Use Patterns and Formulations

    Application Sites: Fruit, vegetable and ornamental plants (including turf grasses), vegetable and field crop seeds, bananas, propagules of sweet potatoes, tree seedlings, bulbs and cuttings of ornamental bulbs, soil, textiles, polyurethane, wood pulp; and sites around homes, airports, seedling nurseries.

    Types of Formulations: Dusts, Wettable Powders and Flowable Suspensions.

    Types and Methods of Application: Dusting, Spraying and Dipping.

Thiram 785

Application Rates: See Use Patterns in EPA Compendium of Registered Pesticides, Vol. II, Fungicides and Nematicides, Part I, Pages T-30-00-01 through T-30-00-09.

Usual Carriers: Marl, Talc, Clays, Petroleum Oil, Graphite, Vermiculite, Mineral Oil, Charcoal and Water.

3. Scientific Findings

Chemical Characteristics:

    Physical State: Powder (Micromilled)
    Color: Cream to White
    Odor:
    Boiling Point: Not Applicable
    Melting Point: Range 155-156°C
    Flash Point: > 300° F

Toxicity Characteristics:

All toxicological data reviewed by the Agency were found to be lacking in information for evaluation, such as identification of compound tested, dose response information, individual animal information, pathology reports, etc.

Physiological and Biochemical Behavioral Characteristics:

Foliar Absorption: The available data do not provide direct evidence that thiram is or is not absorbed by roots or aerial portions of plants.

Translocation: There is inadequate data to conclude that thiram is translocated in plant tissue, but indirect evidence exists to indicate thiram or a degradate of thiram may enter plant tissue.

Mechanism of Pesticidal Action: Not understood as a fungicide. As a repellant to rodents.

Metabolism and Persistence in Plants and Animals: The metabolism of thiram in plants and animals is not adequately understood.

Environmental Characteristics:

Adsorption and Leaching in Basic Soil Types: Inadequate data.

Microbial Breakdown: Inadequate data.

Loss from Photodecomposition and/or Volatilization: No data.

Bioaccumulation: No data.

Resultant Average Persistence: No data.

Ecological Characteristics:

Hazards to fish and wildlife

    Rainbow trout   96-hr LC_{50} = 0.130
    Bluegill sunfish   96-hr LC_{50} = 0.044

(Characterized as "very highly toxic" to both cold water and warm water fish.)

Thiram is "moderately toxic" to birds. There is insufficient information to fully characterize the toxicity of thiram to mammals. Generally, the subcutaneous toxicity is "high," the acute oral toxicity is slight to "moderate," and in some species (mouse) "practically non-toxic." Thiram is characterized as "relatively non-toxic" to honeybees and predaceous ladybird beetles.

Potential Problems Related to Endangered Species:

Additional data (estimated environmental concentrations, persistence, avian reproduction studies, accumulation) are required to complete the endangered species assessment for thiram.

Efficacy Review Results:

No efficacy reviews were made.

Tolerance Assessments:

1. List of Crops and Tolerances:

The following table lists the present status for tolerances in parts per million for residues of thiram:

Raw Agricultural Commodity	U.S.	Canada	Mexico	Codex
Apples	7	0.1	7	3^b
Bananas	7^a	1.0	-	1^c
Celery	7	-	7	5^b
Onions (dry bulb)	0.5	-	0.5	-
Peaches	7	7	7	3^b
Strawberries	7	7	7	3^b
Tomatoes	7	7	7	3^b

a. 7 parts per million in or on bananas (from preharvest and postharvest application) of which not more than 1 part per million shall be in the pulp after peel is removed and discarded.

b. A temporary Codex MRL for the residues of dithiocarbamates (of which thiram is a member) expressed n terms of mg CS_2/kg, has been established.

c. Temporary Codex MRLs of 1 and 0.1 ppm have been established for total dithiocarbamates residues (including thiram expressed as ppm CS_2) in or on whole bananas and banana pulp, respectively.

2. Seed Applications:

No tolerances have been established for thiram residues in or on any crop for which thiram is registered solely for seed treatment, because heretofore seed treatment uses have been considered to be nonfood uses. These crops include: barley, beans (dry and succulent), lima beans, beets, broccoli, brussels sprouts, cabbage, cantaloupe, carrots, castor bean, cauliflower, collards, corn (sweet and field), cotton, cowpeas, cucumber, eggplant, endive, flax, forage-fodder grasses, kale, kohlrabi, lentils, lettuce, millet, muskmelons, mustard, oats, okra, onion, peanuts, peppers, pumpkins, radish, rice, rye, safflower, sesame, small-seeded legumes, sorghum, soybeans, spinach, sugar beets, sunflower, swiss chard, tomato, turnips, watermelon, and wheat.

Results of Tolerance Assessment:

Insufficient data are available to assess the adequacy of the tolerance for thiram in or on all thiram-treated commodities having such tolerances: apples, bananas, celery, onions (dry bulb), peaches, strawberries, and tomatoes (40 CFR 180.132). Note that either green onions must be deleted from thiram labels or residue data and a tolerance proposal must be submitted. Also note that the in-furrow treatment for cotton must either be removed from thiram labels or residue data and tolerance proposals must be submitted for forage, seed, and processed products. In addition, many seed treatment uses are registered for thiram on crops not having tolerances for thiram; the continued "nonfood" classification of these uses is contingent upon the receipt of plant metabolism studies demonstrating that thiram residues of concern are not translocated into food/feed crops grown from thiram-treated seed. If residues of concern are translocated, then residue data and tolerance proposals must be submitted for all of these crops (or at least all of the representative commodity members of each involved crop group). Finally, tolerances for thiram animal products have not been established; if the requested animal metabolism studies reveal that thiram residues of concern are transferred to animals, then animal residue data (feeding studies) and appropriate tolerance proposals will be required for ruminants and poultry. It is imperative that the metabolism of thiram in plants and animals be elucidated, since many of the above-noted data gaps are dependent upon the outcome of the metabolism studies. Refer to the appropriate preceding sections for details of data gaps. The data are insufficient to allow the establishment of any crop group tolerances.

No ADI has been established for thiram. The TMRC is 0.7380 mg/day based on a 1.5 kg diet and the relevant tolerances (40 CFR 180.132) and food factors.

Problems that are Known to Have Occurred with Use of the Chemical.

Two workers exposure effects have been identified:

1. Illness in pine seedling planters and handlers resulting from handling thiram-treated pine seedlings without protective clothing (the illness resulted from the ingestion of alcohol after such an exposure).

    Note: Thiram is the methyl analog of Antabuse, bis-(diethylthiocarbomyl sulfide), a drug used in rehabilitating alcoholics.

2. Skin rashes of the hands and head resulting from exposure to thiram in handling thiram-treated pine seedlings.

These health effects have been reduced by restrictions that require workers to wear gloves and protective clothing when handling thiram products.

4. Summary of Regulatory Position and Rationale

Based on historical use experience (human health effects reports) and the benefits from the pesticidal uses, the Agency has determined to allow the registration of thiram products to continue for existing use-patterns until the hazards are better defined by the data requirements under the Thiram Registration Standard.

5. Summary of Major Data Gaps

All toxicology data, both acute and chronic studies.
Product chemistry data
Residue chemistry studies
Environmental fate
   Hydrolysis studies
   Photodegradation studies
   Metabolism studies
   Mobility studies
   Dissipation studies
   Accumulation studies
Re-entry
   Foliar dissipation studies
   Soil dissipation studies
   Dermal exposure studies
   Inhalation exposure studies
Avian reproduction studies
Field studies with mammals and birds
Aquatic organism studies
Non-target insect studies

6. Contact Person

      Eugene M. Wilson
      EPA
      Office of Pesticide Programs
      Registration Division (TS-767-C)
      Crystal Mall #2
      1921 Jefferson Davis Hwy.
      Arlington, VA 22202
      Telephone (703) 557-1900

DISCLAIMER: The information presented in this Chemical Information Fact Sheet is for informational purposes only and may not be used to fulfill data requirements for pesticide registration and reregistration.

TPTH

Date Issued: September 30, 1984
Fact Sheet Number: 39

1. Description of chemical:

    Generic Name: triphenyltin hydroxide
    Common name: fentin hydroxide (BSI, ISO), triphenyltin hydroxide (USA, S. Africa)
    Trade name: Du-ter®, Duter®, Haitin, TPTH, TPTOH, Suzu H®, Supertin, Tubotin®
    EPA Shaughnessy Code: 083601
    Chemical Abstracts Service (CAS) Number: 76-87-9
    Year of Initial Registration: 1971
    Pesticide Type: Fungicide
    Chemical family: Organotin
    U.S. and Foreign Producers: M & T Chemicals (U.S.A.), Philips-Duphar (Netherlands), Nitto-Kasei Co. (Japan)

2. Use patterns and formulations:

    Application sites: To control early and late blight on potatoes, leaf spot on sugar beets and peanuts, scab and several other diseases on pecans, leaf spot and alternaria blight on carrots, and to suppress spider mites on peanuts.
    Types of formulations: Wettable powders and flowable suspensions
    Types and Methods of Application: Aerial and ground sprays, application through irrigation systems.
    Application Rates: 1.5 to 12 ounces a.i./acre
    Usual carriers: Water, Surfactants, spreaders, or stickers should not be used because excessive phytotoxicity can result. Not to be used with oil sprays.

3. Science Findings:

    Summary science statement:

    TPTH is very highly toxic (Category I) when it reaches the eyes, or when inhaled or absorbed through the skin. Because the major routes of human exposure are skin contact and inhalation, this high toxicity is cause for concern. TPTH produces birth defects in laboratory animals, can damage the immunological systems of exposed animals, produces lesions in the uteruses of female rats and has very high subacute inhalation toxicity. It is not a mutagen.

Chemical characteristics:

    Physical state: solid, fine powder
    Color: white to off-white
    Odor: none
    Vapor pressure: non-volatile
    Melting point: 118-120° C (technical)
    Flammability: 400° C
    Octanol/water partition coefficient: K=1270 and 1370
    Stability: decomposes at about 80°C to bis-triphenyltin oxide
              stable at pH values of 5, 7, and 9 for >30 days

Solubility:			
water - 8 ppm	Benzene - 41 g/l
ether - 28 g/l	Ethanol - 10 g/l
1,2-Dichloromethane - 74 g/l	Acetone - 70 g/l
Methylene chloride - 171 g/l	

    Unusual handling characteristics: None reported (No data on explodability or corrosion characteristics)

Toxicological characteristics:

    Acute Effects:

    Acute Oral LD_{50} - 165 mg/kg (male rats), 156 mg/kg (female rats) (Categor
    Acute Dermal LD_{50} - 127 mg/kg (male rabbits) (Category I)
    Dermal Irritation - Primary Skin Irritation (PSI) = 2.8 (Category III)
    Acute Inhalation Toxicity - 60.3 ug/l (Category I)
    Primary Eye Irritation - Corrosive (Category I)

    Major Routes of Exposure: Dermal, inhalation

Chronic Effects:

Oncogenicity - Caused pathological lesions on the uteruses of female rats at lowest dose tested (LDT). Controversy as to whether this is an oncogenic effect has not yet been resolved.

Teratology - Caused hydrocephalus and hydronephrosis at $\leq$ 1.25 mg/kg (lowest level tested). Other effects: abortions, decreased body weight gain, decreased % live fetuses, decreased fetal weight, increased resorptions.

Mutagenicity - Not a mutagen

Immunotoxicity - Effects spleen weight and IgM AFC spleen cells and spleen cell response to mitogens at 2.5 mg/kg/day (LDT). Decreased leukocyte counts were observed at most dose levels, including LDT. Study does not show a No Observed Effects Level (NOEL).

Subacute Inhalation - Effects were noted at 0.0011 mg/liter (LDT), including alopecia, nasal discharge, red ears, ptosis, piloerection, and epithelial hyperplasia of the skin. The histopathology report has not been completed: the effects at lower dose levels have not been evaluated.

Physiological and Biochemical Behavioral Characteristics:

Translocation: Does not translocate

Environmental Characteristics:

Absorption and leaching characteristics: Relatively immobile in sandy loam, clay loam, and silty clay loam soils.

Loss from photodegradation and/or volatilization: No information is available on photodegradation or volatilization. TPTH has low vapor pressure, so little volatilization is expected.

Resultant average persistence: Half-life of 1 to 3 months in sandy and silt loam soils, 126 days in flooded silt loam.

Half-life in Water: Stable to hydrolysis for 30 days at 21° C, loss of approx. 16% at 32°C

Ecological characteristics:

Hazards to Birds: Cannot be estimated without more data

Hazards to Fish and Aquatic Invertebrates: Any use pattern that would result in contamination of aquatic systems through spray drift or runoff could result in high risk to populations of fish and aquatic invertebrates, because of the very high toxicity of TPTH to aquatic organisms. More data on persistence and chronic effects are needed to complete the hazard evaluation.

Potential Problems with Endangered Species: Cannot be estimated without more data. Aquatic species would presumably be at high risk if exposed.

Tolerance Reassessment:

List of crops and tolerances: (CFR 180.236)

	(PPM)		(PPM)
Beets, sugar, roots	0.1N*	Carrots	0.1N
Cattle, kidney	0.05N	Cattle, liver	0.05N
Goats, kidney	0.05N	Goats, liver	0.05N
Hogs, kidney	0.05N	Hogs, liver	0.05N
Horses, kidney	0.05N	Horses, liver	0.05N
Peanuts	0.05N	Peanuts, hulls	0.4
Pecans	0.05N	Potatoes	0.05N
Sheep, kidney	0.05N	Sheep, liver	0.05N

*"N" stands for "Negligible Residues".

List of food contact uses: Pecans, peanuts, potatoes, carrots

Results of tolerance assessment: No ADI can be set at this time

Problems known to have occurred from use: PIMS file contains eight entries, according to its index. We cannot retrieve the data at this time (Sept. 1984).

4. Summary of Regulatory Position and Rationale:

Use Classification: Reclassified (by the Registration Standard) as a Restricted Use chemical because of toxicity and teratogenic effects. Use, Formulation or Geographic Restrictions: Manufacturing use products may only be formulated into end-use products intended for use as a fungicide on pecan trees, peanuts, carrots, potatoes, sugar beets, and tobacco, or as anti-fouling paint, or spider mite suppressants on peanuts."

Unique Label warning statements:

a. Hazards to Humans Statements

Labels of manufacturing-use and formulated end-use products (EUPs) must bear the statements:

"DANGER - Fatal if inhaled. Corrosive, causes irreversible eye damage. May be harmful or fatal if swallowed or absorbed through the skin. Do not get in eyes, or on skin. Do not breathe dust, vapor, or spray mist. When handling either products containing TPTH or spray-diluted mixtures, wear protective clothing (long pants, long sleeve shirt, impermeable gloves, hat, boots, and a pesticide respirator jointly approved by the Mining Enforcement and Safety Administration and the National Institute for Occupational Safety and Health). When handling concentrated products, wear a face shield. Wash thoroughly with soap and water after handling and before eating or smoking. Remove contaminated clothing and wash before reuse. Do not enter treated areas for at least 24 hours after treatment."; and

"The United States Environmental Protection Agency has determined that triphenyltin hydroxide causes birth defects in laboratory animals. Exposure to triphenyltin hydroxide during pregnancy should be avoided."

The word "POISON" (in red letters) and a skull and crossbones must appear in close proximity to the word "DANGER".

b. Statements of Practical Treatment

Labels of manufacturing-use and end-use products must bear the statements:

"If on skin: Wash with plenty of soap and water."

"If inhaled: Remove victim to fresh air. If not breathing, give artificial respiration, preferably mouth-to-mouth. Get medical attention."

"If in eyes: Flush with plenty of water. Call a physician."

"If swallowed: Do not induce vomiting. Drink promptly a large quantity of milk, egg whites, gelatin solution, or if these are not available, drink large quantities of water. Call a physician or Poison Control Center."

c. Environmental Hazard Statement

The following specific statements must appear on the labels of all manufacturing use products:

"This pesticide is toxic to fish and wildlife. Do not discharge into lakes, streams, ponds, estuaries, oceans, or public waters unless this product is specifically identified and addressed in a NPDES permit. Do not discharge effluent containing this product to sewer systems without previously notifying the sewage treatment plant authority. For guidance contact your State Water Board or Regional Office of the EPA."

All labels of EUPs intended for outdoor use must bear this statement:

"This pesticide is toxic to fish and wildlife. Do not apply directly to water or wetlands. Drift or runoff from treated areas may be hazardous to aquatic organisms in neighboring areas. Cover or incorporate spills. Do not contaminate water by cleaning of equipment or disposal of wastes."

d. Disposal Statements

All labels of manufacturing use or formulated end-use products (EUPs) bear this statement, under the headline "STORAGE AND DISPOSAL":

"Do not contaminate water, food, or feed by storage or disposal. Pesticide wastes are acutely hazardous. Improper disposal of excess pesticide, spray mixture, or rinsate is a violation of Federal law. If these wastes cannot be disposed of by use according to label instructions, contact your State Pesticide or Environmental Control Agency, or the Hazardous Waste representative at the nearest EPA Regional Office for guidance."

The statements required by this standard must appear on the labels of all MUPs and EUPs released for shipment after March 30, 1984. After review of data to be submitted under this standard, the Agency may impose additional label requirements.

Summary of risk/benefit analysis: No risk/benefit analysis per se has been done. The Agency has determined that TPTH meets the risk criteria in 40 CFR §162.11(a), primarily because TPTH has produced teratogenic effects in laboratory animals. TPTH will be placed in the Special Review process, during which there will be an analysis of its risks and benefits.

5. Summary of major data gaps

Dates when major data gaps are due to be filled.

Data Requested	Due date
Description of manufacturing process	April 1, 1985*
Discussion of formation of unintentional ingredients	April 1, 1985*
Preliminary analysis	April 1, 1985*
Certification of limits	April 1, 1985*
Analytical methods and data for enforcement of limits	April 1, 1985*
Vapor pressure	April 1, 1985
Plant residues	April 1, 1985
Animal residues	April 1, 1985

Data Requested (continued)	Due Date
Storage stability data	April 1, 1985*
Magnitude of the residue for each food use	April 1, 1985
Photodegradation	April 1, 1985
Metabolism studies in lab	November 1, 1986
Mobility studies	April 1, 1985
Dissipation studies in field	November 1, 1986
Accumulation studies in rotational crops	November 1, 1986
Accumulation studies in fish	April 1, 1985
Reentry protection	November 1, 1986
90-day feeding (rodent)	January 1, 1986
90-day feeding (non-rodent)	January 1, 1986
21-day dermal	April 1, 1985
90-day inhalation	April 1, 1985**
Chronic toxicity	April 1, 1985***
Oncogenicity	April 1, 1985***
Teratogenicity	April 1, 1985****
Reproduction (2-generation)	September 1, 1986
Chromosomal aberration	September 1, 1986
Other genotoxic effects	September 1, 1986
General metabolism	February 1, 1986
Avian dietary toxicity	April 1, 1985
Avian Reproduction	April 1, 1986
Coldwater fish acute toxicity	April 1, 1985
Density, bult density, or specific gravity	April 1, 1985*
Oxidizing/reducing action	April 1, 1985*
Explodability	April 1, 1985*
Corrosion	April 1, 1985*

*Product-specific data required for manufacturing use products containing TPTH.
**If the registrant commits to conducting new studies, the deadline is January 1, 1986.
***If the registrant commits to conducting new studies, the deadline is November 1, 198
****If the registrant commits to conducting new studies, the deadline is November 1, 19

6. <u>Contact person at EPA:</u> Henry Jacoby, U.S. Environmental Protection Agency, TS-767-C, 401 M Street SW, Washington, DC 20460
(703) 557-1900

DISCLAIMER: The information presented in this Chemical Information Fact Sheet is for informational purposes only and may not be used to fulfill data requirements for pesticide registration and reregistration.

TRIMETHACARB

Date Issued: September 30, 1985
Fact Sheet Number: 76

1. DESCRIPTION OF CHEMICAL

   - Generic Name: 2,3,5- and 3,4,5-trimethylphenyl methylcarbamate
   - Common Name: Trimethacarb
   - Trade Names: Broot® and Landrin®
   - EPA Shaughnessy Codes: 102401 (3,4,5-isomer)
     102402 (2,3,5-isomer)
   - Chemical Abstracts Service (CAS) Numbers:
     2686-99-9 (3,4,5-isomer)
     3971-89-9 (2,3,5-isomer)
   - Year of Initial Registration: 1973
   - Pesticide Type: Insecticide
   - Chemical Family: Carbamate
   - U.S. Producer: Union Carbide Agricultural Products Company, Inc.

2. USE PATTERNS AND FORMULATIONS

   - Application sites: Field corn and popcorn
   - Types of formulations: 15% granular
     50% wettable powder
   - Types and methods of applications: ground application as a granular or spray
   - Application rates: 0.9 - 1.35 lb active ingredient/acre

3. SCIENCE FINDINGS

   Chemical Characteristics

   - Technical trimethacarb is a buff to brown crystalline powder with a mild ester odor. The melting point is 105 - 114°C.
   - Trimethacarb is soluble in water to 58 ppm at 23°C and hydrolyzes to trimethylphenols, CO_2, and methylamine at pH values greater than pH 8.

   Toxicology Characteristics

   - Acute oral: 125 mg/kg (rat), Toxicity Category II
   - Acute dermal: 2 gm/kg (rabbit), Toxicity Category III
   - Primary Eye Irritation: slight eye irritation (rabbit), Toxicity Category III

Toxicology Characteristics

- Acute Inhalation: Data gap
- Primary Skin Irritation: Non irritant, Toxicity Category IV
- Skin Sensitization: Data gap
- Major Routes of Exposure: Human exposure from trimethacarb applications is from handling, mixing and application. Major exposure is expected to be mainly dermal. Because of trimethacarb's low dermal toxicity protective clothing is not required.
- Oncogenicity: Data Gap
- Metabolism: Data Gap
- Teratology: Data Gap
- Reproduction: Data Gap
- Mutagenicity: Data Gap

Physiological and Biochemical Characteristics

- Mechanism of pesticidal action: Cholinesterase inhibition
- Metabolism and persistence in plants and animals: systemic in plants. Like other carbamates, it is metabolized rapidly in animals.

Environmental Characteristics

- Available data are insufficient to fully assess the environmental fate of trimethacarb. Data gaps exist for all required studies.

Ecological Characteristics

- Avian acute oral toxicity: Data Gap
- Avian dietary toxicity: 2300 ppm for Mallard Duck, 1500 ppm for Ring-Neck Pheasant, and 2000 ppm for Japanese Quail (slightly toxic)
- Freshwater Fish toxicity: Cold water species (rainbow trout) -- 1.0 mg/l (highly toxic); warm water species (blue gill) -- 11.6 mg/l (slightly toxic).
- Aquatic freshwater invertebrate toxicity: Data Gap
- Additional data are required to fully characterize the ecological effects of trimethacarb.

Tolerance Assessment

- Established tolerances are published in 40 CFR 180.305 to cover the sum of the residues of both components (3,4,5 and 2,3,5-trimethylphenyl methylcarbamate) of trimethacarb and they are:

Commodity	Parts Per Million
field corn	0.1
popcorn	0.1
corn fodder	0.1
corn forage	0.1

- The Agency is unable to complete a full tolerance assessment for the established tolerances because of certain residue chemistry and significant toxicology data gaps.

- The residue data for trimethacarb in or on corn grain, corn fodder and forage are not adequate to support the established tolerances. Additional data are required.

- Data are required to determine whether food additive tolerances are needed for residues in processed products of corn.

- Residue data must be submitted and a tolerance must be proposed for residues in or on corn silage.

- A pregrazing restriction must be proposed for corn forage.

- Based on the established tolerances the theoretical maximum residue contribution (TMRC) for trimethacarb residues in the human diet is 0.0016 mg/day for a 60 kg person with a 1.5 kg diet. However, no acceptable chronic toxicology exists for calculation of the acceptable daily intake (ADI).

4. Summary of Regulatory Position and Rationale

- The Agency has determined that it should continue to allow the registration of trimethacarb. None of the criteria for unreasonable adverse effects listed in the regulations (§162.11(a)) have been met or exceeded. However, because of gaps in the data base a full risk assessment cannot be completed.

- For end-use products, the product label must include a crop rotation statement that permits crop rotation only in trimethacarb treated soil with those crops registered for trimethacarb. This restriction is imposed until the Agency has adequate accumulation data from typical rotated crops or tolerances are established on these crops.

- Also, a full tolerance reassessment cannot be completed because of major residue chemistry and toxicology data gaps. Until these gaps are filled trimethacarb will not be registered for any significant new uses.

- Available data are insufficient to fully assess the environmental fate of and the ecological effects from trimethacarb.

5. Summary of Major Data Gaps

- The full complement of chronic toxicology requirements: chronic feeding, oncogenicity, reproduction, teratology and mutagenicity.

- The full complement of environmental fate data requirements: degradation (hydrolysis and photodegradation in water), soil metabolism, mobility, dissipation and accumulation.

- Additional data are required : acute avian oral toxicity and freshwater invertebrate acute toxicity.

- Additional data are necessary to characterize the distribution and metabolism of trimethacarb in mature corn plants.

Contact Person at EPA

Jay S. Ellenberger
Product Manager (12)
Insecticide-Rodenticide Branch
Registration Division (TS-767C)
Office of Pesticide Programs
Environmental Protection Agency
401 M. Street, S.W.
Washington, D.C. 20460

Office location and telephone number:
Room 202, Crystal Mall Building #2
1921 Jefferson Davis Highway
Arlington, VA 22202
703-557-2386

DISCLAIMER: THE INFORMATION PRESENTED IN THIS CHEMICAL INFORMATION FACT SHEET IS FOR INFORMATIONAL PURPOSES ONLY AND NOT TO BE USED TO FULFILL DATA REQUIREMENTS FOR PESTICIDE REGISTRATION AND REREGISTRATION.

VITAMIN D_3

Date Issued: December 1, 1984
Fact Sheet Number: 43

1. Description of chemical:

    Generic name: Cholecalciferol, Activated 7-dehydrocholesterol, oleovitamin D_3, 9,10-secocholesta-5,7,10(19)-trien-3B-ol [NEW CHEMICAL]
    Common name:
    Trade name: Vitamin D_3
    EPA Shaughessy code: 208700
    Chemical Abstracts Service (CAS) number: 434-16-2
    Year of initial registration: 1984
    Pesticide type: rodenticide
    Chemical family: sterol
    U.S. and foreign producers: Phillips-Duphar (Netherlands)
    Active ingredient
    Bell Laboratories (Madison, WI)
    Formulated product

2. Use patterns and formulations:

    Application sites: in and around buildings, inside of transport vehicles
    Pest species: norway rats, roof rats, and house mice
    Type of formulations: granular (0.075% bait)
    Type and method of application: topical, hand application
    Application rates: 2-8 oz./15-30 ft. (rats)
    1/4-1 oz./8-12ft. (mice)
    Usual carriers: Confidential Business Information

3. Science findings:

    Summary science statement:

    Adequate studies are available to assess the acute toxicological hazards of technical and formulated Vitamin D_3. No toxicological hazards of concern were identified. Available studies indicate that Vitamin D_3 is of low toxicity to birds; studies on fish are inapplicable because Vitamin D_3 is virtually insoluble in water. The registration of this new active ingredient is conditioned on submittal of additional efficacy data by January 1976.

Chemical characteristics:

Technical Vitamin D_3 is a solid resin. The empirical formula is $C_{27}H_{44}O$, and the molecular weight is 384.62. The melting point is is 84-85°C. Vitamin D_3 is practically soluble in water, soluble in the usual organic solvents, and slightly soluble in vegetable oils.

Toxicological characteristics:

Currently available toxicology studies on Vitamin D_3 are as follows:

- Oral LD_{50} in rats: 352 mg/kg and 42 mg/kg for males and 619 mg/kg for females.
- Dermal LD_{50} in rabbits: >2000 mg/kg.
- Primary dermal irritation: not required because technical is a solid resin.
- Primary eye irritation: not required because technical is a solid resin.
- Inhalation LC_{50}: not required because technical is a solid resin.
- Acute, 60-day, Delayed Toxicity Study in rats: Acute dose, equal to accidental exposure, produced no elevated serum calcium levels and no abnormal long bone growth in young rats.
- Teratology in rats: not required because technical is a solid resin and Vitamin D_3, is a dietary supplement in the adult female diet.

Adequate studies are available to assess the acute toxicological effects of Vitamin D_3. No toxicological hazards of concern have been identified in the studies reviewed for this new pesticide.

Physiological and biochemical behavioral characteristics:

The registrant submitted a volume on the metabolism and function of Vitamin D_3. However, because Vitamin D_3 is a non-food use, is applied topically, and in small amounts, we did not review this submission of such imformation.

Environmental characteristics:

Because of the use pattern of this chemical (in and around buildings and inside of transport vehicles, we did not request any enviornmenal fate data.

Ecological characteristics:

Based on studies available to assess hazards to wildlife and aquatic organisms, Vitamin D_3 is characterized as being of low toxicity to birds. Because the chemical is virtually insoluable in water, we requested no aquatic toxicity data.

Results of currently available studies are as follows:

- avian oral LD_{50}: >2000 mg/kg (mallard duck)
- avian dietary LC_{50}: 4000 ppm (mallard duck)
  2000 ppm (bobwhite quail)
- fish LC_{50}: not required because technical is virtually insoluable in water.
- aquatic invertebrate LC_{50}: not required because technical is virtually insoluable in water.

Product Performance characteristics:

Because the Agency had suspended the efficacy requirements for this type of product, the company did not complete all the pre-suspension data. Subsequently, the Agency notified the company that it had re-instituted these data. The company will have a reasonable period of time (until January 1, 1986) to supply the missing laboratory and field data.

4. Summary of regulatory position and rationale:

The Agency has placed the one registered formulation in Toxicity Category III (CAUTION) and has classified this use pattern (in and around buildings and inside of transport vehicles for norway rats, roofs rats, and house mice) as "Unclassified". Such a product can be sold over-the-counter. The Agency has not identified a potential for adverse effects for man or the environment, based on the submitted toxicology and fish and wildlife data. Because of re-institution of the efficacy data requirements, the registration will be conditioned on the submittal of additional, acceptable laboratory and field efficacy data within fourteen months.

5. Summary of major data gaps:

-LD-50 tests on target species
-Laboratory, choice-test, efficacy data
  for target species
-Field efficacy data

All studies are to be submitted to the Agency by January 1,

Vitamin D_3 805

6. Contact Person at EPA

William H. Miller
Product Manager (16)
Registration Division (TS-767C)
Insecticide-Rodenticide Branch
Environmental Protection Agency
Washington, DC 20460

Tel. No. (703) 557-2600

Disclaimer: The information presented in this Chemical Information Fact Sheet is for informational purposes only and may not be used to fulfill data requirements for pesticide registration and reregistration.

WOOD PRESERVATIVES

Date Issued: July 11, 1984
Fact Sheet Number: 31

INTRODUCTION

The Environmental Protection Agency issued its final regulatory position July 11, 1984 on the use of creosote, inorganic arsenicals and pentachlorophenol (and its salts) as wood preserving pesticides. This regulatory action was taken under the authority of the Federal Insecticide, Fungicide and Rodenticide Act (FIFRA). EPA's mandate under FIFRA is to prevent unreasonable adverse effects associated with exposure to these three wood preservatives and products treated with them, while still allowing the benefits of their use to continue. This fact sheet provides the details of the regulatory action and information on the health and safety concerns that prompted it.

USES

Wood preservatives protect wood from decay and increase the life expectancy of wood by a factor of five or more over that of untreated wood. Pressure treated wood containing the pesticides includes railroad ties, construction lumber, plywood, timbers, foundation materials, fence posts, and utility poles, as well as landscape materials. The preservatives also are sprayed or brushed on fence posts, lumber used around the home and yard such as wooden fences, decks, playground equipment, and lawn furniture, and for millwork, plywood and particleboard. In many instances, creosote, pentachlorophenol and inorganic arsenicals are used as alternatives to each other in treating wood.

In 1982, these three wood preservatives comprised roughly one-third of the 2.7 billion pounds of pesticides produced for both agricultural and industrial uses. Creosote, pentachlorophenol, and the inorganic arsenicals account for over 97% of the wood preservatives used in this country. Creosote, due to its density, accounts for most of the poundage used, but the inorganic arsenicals are found on more treated wood products. The major use of creosote is on railroad ties and utility poles. Pentachlorophenol is used primarily for poles, posts, fences, crossarms and logs for log homes. Sodium and potassium pentachlorophenate are used for preservation of freshly peeled poles or posts during air seasoning prior to pressure-treatment and for green lumber. Inorganic arsenicals are most commonly used on treated wood found outside the home (e.g., decks). The inorganic arsenicals include chromated copper arsenate (CCA), ammoniacal copper arsenate (ACA) and fluor chrome arsenic phenol (FCAP).

REGULATORY HISTORY

On October 18, 1978, the Environmental Protection Agency initiated a special review (Rebuttable Presumption Against Registration or RPAR) of the pesticidal uses of creosote, pentachlorophenol (and its salts), and the inorganic arsenicals based on health concerns. All three wood preservatives exceeded EPA's risk criteria for tumor production. In addition, creosote and the inorganic arsenicals were associated with genetic changes, and pentachlorophenol and the inorganic arsenicals were linked to defects in the offspring of laboratory animals.

In 1981, EPA proposed a set of regulatory actions for safe and proper commercial and domestic application of wood preservatives and recommended regulatory action for safe use and handling of treated wood by consumers. These proposals were reviewed by several government agencies and other interested parties, including industry and environmental groups.

The Agency held a public meeting on April 14, 1983, to give the public an opportunity to comment on a modified proposal to conclude the RPAR. Industry, environmental groups and other interested parties submitted comments as a result of that meeting.

In reaching the final regulatory position, the benefits of using wood preservatives and the availability and efficacy of alternatives (including other chemicals and non-wood products) were taken into account. The risks to applicators during the application process as well as the risks to the general public resulting from using, handling or disposing of the treated wood were examined.

In general, the risks to persons applying these chemicals are more significant than for persons who are only exposed to treated wood. Common uses of treated wood such as decks, lawn furniture or playground equipment do not pose high risks of adverse effects, and there are practical steps that can be taken to reduce exposure to already-treated wood. The consumer information materials developed in connection with this decision point out that the use of appropriate sealers will reduce dermal exposure or inhalation exposure from indoor uses of wood treated with pentachlorophenol or creosote.

ACTIONS

The Environmental Protection Agency has placed several restrictions on the application of the wood preservatives creosote, pentachlorophenol and its salts (including sodium pentachlorophenate) and the inorganic arsenicals. The actions are described below.

1. RESTRICTIONS ON WHO MAY APPLY WOOD PRESERVATIVES

All three chemicals are now classified for restricted use only by certified applicators (or someone under their direct supervision) except for the brush-on treatment of the inorganic arsenicals where use will be for commercial construction purposes only and not for household use. This action will reduce the general availability of wood preservatives to home and farm users, but an individual who chooses to use these products may do so by obtaining training and certification from the appropriate State agency. This provision is designed to ensure that restricted pesticides are properly applied by persons trained to safely handle these materials.

2. USE AND APPLICATION OF WOOD PRESERVATIVES: LABEL CHANGES

The following are the significant label changes that will be required:

The label must state: "Restricted Use Pesticide: For sale and use only by certified applicators or by persons under their direct supervision and only for those uses covered by the certified applicator's certification."

All pentachlorophenol products must include a warning that exposure to women during pregnancy should be avoided because pentachlorophenol has been shown to cause defects in the offspring of laboratory animals.

Protective clothing requirements will be specified. These will include use of impermeable gloves for applying the preservatives and in all situations where dermal contact is expected (e.g., handling freshly treated wood and manually opening cylinders to pressure treatment equipment). In certain situations such as spraying the chemicals and working around pressure treatment equipment, additional clothing is required. Such clothing would include overalls, jacket, boots, respirators, goggles and head covering. Applicators at commercial sites must leave their protective clothing at the plant. For home and farm use, non-disposable protective clothing should be laundered separately from other clothing.

All exposed arsenic treatment plant workers will be required to wear a respirator if the level of ambient arsenic is unknown or exceeds a Permissible Exposure Limit (PEL) of 10 ug/m^3 averaged over an 8 hour work day. This PEL is the same as the standard required by the Occupational Safety and Health Administration.

To reduce exposure to pentachlorophenol, closed systems for mixing and emptying powdered/prilled (granular) formulations will be required within three years. This phase-in period will allow small wood treatment operations which often lack closed systems to comply without undue economic hardship. Closed systems for mixing powdered inorganic arsenicals will be immediately required because of an unacceptably high risk of tumor production.

Applicators may not eat, drink, or use tobacco products during the application process, which may expose them to the treatment formulations. They must wash thoroughly after skin contact and before eating, drinking, smoking or using restrooms.

Pentachlorophenol and creosote may not be applied in homes. Pentachlorophenol may not be applied to wood intended for use in interiors, except for millwork (with outdoor surfaces) and support structures which are in contact with the soil in barns, stables and similar sites and which are subject to decay or insect infestation. A sealer must be applied, however.

Creosote may not be applied to wood intended for use in interiors except for those support structures which are in contact with the soil in barns, stables and similar sites and which are subject to decay or insect infestation. A sealer must be applied.

The application of pentachlorophenol to logs for construction of log homes is prohibited.

If creosote or pentachlorophenol is applied to wood intended for use where it will frequently contact bare skin (e.g., on outdoor furniture), two coats of an appropriate sealer must be applied. Urethane, epoxy, and shellac are acceptable sealers for all creosote-treated wood. Urethane, shellac, latex epoxy enamel, and varnish are acceptable sealers for pentachlorophenol-treated wood. Arsenically-treated wood without sealers is safe for frequent contact because absorption through the skin is negligible.

Pentachlorophenol or creosote should not be used where there may be contamination of feed, food, drinking or irrigation water.

3. USE OF PRESSURE-TREATED WOOD: CONSUMER AWARENESS PROGRAM

The Consumer Awareness Program will require wood pressure-treaters to send Consumer Information Sheets to all places where treated wood is sold to instruct consumers about handling procedures, such as the use of protective gloves and coveralls and face masks when sawing treated wood products. The information sheet will recommend against the use of wood treated with any of the three preservatives in proximity to food, feed and public drinking water.

In interiors of farm buildings where domestic animals are unlikely to lick or bite the wood, pentachlorophenol and creosote may be used if two coats of an approved sealant are applied. In general, shavings, sawdust, and the treated wood itself should not be used for bedding, brooding facilities, food containers, etc.

The information sheet will also recommend that consumers avoid frequent and prolonged skin contact with pentachlorophenol and creosote-treated wood such as treated lawn chairs and other outdoor furniture, unless two coats of an effective sealer have been applied. All treated wood must be visibly clean and free of surface residue for use on patios, decks and walkways.

The use of pentachlorophenol pressure-treated wood in a home is prohibited except for laminated beams or building structures in contact with the ground, provided that two coats of an appropriate sealer are applied. The use of creosote-treated wood inside the home is prohibited. Wood pressure-treated with arsenical preservatives may be used inside residences if dust is vacuumed from the wood surface.

Treated wood may be disposed of by ordinary trash collection or by burial. It should not be burned in a fireplace or open fire because of the toxic fumes or ashes that may be produced. (Commercial users such as railroad workers may use industrial incinerators to burn railroad ties--at these higher temperatures, toxic chemicals are broken down).

4. DIOXIN CONTENT OF PENTACHLOROPHENOL

Registrants of pentachlorophenol will be required to limit immediately the dioxin contaminant (hexachlorodibenzo-p-dioxin or HxCDD) in pentachlorophenol and its salts to 15 ppm and reduce that level to 1 ppm or less within 18 months. HxCDD causes tumors in rats and mice. The highly toxic dioxin TCDD has not been found in pentachlorophenol and no TCDD wil be allowed in the product.

5. DATA REQUIREMENTS

Manufacturers who market or intend to market wood preservatives must submit the following information:

For technical product registrations of pentachlorophenol and sodium pentachlorophenate, a description of the manufacturing process, including any changes to lower HxCDD; product identity; data on the analysis and certification of product ingredients; and information on the technical feasibility and costs of reducing HxCDD lower than the 1 ppm limit.

Epidemiology and exposure monitoring studies of creosote treatment plant workers so that, upon receipt of the data, the Agency can quantitatively estimate the potential risk of cancer.

Glossary

ADI	acceptable daily intake
a.i.	active ingredient
EP	end-use product
EUP	end-use product
LDT	lowest dose tested
LEL	lowest effects level
LOEL	lowest observed effect level
MATC	maximum acceptable toxic concentration
mbyp	meat byproduct
MP	manufacturing-use product
MPI	maximum permitted intake
NOEL	no observed effect level
NPDES	National Pollution Discharge Elimination System
PADI	provisional acceptable daily intake
PAI	pure active ingredient
PGAI	pure grade active ingredient
PGI	pre-gazing interval
PHI	pre-harvest interval
PIMS	Pesticide Incident Monitoring System
PIS	primary irritation score
PLD	provisional listing dose
ppb	parts per billion
ppm	parts per million
PSI	primary skin irritant
TMRC	theoretical maximum residue concentration or contribution

Numerical List of Pesticide Fact Sheets

Number	Chemical Name
1	ALIETTE
2	CRYOLITE
3.1	ETHOPROP
4	NALED
5.1	TERBUFOS
6	EPTC
7	BUTYLATE
8	DICAMBA
9	DIURON
10	FENAMINOSULF
11	FORMETANATE HCL
12	ANILAZINE
13	DCNA
14.1	FENSULFOTHION
15	CHLOROBENZILATE
16	DICOFOL
17	AROSURF
18	THIODICARB
19	ALDICARB
20	AMITROLE
21	CARBARYL
22.1	FONOFOS
23	SIMAZINE
24	CARBOFURAN
25	CARBOPHENOTHION
26	DAMINOZIDE
27	HELIOTHIS NPV
28	LINURON

Number	Chemical Name
29	THIRAM
30	THRICHLORFON*
31	WOOD PRESERVATIVES
32	BRONOPOL
33	DANTOCHLOR
34	PHORATE
35	CAPTAFOL
36	CHLOROTHALONIL
37	CHLORPYRIFOS
38	POTASSIUM BROMIDE
39	TPTH
40	ALACHLOR*
41	CYANAZINE
42	VITAMIN D_3
43	DISULFOTON
44	PROPACHLOR
45	DEMETON
46	FENARIMOL*
47	GLYCOSERVE
48	PICLORAM
49	NAPTALAM
50	PENDIMETHALIN
51	SULFURYL FLUORIDE
52	FLUCHLORALIN
53	METRIBUZIN
54	NITRAPYRIN
55	DIPROPETRYN
56	CYHEXATIN
57	CHLORPYRIFOS METHYL
58	ETHALFLURALIN
59	ACTELLIC
60	NORFLURAZON
61	SODIUM OMADINE
62	CLIPPER (PACLOBUTRAZOL)
63	ARSENAL
64	BENTAZON AND SODIUM BENTAZON
65.1	DINOCAP
66**	
67	3,5-DIBROMO
68.1	DIFLUBENZURON
69**	
70	PRONAMIDE
71	METSULFURON METHYL
72	MONOCROTOPHOS
73	LINDANE
74	PERFLUIDONE
75	CAPTAN

Number	Chemical Name
76	TRIMETHACARB
77	LINALOOL
78	FENOXYCARB
79	SODIUM AND CALCIUM HYPOCHLORITES
80	POTASSIUM PERMANGANATE
81	FLURIDONE
82	CHLORIMURON ETHYL
83	IMAZAQUIN
84	THIOPHANATE ETHYL
85	HYBREX
86	FLUVALINATE
87	COPPER SULFATE
88	FLUOMETURON
89	AVERMECTIN
90	COMMAND
91	ARSENIC ACID
92	THIOPHANATE METHYL
93	BACILLUS THURINGIENSIS*
94.1	2,4-DICHLOROPHENOXY
95	1,3-DICHLOROPROPENE
96	DIAZINON
97*	
98	METHYL BROMIDE
99	PROPARGITE
100	AZINPHOS-METHYL
101	PHOSMET
102	CARBON TETRACHLORIDE
103	CADMIUM
104	TERBUTRYN
105	CYROMAZINE
106	METOLACHLOR
107.1	HEPTACHLOR
108	ALDRIN
109	CHLORDANE
110	ARSENIC TRIOXIDE
111	CALCIUM ARSENATE
112	LEAD ARSENATE
113	SODIUM ARSENITE
114	SODIUM ARSENATE
115	ALDOXYCARB
116	PARATHION
117	METHYL PARATHION
118	ALUMINUM AND MAGNESIUM PHOSPHIDES*
119	FENBUTATIN-OXIDE
120	METHIOCARB
121	PROMETRYN
122	DICHLOBENIL

Number	Chemical Name
123	PROPHAM
124	NABAM
125	MANCOZEB
126	ISOMATE-M
127	EPN
128	LACTOFEN
129	OXAMYL
130	DINOSEB
131	PARAQUAT
132	SODIUM SALT OF FOMESAFEN
133	BROMINATED SALICYLANILIDE
134*	
135	DODINE
136	DIPHENAMID
137	TEBUTHIURON
138	ISAZOPHOS
139	ETHYLENETHIOUREA (ETU)

*Fact Sheet not currently available.
**Number not in use.

Common Name Index

Aldicarb - 6
Aldoxycarb - 19
Aldrin - 24
Amitrole - 34
Anilazine - 43
Anilinocadmium Dilactate - 101
Arsenic Acid - 57
Arsenic Trioxide - 64
Azinphos-methyl - 72
Bentazon - 80
1,3-Bis(hydroxymethyl)-5,5-dimethylhydantoin - 422
Bleach - 708
Bromomethane - 506
Bronopol-Boots - 91
Butylate - 95
2-(sec-Butyl)-4,6-dinitrophenol - 294
Cadmium Carbonate - 101
Cadmium Chloride - 101
Cadmium Sebacate - 101
Cadmium Succinate - 101
Calcium Arsenate - 106
Captafol - 109
Captan - 120
Carbaryl - 130
Carbofuran - 140
Carbophenothion - 152
Carbon Tetrachloride - 149
Chlordane - 157

Chlorimuron ethyl - 164
Chlorobenzilate - 170
Chlorothalonil - 174
Chlorpyrifos - 180
Chlorpyrifos-methyl - 188
Copper Sulfate - 200
Cryolite - 207
Cyanazine - 211
Cyhexatin - 220
Cyromazine - 226
2,4-D - 268
Daminozide - 232
Demeton-O - 243
Demeton-S - 243
Diazinon - 247
Dicamba - 254
Dichlobenil - 260
1,3-Dichloro-5-ethyl-5-methylhydantoin - 236
Dicloran - 239
Dicofol - 281
Diflubenzuron - 286
Dinocap - 292
Dinoseb - 294
Diphenamid - 298
Dipropetryn - 307
Disulfoton - 314
Diuron - 324
Dodine - 329

Common Name Index 817

DPX-F6025 - 164
EPN - 336
EPTC - 346
Ethalfluralin - 351
Ethoprop - 354
Ethylenethiourea - 360
Fenaminosulf - 364
Fenbutatin-oxide 367
Fenoxycarb - 373
(ISO) Fenridazone-potassium - 440
Fensulfothion - 378
Fentin hydroxide - 789
Fluchloralin - 385
Fluridone - 400
Fluvalinate - 405
Fonofos - 413
Formetanate hydrochloride - 417
Fosetyl-Al - 30
Heliothis - 428
Heptachlor - 433
Imazaquin - 447
Isazophos - 453
Lactofen - 462
Lead Arsenate - 469
Linalool - 473
Lindane - 475
Linuron - 484
Mancozeb - 491
Mercaptophos - 243
Mercaptophos teolevy (USSR) - 243
Methiocarb - 500
Methyl Bromide - 506
Methyl Parathion - 513
Metolachlor - 522
Metribuzin - 531
Metsulfuron methyl - 541
Monocrotophos - 547
Nabam - 555
Naled - 563
Naptalam - 571

Nitrapyrin - 578
Norflurazon - 583
Oriental Fruit Moth Pheromone - 458
Oxamyl - 590
Paclobutrazol - 192
Paraquat - 596
Parathion - 606
Pendimethalin - 614
Perfluidone - 625
Phorate - 632
Phosmet - 637
Picloram - 646
Pirimiphos-methyl - 1
Potassium Bromide - 651
Potassium Permanganate - 655
Prometryn - 658
Pronamide - 666
Propachlor - 675
Propargite - 683
Propham - 693
RH-315 - 666
Simazine - 703
Sodium arsenate - 712
Sodium arsenite - 715
Sodium bentazon - 80
Sodium omadine - 719
Sodium salt of fomesafen - 722
TBS - 89, 252
Tebuthiuron - 735
Terbufos - 744
Terbutryn - 750
Thiodicarb - 759
Thiophanate - 769
Thiophanate ethyl - 769
Thiophanate methyl - 773
Thiram - 784
Tribromsalan - 89, 252
Trimethacarb - 798
Triphenyltin hydroxide - 789

Generic Name Index

Acid orthoarsenate - 469
Activated 7-dehydrocholesterol - 802
Aluminum tris (O-ethyl phosphonate - 30
4-Amino-6(1,1-dimethylethyl)-3-(methylthio)-1,2,4-triazin-5(4H) - 531
4-Amino-3,5,6-trichloropicolinic acid - 646
Amitrole - 34
Arsenic acid - 57
Arsenious oxide - 64
Avermectin B_1 - 67
Basic Orthoarsenate - 469
Bis-(dimethylthiocarbamyl) disulfide - 784
2,4-Bis(isopropylamino)-6-(methylthio)-s-triazine - 658
Bis[tris(2-methyl-2-phenylpropyl)-tin] oxide - 367
2-Bromo-2-nitropropane-1,3-diol - 91
2-(tert-Butylamino)-4-(ethylamino)-6-(methylthio-s-triazine - 750
Butanedioic acid mono(2,2-dimethylhydrazine) - 232
Butylate - 95
2-(p-tert-Butylphenoxy)cyclohexyl 2-propynyl sulfite - 683
Calcium orthoarsenate - 106

1-(Carboethoxy)ethyl-5-[2-chloro-4-(trifluoromethyl)phenoxy] - 2-nitrobenzoate - 462
Carbon tetrachloride - 149
Chlorinated hydantoin derivative -236
2-Chloro-4,6-bis(ethylamino)-s-triazine - 703
2-Chloro-N-(2-ethyl-6-methylphenyl) N-(2-methoxy-1-methylethyl)acetamide - 522
2-Chloro-N-isopropyl acetanilide - 675
4-Chloro-5-(methylamino)-2-(alpha, alpha, alpha-trifluoromethyl-m-tolyl)-3(2H)-pyridazinone - 583
O-(5-Chloro-1-[methylethyl]-1H-1,2,4-triazol-3-yl) O,O-diethyl phosphorothioate - 453
N-[[(4-Chlorophenyl)amino]-carbonyl] -2,6-difluoro-benzamide - 286
S-[[(p-Chlorophenyl)thio] methyl] - O,O-diethyl phosphorodithioate - 152
2-Chloro-6-(trichloromethyl) pyridine - 578

Generic Name Index 819

5-[2-Chloro-4-(trifluoromethyl)-
  phenoxy] -N-(methylsulfonyl)-
  2-nitrobenzamide - 722
N-[2-Chloro-4-(trifluoromethyl)-
  phenyl] -DL-valine(±)-cyano-
  (3-phenoxyphenyl) methyl
  ester - 405
Cholecalciferol - 802
Copper Sulfate - 200
Cyanazine - 211
N-Cyclopropyl-1,3,5-triazine-
  2,4,6 triamine - 226
3,5 Dibromosalicylanilide - 89, 252
4',5 Dibromosalicylanilide - 89, 252
1,2-Dibromo-2,2-dichloroethyl
  dimethyl phosphate - 563
3,6-Dichloro-o-anisic acid - 254
2,6-Dichlorobenzonitrile - 260
2,4-Dichloro-6-(O-chloroanilino)-s-
  triazine - 43
3,5-Dichloro-N-(1,1-dimethyl-2-
  propynyl) benzamide - 666
2,6-Dichloro-4-nitroaniline - 239
(2,4-Dichlorophenoxy) acetic
  acid - 268
3-(3,4-Dichlorophenyl) 1,1-
  dimethylurea - 324
3-(3,4-Dichlorophenyl)-1-methoxy-
  1-methylurea - 484
1,3-Dichloropropene - 273
Dicofol - 281
O-[2-(Diethylamino)-6-methyl-4-
  pyrimidinyl] O,O-dimethyl
  phosphorodithioate - 1
O,O-Diethyl S-[2-(ethylthio)ethyl]
  phosphorodithioate - 314
O,O-Diethyl O-[2(ethylthio) ethyl]
  phosphorothioate and O,O-
  diethyl S-[2(ethylthio)ethyl]
  phosphorothioate mixture - 243
O,O-Diethyl S-[(ethylthio)methyl] -
  phosphorodithioate - 632
O,O-Diethyl O-(2-isopropyl-6-methyl-
  4-pyrimidinyl) phosphorothioate - 247
O,O-Diethyl O-[p-(methylsulfinyl)
  phenyl] phosphorothioate - 378
O,O-Diethyl-O-p-nitrophenyl
  phosphorothioate - 606

Diethyl 4,4'-o-phenylenebis[3-
  thioallophanate - 769
O,O-Diethyl O-(3,5,6-trichloro-2-
  pyridyl) phosphorothioate - 180
2,3-Dihydro-2,2-dimethyl-7-benzo-
  furanyl methylcarbamate - 140
2-[4,5-Dihydro-4-methyl-4-(1-
  methylethyl)-5-oxo-1H-imidazol-
  2-yl] -3-quinoline carboxylic
  acid - 447
p-(Dimethylamino)benzenediazo
  sodium sulfonate - 364
1,1'-Dimethyl-4,4'-bipyridinium
  ion - 596
N,N-Dimethyldiphenylacetamide - 298
N-[5-(1,1-Dimethylethyl)-1,3,4-thia-
  diazol-2-yl] -N,N'-dimethylurea - 735
S-[[(1,1-Dimethylethyl)thio] methyl]
  O,O-diethyl phosphorodithioate - 744
N,N-Dimethyl-N'-[3-[[(methylamino)-
  carbonyl] oxy] phenyl] methanimid-
  amide monohydrochloride - 417
O,O-Dimethyl-O-4-nitrophenyl
  phosphorothioate - 513
O,O-Dimethyl-s-[(4-oxo-1,2,3-
  benzotriazin-3(4H)-yl)methyl]
  phosphorodithioate - 72
N-(1,1-Dimethylpropynyl)-3,5-
  dichlorobenzamide - 666
Dimethyl N,N'-thiobis(methylimino)
  carbonyloxy bis ethanimido-
  thioate - 759
O,O-Dimethyl O-(3,5,6-trichloro-2-
  pyridyl) phosphorothioate - 188
2,4-Dinitro-6-octylphenol crotonate
  and 2,6-dinitro-4-octylphenol
  crotonate mixture - 292
Disodium ethylene bisdithiocarbamate - 555
2-8-Dodecen-1-ol - 458
E-8-Dodecen-1-yl-acetate - 458
2-8-Dodecen-1-yl-acetate - 458
Ethyl 2-[[[[(4-chloro-6-methoxy-
  pyrimidin-2-yl)amino] carbonyl] -
  amino] sulfonyl] benzoate - 164
Ethyl 4,4'-dichlorobenzilate - 170
O-Ethyl S,S-dipropyl phosphorodithioate - 354
S-Ethyl dipropylthiocarbamate - 346

N-Ethyl-N-(2-methyl-2-propenyl)-
2,6-dinitro-4-(trifluoromethyl)
benzamine - 351
O-Ethyl O-p-nitrophenyl
phenylphosphonothioate - 336
Ethyl [2-(4-phenoxyphenoxy)ethyl]
carbamate - 373
O-Ethyl S-phenyl ethylphosphonodithioate - 413
N-(1-Ethylpropyl)-3,4-dimethyl-
2,6-denitrobenzenamine - 614
2(Ethylthio)-4,6-bis(isopropylamino)-
s-triazine - 307
Fluchloralin - 385
Fluometuron - 392
Heliothis zea - 428
1,4,5,6,7,8,8-Heptachloro-3a,4,7,7a-
tetrahydro-4,7-methano-1H
indene - 433
1,2,3,4,5,6-Hexachlorocyclohexane,
gamma isomer - 475
1,2,3,4,10,10-Hexachloro-1,4,4a,5,8,8a-
hexahydroexo-1,4-endo-5,8-dimeth-
anonaphthalene - 24
Hexakis(2-methyl-2-phenylpropyl)-
distannoxane - 367
3-Hydroxy-N-methyl-cis-crotonamide,
dimethyl phosphate - 547
1-Hydroxy-2(1H)-pyridinethione,
sodium salt - 719
Isopropyl carbanilate - 693
Linalool - 473
Manganese ethylene bisdithiocarbamate - 491
N-(Mercaptomethyl)phthalimide
S-(O,O-dimethylphosphorodi-
thioate) - 637
Methyl N',N'-dimethyl-N-[(methylcarbamoyl)oxy]-1-oxamimidate - 590
3-(1-Methylethyl)-1H-2,1,3-benzo-
thiadiazin-4(3H)-one 2,2-diox-
ide - 80
3-(1-Methylethyl)-1H,2,1,3-benzo-
thiadiazin-4(3H)-one 2,2-dioxide,
sodium salt of - 80
Methyl-2-[[[[(4-methoxy-6-methyl-
1,3,5-triazin-2-yl)amino] carbonyl] -
amino] sulfonyl] benzoate - 541

2-Methyl-2-(methylsulfonyl)pro-
panol-O-(methylamino carbonyl
oximel) - 19
2-Methyl-2-(methylthio)propion-
aldehyde - 6
1-Methyl-3-phenyl-5-[3-(trifluoromethyl)phenyl] -4[1H] -pyridi-
none - 400
4-Methylthio-3,5-xylylmethyl-
carbamate - 500
Naptalam - 571
Naptalam sodium - 571
1-Napthyl N-methylcarbamate - 130
1,2,4,5,6,7,8,8-Octachloro-2,3,3a-
4,7,7a-hexahydro-4,7-methano-
indene - 157
Oleovitamin D_3 - 802
Orthoarsenic acid - 57
Potassium bromide - 651
Potassium 1-(4-chlorophenyl)-1,4-
dihydro-6-methyl-4-oxypyridazine-
3-carboxylate - 440
Poly(oxy-1,2-ethanediyl), alpha-
isooctadecyl-omega-hydroxy - 48
9,10-Secocholesta-5,7,10(19)-trien-
3B-ol - 802
Sodium aluminofluoride - 207
Sodium and calcium hypo-
chlorites - 708
Sodium fluoaluminate - 207
Sodium metaarsenite - 715
Sodium orthoarsenate - 712
Substituted dimethyl hydantoin - 422
Succinic acid 2,2-dimethyl-
hydrazine - 232
Sulfuryl fluoride - 731
cis-N-[1,1,2,2-Tetrachloroethyl)thio] -4-
cyclohexene-1,2-dicarboximide - 109
Tetrachloroisophthalonitrile - 174
3,4',5 Tribromosalicylanilide - 89, 252
N-Trichloromethylthio-4-cyclohexene-
1,2-dicarboximide - 120
Tricyclohexylhydroxystannane - 220
1,1,1-Trifluoro-N-[2-methyl-4-(phenyl
sulfonyl)phenyl] -methanesul-
fonamide - 625
2,3,5- and 3,4,5-Trimethylphenyl
methylcarbamate - 798
Triphenyltin hydroxide - 789

Trade Name Index

Includes all names listed on each individual fact sheet under the headings *Trade Names* or *Trade and Other Names*.

A-831010 - 298
AAtak - 784
AC 5223 - 329
AC 87,844 - 773
AC 92553 - 614
Acaraben - 170
Acarin - 281
Acarstin - 220
Actellic - 1
Actor Cekuquat - 596
Affirm - 67
AG500 - 247
Alanap-L - 571
Alar - 232
Aldrex - 24
Aldrex 30 - 24
Aldrine - 24
Aldrite - 24
Aldrosol - 24
Alfanox - 484
Alfa-tox - 247
Algae-A-Way - 703
Algaecide - 703
Algidize - 703
Algi-eater - 703
Algi-gon - 703
Aliette - 30
Alkron - 606
Alleron - 606

Allisan - 239
Altox - 24
Amdon - 646
Aminotriazole 90 - 34
Amino Triazole Weed Killer - 34
Aminozide - 232
Amitrol T - 34
Amizine - 703
Amizole - 34
Alphamite - 606
Apodrin - 547
Appa - 637
Aquazine - 703
Arasan - 784
Arosurf MSF - 48
AT - 34
ATA - 34
Atomicide - 703
Azodrin - 547
B-622 - 43
B-995 - 232
B-nine - 232
Banex - 254
Bangald - 24
Banvel - 254
Basagran Postemergence Herbicide - 80
Basalin - 385
Basanite - 294
Bay 9027 - 72

Bay 17147 - 72
Bay 25141 - 378
Bay 37344 - 500
Bay 78537 - 140
Bayer 5072 - 364
Bayer 22555 - 364
Belt - 157
Bentazon Manufacturer's
  Concentrate - 80
Benzinoform - 149
Bexton - 675
Bilobron - 547
Biotrol VHZ - 428
Birgin - 693
Bladan - 606
Bladan M - 513
Bladex 4-WDS or 4L - 211
Bladex 15G - 211
Bladex 80 WP or 80W - 211
Bluestone - 200
Botran - 239
Bravo - 174
Brom-O-Gas - 506
Broot - 798
Brovomil - 174
Brush Buster - 254
Caldon - 294
Caparol - 658
Carbax - 281
Carbona - 149
Carfene - 72
Carpene - 329
Carzol - 417
CAS 56-235 - 149
Casoron - 260
Cekiuron - 324
Cekumethion - 513
Cekusan - 703
Celfume - 506
Cercobin - 769
Cercobin M - 773
CGA-12223 - 453
CGA-24705 - 522
Chem Bam - 555
Chem-Hoe - 693
Chemolimpex - 346
Chemox General & PE - 294
Chempar Amitrole - 34

Chem Pels C - 715
Chemsect - 294
Chem-Sen 56 - 715
Chlordan - 157
Chlor-Kil - 157
Chlorothalonil Technical - 174
Chlortox - 157
Cimacide - 703
CL 7521 - 329
Clarosan - 750
Cleary's 3336 - 769
Clipper 50 WP - 192
Clorox - 708
Cobra - 462
Code 133 - 260
Comite - 683
Compound 118 - 24
Copper Sulfate - 200
Corodane - 157
Corothion - 606
Cotion-Methyl - 72
Cotoran - 392
Cototar - 307
Counter - 744
CP 31393 - 675
Crisodrin - 547
Crisquat - 596
Curaterr - 140
Curitan - 329
Cycosin - 773
Cyprex - 329
Cytox 2013 - 329
Cytrol - 34
D 1221 - 140
Daconil - 174
Dailon - 324
Dalf - 513
Dantochlor - 236
Dasanit - 378
2,6-DBN - 260
DCNA - 239
Decabane - 260
Decofol - 281
Delsan - 784
Desiccant L-10 - 57
Destun - 625
Devithion - 513
Dexon - 364

Dextrone - 596
Dextrone X - 596
Dexuron - 596
Diater - 324
Diazoben - 364
Dibrom - 563
4,6-Dichloro-N-(2-chlorophenyl)-
1,3,5-triazin-2-amine - 43
Difenamid - 298
Difolatan - 109
Dikar - 292
O,O-Dimethyl O-(4-nitrophenyl) phosphorothioate - 513
O,O-Dimethyl O-(p-nitrophenyl) phosphorothioate - 513
Dimethyl Parathion - 513
Dimilin - 286
Dinitrex - 294
Dinitro - 294
Dinitro-3 - 294
Dinitro General - 294
Dinoseb (F-ISO) - 294
Di-on - 324
Direz - 43
Di-Syston - 314
Dithane A-40 - 555
Dithane D-14 - 555
Dithane M-45 - 491
Ditranil - 239
Diurol - 324
Diurox - 324
DNBP - 294
DNOSBP - 294
Dodecylguanidine acetate - 329
Dodecylguanidine hydrochloride (DGH) - 329
Dodecylguanidine monoacetate - 329
Dodecylguanidine terephthalate (DGT) - 329
Dodine acetate - 329
Doguanidine - 329
Domatol - 34
Dowco 213 - 220
Dowfume - 506
Dowfume 75 - 149
Dow Telone - 273
DPX 1410 - 590
Drexel Diuron 4L - 324

Drexel Dynamite 3 - 294
Drinox - 24
Drinox H-34 - 433
DSE - 555
Dual - 522
DuPont Ally Herbicide - 541
Dupont Classic Herbicide - 164
DuPont Escort Herbicide - 541
Durotex 7487-A - 329
Dursban - 180
Duter - 789
Du-ter - 789
Dyfonate - 413
Dymid - 298
Dynamite - 294
Dynanap - 294
Dynex - 324
Dyrene - 43
D-Z-N Diazinon 14G - 247
E601 - 513
E-3314 - 433
Elcar - 428
Elgetol 318 - 294
Embafume - 506
Enide - 298
ENT 4705 - 149
ENT 27164 - 140
Ent-26,058 - 43
Entoma 15949 - 24
Eptam - 346
Eradicane - 346
Esgram - 596
Ethyl Parathion - 606
Evital - 583
Exagamma - 475
Fenbutatin oxyde - 367
Flex - 722
Flukoide - 149
FMC 10242 - 140
Folcid - 109
Folidol E-605 - 606
Folidol M - 513
Fore - 491
Forlin - 475
Fosferno 50 - 606
Fosferno M50 - 513
Framed - 703
Fungo - 773

Fungo 70 - 773
Furadan - 140
G-34161 - 658
Gallogamma - 475
Gammaphex - 475
Gammex - 475
Gebutox - 294
Gesagard - 658
Gesapun - 703
Gesatop - 703
Gexane - 475
Glore Phos 36 - 547
Glycoserve - 422
Gold Crest C-50 - 157
Gold Crest C-100 - 157
Gold Crest H-60 - 433
Goldquat 276 - 596
Gramanol - 596
Grammapoz - 475
Grammexane - 475
Gramoxone - 596
Gramuron - 596
Grazon - 646
Groutcide - 174
GS 14260 - 750
GS-16068 - 307
Gusathion - 72
Gusation-M - 72
Guthion - 72
Gypsine - 469
H-133 - 260
H-321 - 500
Haipen - 109
Haitin - 789
Halon 104 - 149
HCl-CO8684 - 43
Hebaxon - 596
Hel-Fire - 294
Heptachlore - 433
1,4,5,6,7,8,8-Heptachloro-3a,4,7,7a-
 tetrahydro-4,7-methanoindene - 433
Heptagran - 433
Heptalube - 433
Heptamul - 433
Heptox - 433
Herbadox - 614
Herboxone - 596
HHDN - 24

Hi Yield H-10 - 57
Hi Yield Synergized H-10 - 57
Hybrex - 440
Igran - 750
Imidan - 637
Inexit - 475
Isomate-M - 458
Karathane - 292
Karmex - 324
Kayafume - 506
Kelthane - 281
p,p' Kelthane - 281
Kelthane A - 281
Kemate - 43
Kemolate - 637
Kerb - 666
Kilex - 157
Kill-All - 715
Kiloseb - 294
Klean Krop - 294
Kryocide - 207
Kwell - 475
Kylar - 232
Kypchlor - 157
L-34-314 - 298
Laddock Postemergence Herbicide - 80
Landrin - 798
Lanex - 392
Larvadex Technical and Premix - 226
Larvin - 759
Lead Arsenate - 469
Lesan - 364
Lexone - 531
Linalool 925 - 473
Lindafor - 475
Lindagrain - 475
Lindagram - 475
Lindagranox - 475
Lindalo - 475
Lindamul - 475
Lindapoudre - 475
Lindaterra - 475
Lindex - 475
Lindust - 475
Lintox - 475
Linurex - 484
Logic 373
Londax - 484

Lorox - 484
Lorsban - 180
M-410 - 157
Manzate 200 - 491
Marvik - 405
Mediben - 254
Melprex - 329
Mercaptodimethur - 500
Mercuram - 784
Merpafol - 109
Merpan - 120
Mesurol - 500
Metacide - 513
Metaphos - 513
Methiocarbe - 500
Meth-O Gas - 506
Metiltriazotion - 72
Metmercapturnon - 500
Metron - 513
Mibol - 281
Micromite - 286
Mildex - 292
Mildothane - 773
Mitigan - 281
Mocap - 354
Mold-Ex - 174
Monocil 40 - 547
Monocron - 547
Myacide AS - 91
Myacide B10 - 91
Nabasan - 555
Neostanox - 367
NF44 - 773
NIA 10242 - 140
Niran - 157, 606
Nitropone C - 294
Nitrox 80 - 513
Nomersan - 784
Nopcocide N-96 - 174
Novigram - 475
N-Serve - 578
Nuracron - 547
Octachlor - 157
1,2,4,5,6,7,8,8-Octachloro-3a,4,7,7a-
   tetrahydro-4,7-methanoindan - 157
Octalene - 24
Omadine sodium - 719
Omite - 683

Ontrack - 522
Ortho 5865 - 109
Orthocide - 120
Orthophos - 606
Osadan - 367
Panthion - 606
Paracol Paraquat CL - 596
Paramar - 606
Paraphos - 606
Parataf - 513
Parathene - 606
Parathion-methyl - 513
Paratox - 513
Parawet - 606
Partron M - 513
Parzate - 555
Penite - 715
Pennant - 522
Penncap M - 513
Penticklor - 157
Phoskil - 606
Phthalofos - 637
Pillardrin - 547
Pillarquat - 596
Pillarxone - 596
Plantdrin - 547
Plictran - 220
PMP - 637
Poly Brand Desiccant - 57
Polyram-Alltra - 784
Pomarsol - 784
Potassium Bromide - 651
PP148 (dichloride) - 596
PPG-844 - 462
Prebane - 750
Premerge Plus with Dinitro - 294
Prentox - 157
Primatol Q - 658
Primatol S - 703
Princep - 703
Prodalumnol Double - 715
Prolate - 637
Prometrex - 658
Prometryne - 658
Prowl - 614
Purex - 708
R-1504 - 637
R-1582 - 72

Rampart - 632
Ramrod - 675
Rasayaldrin - 24
Reflex - 722
Reldan - 188
Resisan - 239
Rhodiatox - 606
Ro 13-5223 - 373
S-767 - 378
SADH - 232
Sancap - 307
Sanspor - 109
Sarolex - 247
Scepter - 447
SD 14114 - 367
Security - 469
Seedrin Liquid - 24
Sencor - 531
Sevin - 130
Short-stop - 750
Silvanol - 475
Simadex - 703
Simanex - 703
Sim-trol - 703
Sinox - 294
Sodium 1-hydroxypyridine -2-thione - 719
Sodium 2-mercaptopyridine - 719
Sodium 2-pyridinethiol 1-oxide - 719
Solicam - 583
Sonar - 400
Soprathion - 606
Spatrete - 784
Spectracide - 247
Spot-kleen - 773
Spring Bak - 555
SR-406 - 120
Standak - 19
Stathion - 606
Stomp - 614
Subitex - 294
Sulfenimide - 109
Supertin - 789
Susrin - 547
Sutan - 95
Suzu H - 789
Sweep - 596
Syllit - 329

Synchlor - 157
Systox - 243
Talbot - 469
TD-1771 - 773
Tekwaisa - 513
Telone II - 273
Temasept IV - 89, 252
Temik - 6
Terbutrex - 750
Terbutryne - 750
Termi-Ded - 157
Termil - 174
Terracur P - 378
Terr-O-Gas 100 - 506
Tersan - 784
Tetramethylthiuram disulfide - 784
Thimer - 784
Thimet - 632
Thiooxamyl - 590
Thiophanate-methyl - 773
Thiophos - 606
Thiramid - 784
Thirasan - 784
Thiuramin - 784
TM-95 - 784
TMTD - 784
TMTDS - 784
Topiclor 20 - 157
Topsin - 769
Topsin M - 773
Tordon - 646
Torque - 367
TPTH - 789
TPTOH - 789
Trametan - 784
Triampa - 784
Triasyn - 43
Trigard - 226
Triherbide - 693
Tripomol - 784
Trithion - 152
Triumph - 453
Tsitrex - 329
Tuads - 784
Tubotin - 789
Turf-Cal - 106
Ulvair - 547
Unicrop DNBP - 294

Unidron - 324
Uniroyal D014 - 683
Urox - 324
Ustinex - 34
Vancide - 784
Vancide 89 - 120
Velsicol 58-CS-11 - 254
Velsicol 104 - 433
Velsicol 168 - 157
Velsicol 1068 - 157
Vendex - 367
Venturol - 329
Vertac Dinitro Weed Killer 5 - 294
Vertac General and Selective Weed Killer - 294

Vigilante - 286
Vikane - 731
Viron/H - 428
Vitamin D_3 - 802
Vondodine - 329
Vonduron - 324
Vorox - 34
Vydate - 590
Weedazole - 34
White Arsenic - 64
Wofatox - 513
X-All - 34
Yaltox - 140
Zorial - 583

Other Noyes Publications

PESTICIDE WASTE DISPOSAL TECHNOLOGY

Edited by
James S. Bridges and Clyde R. Dempsey
U.S. Environmental Protection Agency

Pollution Technology Review No. 148

This book attempts to define practical solutions to pesticide users' disposal problems.

A major agreement must be reached on what can be done, legally and technically, to deal with the difficulties of proper pesticide-related waste disposal, and who should share in the cost of a clean environment. Pesticide commerce and use are regulated under the Federal Insecticide, Fungicide and Rodenticide Act and by state laws and rules. However, once applications of pesticides are completed, any excess pesticide concentrate, unapplied diluted pesticide, and discarded pesticide containers may be regulated as wastes, some of which may be considered hazardous. Although past disposal problems and future policy changes are of major importance, the primary focus must be the solutions to the existing problems facing the pesticide user industry today.

The book is presented in three parts. Part I covers disposal needs; federal/state regulatory requirements; pesticide degradation properties; disposal technology options, including physical treatment, biological treatment, chemical treatment, land application and incineration options; storage, handling, and shipments of pesticide wastes; and empty pesticide container disposal programs.

Part II addresses issues regarding the effectiveness of current state-of-the-art capabilities, identifies emerging techniques or technologies that may be applicable along with technologies being applied in other areas, and describes the need for research efforts capable of providing results in a three-to-five year time frame as they pertain to the treatment/disposal of dilute pesticide wastewater. Twelve technologies are discussed in some detail.

Part III includes industry's role in users' waste disposal, on-site demonstration projects, users' waste minimization/reuse and users' treatment/storage/disposal.

The condensed contents given below lists **parts and selected chapter titles.**

PART I
1. Applicator Disposal Needs
2. An Agricultural Aviator's Perspective
3. A Ground Applicator's Perspective
4. A Private Applicator's Perspective
5. Federal Regulation of Pesticide Disposal
6. California Regulatory Requirements
7. Pesticide Degradation Properties
8. Physical Treatment Options
9. Biological Methods
10. Land Disposal of Pesticide Rinsate
11. Storage, Handling and Shipment of Pesticide Waste—Regulatory Requirements
12. Empty Pesticide Container Management

PART II
13. Pesticide Rinsewater Recycling
14. Granular Carbon Adsorption
15. UV-Ozonation
16. Small-Scale Incineration
17. Solar Photo-Decomposition
18. Chemical Degradation
19. Evaporation, Photodegradation, and Biodegradation in Containment Devices
20. Genetically Engineered Products
21. Leach Fields
22. Acid and Alkaline Trickling Filter Systems
23. Organic Matrix Adsorption and Microbial Degradation
24. Evaporation and Biological Treatment with Wicks

PART III
25. Industry's Role in Users' Waste Disposal
26. Carbon Adsorption Treatment of Rinsewater
27. Treatment of Pesticide-Containing Soil
28. Biological and Chemical Disposal Systems for Waste Pesticide Solutions
29. Wastewater Recycling
30. Dealing with Emergencies
31. Reconditioning Containers
32. Recycling Metal Containers
33. Disposal of 55-Gallon All-Plastic Drums
34. Off-Site Disposal of Pesticides and Pesticide Containers

ISBN 0-8155-1157-4 (1988)　　　　7"x10"　　　　331 pages

Other Noyes Publications

PESTICIDE MANUFACTURING AND TOXIC MATERIALS CONTROL ENCYCLOPEDIA 1980

Edited by Marshall Sittig

Chemical Technology Review No. 168
Environmental Health Review No. 3
Pollution Technology Review No. 69

This book contains a total of 514 pesticide materials arranged in an alphabetical and encyclopedic fashion by the common or generic name of each pesticide. It is a thorough revision of our previous *Pesticides Process Encyclopedia* published in 1977, plus additional material relative to toxic materials control. The data on manufacturing processes were drawn primarily from the patent literature, while the data on product toxicity, emissions and product use came mostly from published and unpublished reports released by the Environmental Protection Agency.

This book is definitely of interest to pesticide manufacturers, chemical raw material suppliers, formulators, growers, farmers and food processors. It should also prove useful to chemists, lawyers, industrial hygienists and environmentalists.

It contains much useful extrinsic information, e.g. *allowable tolerance limits, animal and human toxicities,* and similar data not easily ascertained elsewhere.

The use of pesticides leads to healthier plants and bigger crops, and exports of pesticides could provide fast growth for U.S. producers in the coming years.

An indication of the comprehensive nature of this one-volume encyclopedia is given here:

INTRODUCTION
What Is a Pesticide?
Pesticide Manufacture
Pollution Problems
Pesticide Formulations
   Dusts & Wettable Powders
   Granules
   Liquid Formulations
   Packing & Storage
Pesticide Applications

TOXIC MATERIALS CONTROL
Safe Work Practices
Pollution Control in Manufacture
Restrictions on Exposure & Use
Concentrations in Air/Water
Registration
Residue Tolerances

ENVIRONMENTALLY ACCEPTABLE ALTERNATIVES
Biodegradable Pesticides
Physical Control of Toxic Pesticides

Controlled Release Pesticides
Ultra-Low Volume Application
Undesigned Pesticides
Biological Controls
Pheromones
Integrated Pest Management

DATA ON 514 INDIVIDUAL PESTICIDES:
Acephate
Acrolein
Acrylonitrile
Alachlor
Aldicarb
Aldoxycarb
Aldrin
Allethrin
Allidochlor
Allyl Alcohol
Aluminum Phosphide
Ametryne
Aminocarb
Amitraz
AMS
Ancymidol
Anilazine
Anthraquinone
ANTU
Arsenic Acid
Asulam
Atrazine
Azinphos-Ethyl
Azinphos-Methyl
Aziprotryn
Bacillus Thuringiensis
Barban
Benazolin
Bendiocarb
Benfluralin
Benodalin
Benomyl
Bensulide
Bentazon
Benzene Hexachloride
Benzoximate
Benzoylprop-Ethyl
Benzthiazuron
S-Benzyl Di-sec-butylthiocarbamate
Bifenox
plus 474 other pesticides

RAW MATERIALS INDEX
TRADE NAMES INDEX

810 pages

ISBN 0-8155-0814-X